Introduction to the Physics and Techniques of Remote Sensing

WILEY SERIES IN REMOTE SENSING

Jin Au Kong, Editor

Introduction to the Physics and Techniques of Remote Sensing

Second Edition

Charles Elachi
Jakob van Zyl

A JOHN WILEY & SONS, INC., PUBLICATION

The images on the cover show how complimentary information is gained with remote sensing in different parts of the electromagnetic spectrum. Both images are of the Nile River, near the Fourth Cataract in Sudan. The top image is a color infrared photograph taken from Space Shuttle Columbia in November 1995. The bottom image was acquired by the Spaceborne Imaging Radar C/X-Band Synthetic Aperture Radar (SIR-C/X-SAR) aboard Space Shuttle Endeavour in April 1994. The thick, white band in the top right of the radar image is an ancient channel of the Nile that is now buried under layers of sand. This channel cannot be seen in the infrared photograph and its existence was not known before this radar image was processed. (Courtesy of NASA/JPL-Caltech)

For general information on our other products and services or for technical support, please contact our Customer Care Department within the United States at (800) 762-2974, outside the United States at (317) 572-3993 or fax (317) 572-4002.

Wiley also publishes its books in a variety of electronic formats. Some content that appears in print may not be available in electronic format. For information about Wiley products, visit our web site at www.wiley.com.

Library of Congress Cataloging-in-Publication Data is available.

Printed in the United States of America.

ISBN-13 978-0-471-47569-9
ISBN-10 0-471-47569-6

10 9 8 7 6 5 4 3 2 1

To Valerie,
Joanna and Lauren (CE)
and Kalfie and Jorkie (JvZ)

Contents

Preface

The advent of the space age opened a whole new dimension in our ability to observe, study, and monitor planetary (including Earth) surfaces and atmospheres on a global and continuous scale. This led to major developments in the field of remote sensing, both in its scientific and technical aspects. In addition, recent technological developments in detectors and digital electronics opened the whole electromagnetic spectrum to be used for detecting and measuring ever finer details of the emitted and reflected waves that contain the "fingerprints" of the medium with which they interact. Spaceborne imaging spectrometers from the visible to the far infrared are being developed to acquire laboratory quality spectra for each observed surface pixel or atmospheric region, thus allowing direct identification of the surface or atmospheric composition. Multispectral polarimetric and interferometric imaging radars will provide detailed maps of the surface morphology, its motion, and the near-subsurface structure, as well as three-dimensional maps of precipitation regions in the atmosphere. Active microwave sensors are being used to monitor, on a global basis, the dynamics of the ocean: its topography, currents, near-surface wind, and polar ice. Passive and active atmospheric sounders provide detailed profiles of the atmosphere and ionosphere characteristics: temperature, pressure, wind velocity, and electron content. Large-scale multispectral imagers provide repetitive global images of the surface biomass cover and monitor our planet's environmental changes resulting from natural causes as well as from the impact of human civilization.

These capabilities are also being applied more extensively in exploring the planets in our solar system with flyby and orbiting spacecraft. All the major bodies in the solar system, with the exception of Pluto, have been visited and explored. The surface of Venus has been mapped globally by radar, and Mars has been explored with orbiters and rovers. Jupiter and Saturn, as well as their satellites, have been mapped by sophisticated orbiters.

The next decade will also see major advances in our use of spaceborne remote sensing techniques to understand the dynamics of our own planet and its environment. A number of international platforms continuously monitor our planet's surface and atmosphere using multispectral sensors, allowing us to observe long-term global and regional changes. More sophisticated systems will be deployed in the next decade to globally measure ocean salinity, soil moisture, gravity field changes, surface tectonic motion, and so on. These systems will make full use of new developments in technology, information handling, and modeling.

Remote sensing is a maturing discipline that calls on a wide range of specialties and crosses boundaries between traditional scientific and technological disciplines. Its multi-disciplinary nature requires its practitioner to have a good basic knowledge in many areas of science and requires interaction with researchers in a wide range of areas such as electromagnetic theory, spectroscopy, applied physics, geology, atmospheric sciences, agronomy, oceanography, plasma physics, electrical engineering, and optical engineering.

The purpose of this text is to provide the basic scientific and engineering background for students and researchers interested in remote sensing and its applications. It addresses (1) the basic physics involved in wave–matter interactions, which is the fundamental element needed to fully interpret the data; (2) the techniques used to collect the data; and (3) the applications to which remote sensing is most successfully applied. This is done keeping in mind the broad educational background of interested readers. The text is self-comprehensive and requires the reader to have the equivalent of a junior level in physics, specifically introductory electromagnetic and quantum theory.

The text is divided into three major parts. After the introduction, Chapter 2 gives the basic properties of electromagnetic waves and their interaction with matter. Chapters 3 through 7 cover the use of remote sensing in solid (including ocean) surface studies. Each chapter covers one major part of the electromagnetic spectrum (visible/near infrared, thermal infrared, passive microwave, and active microwave, respectively). Chapters 8 through 12 cover the use of remote sensing in the study of atmospheres and ionospheres. In each chapter, the basic interaction mechanisms are covered first. This is followed by the techniques used to acquire, measure, and study the information (waves) emanating from the medium under investigation. In most cases, a specific advanced sensor flown or under development is used for illustration.

The text is generously illustrated and includes many examples of data acquired from spaceborne sensors. As a special feature, sixteen of the illustrations presented in the text are reproduced in a separate section of color plates.

This book is based on an upper undergraduate and first-year graduate course that we teach at the California Institute of Technology to a class that consists of students in electrical engineering, applied physics, geology, planetary science, astronomy, and aeronautics. It is intended to be a two-quarter course. This text is also intended to serve engineers and scientists involved in all aspects of remote sensing and its applications.

This book is a result of many years of research, teaching, and learning at Caltech and the Jet Propulsion Laboratory. Through these years, we have collaborated with a large number of scientists, engineers, and students who helped in developing the basis for the material in this book. We would like to acknowledge all of them for creating the most pleasant atmosphere for work and scientific "enjoyment." To name all of them would lead to a very long list. We would also like to acknowledge numerous researchers at JPL who were kind enough to read and provide suggestions on how to improve the text—they include M. Abrams, M. Chahine, J. Curlander, M. Freilich, D. McCleese, J. Waters and H. Nair—as well as all of our students at Caltech, who hopefully became interested enough in this field to carry the banner into the next century.

We also would like to acknowledge the secretaries and artists who typed the text of the First Edition, improved the grammar, and did the artwork, in particular, Clara Sneed, Susan Salas, and Sylvia Munoz. Of course, with the advances in "office technology" we had to type the changes for the Second Edition ourselves.

CHARLES ELACHI
JAKOB VAN ZYI

Pasadena, California
May 2005

1

Introduction

Remote sensing is defined as the acquisition of information about an object without being in physical contact with it. Information is acquired by detecting and measuring changes that the object imposes on the surrounding field, be it an electromagnetic, acoustic, or potential. This could include an electromagnetic field emitted or reflected by the object, acoustic waves reflected or perturbed by the object, or perturbations of the surrounding gravity or magnetic potential field due to the presence of the object.

The term "remote sensing" is most commonly used in connection with electromagnetic techniques of information acquisition. These techniques cover the whole electromagnetic spectrum from low-frequency radio waves through the microwave, submillimeter, far infrared, near infrared, visible, ultraviolet, x-ray, and gamma-ray regions of the spectrum.

The advent of satellites is allowing the acquisition of global and synoptic detailed information about the planets (including the Earth) and their environments. Sensors on Earth-orbiting satellites provide information about global patterns and dynamics of clouds, surface vegetation cover and its seasonal variations, surface morphologic structures, ocean surface temperature, and near-surface wind. The rapid wide coverage capability of satellite platforms allows monitoring of rapidly changing phenomena, particularly in the atmosphere. The long duration and repetitive capability allows the observation of seasonal, annual, and longer-term changes such as polar ice cover, desert expansion, and tropical deforestation. The wide-scale synoptic coverage allows the observation and study of regional and continental-scale features such as plate boundaries and mountain chains.

Sensors on planetary probes (orbiters, flybys, surface stations, and rovers) are providing similar information about the planets and objects in the solar system. By now, all the planets in the solar system, except for Pluto, have been visited by one or more spacecraft. The comparative study of the properties of the planets is providing new insight into the formation and evolution of the solar system.

Introduction to the Physics and Techniques of Remote Sensing. By C. Elachi and J. van Zyl

1-1 TYPES AND CLASSES OF REMOTE SENSING DATA

The type of remote sensing data acquired is dependent on the type of information being sought, as well as on the size and dynamics of the object or phenomena being studied. The different types of remote sensing data and their characteristics are summarized in Table 1-1. The corresponding sensors and their role in acquiring different types of information are illustrated in Fig. 1-1.

Two-dimensional images and three-dimensional perspectives are usually required when high-resolution spatial information is needed, such as in the case of surface cover and structural mapping (Figs. 1-2, 1-3), or when a global synoptic view is instantaneously required, such as in the case of meteorological and weather observations (Fig. 1-4). Two-dimensional images can be acquired over wide regions of the electromagnetic spectrum (Fig. 1-5) and with a wide selection of spectral bandwidths. Imaging sensors are available in the microwave, infrared (IR), visible, and ultraviolet parts of the spectrum using electronic and photographic detectors. Images are acquired by using active illumination, such as radars or lasers; solar illumination, such as in the ultraviolet, visible, and near infrared;

TABLE 1-1. Types of Remote Sensing Data

Important type of information needed	Type of sensor	Examples of sensors
High spatial resolution and wide coverage	Imaging sensors, cameras	Large-format camera (1984), Seasat imaging radar (1978), Magellan radar mapper (1989), Mars Global Surveyor Camera (1996), Mars Rover Camera (2004)
High spectral resolution over limited areas or along track lines	Spectrometers, spectroradiometers	Shuttle multispectral imaging radiometer (1981), Hyperion (2000)
Limited spectral resolution with high spatial resolution	Multispectral mappers	Landsat multispectral mapper and thematic mapper (1972–1999), SPOT (1986–2002), Galileo NIMS (1989)
High spectral and spatial resolution	Imaging spectrometer	Spaceborne imaging spectrometer (1991), ASTER (1999), Hyperion (2000)
High-accuracy intensity measurement along line tracks or wide swath	Radiometers, scatterometers	Seasat (1978), ERS-1/2 (1991, 1997), NSCAT (1996), QuikSCAT (1999), SeaWinds (2002) scatterometers
High-accuracy intensity measurement with moderate imaging resolution and wide coverage	Imaging radiometers	Electronically scanned microwave radiometer (1975), SMOS (2007)
High-accuracy measurement of location and profile	Altimeters, sounders	Seasat (1978), GEOSAT (1985), TOPEX/Poseidon (1992), and Jason (2001) altimeter, Pioneer Venus orbiter radar (1979), Mars orbiter altimeter (1990)
Three-dimensional topographic mapping	Scanning altimeters and interferometers	Shuttle Radar Topography Mission (2000)

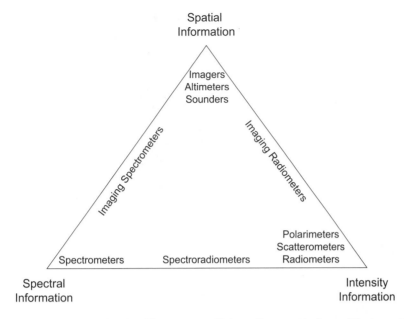

Figure 1-1. Diagram illustrating the different types of information sought after and the type of sensor used to acquire this information. For instance, spectral information is acquired with a spectrometer. Two-dimensional surface spatial information is acquired with an imager such as a camera. An imaging spectrometer also acquires data for each pixel in the image spectral information.

or emission from the surface, such as in thermal infrared, microwave emission (Fig. 1-6), and x and gamma rays.

Spectrometers are used to detect, measure, and chart the spectral content of the incident electromagnetic field (Figs. 1-7, 1-8). This type of information plays a key role in identifying the chemical composition of the object being sensed, be it a planetary surface or atmosphere. In the case of atmospheric studies, the spatial aspect is less critical than the spectral aspect due to the slow spatial variation in the chemical composition. In the case of surface studies, both spatial and spectral information are essential, leading to the need for imaging spectrometers (Figs. 1-9, 1-10). The selection of the number of spectral bands, the bandwidth of each band, the imaging spatial resolution, and the instantaneous field of view leads to trade-offs based on the object being sensed, the sensor data-handling capability, and the detector technological limits.

In a number of applications, both the spectral and spatial aspects are less important, and the information needed is contained mainly in the accurate measurement of the intensity of the electromagnetic wave over a wide spectral region. The corresponding sensors, called radiometers, are used in measuring atmospheric temperature profiles and ocean surface temperature. Imaging radiometers are used to spatially map the variation of these parameters (Fig. 1-11). In active microwave remote sensing, scatterometers are used to accurately measure the backscattered field when the surface is illuminated by a signal with a narrow spectral bandwidth (Fig. 1-12). One special type of radiometer is the polarimeter, in which the key information is embedded in the polarization state of the transmitted, reflected, or scattered wave. The polarization characteristic of reflected or scattered sunlight provides information about the physical properties of planetary atmospheres.

In a number of applications, the information required is strongly related to the three-

Figure 1-2. Landsat MSS visible/near IR image of the Imperial Valley area in California.

dimensional spatial characteristics and location of the object. In this case, altimeters and interferometric radars are used to map the surface topography (Figs. 1-13 to 1-16), and sounders are used to map subsurface structures (Fig. 1-17) or to map atmospheric parameters (such as temperature, composition, or pressure) as a function of altitude (Fig. 1-18).

1-2 BRIEF HISTORY OF REMOTE SENSING

The early development of remote sensing as a scientific field is closely tied to developments in photography. The first photographs were reportedly taken by Daguerre and Niepce in 1839. The following year, Arago, Director of the Paris Observatory, advocated the use of photography for topographic purposes. In 1849, Colonel Aimé Laussedat, an officer in the French Corps of Engineers, embarked on an exhaustive program to use photography in topographic mapping. By 1858, balloons were being used to make photographs of large areas. This was followed by the use of kites in the 1880s and pigeons in the early 1900s to carry cameras to many hundred meters of altitude. The advent of the

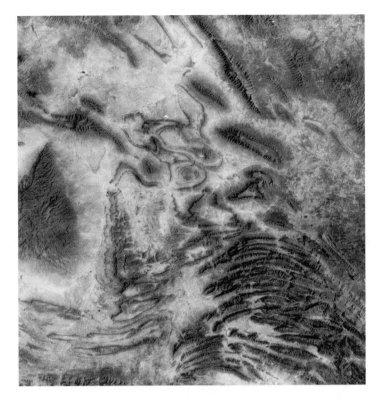

Figure 1-3. Folded mountains in the Sierra Madre region, Mexico (Landsat MSS).

airplane made aerial photography a very useful tool because acquisition of data over specific areas and under controlled conditions became possible. The first recorded aerial photographs were taken from an airplane piloted by Wilbur Wright in 1909 over Centocelli, Italy.

Color photography became available in the mid-1930s. At the same time, work was continuing on the development of films that were sensitive to near-infrared radiation. Near-infrared photography was particularly useful for haze penetration. During World War II, research was conducted on the spectral reflectance properties of natural terrain and the availability of photographic emulsions for aerial color infrared photography. The main incentive was to develop techniques for camouflage detection.

In 1956, Colwell performed some of the early experiments on the use of special-purpose aerial photography for the classification and recognition of vegetation types and the detection of diseased and damaged vegetation. Beginning in the mid-1960s, a large number of studies of the application of color infrared and multispectral photography were undertaken under the sponsorship of NASA, leading to the launch of multispectral imagers on the Landsat satellites in the 1970s.

At the long-wavelength end of the spectrum, active microwave systems have been used since the early twentieth century and particularly after World War II to detect and track moving objects such as ships and, later, planes. More recently, active microwave sensors have been developed that provide two-dimensional images that look very similar to regular photography, except that the image brightness is a reflection of the scat-

↑ 22:30 28SE76 13A-Z 0006-1640 FULL DISC IR

Figure 1-4. Infrared image of the western hemisphere acquired from a meterological satellite.

tering properties of the surface in the microwave region. Passive microwave sensors were also developed to provide "photographs" of the microwave emission of natural objects.

The tracking and ranging capabilities of radio systems were known as early as 1889, when Heinrich Hertz showed that solid objects reflected radio waves. In the first quarter of the twentieth century, a number of investigations were conducted in the use of radar systems for the detection and tracking of ships and planes and for the study of the ionosphere.

Radar work expanded dramatically during World War II. Today, the diversity of applications for radar is truly startling. It is being used to study ocean surface features, lower- and upper-atmospheric phenomena, subsurface and surface land structures, and surface cover. Radar sensors exist in many different configurations. These include altimeters to provide topographic measurements, scatterometers to measure surface roughness, and polarimetric and interferometric imagers.

In the mid-1950s, extensive work took place in the development of real-aperture airborne imaging radars. At about the same time, work was ongoing in developing synthetic-aperture imaging radars (SAR), which use coherent signals to achieve high-reso-

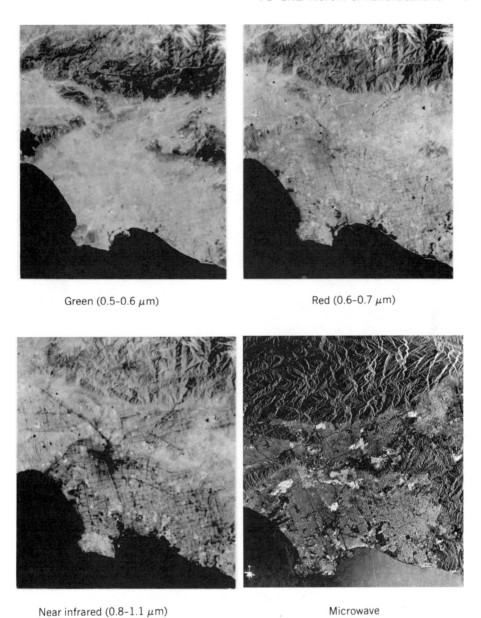

Green (0.5–0.6 μm) Red (0.6–0.7 μm)

Near infrared (0.8–1.1 μm) Microwave

Figure 1-5. Multispectral satellite images of the Los Angeles basin acquired in the visible, infrared, and microwave regions of the spectrum. See color section.

lution capability from high-flying aircraft. These systems became available to the scientific community in the mid-1960s. Since then, work has continued at a number of institutions to develop the capability of radar sensors to study natural surfaces. This work led to the orbital flight around the Earth of the Seasat SAR (1978) and the Shuttle Imaging Radar (1981, 1984). Since then, several countries have flown orbital SAR systems.

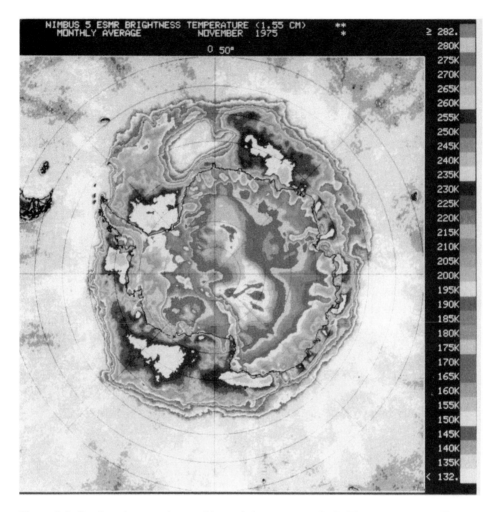

Figure 1-6. Passive microwave image of Antarctic ice cover acquired with a spaceborne radiometer. The color chart corresponds to the surface brightness temperature. See color section.

The most recently introduced remote sensing instrument is the laser, which was first developed in 1960. It is mainly being used for atmospheric studies, topographic mapping, and surface studies by fluorescence.

There has been great progress in spaceborne remote sensing over the past three decades. Most of the early remote sensing satellites were developed exclusively by government agencies in a small number of countries. Now nearly 20 countries are either developing or flying remote sensing satellites, and many of these satellites are developed, launched and operated by commercial firms. In some cases, these commercial firms have completely replaced government developers, and the original developers in the governments now are simply the customers of the commercial firms.

The capabilities of remote sensing satellites have also dramatically increased over the past two decades. The number of spectral channels available has grown from a few to more than 200 in the case of the Hyperion instrument. Resolutions of a few meters or less are now available from commercial vendors. Synthetic-aperture radars are now ca-

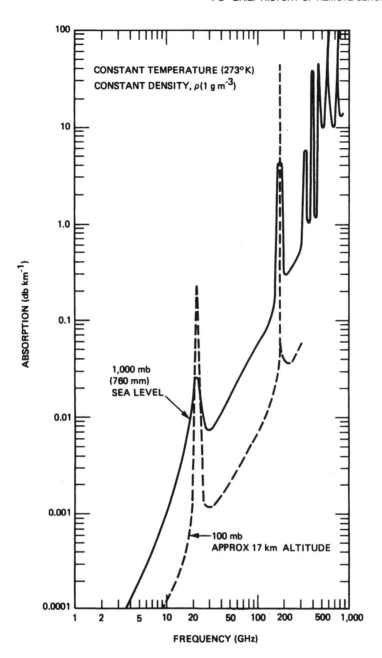

Figure 1-7. Absorption spectrum of H_2O for two pressures (100 mbars and 1000 mbars), at constant temperature of 273° K. (From Chahine et al., 1983.)

pable of collecting images on demand in many different modes. Satellites are now acquiring images of other planets in more spectral channels and with better resolutions than what was available for the Earth two decades ago. And as the remote sensing data have become more available, the number of applications has grown. In many cases, the limitation now has shifted from the technology that acquires the data to the techniques

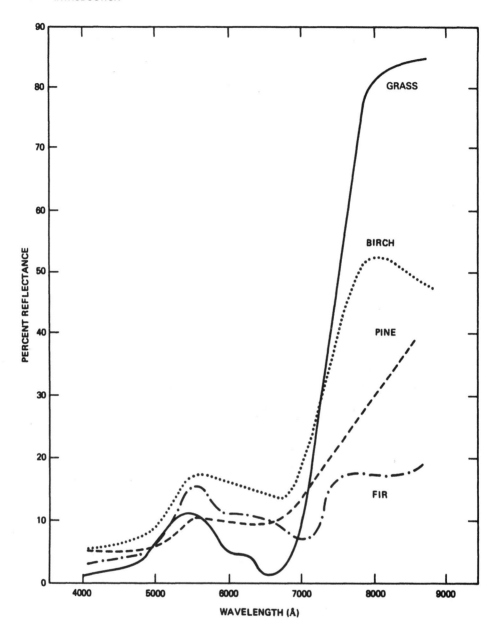

Figure 1-8. Spectral signature of some vegetation types. (From Brooks, 1972.)

and training needed to optimally exploit the information embedded in the remote sensing data.

1-3 REMOTE SENSING SPACE PLATFORMS

Up until 1946, remote sensing data were mainly acquired from airplanes or balloons. In 1946, pictures were taken from V-2 rockets. The sounding rocket photographs proved

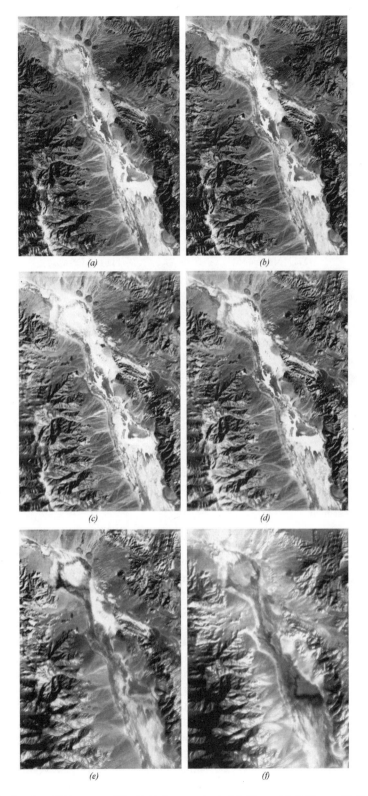

Figure 1-9. Landsat TM images of Death Valley acquired at 0.48 μm (*a*), 0.56 μm. (*b*), 0.66 μm (*c*), and 0.83 μm (*d*), 1.65 μm (*e*), and 11.5 μm (*f*).

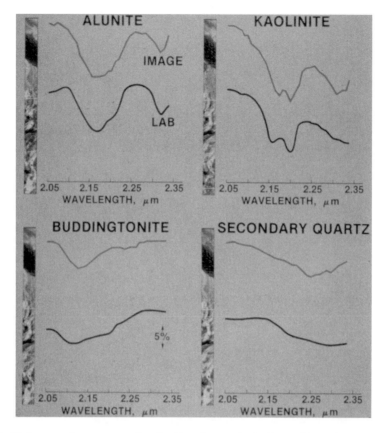

Figure 1-10. Images of an area near Cuprite, Nevada, acquired with an airborne imaging spectrometer. The image is shown to the left. The spectral curves derived from the image data are compared to the spectral curves measured in the laboratory using samples from the same area. (Courtesy of JPL.) See color section.

invaluable in illustrating the potential value of photography from orbital altitudes. Systematic orbital observations of the Earth began in 1960 with the launch of Tiros I, the first meteorological satellite, using a low-resolution imaging system. Each Tiros spacecraft carried a narrow-angle TV, five-channel scanning radiometer, and a bolometer.

In 1961, orbital color photography was acquired by an automatic camera in the unmanned MA-4 Mercury spacecraft. This was followed by photography acquired during the Mercury, Gemini, Apollo, and Skylab missions. On Apollo 9, the first multispectral images were acquired to assess their use for Earth resources observation. This was followed by the launch in 1972 of the first Earth Resources Technology Satellite (ERTS-1, later renamed Landsat-1), which was one of the major milestones in the field of Earth remote sensing. ERTS-1 was followed by the series of Landsat missions.

Earth orbital spacecraft are also used to acquire remote sensing data other than regular photography. To name just a few, the Nimbus spacecraft carry passive microwave radiometers, infrared spectrometers, and infrared radiometers. The Synchronous Meteorological Satellite (SMS) carried visible and IR spin-scan cameras. Skylab (1972) carried a

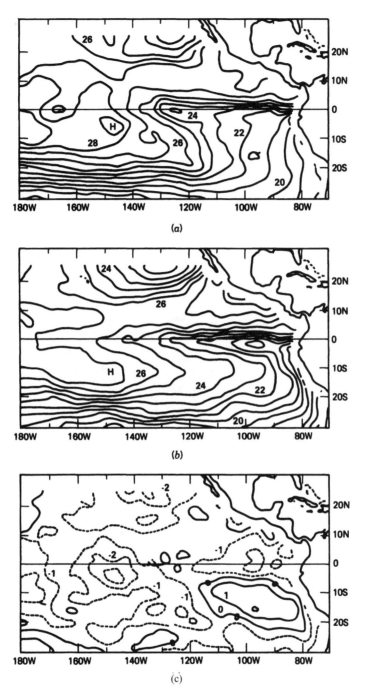

Figure 1-11. Sea surface temperature derived from ship observations (*a*) and from the Seasat Multi-spectral Microwave Radiometer (*b*). (*c*) shows the difference. (From Liu, 1983.)

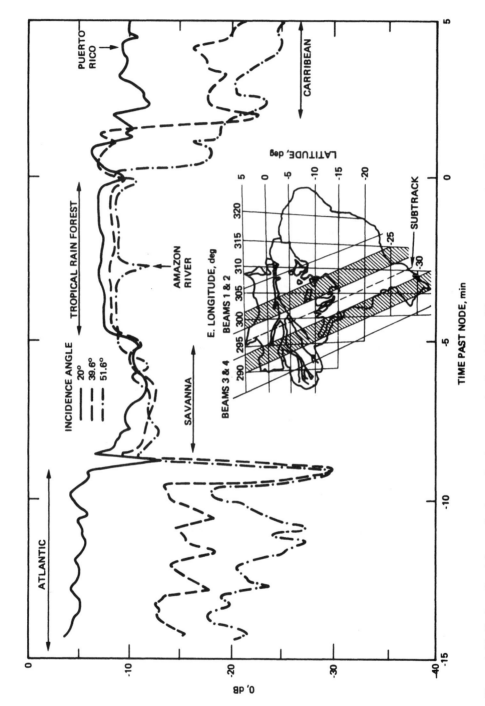

Figure 1-12. Backscatter data acquired over the Amazon region (insert). The different curves correspond to different incidence angles. Data was acquired by the Seasat Scatterometer at 14.6 GHz and at VV polarization. (Bracalante et al., 1980.)

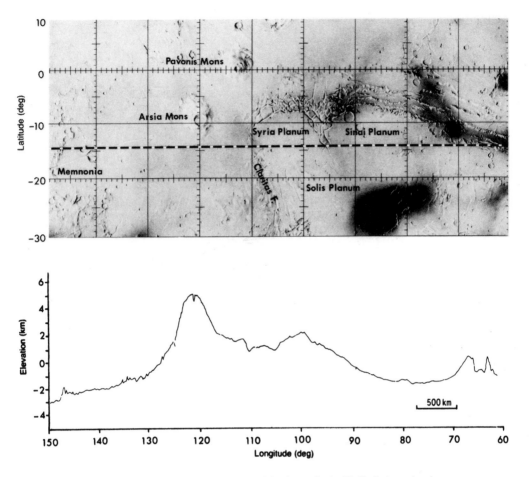

Figure 1-13. Profile of Tharsis region (Mars) acquired with Earth-based radar.

radiometer and a radar scatterometer. Seasat (1978) carried an imaging radar, a scatterometer, and an altimeter.

In the 1980s and 1990s, the Space Shuttle provided an additional platform for remote sensing. A number of shuttle flights carried imaging radar systems. In particular, the Shuttle Radar Topography Mission, flown on the Space Shuttle in 2000, allowed global mapping of the Earth's topography.

Remote sensing activity was also expanding dramatically using planetary spacecraft. Images were acquired of the surfaces of the Moon, Mercury, Venus, Mars, and the Jovian and Saturnian satellites, and of the atmospheres of Venus, Jupiter, Saturn, and Uranus. Other types of remote sensors, such as radar altimeters, sounders, gamma-ray detectors, infrared radiometers, and spectrometers were used on a number of planetary missions.

The use of orbiting spacecraft is becoming a necessity in a number of geophysical disciplines because they allow the acquisition of global and synoptic coverage with a relatively short repetitive period. These features are essential for observing dynamic atmospheric, oceanic, and biologic phenomena. The global coverage capability is also essential in a number of geologic applications where large-scale structures are being investigated.

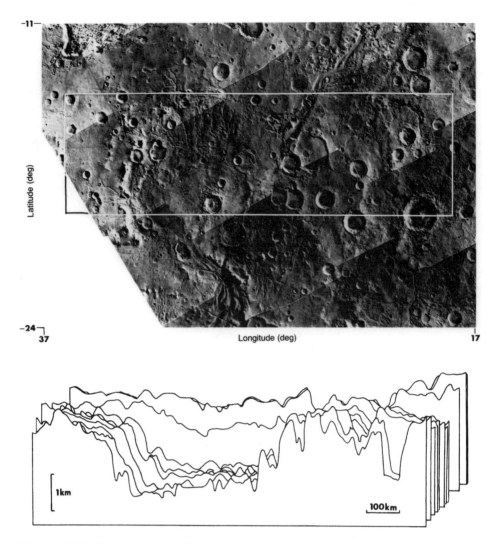

Figure 1-14. Profiles of an unnamed impact basin on Mars using Earth-based radar. The set of profiles shown correspond to the box overlay on the figure.

In addition, planetary rovers are using remote sensing instruments to conduct close-up analysis of planetary surfaces.

1-4 TRANSMISSION THROUGH THE EARTH AND PLANETARY ATMOSPHERES

The presence of an atmosphere puts limitations on the spectral regions that can be used to observe the underlying surface. This is a result of wave interactions with atmospheric and ionospheric constituents, leading to absorption or scattering in specific spectral regions (Fig. 1-19).

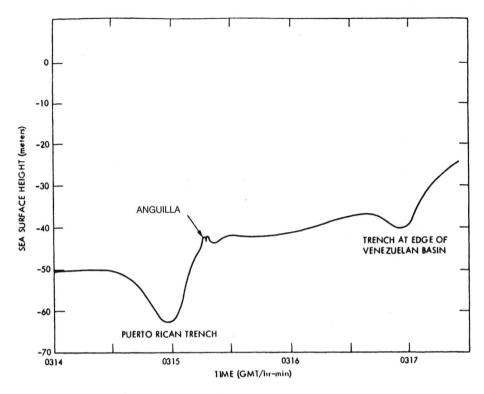

Figure 1-15. Sea surface height over two trenches in the Caribbean acquired with the Seasat altimeter (Townsend, 1980).

At radio frequencies below 10 MHz, the Earth's ionosphere blocks any transmission to or from the surface. In the rest of the radio frequency region, up to the low microwave (10 GHz), the atmosphere is effectively transparent. In the rest of the microwave region, there are a number of strong absorption bands, mainly associated with water vapor and oxygen.

In the submillimeter and far-infrared regions, the atmosphere is almost completely opaque, and the surface is invisible. This opacity is due mainly to the presence of absorption-spectral bands associated with the atmospheric constituents. This makes the spectral region most appropriate for atmospheric remote sensing.

The opacity of the atmosphere in the visible and near infrared is high in selected bands in which the high absorption coefficients are due to a variety of electronic and vibrational processes mainly related to the water vapor and carbon dioxide molecules. In the ultraviolet, the opacity is mainly due to the ozone layer in the upper atmosphere.

The presence of clouds leads to additional opacity due to absorption and scattering by cloud drops. This limits the observation capabilities in the visible, infrared, and submillimeter regions. In the microwave and radio frequency regions, clouds are basically transparent.

In the case of the other planets, more extreme conditions are encountered. In the case of Mercury and the Moon, no significant atmosphere exists, and the whole electromagnetic spectrum can be used for surface observation. In the case of Venus and Titan, the con-

Figure 1-16. Shaded relief display of the topography of California measured by Shuttle Radar Topography Mission using an interferometric SAR.

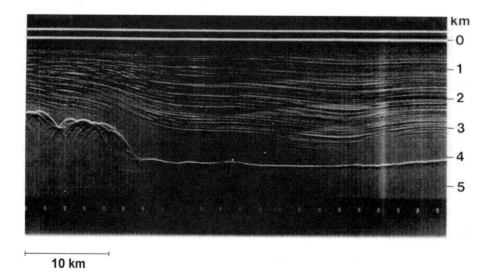

10 km

Figure 1-17. Subsurface layering in the ice cover and bedrock profile acquired with an airborne electromagnetic sounder over a part of the Antarctic ice sheet.

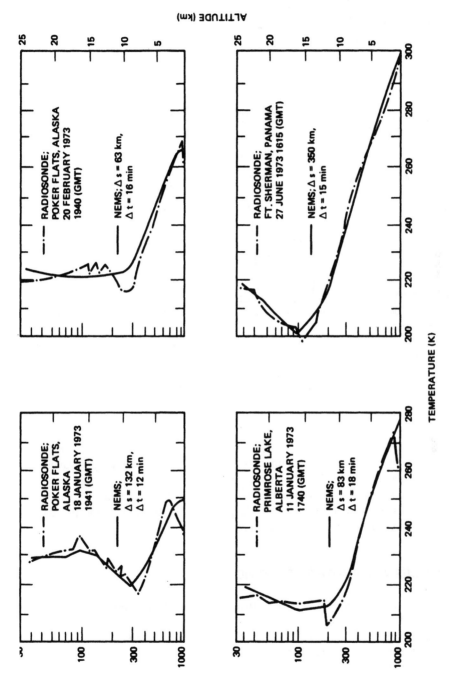

Figure 1-18. Comparison of temperature profiles acquired with a microwave sounder (NEMS) and radiosounder. The spatial and temporal differences, Δs and Δt, between the two measurements are indicated (Waters et al., 1975).

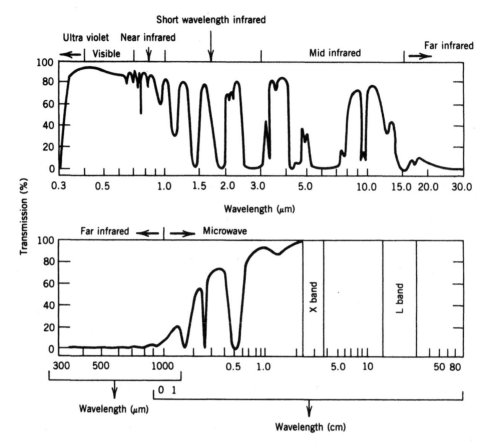

Figure 1-19. Generalized absorption spectrum of the Earth's atmosphere at the zenith. The curve shows the total atmospheric transmission.

tinuous and complete cloud coverage limits surface observation to the longer-wavelength regions, particularly radio frequency and microwave bands. In the case of Mars, the tenuous atmosphere is essentially transparent across the spectrum even though a number of absorption bands are present. In the case of the giant planets, the upper atmosphere is essentially all that can be observed and studied remotely.

REFERENCES AND FURTHER READING

Bracalante, E. M. et al. The SASS scattering coefficient algorithm. *IEEE Journal of. Oceanic Engineering,* **OE-5,** 145–153, 1980.

Chahine, M. et al. Interaction mechanisms within the atmosphere. Chapter 5 in *Manual of Remote Sensing,* American Society of Photogrammetry, Falls Church, VA, 1983.

Colwell, R. N. (Ed.). *Manual of Remote Sensing.* American Society of Photogrammetry, Falls Church, VA, 1983.

Kramer, H. J. *Observation of the Earth and its Environment—Survey of Missions and Sensors,* Springer-Verlag, Berlin, Germany, 2002.

Liu, C. T. Tropical Pacific sea surface temperatures measured by Seasat microwave radiometer and by ships. *Journal of Geophysical Research,* **88,** 1909–1916, 1983.

Townsend, W. F. The initial assessment of the performance achieved by the Seasat radar altimeter. *IEEE Journal of Oceanic Engineering,* **OE-5,** 80–92, 1980.

Waters, J. et al. Remote sensing of atmospheric temperature profiles with the Nimbus 5 microwave spectrometer. *Journal of Atmospheric Sciences,* **32**(10), 1975.

2

Nature and Properties of Electromagnetic Waves

2-1 FUNDAMENTAL PROPERTIES OF ELECTROMAGNETIC WAVES

Electromagnetic energy is the means by which information is transmitted from an object to a sensor. Information could be encoded in the frequency content, intensity, or polarization of the electromagnetic wave. The information is propagated by electromagnetic radiation at the velocity of light from the source directly through free space, or indirectly by reflection, scattering, and reradiation to the sensor. The interaction of electromagnetic waves with natural surfaces and atmospheres is strongly dependent on the frequency of the waves. Waves in different spectral bands tend to excite different interaction mechanisms such as electronic, molecular, or conductive mechanisms.

2-1-1 Electromagnetic Spectrum

The electromagnetic spectrum is divided into a number of spectral regions. For the purpose of this text, we use the classification illustrated in Figure 2-1.

The radio band covers the region of wavelengths longer than 10 cm (frequency less than 3 GHz). This region is used by active radio sensors such as imaging radars, altimeters, and sounders, and, to a lesser extent, passive radiometers.

The microwave band covers the neighboring region, down to a wavelength of 1 mm (300 GHz frequency). In this region, most of the interactions are governed by molecular rotation, particularly at the shorter wavelengths. This region is mostly used by microwave radiometers/spectrometers and radar systems.

The infrared band covers the spectral region from 1 mm to 0.7 μm. This region is subdivided into subregions called submillimeter, far infrared, thermal infrared, and near infrared. In this region, molecular rotation and vibration play important roles. Imagers, spectrometers, radiometers, polarimeters, and lasers are used in this region for remote

Introduction to the Physics and Techniques of Remote Sensing. By C. Elachi and J. van Zyl
Copyright © 2006 John Wiley & Sons, Inc.

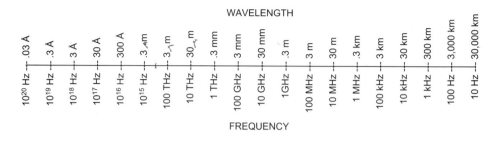

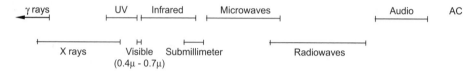

Figure 2-1. Electromagnetic spectrum.

sensing. The same is true in the neighboring region, the visible region (0.7–0.4 μm) where electronic energy levels start to play a key role.

In the next region, the ultraviolet (0.4 μm to 300 Å), electronic energy levels play the main role in wave–matter interaction. Ultraviolet sensors have been used mainly to study planetary atmospheres or to study surfaces with no atmospheres because of the opacity of gases at these short wavelengths.

X-rays (300 Å to 0.3 Å) and gamma rays (shorter than 0.3 Å) have been used to an even lesser extent because of atmospheric opacity. Their use has been limited to low-flying aircraft platforms or to the study of planetary surfaces with no atmosphere (e.g., the Moon).

2-1-2 Maxwell's Equations

The behavior of electromagnetic waves in free space is governed by Maxwell's equations:

$$\nabla \times \mathbf{E} = -\frac{\partial \mathbf{B}}{\partial t} \tag{2-1}$$

$$\nabla \times \mathbf{H} = \frac{\partial \mathbf{D}}{\partial t} + \mathbf{J} \tag{2-2}$$

$$\mathbf{B} = \mu_0 \mu_r \mathbf{H} \tag{2-3}$$

$$\mathbf{D} = \varepsilon_0 \varepsilon_r \mathbf{E} \tag{2-4}$$

$$\nabla \cdot \mathbf{E} = 0 \tag{2-5}$$

$$\nabla \cdot \mathbf{B} = 0 \tag{2-6}$$

where
$\mathbf{E}$ = electric vector
$\mathbf{D}$ = displacement vector

$\mathbf{H}$ = magnetic vector
$\mathbf{B}$ = induction vector
μ_0, ε_0 = permeability and permittivity of vacuum
μ_r, ε_r = relative permeability and permittivity

Maxwell's concept of electromagnetic waves is that a smooth wave motion exists in the magnetic and electric force fields. In any region in which there is a temporal change in the electric field, a magnetic field appears automatically in that same region as a conjugal partner and vice versa. This is expressed by the above coupled equations.

2-1-3 Wave Equation and Solution

In homogeneous, isotropic, and nonmagnetic media, Maxwell's equations can be combined to derive the wave equation:

$$\nabla^2 \mathbf{E} - \mu_0 \varepsilon_0 \mu_r \varepsilon_r \frac{\partial^2 \mathbf{E}}{\partial t^2} = 0 \tag{2-7}$$

or, in the case of a sinusoidal field,

$$\nabla^2 \mathbf{E} + \frac{\omega^2}{c_r^2} \mathbf{E} = 0 \tag{2-8}$$

where

$$c_r = \frac{1}{\sqrt{\mu_0 \varepsilon_0 \mu_r \varepsilon_r}} = \frac{c}{\sqrt{\mu_r \varepsilon_r}} \tag{2-9}$$

Usually, $\mu_r = 1$ and ε_r varies from 1 to 80 and is a function of the frequency. The solution for the above differential equation is given by

$$\mathbf{E} = \mathbf{A} e^{i(kr - \omega t + \phi)} \tag{2-10}$$

where $\mathbf{A}$ is the wave amplitude, ω is the angular frequency, ϕ is the phase, and k is the wave vector in the propagation medium ($k = 2\pi \sqrt{\varepsilon_r}/\lambda$, λ = wavelength = $2\pi c/\omega$, c = speed of light in vacuum). The wave frequency v is defined as $v = \omega/2\pi$.

Remote sensing instruments exploit different aspects of the solution to the wave equation in order to learn more about the properties of the medium from which the radiation is being sensed. For example, the interaction of electromagnetic waves with natural surfaces and atmospheres is strongly dependent on the frequency of the waves. This will manifest itself in changes in the amplitude [the magnitude of $\mathbf{A}$ in Equation (2-10)] of the received wave as the frequency of the observation is changed. This type of information is recorded by multispectral instruments such as the LandSat Thematic Mapper and the Advanced Advanced Spaceborne Thermal Emission and Reflection Radiometer. In other cases, one can infer information about the electrical properties and geometry of the surface by observing the polarization [the vector components of $\mathbf{A}$ in Equation (2-10)] of the received waves. This type of information is recorded by polarimeters and polarimetric radars. Doppler lidars and radars, on the other hand, measure the change in frequency between the transmitted and received waves in order to infer the velocity with which an object or medium is moving. This information is contained in the angular frequency ω of the wave shown in Equation (2-10). The quantity $kr - \omega t + \phi$ in Equation (2-10) is known as the *phase* of the wave. This phase changes by 2π every time the wave moves through a distance equal to the wavelength λ. Measuring the phase of a wave therefore provides an extremely accurate way to measure the distance that the wave actually travelled. Interferometers exploit this property of the wave to accurately measure differences in the path length between a source and two collectors, allowing one to significantly increase the resolution with which the position of the source can be established.

2-1-4   Quantum Properties of Electromagnetic Radiation

Maxwell's formulation of electromagnetic radiation leads to a mathematically smooth wave motion of fields. However, at very short wavelengths, it fails to account for certain significant phenomena that occur when the wave interacts with matter. In this case, a quantum description is more appropriate.

The electromagnetic energy can be presented in a quantized form as bursts of radiation with a quantized radiant energy Q, which is proportional to the frequency v:

$$Q = hv \qquad (2\text{-}11)$$

where h = Planck's constant = 6.626×10^{-34} joule second. The radiant energy carried by the wave is not delivered to a receiver as if it is spread evenly over the wave, as Maxwell had visualized, but is delivered on a probabilistic basis. The probability that a wave train will make full delivery of its radiant energy at some place along the wave is proportional to the flux density of the wave at that place. If a very large number of wave trains are co-existent, then the overall average effect follows Maxwell's equations.

2-1-5   Polarization

An electromagnetic wave consists of a coupled electric and magnetic force field. In free space, these two fields are at right angles to each other and transverse to the direction of propagation. The direction and magnitude of only one of the fields (usually the electric field) is sufficient to completely specify the direction and magnitude of the other field using Maxwell's equations.

The polarization of the electromagnetic wave is contained in the elements of the vector amplitude **A** of the electric field in Equation (2-10). For a transverse electromagnetic wave, this vector is orthogonal to the direction in which the wave is propagating, and, therefore, we can completely describe the amplitude of the electric field by writing **A** as a two-dimensional complex vector:

$$\mathbf{A} = a_h e^{i\delta_h}\hat{\mathbf{h}} + a_v e^{i\delta_v}\hat{\mathbf{v}} \qquad (2\text{-}12)$$

Here, we denote the two orthogonal basis vectors as $\hat{\mathbf{h}}$ for *horizontal* and $\hat{\mathbf{v}}$ for *vertical*. Horizontal polarization is usually defined as the state in which the electric vector is perpendicular to the plane of incidence. Vertical polarization is orthogonal to both horizontal polarization and the direction of propagation, and corresponds to the case in which the electric vector is in the plane of incidence. Any two orthogonal basis vectors could be used to describe the polarization, and in some cases the right- and left-handed circular basis is used. The amplitudes, a_h and a_v, and the relative phases, δ_h and δ_v, are real numbers. The polarization of the wave can be thought of as that figure that the tip of the electric field would trace over time at a fixed point in space. Taking the real part of Equation (2-12), we find that the polarization figure is the locus of all the points in the h–v plane that have the coordinates $E_h = a_h \cos \delta_h$, $E_v = a_v \cos \delta_v$. It can easily be shown that the points on the locus satisfy the expression

$$\left(\frac{E_h}{a_h}\right)^2 + \left(\frac{E_v}{a_v}\right)^2 - 2\frac{E_h}{a_h}\frac{E_v}{a_v}\cos(\delta_h - \delta_v) = \sin^2(\delta_h - \delta_v) \qquad (2\text{-}13)$$

This is the expression of an ellipse, shown in Figure 2-2. Therefore, in the general case, electromagnetic waves are *elliptically* polarized. In tracing the ellipse, the tip of the electric field can rotate either clockwise, or counterclockwise; this direction is denoted by the *handedness* of the polarization. The definition of handedness accepted by the Institute of Electrical and Electronics Engineers (IEEE), is that a wave is said to have *right-handed* polarization if the tip of the electric field vector rotates *clockwise* when the wave is viewed receding from the observer. If the tip of the electric field vector rotates *counterclockwise* when the wave is viewed in the same way, it has a *left-handed* polarization. It is worth pointing out that in the optical literature a different definition of handedness is often encountered. In that case, a wave is said to be right-handed (left-handed) polarized when the wave is view approaching the observer, and tip of the electric field vector rotates in the clockwise (counterclockwise) direction.

In the special case in which the ellipse collapses to a line, which happens when $\delta_h - \delta_v = n\pi$ with n any integer, the wave is said to be *linearly* polarized. Another special case is encountered when the two amplitudes are the same ($a_h = a_v$) and the relative phase difference $\delta_h - \delta_v$ is either $\pi/2$ or $-\pi/2$. In this case, the wave is *circularly* polarized.

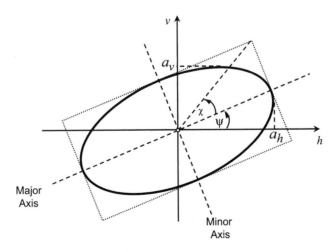

Figure 2-2. Polarization ellipse.

The polarization ellipse (see Fig. 2-2) can also be characterized by two angles known as the ellipse orientation angle (ψ in Fig. 2-2, $0 \leq \psi \leq \pi$) and the ellipticity angle, shown as χ ($-\pi/4 \leq \chi \leq \pi/4$) in Figure 2-2. These angles can be calculated as follows:

$$\tan 2\psi = \frac{2a_h a_v}{a_h^2 - a_v^2} \cos(\delta_h - \delta_v); \qquad \sin 2\chi = \frac{2a_h a_v}{a_h^2 + a_v^2} \sin(\delta_h - \delta_v) \qquad (2\text{-}14)$$

Note that linear polarizations are characterized by an ellipticity angle $\chi = 0$.

So far, it was implied that the amplitudes and phases shown in Equations (2-12) and (2-13) are constant in time. This may not always be the case. If these quantities vary with time, the tip of the electric field vector will not trace out a smooth ellipse. Instead, the figure will in general be a noisy version of an ellipse that after some time may resemble an "average" ellipse. In this case, the wave is said to be *partially* polarized, and it can be considered that part of the energy has a deterministic polarization state. The radiation from some sources, such as the sun, does not have any clearly defined polarization. The electric field assumes different directions at random as the wave is received. In this case, the wave is called *randomly* polarized or *unpolarized.* In the case of some man-made sources, such as lasers and radio/radar transmitters, the wave usually has a well-defined polarized state.

Another way to describe the polarization of a wave, particularly appropriate for the case of partially polarized waves, is through the use of the *Stokes parameters* of the wave. For a monochromatic wave, these four parameters are defined as

$$\begin{aligned} S_0 &= a_h^2 + a_v^2 \\ S_1 &= a_h^2 - a_v^2 \\ S_2 &= 2a_h a_v \cos(\delta_h - \delta_v) \\ S_3 &= 2a_h a_v \sin(\delta_h - \delta_v) \end{aligned} \qquad (2\text{-}15)$$

Note that for such a fully polarized wave, only three of the Stokes parameters are independent, since $S_0^2 = S_1^2 + S_2^2 + S_3^2$. Using the relations in Equations (2-14) between the ellipse orientation and ellipticity angles and the wave amplitudes and relative phases, it can be shown that the Stokes parameters can also be written as

$$\begin{aligned} S_1 &= S_0 \cos 2\chi \cos 2\psi \\ S_2 &= S_0 \cos 2\chi \sin 2\psi \\ S_3 &= S_0 \sin 2\psi \end{aligned} \qquad (2\text{-}16)$$

The relations in Equations (2-16) lead to a simple geometric interpretation of polarization states. The Stokes parameters S_1, S_2, and S_3 can be regarded as the Cartesian coordinates of a point on a sphere, known as the Poincaré sphere, of radius S_0 (see Fig. 2-3). There is, therefore, a unique mapping between the position of a point on the surface of the sphere and a polarization state. Linear polarizations map to points on the equator of the Poincaré sphere, whereas the circular polarizations map to the poles (Fig. 2-4).

In the case of partially polarized waves, all four Stokes parameters are required to fully describe the polarization of the wave. In general, the Stokes parameters are related by $S_0^2 \geq S_1^2 + S_2^2 + S_3^2$, with equality holding only for fully polarized waves. In the extreme

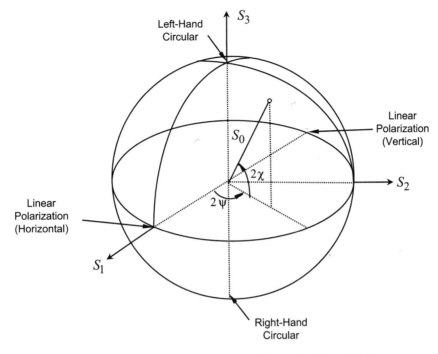

Figure 2-3. Polarization represented as a point on the Poincaré sphere.

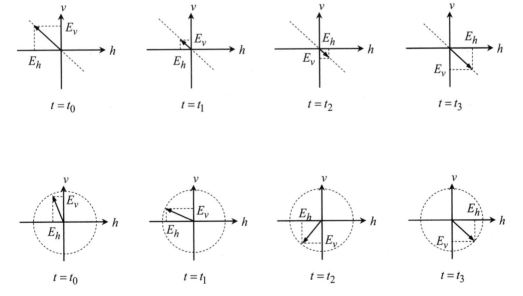

Figure 2-4. Linear (upper) and circular (lower) polarization.

case of an unpolarized wave, the Stokes parameters are $S_0 > 0$; $S_1 = S_2 = S_3 = 0$. It is always possible to describe a partially polarized wave by the sum of a fully polarized wave and an unpolarized wave. The magnitude of the polarized wave is given by $\sqrt{S_1^2 + S_2^2 + S_3^2}$, and the magnitude of the unpolarized wave is $S_0 - \sqrt{S_1^2 + S_2^2 + S_3^2}$. Finally, it should be pointed out that the Stokes parameters of an unpolarized wave can be written as the sum of two fully polarized waves:

$$
\begin{pmatrix} S_0 \\ 0 \\ 0 \\ 0 \end{pmatrix} = \frac{1}{2} \begin{pmatrix} S_0 \\ S_1 \\ S_2 \\ S_3 \end{pmatrix} + \frac{1}{2} \begin{pmatrix} S_0 \\ -S_1 \\ -S_2 \\ -S_3 \end{pmatrix} \tag{2-17}
$$

These two fully polarized waves have orthogonal polarizations. This result shows that when an antenna with a particular polarization is used to receive unpolarized radiation, the amount of power received by the antenna will be only that half of the power in the unpolarized wave that aligns with the antenna polarization. The other half of the power will not be absorbed because its polarization is orthogonal to that of the antenna.

The polarization states of the incident and reradiated waves play an important role in remote sensing. They provide an additional information source (in addition to the intensity and frequency) that may be used to study the properties of the radiating or scattering object. For example, at an incidence angle of 37° from vertical, an optical wave polarized perpendicular to the plane of incidence will reflect about 7.8% of its energy from a smooth water surface, whereas an optical wave polarized in the plane of incidence will not reflect any energy from the same surface. All the energy will penetrate into the water. This is the Brewster effect.

2-1-6 Coherency

In the case of a monochromatic wave of certain frequency v_0, the instantaneous field at any point P is well defined. If the wave consists of a large number of monochromatic waves with frequencies over a bandwidth ranging from v_0 to $v_0 + \Delta v$, then the random addition of all the component waves will lead to irregular fluctuations of the resultant field.

The coherency time Δt is defined as the period over which there is strong correlation of the field amplitude. More specifically, it is the time after which two waves at v and $v + \Delta v$ are out of phase by one cycle; that is, it is given by

$$
v\Delta t + 1 = (v + \Delta v)\Delta t \rightarrow \Delta v \Delta t = 1
$$
$$
\rightarrow \Delta t = \frac{1}{\Delta v} \tag{2-18}
$$

The coherence length is defined as

$$
\Delta l = c\Delta t = \frac{c}{\Delta v} \tag{2-19}
$$

Two waves or two sources are said to be coherent with each other if there is a systematic relationship between their instantaneous amplitudes. The amplitude of the resultant field varies between the sum and the difference of the two amplitudes. If the two waves are incoherent, then the power of the resultant wave is equal to the sum of the power of the two constituent waves. Mathematically, let $E_1(t)$ and $E_2(t)$ be the two component fields at a certain location. Then the total field is

$$E(t) = E_1(t) + E_2(t) \tag{2-20}$$

The average power is

$$
\begin{aligned}
P \sim \overline{[E(t)]^2} &= \overline{[E_1(t) + E_2(t)]^2} \\
&= \overline{E_1(t)^2} + \overline{E_2(t)^2} + 2\,\overline{E_1(t)E_2(t)}
\end{aligned}
\tag{2-21}
$$

If the two waves are incoherent relative to each other, then $\overline{E_1(t)E_2(t)} = 0$ and $P = P_1 + P_2$. If the waves are coherent, then $\overline{E_1(t)E_2(t)} \neq 0$. In the latter case, we have

$$P > P_1 + P_2 \quad \text{in some locations}$$
$$P < P_1 + P_2 \quad \text{in other locations}$$

This is the case of optical interference fringes generated by two overlapping coherent optical beams. The bright bands correspond to where the energy is above the mean and the dark bands correspond to where the energy is below the mean.

2-1-7 Group and Phase Velocity

The phase velocity is the velocity at which a constant phase front progresses (see Fig. 2-5). It is equal to

$$v_p = \frac{\omega}{k} \tag{2-22}$$

If we have two waves characterized by $(\omega - \Delta\omega, k - \Delta k)$ and $(\omega + \Delta\omega, k + \Delta k)$, then the total wave is given by

$$
\begin{aligned}
E(z, t) &= A e^{i[(k-\Delta k)z - (\omega - \Delta\omega)t]} + A e^{i[(k+\Delta k)z - (\omega + \Delta\omega)t]} \\
&= 2A e^{i(kz - \omega t)} \cos(\Delta k z - \Delta\omega t)
\end{aligned}
\tag{2-23}
$$

In this case, the plane of constant amplitude moves at a velocity v_g, called the group velocity:

$$v_g = \frac{\Delta\omega}{\Delta k} \tag{2-24}$$

As $\Delta\omega$ and Δk are assumed to be small, we can write

$$v_g = \frac{\partial\omega}{\partial k} \tag{2-25}$$

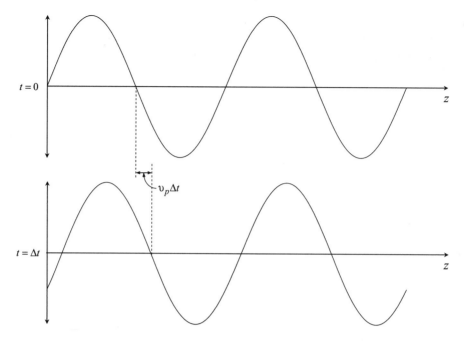

Figure 2-5. Phase velocity.

This is illustrated in Figure 2-6. It is important to note that v_g represents the velocity of propagation of the wave energy. Thus, the group velocity v_g must be equal to or smaller than the speed of light c. However, the phase velocity v_p can be larger than c.

If the medium is nondispersive, then

$$\omega = ck \qquad (2\text{-}26)$$

This implies that

$$v_p = \frac{\omega}{k} = c \qquad (2\text{-}27)$$

$$v_g = \frac{\partial \omega}{\partial k} = c \qquad (2\text{-}28)$$

However, if the medium is dispersive (i.e., ω is a nonlinear function of k), such as in the case of ionospheres, then the two velocities are different.

2-1-8 Doppler Effect

If the relative distance between a source radiating at a fixed frequency v and an observer varies, the signal received by the observer will have a frequency v', which is different than v. The difference, $v_d = v' - v$, is called the Doppler shift. If the source–observer distance is decreasing, the frequency received is higher than the frequency transmitted, lead-

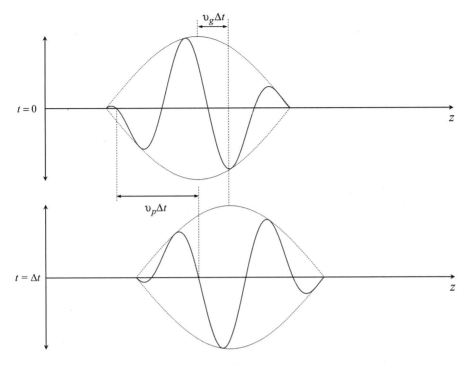

Figure 2-6. Group velocity.

ing to a positive Doppler shift ($v_d > 0$). If the source–observer distance is increasing, the reverse effect occurs (i.e., $v_d < 0$) and the Doppler shift is negative.

The relationship between v_d and v is

$$v_d = v \frac{v}{c} \cos \theta \qquad (2\text{-}29)$$

where v is the relative speed between the source and the observer, c is the velocity of light, and θ is the angle between the direction of motion and the line connecting the source and the observer (see Fig. 2-7). The above expression assumes no relativistic effects ($v \ll c$), and it can be derived in the following simple way.

Referring to Figure 2-8, assume an observer is moving at a velocity v with an angle θ relative to the line of propagation of the wave. The lines of constant wave amplitude are separated by the distance λ (i.e., wavelength) and are moving at velocity c. For the observer, the apparent frequency v' is equal to the inverse of the time period T' that it takes the observer to cross two successive equiamplitude lines. This is given by the expression

$$cT' + vT' \cos \theta = \lambda \qquad (2\text{-}30)$$

which can be written as

$$\frac{c}{v'} + \frac{v \cos \theta}{v'} = \frac{c}{v} \ \rhd \ v' = v + v \frac{v}{c} \cos \theta = v + v_d \qquad (2\text{-}31)$$

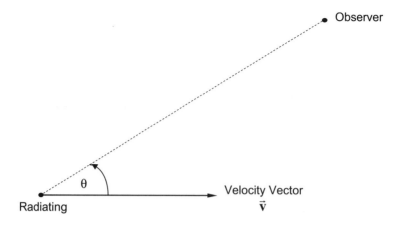

Figure 2-7. Doppler geometry for a moving source, fixed observer.

The Doppler effect also occurs when the source and observer are fixed relative to each other but the scattering or reflecting object is moving (see Fig. 2-9). In this case, the Doppler shift is given by

$$v_d = v \frac{v}{c} (\cos \theta_1 + \cos \theta_2) \qquad (2\text{-}32)$$

and if the source and observer are collocated (i.e., $\theta_1 = \theta_2 = \theta$), then

$$v_d = 2v \frac{v}{c} \cos \theta \qquad (2\text{-}33)$$

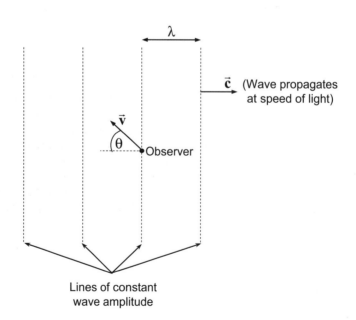

Figure 2-8. Geometry illustrating wave fronts passing by a moving observer.

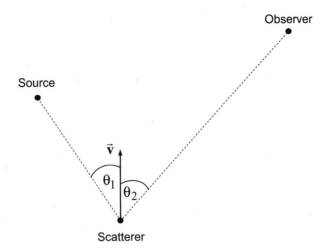

Figure 2-9. Doppler geometry for a moving scatterer with fixed source and observer.

The Doppler effect is used in remote sensing to measure target motion. It is also the basic physical effect used in synthetic-aperture imaging radars to achieve very high resolution imaging.

2-2 NOMENCLATURE AND DEFINITION OF RADIATION QUANTITIES

2-2-1 Radiation Quantities

A number of quantities are commonly used to characterize the electromagnetic radiation and its interaction with matter. These are briefly described below and summarized in Table 2-1.

Radiant energy. The energy carried by an electromagnetic wave. It is a measure of the capacity of the wave to do work by moving an object by force, heating it, or changing its state. The amount of energy per unit volume is called radiant energy density.

Radiant flux. The time rate at which radiant energy passes a certain location. It is closely related to the wave *power,* which refers to the time rate of doing work. The term *flux* is also used to describe the time rate of flow of quantized energy elements such as photons. Then the term *photon flux* is used.

Radiant flux density. Corresponds to the radiant flux intercepted by a unit area of a plane surface. The density for flux incident upon a surface is called irradiance. The density for flux leaving a surface is called exitance or emittance.

Solid angle. The solid angle Ω subtended by area A on a spherical surface is equal to the area A divided by the square of the radius of the sphere.

Radiant intensity. The radiant intensity of a point source in a given direction is the radiant flux per unit solid angle leaving the source in that direction.

Radiance. The radiant flux per unit solid angle leaving an extended source in a given direction per unit projected area in that direction (see Fig. 2-10). If the radiance does not change as a function of the direction of emission, the source is called Lambertian.

A piece of white matte paper, illuminated by diffuse skylight, is a good example of a Lambertian source.

Hemispherical reflectance. The ratio of the reflected exitance (or emittance) from a plane of material to the irradiance on that plane.

Hemispherical transmittance. The ratio of the transmitted exitance, leaving the opposite side of the plane, to the irradiance.

Table 2-1. Radiation Quantities

Quantity	Usual symbol	Defining equation	Units
Radiant energy	Q		joule
Radiant energy density	W	$W = \dfrac{dQ}{dV}$	joule/m^3
Radiant flux	Φ	$\Phi = \dfrac{dQ}{dt}$	watt
Radiant flux density	E (irradiance) M (emittance)	$E, M = \dfrac{d\Phi}{dA}$	watt/m^2
Radiant intensity	I	$I = \dfrac{d\Phi}{d\Omega}$	watt/steradian
Radiance	L	$L = \dfrac{dI}{dA \cos\theta}$	watt/steradian m^2
Hemispherical reflectance	ρ	$\rho = \dfrac{M_r}{E}$	
Hemispherical absorptance	α	$\alpha = \dfrac{M_a}{E}$	
Hemispherical transmittance	τ	$\tau = \dfrac{M_t}{E}$	

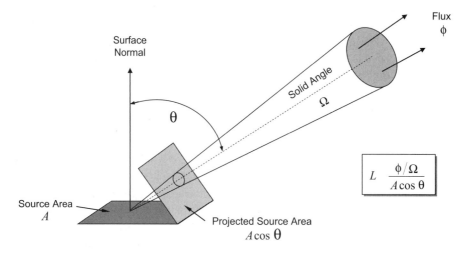

$$L = \frac{\phi/\Omega}{A \cos\theta}$$

Figure 2-10. Concept of radiance.

Hemispherical absorptance. The flux density that is absorbed over the irradiance. The sum of the reflectance, transmittance, and absorptance is equal to one.

2-2-2 Spectral Quantities

Any wave can be considered as being composed of a number of sinusoidal component waves or spectral components, each carrying a part of the radiant flux of the total wave form. The spectral band over which these different components extend is called the spectral width or bandwidth of the wave. The manner with which the radiation quantities are distributed among the components of different wavelengths or frequencies is called the spectral distribution. All radiance quantities have equivalent spectral quantities that correspond to the density as a function of the wavelength or frequency. For instance, the spectral radiant flux $\Phi(\lambda)$ is the flux in a narrow spectral width around λ divided by the spectral width:

$$\Phi(\lambda) = \frac{\text{Flux in all waves in the band } \lambda - \Delta\lambda \text{ to } \lambda + \Delta\lambda}{2\Delta\lambda}$$

To get the total flux from a wave form covering the spectral band from λ_1 to λ_2, the spectral radiant flux must be integrated over that band:

$$\Phi(\lambda_1 \text{ to } \lambda_2) = \int_{\lambda_1}^{\lambda_2} \Phi(\lambda)d\lambda \tag{2-34}$$

2-2-3 Luminous Quantities

Luminous quantities are related to the ability of the human eye to perceive radiative quantities. The relative effectiveness of the eye in converting radiant flux of different wavelengths to visual response is called the spectral luminous efficiency $V(\lambda)$. This is a dimensionless quantity that has a maximum of unity at about 0.55 μm and covers the spectral region from 0.4 to 0.7 μm (see Fig. 2-11). $V(\lambda)$ is used as a weighting function in relating radiant quantities to luminous quantities. For instance, luminous flux Φ_v is related to the radiant spectral flux $\Phi_e(\lambda)$ by

$$\Phi_v = 680 \int_0^\infty \Phi_e(\lambda)V(\lambda)d\lambda \tag{2-35}$$

where the factor 680 is to convert from radiant flux units (watts) to luminous flux units (lumens).

Luminous quantities are also used in relation to sensors other than the human eye. These quantities are usually referenced to a standard source with a specific blackbody temperature. For instance, standard tungsten lamps operating at temperatures between 3200°K and 2850°K are used to test photoemissive tubes.

2-3 GENERATION OF ELECTROMAGNETIC RADIATION

Electromagnetic radiation is generated by transformation of energy from other forms such as kinetic, chemical, thermal, electrical, magnetic, or nuclear. A variety of transformation

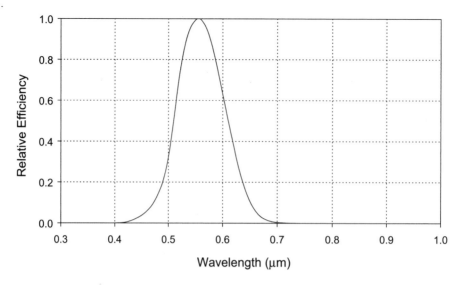

Figure 2-11. Spectral luminous efficiency $V(\lambda)$.

mechanisms lead to electromagnetic waves over different regions of the electromagnetic spectrum. In general, the more organized (as opposed to random) the transformation mechanism is, the more coherent (or narrower in spectral bandwidth) is the generated radiation.

Radio frequency waves are usually generated by periodic currents of electric charges in wires, electron beams, or antenna surfaces. If two short, straight metallic wire segments are connected to the terminals of an alternating current generator, electric charges are moved back and forth between them. This leads to the generation of a variable electric and magnetic field near the wires and to the radiation of an electromagnetic wave at the frequency of the alternating current. This simple radiator is called a dipole antenna.

At microwave wavelengths, electromagnetic waves are generated using electron tubes that use the motion of high-speed electrons in specially designed structures to generate a variable electric/magnetic field, which is then guided by waveguides to a radiating structure. At these wavelengths, electromagnetic energy can also be generated by molecular excitation, as is the case in masers. Molecules have different levels of rotational energy. If a molecule is excited by some means from one level to a higher one, it could drop back to the lower level by radiating the excess energy as an electromagnetic wave.

Higher-frequency waves in the infrared and the visible spectra are generated by molecular excitation (vibrational or orbital) followed by decay. The emitted frequency is exactly related to the energy difference between the two energy levels of the molecules. The excitation of the molecules can be achieved by a variety of mechanisms such as electric discharges, chemical reactions, or photonic illumination.

Molecules in the gaseous state tend to have well-defined, narrow emission lines. In the solid phase, the close packing of atoms or molecules distorts their electron orbits, leading to a large number of different characteristic frequencies. In the case of liquids, the situation is compounded by the random motion of the molecules relative to each other.

Lasers use the excitation of molecules and atoms and the selective decay between energy levels to generate narrow-bandwidth electromagnetic radiation over a wide range of the electromagnetic spectrum ranging from UV to the high submillimeter.

Heat energy is the kinetic energy of random motion of the particles of matter. The random motion results in excitation (electronic, vibrational, or rotational) due to collisions, followed by random emission of electromagnetic waves during decay. Because of its random nature, this type of energy transformation leads to emission over a wide spectral band. If an ideal source (called a blackbody) transforms heat energy into radiant energy with the maximum rate permitted by thermodynamic laws, then the spectral emittance is given by Planck's formula as

$$S(\lambda) = \frac{2\pi hc^2}{\lambda^5} \frac{1}{e^{ch/\lambda kT} - 1} \tag{2-36}$$

where h is Planck's constant, k is the Boltzmann constant, c is the speed of light, λ is the wavelength, and T is the absolute temperature in degrees Kelvin. Figure 2-12 shows the spectral emittance of a number of blackbodies with temperatures ranging from 2000° (temperature of the Sun's surface) to 300°K (temperature of the Earth's surface). The spectral emittance is maximum at the wavelength given by

$$\lambda_m = \frac{a}{T} \tag{2-37}$$

where $a = 2898$ μm°K. The total emitted energy over the whole spectrum is given by the Stefan–Boltzmann law:

$$S = \sigma T^4 \tag{2-38}$$

where $\sigma = 5.669 \times 10^{-8}$ Wm^{-2}K^{-4}. Thermal emission is usually unpolarized and extends through the total spectrum, particularly at the low-frequency end. Natural bodies are also characterized by their spectral emissivity $\varepsilon(\lambda)$, which expresses the capability to emit radiation due to thermal energy conversion relative to a blackbody with the same temperature. The properties of this emission mechanism will be discussed in more detail in Chapters 4 and 5.

Going to even higher energies, waves in the gamma-ray regions are mainly generated in the natural environment by radioactive decay of uranium (U), thorium (Th), and potassium 40 (^{40}K). The radioisotopes found in nature, ^{238}U and ^{232}Th, are long-lived alpha emitters and parents of individual radioactive decay chains. Potassium is found in almost all surfaces of the Earth, and its isotope ^{40}K, which makes up 0.12% of natural potassium, has a half-life of 1.3 billion years.

2-4 DETECTION OF ELECTROMAGNETIC RADIATION

The radiation emitted, reflected, or scattered from a body generates a radiant flux density in the surrounding space that contains information about the body's properties. To measure the properties of this radiation, a collector is used, followed by a detector.

The collector is a collecting aperture that intercepts part of the radiated field. In the microwave region, an antenna is used to intercept some of the electromagnetic energy. Examples of antennas include dipoles, an array of dipoles, or dishes. In the case of dipoles, the surrounding field generates a current in the dipole with an intensity proportional to the

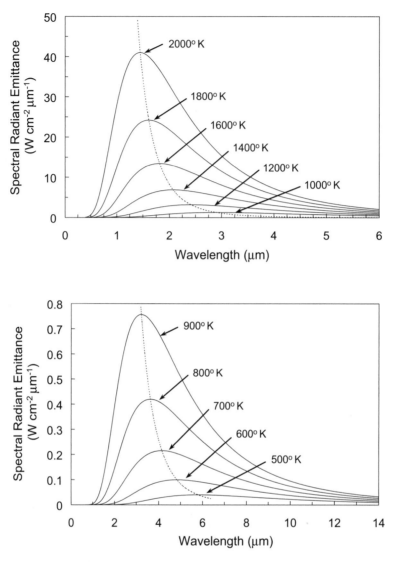

Figure 2-12. Spectral radiant emittance of a blackbody at various temperatures. Note the change of scale between the two graphs.

field intensity and a frequency equal to the field frequency. In the case of a dish, the energy collected is usually focused onto a limited area where the detector (or waveguide connected to the detector) is located.

In the IR, visible, and UV regions, the collector is usually a lens or a reflecting surface that focuses the intercepted energy onto the detector. Detection then occurs by transforming the electromagnetic energy into another form of energy such as heat, electric current, or state change.

Depending on the type of the sensor, different properties of the field are measured. In the case of synthetic-aperture imaging radars, the amplitude, polarization, frequency, and phase of the fields are measured at successive locations along the flight line. In the case of optical spectrometers, the energy of the field at a specific location is measured as a function of wavelength. In the case of radiometers, the main parameter of interest is the total radiant energy flux. In the case of polarimeters, the energy flux at different polarizations of the wave vector is measured.

In the case of x-ray and gamma-ray detection, the detector itself is usually the collecting aperture. As the particles interact with the detector material, ionization occurs, leading to light emission or charge release. Detection of the emitted light or generated current gives a measurement of the incident energy flux.

2-5 INTERACTION OF ELECTROMAGNETIC WAVES WITH MATTER: QUICK OVERVIEW

The interaction of electromagnetic waves with matter (e.g., molecular and atomic structures) calls into play a variety of mechanisms that are mainly dependent on the frequency of the wave (i.e., its photon energy) and the energy level structure of the matter. As the wave interacts with a certain material—be it gas, liquid, or solid—the electrons, molecules, and/or nuclei are put into motion (rotation, vibration, or displacement), which leads to exchange of energy between the wave and the material. This section gives a quick simplified overview of the interaction mechanisms between waves and matter. Detailed discussions are given later in the appropriate chapters throughout the text.

Atomic and molecular systems exist in certain stationary states with well-defined energy levels. In the case of isolated atoms, the energy levels are related to the orbits and spins of the electrons. These are called the electronic levels. In the case of molecules, there are additional rotational and vibrational energy levels that correspond to the dynamics of the constituent atoms relative to each other. Rotational excitations occur in gases where molecules are free to rotate. The exact distribution of the energy levels depends on the exact atomic and molecular structure of the material. In the case of solids, the crystalline structure also affects the energy level distribution.

In the case of thermal equilibrium, the density of population N_i at a certain level i is proportional to (Boltzmann's law):

$$N_i \sim e^{-E_i/kT} \tag{2-39}$$

where E_i is the level energy, k is Boltzmann's constant, and T is the absolute temperature. At absolute zero, all the atoms will be in the ground state. Thermal equilibrium requires that a level with higher energy be less populated than a level of lower energy (Fig. 2-13).

To illustrate, for $T = 300°K$, the value for kT is 0.025 eV (one eV is 1.6×10^{-19} joules). This is small relative to the first excited energy level of most atoms and ions, which means that very few atoms will be in the excited states. However, in the case of molecules, some vibrational and many rotational energy levels could be even smaller than kT, thus allowing a relatively large population in the excited states.

Let us assume that a wave of frequency v is propagating in a material in which two of the energy levels i and j are such that

$$hv = E_j - E_i \tag{2-40}$$

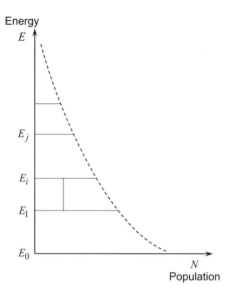

Figure 2-13. Curve illustrating the exponential decrease of population as a function of the level energy for the case of thermal equilibrium.

This wave would then excite some of the population of level i to level j. In this process, the wave loses some of its energy and transfers it to the medium. The wave energy is absorbed. The rate p_{ij} of such an event happening is equal to

$$p_{ij} = B_{ij}\varepsilon_v \qquad (2\text{-}41)$$

where ε_v is the wave energy density per unit frequency and B_{ij} is a constant determined by the atomic (or molecular) system. In many texts, p_{ij} is also called transition probability.

Once exited to a higher level by absorption, the atoms may return to the original lower level directly by spontaneous or stimulated emission, and in the process they emit a wave at frequency v, or they could cascade down to intermediate levels and in the process emit waves at frequencies lower than v (see Fig. 2-14). Spontaneous emission could occur any time an atom is in an excited state independent of the presence of an external field. The rate of downward transition from level j to level i is given by

$$p_{ji} = A_{ji} \qquad (2\text{-}42)$$

where A_{ji} is characteristic of the pair of energy levels in question.

Stimulated emission corresponds to downward transition, which occurs as a result of the presence of an external field with the appropriate frequency. In this case, the emitted wave is in phase with the external field and will add energy to it coherently. This results in an amplification of the external field and energy transfer from the medium to the external field. The rate of downward transition is given by

$$p_{ji} = B_{ji}\varepsilon_v \qquad (2\text{-}43)$$

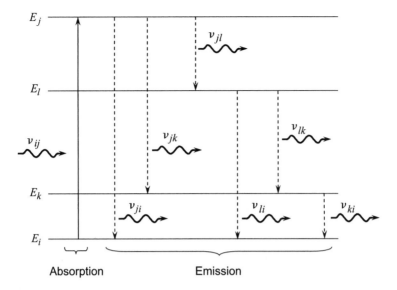

Figure 2-14. An incident wave of frequency v_{ij} is adsorbed due to population excitation from E_i to E_j. Spontaneous emission for the above system can occur at v_{ij} as well as by cascading down via the intermediate levels l and k.

The relationships between A_{ji}, B_{ji}, and B_{ij} are known as the Einstein's relations:

$$B_{ji} = B_{ij} \tag{2-44}$$

$$A_{ji} = \frac{8\pi h v^3 n^3}{c^3} B_{ji} = \frac{8\pi h n^3}{\lambda^3} B_{ji} \tag{2-45}$$

where n is the index of refraction of the medium.

Let us now consider a medium that is not necessarily in thermal equilibrium and where the two energy levels i and j are such that $E_i < E_j$. N_i and N_j are the population in the two levels, respectively. The number of downward transitions from level j to level i is $(A_{ji} + B_{ji}\varepsilon_v)N_j$. The number of upward transitions is $B_{ij}\varepsilon_v N_i = B_{ji}\varepsilon_v N_i$. The incident wave would then lose $(N_i - N_j)B_{ji}\varepsilon_v$ quanta per second. The spontaneously emitted quanta will appear as scattered radiation, which does not add coherently to the incident wave.

The wave absorption is a result of the fact that usually $N_i > N_j$. If this inequality can be reversed, the wave would be amplified. This requires that the population in the higher level is larger than the population in the lower energy level. This population inversion is the basis of laser and maser operations. However, it is not usually encountered in the cases of natural matter/waves interactions, which form the topic of this text. (*Note:* Natural maser effects have been observed in astronomical objects; however, these are beyond the scope of this text.)

The transition between different levels in usually characterized by the lifetime τ. The lifetime of an excited state i is equal to the time period after which the number of excited atoms in this state have been reduced by a factor e^{-1}. If the rate of transition out of the state i is A_i, the corresponding lifetime can be derived from the following relations:

$$dN_i = -A_i N_i\, dt \qquad (2\text{-}46)$$

$$\Rightarrow\ N_i(t) = N_i(0)e^{-A_i t} = N_i(0)e^{-t/\tau i} \qquad (2\text{-}47)$$

$$\Rightarrow\ \tau_i = 1/A_i \qquad (2\text{-}48)$$

If the transitions from i occur to a variety of lower levels j, then

$$A_i = \sum_j A_{ij} \qquad (2\text{-}49)$$

$$\Rightarrow\ \frac{1}{\tau_i} = \sum_j \frac{1}{\tau_{ij}} \qquad (2\text{-}50)$$

2-6   INTERACTION MECHANISMS THROUGHOUT THE ELECTROMAGNETIC SPECTRUM

Starting from the highest spectral region used in remote sensing, gamma- and x-ray inter-actions with matter call into play atomic and electronic forces such as the photoelectric effect (absorption of photon with ejection of electron), Compton effect (absorption of photon with ejection of electron and radiation of lower-energy photon), and pair produc-tion effect (absorption of photon and generation of an electron–positron pair). The photon energy in this spectral region is larger than 40 eV (Fig. 2-15). This spectral region is used mainly to sense the presence of radioactive materials.

In the ultraviolet region (photon energy between 3 eV and 40 eV), the interactions call into play electronic excitation and transfer mechanisms, with their associated spectral bands. This spectral region is used mostly for remote sensing of the composition of the upper layers of the Earth and planetary atmospheres. An ultraviolet spectrometer was flown on the Voyager spacecraft to determine the composition and structure of the upper atmospheres of Jupiter, Saturn, and Uranus.

In the visible and near infrared (energy between 0.2 eV and 3 eV), vibrational and electronic energy transitions play the key role. In the case of gases, these interactions usu-ally occur at well-defined spectral lines, which are broadened due to the gas pressure and temperature. In the case of solids, the closeness of the atoms in the crystalline structure leads to a wide variety of energy transfer phenomena with broad interaction bands. These include molecular vibration, ionic vibration, crystal field effects, charge transfer, and electronic conduction. Some of the most important solid surface spectral features in this wavelength region include the following:

1. The steep fall-off of reflectance in the visible toward the ultraviolet and an absorp-tion band between 0.84 and 0.92 μm associated with the Fe^{3+} electronic transition. These features are characteristic of iron oxides and hydrous iron oxides, collective-ly referred to as limonite.

2. The sharp variation of chlorophyll reflectivity in the neighborhood of 0.75 μm, which has been extensively used in vegetation remote sensing.

3. The fundamental and overtone bending/stretching vibration of hydroxyl (OH) bear-ing materials in the 2.1 to 2.8 μm region, which are being used to identify clay-rich areas associated with hydrothermal alteration zones.

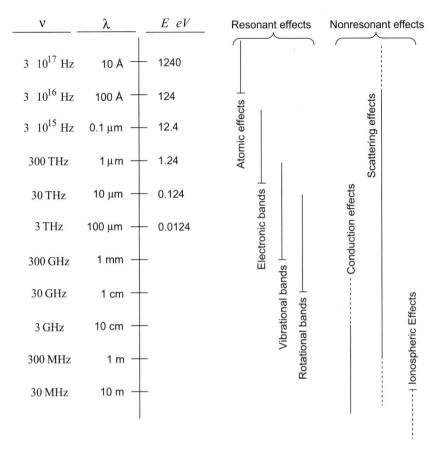

Figure 2-15. Correspondence of spectral bands, photon energy, and range of different wave–matter interaction mechanisms of importance in remote sensing. The photon energy in electron volts is given by $E(eV) = 1.24/\lambda$, where λ is in μm.

In the mid-infrared region (8 to 14 μm), the Si–O fundamental stretching vibration provides diagnostics of the major types of silicates (Fig. 2-16). The position of the restrahlen bands, or regions of metallic-like reflection, are dependent on the extent of interconnection of the Si–O tetrahedra comprising the crystal lattice. This spectral region also corresponds to vibrational excitation in atmospheric gaseous constituents.

In the thermal infrared, the emissions from the Earth's and other planets' surfaces and atmospheres are strongly dependent on the local temperature, and the resulting radiation is governed by Planck's law. This spectral region provides information about the temperature and heat constant of the object under observation. In addition, a number of vibrational bands provide diagnostic information about the emitting object's constituents.

In the submillimeter region, a large number of rotational bands provide information about the atmospheric constituents. These bands occur all across this spectral region, making most planetary atmospheres completely opaque for surface observation. For some gases such as water vapor and oxygen, the rotational band extends into the upper regions of the microwave spectrum.

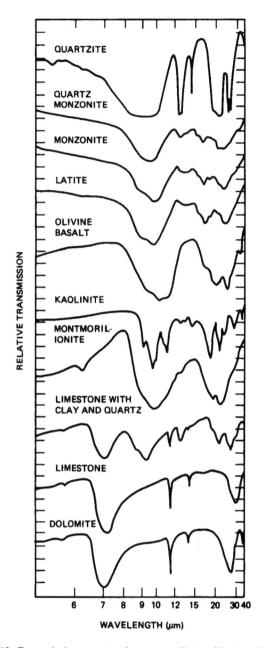

Figure 2-16. Transmission spectra of common silicates (Hunt and Salisbury, 1974).

The interaction mechanisms in the lower-frequency end of the spectrum ($v < 20$ GHz, $\lambda > 1.5$ cm) do not correspond to energy bands of specific constituents. They are, rather, collective interactions that result from electronic conduction and nonresonant magnetic and electric multipolar effects. As a wave interacts with a simple molecule, the resulting displacement of the electrons results in the formation of an oscillating dipole that generates an electromagnetic field. This will result in a composite field moving at a speed low-

TABLE 2-2. Wave–Matter Interaction Mechanisms across the Electromagnetic Spectrum

Spectral region	Main interaction mechanisms	Examples of remote sensing applications
Gamma rays, x-rays	Atomic processes	Mapping of radioactive materials
Ultraviolet	Electronic processes	Presence of H and He in atmospheres
Visible and near infrared	Electronic and vibration molecular processes	Surface chemical composition, vegetation cover, and biological properties
Mid-infrared	Vibrational, vibrational–rotational molecular processes	Surface chemical composition, atmospheric chemical composition
Thermal infrared	Thermal emission, vibrational and rotational processes	Surface heat capacity, surface temperature, atmospheric temperature, atmospheric and surface constituents
Microwave	Rotational processes, thermal emission, scattering, conduction	Atmospheric constituents, surface temperature, surface physical properties, atmospheric precipitation
Radio frequency	Scattering, conduction, ionospheric effect	Surface physical properties, subsurface sounding, ionospheric sounding

er than the speed of light in vacuum. The effect of the medium is described by the index of refraction or the dielectric constant. In general, depending on the structure and composition of the medium, the dielectric constant could be anisotropic or could have a loss term that is a result of wave energy transformation into heat energy.

In the case of an interface between two media, the wave is reflected or scattered depending on the geometric shape of the interface. The physical properties of the interface and the dielectric properties of the two media are usually the major factors affecting the interaction of wave and matter in the microwave and radio frequency part of the spectrum. Thus, remote sensing in this region of the spectrum will mainly provide information about the physical and electrical properties of the object instead of its chemical properties, which are the major factors in the visible/infrared region, or its thermal properties, which are the major factors in the thermal infrared and upper microwave regions (see Table 2-2).

In summary, a remote sensing system can be visualized (Fig. 2-17) as a source of electromagnetic waves (e.g., the sun, a radio source, etc.) that illuminate the object being studied. An incident wave interacts with the object and the scattered wave is modulated by a number of interaction processes that contain the "fingerprints" of the object. In some cases, the object itself is the source and the radiated wave contains information about its properties. A part of the scattered or radiated wave is then collected by a collector, focused on a detector, and its properties measured. An inverse process is then used to infer the properties of the object from the measured properties of the received wave.

EXERCISES

2-1. In order to better visualize the relative scale of the waves' wavelength in different regions of the spectrum, assume that the blue wavelength ($\lambda = 0.4$ μm) is ex-

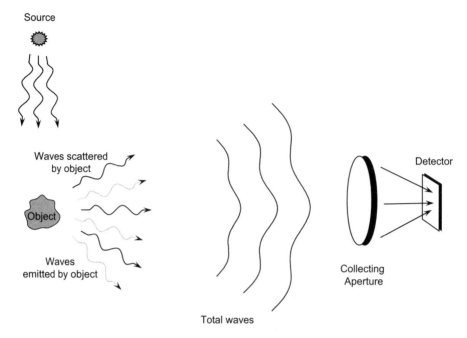

Figure 2-17. Sketch of key elements of a remote sensing system.

panded to the size of a keyhole (1 cm). What would be the wavelength size of other spectra regions in terms of familiar objects?

2-2.   The sun radiant flux density at the top of the Earth's atmosphere is 1.37 kilowatts/m². What is the flux density at Venus (0.7 AU), Mars (1.5 AU), Jupiter (5.2 AU), and Saturn (9.5 AU)? Express these values in kilowatts/m² and in photons/m² sec. Assume λ = 0.4 μm for the sun illumination. (Note that AU = astronomical unit = Earth/sun distance.)

2-3.   Assume that the sun emittance spectrum follows exactly Plank's formula:

$$S(\lambda)\frac{2\pi hc^2}{\lambda^5}\frac{1}{e^{ch/\lambda kT}-1}$$

with T = 6000°K. Calculate the percent of solar energy in the following spectral regions:

- In the UV (λ < 0.4 μm)
- In the visible (0.4 μm < λ < 0.7 μm)
- In the infrared (0.7 μm < λ < 10 μm)
- In the thermal infrared and submillimeter (10 μm < λ < 3 mm)
- In the microwave (λ > 3 mm)

2-4.   The amplitudes of two coexistent electromagnetic waves are given by

$$\mathbf{E}_1 = \hat{\mathbf{e}}_x A \cos(kz - \omega t)$$
$$\mathbf{E}_2 = \hat{\mathbf{e}}_y B \cos(kz - \omega t + \phi)$$

Describe the temporal behavior of the total electric field $\mathbf{E} = \mathbf{E}_1 + \mathbf{E}_2$ for the following cases:

- $A = B, \phi = 0$
- $A = B, \phi = \pi$
- $A = B, \phi = \pi/2$
- $A = B, \phi = \pi/4$

Repeat the exercise for $A = 2B$.

2-5.   The amplitudes of two coexistent electromagnetic waves are given by

$$\mathbf{E}_1 = \hat{\mathbf{e}}_x A \cos\left[\omega\left(\frac{z}{c} - t\right)\right]$$

$$\mathbf{E}_2 = \hat{\mathbf{e}}_y B \cos\left[\omega'\left(\frac{z}{c} - t\right)\right]$$

where c is the speed of light in vacuum. Let $\omega' = \omega + \Delta\omega$ and $\Delta\omega \ll \omega$ Describe the behavior of the power $P \sim |\mathbf{E}_1 + \mathbf{E}_2|^2$ of the composite wave as a function of P_1 and P_2 of each individual wave.

2-6.   A plasma is an example of a dispersive medium. The wavenumber for a plasma is given by

$$k = \frac{1}{c}\sqrt{\omega^2 - \omega_p^2}$$

where c = speed of light and ω_p = plasma angular frequency.

(a) Calculate and plot the phase and group velocity of a wave in the plasma as a function of frequency.

(b) Based on the results of (a), why is a plasma called a *dispersive* medium?

2-7.   A radar sensor is carried on an orbiting satellite that is moving at a speed of 7 km/sec parallel to the Earth's surface. The radar beam has a total beam width of $\theta = 4°$ and the operating frequency is $v = 1.25$ GHz. What is the center frequency and the frequency spread of the echo due to the Doppler shift across the beam for the following cases:

(a) a nadir-looking beam

(b) a 45° forward-looking beam

(c) a 45° backward-looking beam

2-8.   Repeat Problem 2-7 for the case in which the satellite is moving at a velocity of 7 km/sec but at a 5° angle above the horizontal.

2-9.   Plot and compare the spectral emittance of blackbodies with surface temperatures of 6000°K (Sun), 600°K (Venus), 300°K (Earth), 200°K (Mars), and 120 K (Titan). In particular, determine the wavelength for maximum emission for each body.

2-10. A small object is emitting Q watts isotropically in space. A collector of area A is located a distance d from the object. How much of the emitted power is being intercepted by the collector? Assuming that $Q = 1$ kW, and $d = 1000$ km, what size collector is needed to collect 1 milliwatt and 1 microwatt?

2-11. Two coexistent waves are characterized by

$$E_1 = A \cos(\omega t)$$

$$E_1 = A \cos(\omega t + \alpha)$$

Describe the behavior of the total wave $E = E_1 + E_2$ for the cases where $\alpha = 0$, $\alpha = \pi/2$, $\alpha = \pi$, and $\alpha = $ a random value with equal probability of occurrence between 0 and 2π.

REFERENCES AND FURTHER READING

Goetz, A., and L. Rowan. Geologic remote sensing. *Science,* **211,** 781–791, 1981.

Hunt, R., and W. Salisbury. Mid infrared spectral behavior of igneous rocks. U.S. Air Force Cambridge Research Laboratories Report AFCRL-TR-74-0625, 1974.

Papas, C. H. *Theory of Electromagnetic Wave Propagation.* McGraw-Hill, New York, 1965.

Reeves, R. G. (Ed.). *Manual of Remote Sensing,* Chapters 3, 4, and 5. American Society of Photogrammetry, Falls Church, VA, 1975.

Sabins, F. *Remote Sensing: Principles and Interpretation.* Freeman, San Francisco, 1978.

3

Solid Surfaces Sensing in the Visible and Near Infrared

The visible and near-infrared regions of the electromagnetic spectrum have been the most commonly used in remote sensing of planetary surfaces. This is partially due to the fact that this is the spectral region of maximum illumination by the sun and most widely available detectors (electrooptical and photographic). The sensor detects the electromagnetic waves reflected by the surface and measures their intensity in different parts of the spectrum. By comparing the radiometric and spectral characteristics of the reflected waves to the characteristics of the incident waves, the surface reflectivity is derived. This in turn is analyzed to determine the chemical and physical properties of the surface. The chemical composition and crystalline structure of the surface material has an effect on the reflectivity because of the molecular and electronic processes that govern the interaction of waves with matter. The physical properties of the surface, such as roughness and slope, also affect the reflectivity, mainly due to geometric factors related to the source–surface–sensor relative angular configuration.

Thus, information about the surface properties is acquired by measuring the modulation that the surface imprints on the reflected wave (Fig. 3-1) by the process of wave–matter interactions, which will be discussed in this chapter.

3-1 SOURCE SPECTRAL CHARACTERISTICS

By far the most commonly used source of illumination in the visible and near infrared is the sun. In the most simple terms, the sun emits approximately as a hot blackbody at 6000°K temperature. The solar illumination spectral irradiance at the Earth's distance is shown in Figure 3-2. The total irradiance is measured to be approximately 1370 W/m^2 above the Earth's atmosphere. This irradiance decreases as the square of the distance from the sun because of the spherical geometry. Thus, the total solar irradiance at Venus is about twice the value for Earth, whereas it is half that much at Mars (see Table 3-1).

Introduction to the Physics and Techniques of Remote Sensing. By C. Elachi and J. van Zyl

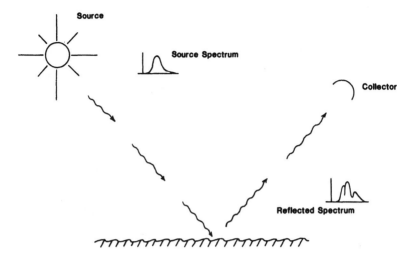

Figure 3-1. The surface spectral imprint is reflected in the spectrum of the reflected wave.

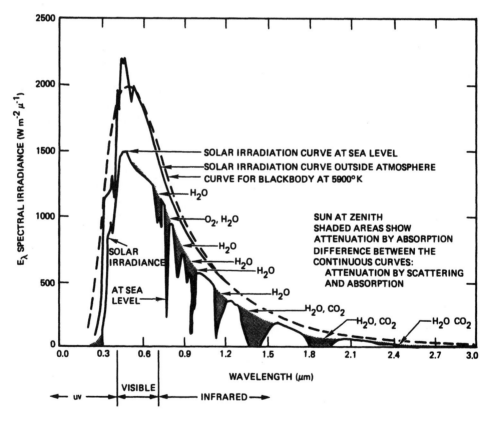

Figure 3-2. Sun illumination spectral irradiance at the Earth's surface. (From Chahine, et al. 1983.)

Table 3-1. Solar Irradiance at the Distance of the Different Planets

Planets	Mean distance to the sun		Solar irradiance (W/m^2)
	(AU)	Million kilometers	
Mercury	0.39	58	9000
Venus	0.72	108	2640
Earth	1	150	1370
Mars	1.52	228	590
Jupiter	5.19	778	50
Saturn	9.51	1427	15
Uranus	19.13	2870	3.7
Neptune	30	4497	1.5
Pluto	39.3	5900	0.9

As the solar waves propagate through a planet's atmosphere, they interact with the atmospheric constituents, leading to absorption in specific spectral regions, which depends on the chemical composition of these constituents. Figure 3-2 shows the sun illumination spectral irradiance at the Earth's surface. Strong absorption bands exist in the near infrared, particularly around 1.9, 1.4, 1.12, 0.95, and 0.76 μm. These are mainly due to the presence of water vapor (H_2O), carbon dioxide (CO_2), and, to a lesser extent, oxygen (O_2). In addition, scattering and absorption lead to a continuum of attenuation across the spectrum. For comparison, the transmittivity of the Martian atmosphere is shown in Figure 3-3. In the near infrared, the Martian atmosphere is opaque only in very narrow spectral bands near 2.7 and 4.3 μm, which are due to CO_2. The absorption band near 2 μm is due to H_2O.

Another important factor in visible and near-infrared remote sensing is the relative configuration of the sun–surface–sensor. Because of the tilt of the Earth's rotation axis relative to the plane of the ecliptic, the sun's location in the sky varies as a function of the seasons and the latitude of the illuminated area. In addition, the ellipticity of the Earth's orbit has to be taken into account (Dozier and Strahler, 1983; Wilson, 1980).

With recent and continuing development, high-power lasers are becoming viable sources of illumination even from orbital altitudes. These sources have a number of advantages, including controlled illumination geometry, controlled illumination timing, which can be used in metrology, and the possibility of very high powers in very narrow spectral bands. However, these advantages have to be compared to the lack of instantaneous wide-spectral coverage and the need for orbiting the laser source with its associated weight and power requirements.

3-2 WAVE–SURFACE INTERACTION MECHANISMS

When an electromagnetic wave interacts with a solid material, there are a number of mechanisms that affect the properties of the resulting wave. Some of these mechanisms operate over a narrow band of the spectral region, whereas others are wide band and thus affect the entire spectrum from 0.3 to 2.5 μm. These interaction mechanisms are summarized in Table 3-2 and discussed in detail in this section. The narrowband interactions are usually associated with resonant molecular and electronic processes. These mechanisms

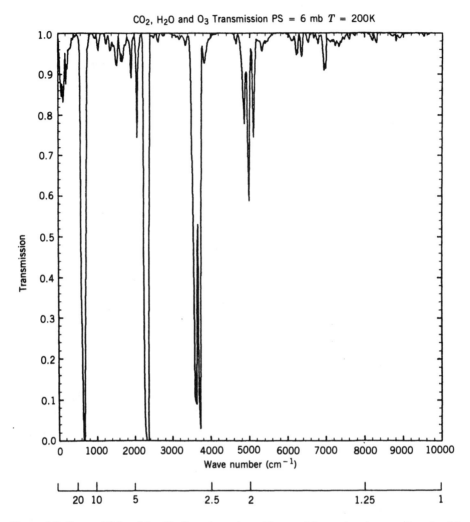

Figure 3-3. Transmittivity of the Martian atmosphere. The model uses band parameters for CO_2, H_2O, and O_3. Surface pressure assumed is 6 mbar. (Courtesy of D. Crisp and D. McCleese.)

are strongly affected by the crystalline structure, leading to splitting, displacement, and broadening of the spectral lines into spectral bands. The wide-band mechanisms are usually associated with nonresonant electronic processes that affect the material index of refraction (i.e., velocity of light in the material).

3-2-1   Reflection, Transmission, and Scattering

When an electromagnetic wave is incident on an interface between two materials (in the case of remote sensing, one of the materials is usually the atmosphere), some of the energy is reflected in the specular direction, some of it is scattered in all directions of the incident medium, and some of it is transmitted through the interface (Fig. 3-4). The transmitted energy is usually absorbed in the bulk material and either reradiated by electronic or thermal processes or dissipated as heat.

When the interface is very smooth relative to the incident wavelength λ (i.e., $\lambda \gg$ interface roughness), the reflected energy is in the specular direction and the reflectivity is given by Snell's law. The reflection coefficient is a function of the complex index of refraction n and the incidence angle. The expression of the reflection coefficient is given by

$$|R_h|^2 = \frac{\sin^2 (\theta - \theta_t)}{\sin^2 (\theta + \theta_t)} \tag{3-1}$$

Table 3-2. Interaction Mechanisms

General physical mechanisms	Specific physical mechanisms	Example
Geometrical and physicaloptics	Dispersive refraction	Chromatic aberration, rainbow
	Scattering	Blue sky, rough surfaces
	Reflection, refraction, and interference	Mirror, polished surfaces, oil film on water, lens coating
	Diffraction grating	Opal, liquid crystals, gratings
Vibrational excitation	Molecular vibration	H_2O, aluminum + oxygen, or silicon + oxygen
	Ion vibration	Hydroxyl ion (OH)
Electronic excitation	Crystal field effects in transition metal compounds	Turquoise, most pigments, some fluorescent materials
	Crystal field effects in transition metal impurities	Ruby, emerald, red sandstone
	Charge transfer between molecular orbits	Magnetite, blue sapphire
	Conjugated bonds	Organic dyes, most plant colors
	Transitions in materials with energy bands	Metallic: copper, silver, gold; semiconductors: silicon, cinnabar, diamond, galena

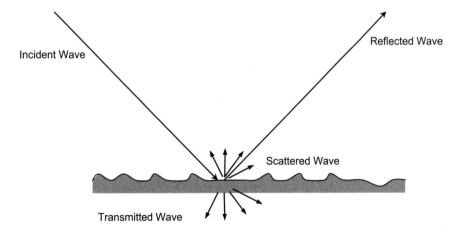

Figure 3-4. Wave interaction with an interface.

for horizontally polarized incidence waves, and

$$|R_\nu|^2 = \frac{\tan^2(\theta - \theta_t)}{\tan^2(\theta + \theta_t)} \tag{3-2}$$

for vertically polarized incidence waves. θ_t is the transmission angle and is given by

$$\sin \theta = n \sin \theta_t \tag{3-3}$$

The behavior of the reflection coefficient as a function of the incidence angle is illustrated in Figure 3-5. One special aspect of the vertically polarized configuration is the presence of a null (i.e., no reflection) at an angle that is given by

$$\tan \theta = n \tag{3-4}$$

and which corresponds to $R_\nu = 0$. This is called the Brewster angle.

In the case of normal incidence, the reflection coefficient is

$$R_h = R_\nu = R = \left(\frac{n-1}{n+1} \right) = \left(\frac{N_r + iN_i - 1}{N_r + iN_i + 1} \right) \tag{3-5}$$

and

$$|R|^2 = \frac{(N_r - 1)^2 + N_i^2}{(N_r + 1)^2 + N_i^2} \tag{3-6}$$

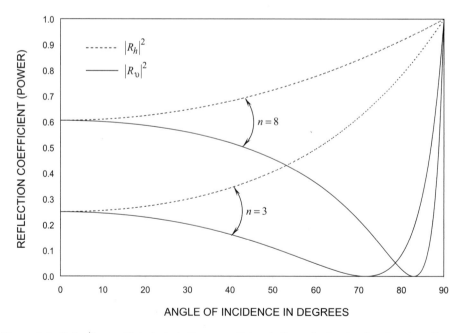

Figure 3-5. Reflection coefficient of a half space with two indices of refraction ($n = 3$ and $n = 8$) as a function of incidence angle. The continuous curve corresponds to horizontal polarization. The dashed curve corresponds to vertical polarization.

where N_r and N_i are the real and imaginary parts of n, respectively. In the spectral regions away from any strong absorption bands, $N_i \ll N_r$ and

$$|R|^2 = \left(\frac{N_r - 1}{N_r + 1} \right)^2 \tag{3-7}$$

However, in strong absorption bands, $N_i \gg N_r$ and

$$|R|^2 \simeq 1 \tag{3-8}$$

Thus, if wide-spectrum light is incident on a polished surface, the reflected light will contain a relatively large portion of spectral energy around the absorption bands of the surface material (Fig. 3-6). This is the *restrahlen* effect.

In actuality, most natural surfaces are rough relative to the wavelength and usually consist of particulates. Thus, scattering plays a major role and the particulates' size distribution has a significant impact on the spectral signature. An adequate description of the scattering mechanism requires a rigorous solution of Maxwell's equations, including multiple scattering. This is usually very complex, and a number of simplified techniques and empirical relationships are used. For instance, if the particles are small relative to the wavelength, then Rayleigh's law is applicable, and it describes adequately the scattering mechanism. In this case, the scattering cross section is a fourth power of the ratio a/λ [i.e., scattering is $\sim(a/\lambda)^4$] of the particles' size over the wavelength. The Rayleigh scattering explains the blue color of the sky; molecules and very fine dust scatter blue light about four times more than red light, making the sky appear blue. The direct sunlight, on the other hand, is depleted of the blue and therefore appears reddish.

In the case of a particulate surface, the incident wave is multiply scattered and some of the energy that is reflected toward the incident medium penetrates some of the particles (Fig. 3-7). Thus, if the material has an absorption band, the reflected energy is depleted of energy in that band. Usually, as the particles get larger, the absorption features become more pronounced even though the total reflected energy is decreased. This is commonly observed in measured spectra, as illustrated in Figure 3-8. This is the usual case in remote sensing in the visible and infrared regions.

In the general case of natural surfaces, the reflectivity is modeled by empirical expressions. One such expression is the Minnaert law (Minnaert, 1941), which gives

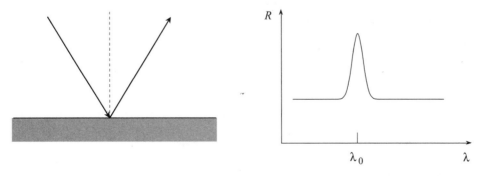

Figure 3-6. For a polished surface, there is an increase in the reflected energy near an absorption spectral line.

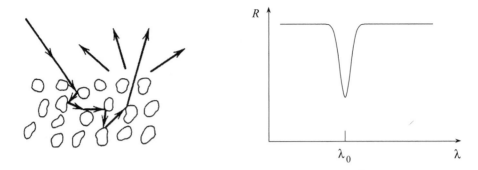

Figure 3-7. In the case of a particulate layer, the volume scattering and resulting absorption leads to a decrease of the reflected (scattered) energy near an absorption spectral line.

$$B \cos \theta_s = B_0 (\cos \theta_i \cos \theta_s)^\kappa \qquad (3\text{-}9)$$

where B is the apparent surface brightness, B_0 is the brightness of an ideal reflector at the observation wavelength, κ is a parameter that describes darkening at zero phase angle, and θ_i and θ_s are the incidence and scattering (or emergence) angles, respectively. For a Lambertian surface, $\kappa = 1$ and

$$B = B_0 \cos \theta_i \qquad (3\text{-}10)$$

which is the commonly used cosine darkening law.

In some special cases, diffraction gratings are encountered in nature. For instance, opal consists of closely packed spheres of silicon dioxide and a little water imbedded in a transparent matrix that has similar composition but slightly different index of refraction. The

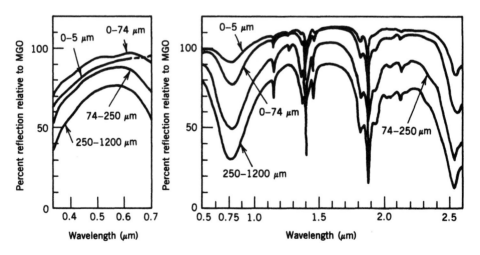

Figure 3-8. Bidirectional reflection spectra of four different particle-size-range samples of the mineral beryl. The smallest particle-size range displays the highest relative reflection, but shows the least contrast in its spectral bands (Hunt, 1977).

spheres have a diameter of about 0.25 μm. Dispersion of white light by the three-dimensional diffraction grating gives rise to spectrally pure colors that glint from within an opal.

The reflection of visible and near-infrared waves from natural surfaces occurs within the top few microns. Thus, surface cover plays an important role. For instance, the iron oxide in desert varnish can quench or even completely mask the spectrum of the underlying rocks.

3-2-2 Vibrational Processes

Vibrational processes correspond to small displacements of the atoms from their equilibrium positions. In a molecule composed of N atoms, there are $3N$ possible modes of motion because each atom has three degrees of freedom. Of these modes of motion, three constitute translation, three (two in the case of linear molecules) constitute rotation of the molecule as a whole, and $3N - 6$ ($3N - 5$ in the case of linear molecules) constitute the independent type of vibrations (Fig. 3-9). Each mode of motion leads to vibration at a clas-

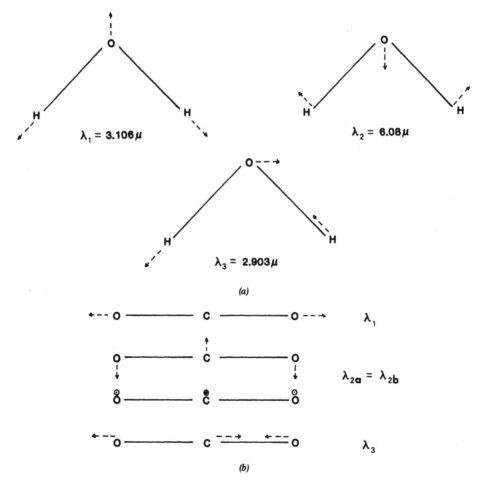

(a)

(b)

Figure 3-9. (a) H_2O molecule fundamental vibrational modes. (b) CO_2 molecule (linear) fundamental vibrational modes.

sical frequency v_i. For a molecule with many classical frequencies v_i, the energy levels are given by

$$E = (n_1 + \tfrac{1}{2})hv_1 + \cdots + (n_{3N-6} + \tfrac{1}{2})hv_{3N-6} \qquad (3\text{-}11)$$

where $(n + \tfrac{1}{2})hv$ are the energy levels of a linear harmonic oscillator, and n_i are vibrational quantum numbers ($n_i = 0, 1, \ldots$). The number and values of the energy levels for a material are thus determined by the molecular structure (i.e., number and type of the constituent atoms, the molecular geometry, and the strength of the bonds).

The transitions between the ground state (all $n_i = 0$) to a state in which only one $n_i = 1$ are called fundamental tones. The corresponding frequencies are $v_1, v_2, \ldots, v_i$. These usually occur in the far-to mid-infrared (> 3 μm). The transitions between the ground state to a state in which only one $n_i = 2$ (or some multiple integer) are called overtones. The corresponding frequencies are $2v_1, 2v_2, \ldots$ (or higher-order overtones). The other transitions are called combination tones, which are combinations of fundamental and overtone transitions. The corresponding frequencies are $lv_i + mv_j$, where l and m are integers. Features due to overtone and combination tones usually appear between 1 and 5 μm.

As an illustration, let us consider the case of the water molecule (Fig. 3-9a). It consists of three atoms ($N = 3$) and has three classical frequencies v_1, v_2, v_3, which correspond to the three wavelengths:

$\lambda_1 = 3.106$ μm, which corresponds to the symmetric OH stretch

$\lambda_2 = 6.08$ μm, which corresponds to the HOH bend

$\lambda_3 = 2.903$ μm, which corresponds to the asymmetric OH stretch

These wavelengths are the fundamentals. The lowest-order overtones correspond to the frequencies $2v_1, 2v_2$, and $2v_3$, and to wavelengths $\lambda_1/2, \lambda_2/2$, and $\lambda_3/2$. An example of a combination tone can be $v = v_3 + v_2$, which has a wavelength given by

$$\frac{1}{\lambda} = \frac{1}{\lambda_3} + \frac{1}{\lambda_2} \rightarrow \lambda = 1.87 \ \mu\text{m}$$

or $v' = 2v_1 + v_3$, which has a wavelength $\lambda' = 0.962$ μm.

In the spectra of minerals and rocks, whenever water is present, two bands appear—one near 1.45 μm (due to $2v_3$) and one near 1.9 μm (due to $v_2 + v_3$). The presence of these bands is usually indicative of the presence of the water molecule. These bands could be sharp, indicating that the water molecules are located in well-defined, ordered sites, or they may be broad, indicating that they occupy unordered or several unequivalent sites. The exact location and appearance of the spectral bands give quite specific information about the way in which the molecular water is associated with the inorganic material. Figure 3-10 illustrates this effect by showing spectra of various material that contains water. The $2v_3$ and $v_2 + v_3$ tones are clearly displayed and the variation in the exact location and spectral shape is clearly visible.

All the fundamental vibrational modes of silicon, magnesium, and aluminum with oxygen occur near 10 μm or at longer wavelengths. Because the first overtone near 5 μm is not observed, one does not expect to detect direct evidence of them at higher-order overtones in the near infrared. What is observed in the near infrared are features due to the overtones and combinations involving materials that have very high fundamental frequen-

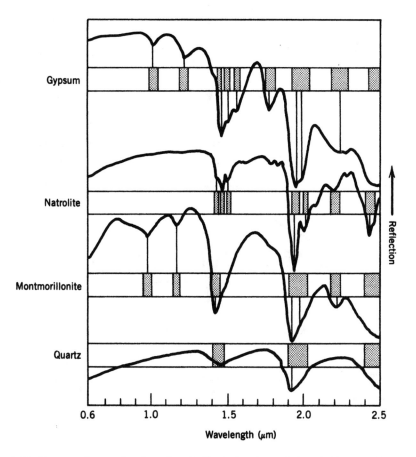

Figure 3-10. Spectra of water-bearing minerals illustrating the variations in the exact location and shape of the spectral lines associated with two of the tones of the water molecule: $2v_3$ near 1.4 μm and $v_2 + v_3$ near 1.9 μm. (From Hunt, 1977.)

cies. There are only a few groups that provide such features and by far the most common of these are the near-infrared bands that involve the OH stretching mode.

The hydroxyl ion, OH, very frequently occurs in inorganic solids. It has one stretching fundamental that occurs near 2.77 μm, but its exact location depends upon the site at which the OH ion is located and the atom attached to it. In some situations, this spectral feature is doubled, indicating that the OH is in two slightly different locations or attached to two different minerals. Al–OH and Mg–OH bending have fundamental wavelengths at 2.2 and 2.3 μm, respectively. The first overtone of OH ($2v$ at about 1.4 μm) is the most common feature present in the near-infrared spectra of terrestrial materials. Figure 3-11 shows examples of spectra displaying the hydroxyl tones.

Carbonate minerals display similar features between 1.6 and 2.5 μm that are combinations and overtones of the few fundamental internal vibrations of the CO_3^{2-} ion.

3-2-3 Electronic Processes

Electronic processes are associated with electronic energy levels. Electrons in an atom can only occupy specific quantized orbits with specific energy levels. The most common

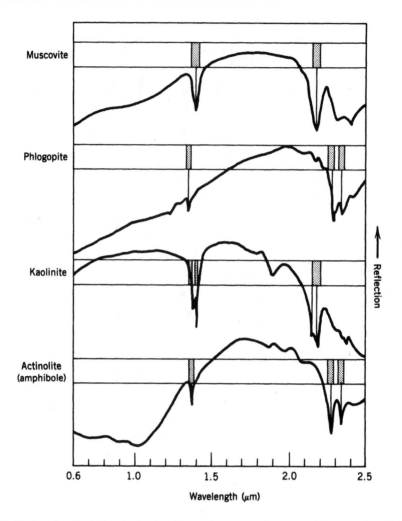

Figure 3-11. Spectra displaying the hydroxyl group tones: overtone near 1.4 μm and combination tones near 2.3 μm. (From Hunt, 1977.)

elements in rocks, namely silicon, aluminum, and oxygen, do not possess energy levels such that transitions between them can yield spectral features in the visible or near infrared. Consequently, no direct information is available on the bulk composition of geological materials. However, indirect information is available as a consequence of the crystal or molecular structure imposing its effect upon the energy levels of specific ions. The most important effects relevant to remote sensing are the crystal field effect, charge transfer, conjugate bonds, and transitions in materials with energy bands.

Crystal Field Effect. One consequence of the binding together of atoms is a change in the state of valence electrons. In an isolated atom, the valence electrons are unpaired and are the primary cause of color. In a molecule and in many solids, the valence electrons of adjacent atoms form pairs that constitute the chemical bonds that hold atoms together. As a result of this pair formation, the absorption bands of the valence electrons are usually displaced to the UV region. However, in the case of transition metal elements such as iron, chromium, copper, nickel, cobalt, and manganese, the atoms have inner shells that remain only partially filled. These unfilled inner shells hold unpaired electrons that have excited states that often fall in the visible spectrum. These states are strongly affected by the electrostatic field that surrounds the atom. The field is determined by the surrounding crystalline structure. The different arrangement of the energy levels for different crystal fields leads to the appearance of different spectra for the same ion. However, not all possible transitions may occur equally strongly. The allowed transitions are defined by the selection rules. In this situation, the most pertinent rule is the one related to the electron spins of the state involved. This rule allows transitions between states of identical spins only. Typical examples of such situations occur in ruby and emerald.

The basic material of ruby is corundum, an oxide of aluminum (Al_2O_3). A few percent of the aluminum ions are replaced by chromium ions (Cr^{3+}). Each chromium ion has three unpaired electrons whose lowest possible energy is a ground state designated $4A_2$ and a complicated spectrum of excited states. The exact position of the excited state is determined by the crystal electric field in which the ion is immersed. The symmetry and strength of the field are determined in turn by the nature of the ions surrounding the chromium and their arrangement. In the case of the ruby, there are three excited states ($2E$, $4T_1$, and $4T_2$) with energy bands in the visible range.

Selection rules forbid a direct transition from $4A_2$ to $2E$, but allow transitions to both $4T_1$ and $4T_2$ (see Fig. 3-12). The energies associated with these transitions correspond to wavelengths in the violet and yellow-green regions of the spectrum. Hence, when white light passes through a ruby, it emerges as a deep red (i.e., depleted of violet, yellow, or green).

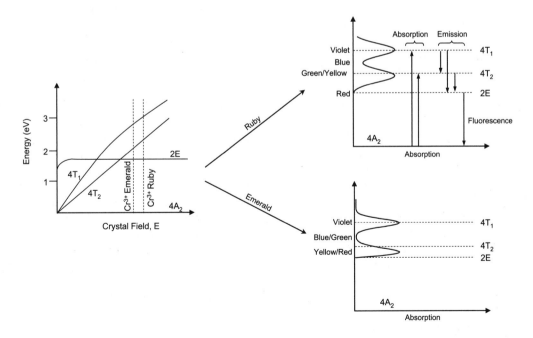

Figure 3-12. Basis for the color characteristics of emerald and ruby.

Because of the selection rules, electrons can return from the 4T states to the $4A_2$ ground state only through the 2E level. The 4T to 2E transition releases infrared waves, but the 2E to $4A_2$ transition gives rise to strong red light emission.

In the case of emerald, the impurity is also Cr^{3+}; however, because of the exact structure of the surrounding crystal, the magnitude of the electric field surrounding a chromium ion is somewhat reduced. This results in lower energy for the 4T states (Fig. 3-12). This shifts the absorption band into the yellow-red, hence the green color of emerald.

Crystal field effects can arise whenever there are ions bearing unpaired electrons in a solid. Aquamarine, jade, and citrine quartz have iron instead of chromium. Blue or green azurite, turquoise, and malachite have copper as a major compound instead of impurity. Similar situations occur in red garnets, in which iron is a major compound.

One of the most important cases in remote sensing involves ferrous ion Fe^{2+}. For a ferrous ion in a perfectly octahedral site, only one spin-allowed transition is in the near infrared. However, if the octahedral site is distorted, the resulting field may lead to splitting of the energy level, thus allowing additional transitions. If the ferrous ions are in different nonequivalent sites, as is the case in olivine, additional characteristic transitions occur. Thus important information on the bulk structure of the mineral can be derived indirectly by using the ferrous ion spectral feature as a probe. Figure 3-13 shows the reflection spectra of several minerals that contain ferrous ion, illustrating the spectral changes resulting from the bulk structure.

The crystal field effect is not only limited to electrons in transition metal ions. If an electron is trapped at a structural defect, such as a missing ion or an impurity, a similar transition energy level occurs. These anomalies are called color centers or F centers (from the German word for color—Farbe). The color center effect is responsible for the purple color of fluorite (CaF_2), in which each calcium ion is surrounded by eight fluorine ions. An F center forms when a fluorine ion is missing from its usual position. In order to maintain electrical neutrality, an electron is trapped in the fluorine location and is bound to the location by the crystal field of the surrounding ions. Within this field, it can occupy a ground state and various excited states.

Charge Transfer. In many situations, paired electrons are not confined to a particular bond between two atoms but move over longer distances. They even range throughout the molecule or even throughout a macroscopic solid. They are then bound less tightly and the energy needed to create an excited state is reduced. The electrons are said to occupy molecular orbits (in contrast to atomic orbits). One mechanism by which molecular orbits can contribute to the spectral characteristics in the visible and NIR regions is the transfer of electric charges from one ion to another. An example of such effect occurs in materials in which iron is present in both of its common values Fe^{2+} and Fe^{3+}. Charge transfers between these forms give rise to colors ranging from deep blue to black, such as in the black iron ore magnetite. A similar mechanism occurs in blue sapphire. In this case, the impurities are Fe^{2+} and Ti^{4+}. An excited state is formed when an electron is transferred from the iron to the titanium, giving both a charge. About 2 eV are needed to drive the charge transfer, creating a broad absorption band that extends from the yellow through the red, leaving the sapphire with a deep blue color.

The spectral features that occur as a result of charge transfers typically are very intense, much more intense than the features corresponding to crystal field effects.

Conjugate Bonds. Molecular orbital transitions play a major role in the spectral response of biological pigments and many organic substances in which carbon, and some-

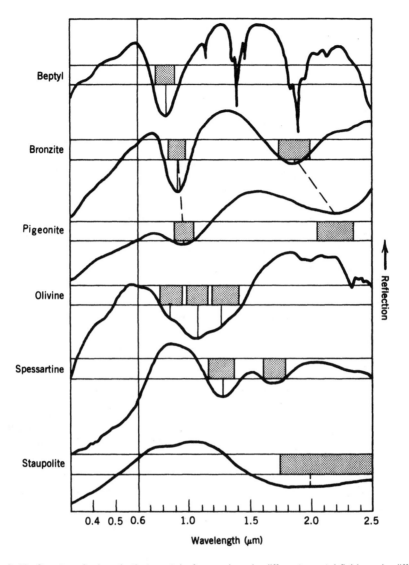

Figure 3-13. Spectra of minerals that contain ferrous ions in different crystal fields or in different sites. All the features indicated (vertical line is band minimum and shaded area shows bandwidth) in these spectra are due to spin-allowed transitions. (From Hunt, 1977.)

time nitrogen, atoms are joined by a system of alternating single and double bonds called conjugate bonds. Because each bond represents a pair of shared electrons, moving a pair of electrons from each double bond to the adjacent single bond reverses the entire sequence of bonds, leading to an equivalent structure. Actually, the best representation of the structure shows all the atoms connected by single bonds with the remaining pairs of bonding electrons distributed over the entire structure in molecular orbitals, which in this instance are called pi-orbitals.

The extended nature of pi-orbitals in a system of conjugate bonds tends to diminish the excitation energy of the electron pairs, allowing absorption in the visible spectrum. A number of biological pigments owe their spectral properties to extended systems of pi-orbitals. Among them are the green chlorophyll of plants and the red hemoglobin of blood.

Materials with Energy Bands. The spatial extent of electron orbitals reaches its maximum value in metals and semiconductors. Here, the electrons are released entirely from attachment to particular atoms or ions and even move freely throughout a macroscopic volume.

In a metal, all the valence electrons are essentially equivalent since they can freely exchange places. The energy levels form a continuum. Thus, a metal can absorb radiation at any wavelength. This might lead someone to conclude that metals should be black. However, when an electron in a metal absorbs a photon and jumps to an excited state, it can immediately reemit a photon of the same energy and return to its original state. Because of the rapid and efficient reradiation, the surface appears reflective rather than absorbent; it has the characteristic metallic luster. The variations in the color of metallic surfaces result from differences in the number of states available at particular energies above the Fermi level. Because the density of states is not uniform, some wavelengths are absorbed and reemitted more efficiently than others.

In the case of semiconductors, there is a splitting of the energy levels into two broad bands with a forbidden gap (Fig. 3-14). The lower energy levels in the valence band are completely occupied. The upper conduction band is usually empty. The spectral response of a pure semiconductor depends on the width of the energy gap. The semiconductor cannot absorb photons with energy less than the gap energy. If the gap is small, all wavelengths in the visible spectrum can be absorbed. Small-gap semiconductors in which reemission is efficient and rapid, such as silicon, have a metal-like luster.

If the gap is large, no wavelengths in the visible spectrum can be absorbed (i.e., the photon energy less than the gap energy) and the material is colorless. Diamond is such a material, with an energy gap of 5.4 eV (i.e., $\lambda = 0.23$ μm). If the energy gap is intermediate, the semiconductor will have a definite color. Cinnebar (HgS) has a band gap of 2.1 eV (i.e., $\lambda = 0.59$ μm). All photons with energy higher than this level (i.e., blue or green) are absorbed, and only the longest visible wavelengths are transmitted; as a result, cinnebar appears red (see Fig. 3-15). Semiconductor materials are characterized by a sharp edge of transition in their spectrum due to the sharp edge of the conduction band. The

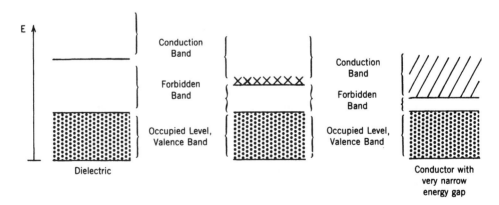

Figure 3-14. Configurations of energy bands for different types of solid materials.

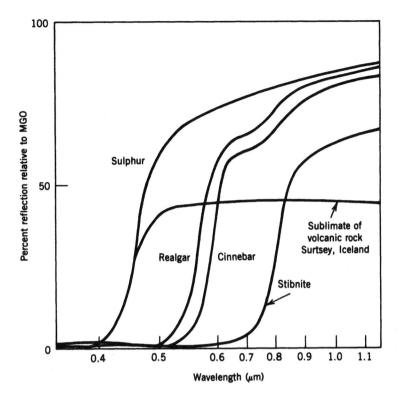

Figure 3-15. Visible and infrared bidirectional reflection spectra of particulate samples of different materials, all of which display sharp absorption edge effect. (From Siegal and Gillespie, © 1980 John Wiley & Sons. Reprinted with permission.)

sharpness of the absorption edge is a function of the purity and crystallinity of the material. For particulate materials, the absorption edge is more sloped.

If the semiconductor is doped with impurity, new intermediate energy levels are available, allowing some large-gap semiconductors to have spectral signatures in the visible spectrum. The technique of doping semiconductors is often used in the manufacture of detector materials. Since the doped layer has a lower transition energy, longer-wavelength radiation can typically be detected. An example is the case in which silicon is doped with arsenic. The resulting detectors have sensitivities that extend well into the infrared region far beyond the normal cutoff wavelength of pure silicon at about 1.1 μm.

3-2-4 Fluorescence

As illustrated in the case of the interaction of light with ruby, energy can be absorbed at one wavelength and reemitted at a different wavelength due to the fact that the excited electrons will cascade down in steps to the ground state. This is called fluorescence. This effect can be used to acquire additional information about the composition of the surface. In the case of sun illumination, it would seem at first glance that it is not possible to detect fluorescence because the emitted fluorescent light cannot be separated from the reflected light at the fluorescence wavelength. However, this can be circumvented due to the fact that the sun spectrum has a number of very narrow dark lines, called Fraunhofer lines, which are due to absorption in the solar atmosphere. These lines have widths ranging from 0.01 to 0.1 μm, and the central intensity of some of them is less than 10% of the surrounding continuum. The fluorescence remote sensing technique consists of measuring to what extent a Fraunhofer "well" is filled up relative to the continuum due to fluorescence energy resulting from excitation by shorter wavelengths (see Fig. 3-16). Thus, by comparing the depth of a Fraunhofer line relative to the immediate continuum in the reflected light and the direct incident light, surface fluorescence can be detected and measured.

Let F_s be the ratio of solar illumination intensity at the center of a certain Fraunhofer line (I_0) to the intensity of the continuum illumination (I_c) immediately next to the line (Fig. 3-16)

$$F_s = \frac{I_0}{I_c} \tag{3-12}$$

and let F_r be the ratio for the reflected light. If the reflecting material is not fluorescent, then

$$F_r = \frac{RI_0}{RI_c} = \frac{I_0}{I_c} = F_s \tag{3-13}$$

where R is the surface reflectivity. If the surface is fluorescent with fluorescence emission near the Fraunhofer line being measured, then an additional intensity I_f is added in the reflected light:

$$F_r = \frac{RI_0 + I_f}{RI_c + I_f} > F_s \tag{3-14}$$

and the intensity of the fluorescence illumination I_f can be derived.

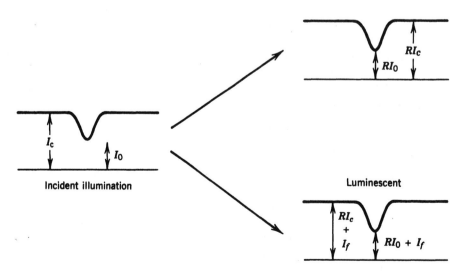

Figure 3-16. Use of Fraunhofer lines to detect luminescent material from their reflection properties.

3-3 SIGNATURE OF SOLID-SURFACE MATERIALS

Solid-surface materials can be classified into two major categories: geologic materials and biologic materials. Geologic materials correspond to the rocks and soils. Biologic materials correspond to the vegetation cover (natural and human-grown). For the purpose of this text, snow cover and urban areas are included in the biologic category.

3-3-1 Signature of Geologic Materials

As discussed earlier, the signature of geologic materials in the visible and near infrared regions is mainly a result of electronic and vibrational transitions. The absorption bands of specific constituents are strongly affected by the crystalline structure surrounding them, their distribution in the host material, and the presence of other constituents. A spectral signature diagram for a variety of geologic materials is shown in Figure 3-17 and

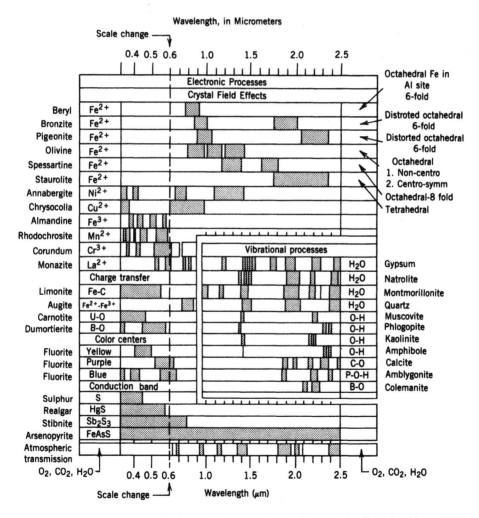

Figure 3-17. Spectral signature diagram of a variety of geologic materials. (Based on Hunt, 1977.)

is based on the work by Hunt (1977). In the case of vibrational processes, the water molecule (H_2O) and hydroxyl (–OH) group play a major role in characterizing the spectral signature of a large number of geologic materials. In the case of electronic processes, a major role is played by the ions of transition metals (e.g., Fe, Ni, Cr, and Co), which are of economic importance. The sulfur molecule illustrates the conduction band effect on the spectral signature of a certain class of geologic materials.

The wide variety of geologic materials with their corresponding composition makes it fairly complicated to uniquely identify a certain material based on a few spectral measurements. The measurement of surface reflectivity at a few spectral bands would lead to ambiguities. The most ideal situation would be to acquire the spectral signature of each pixel element in an image all across the whole accessible spectrum from 0.35 to 3 μm and beyond in the infrared region. This would allow unique identification of the constituents.

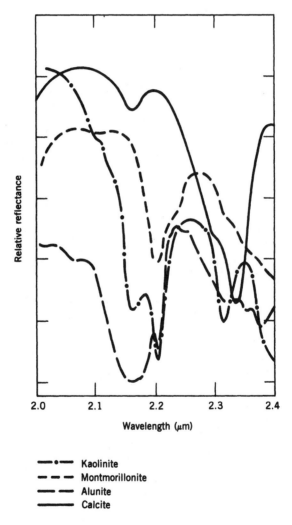

Figure 3-18. High-resolution laboratory spectra of common minerals typically associated with hydrothermal alteration. (From Goetz et al., 1983.)

However, it requires a tremendous amount of data handling capability, as discussed later. A more feasible approach would be to concentrate on diagnostic regions for specific materials. For instance, a detailed spectrum in the 2.0 to 2.4 μm region will allow the identification of the OH group minerals. Figure 3-18 illustrates the detailed structure in this spectral region for minerals typically associated with hydrothermal alterations.

3-3-2   Signature of Biologic Materials

The presence of chlorophyll in vegetation leads to strong absorption at wavelengths shorter than 0.7 μm. In the 0.7 to 1.3 μm region, the strong reflectance is due to the refractive index discontinuity between air and the leaf cell. In the region from 1.3 to 2.5 μm, the spectral reflectance curve of a leaf is essentially the same as that of pure water. Figure 3-19 shows the spectral reflectance of corn, soybean, bare soil, and a variety of foliages.

One of the major objectives in remotely sensing biologic materials is to study their dynamic behavior through a growing cycle and monitor their health. Thus, the variations in their spectral signature as a function of their health is of particular importance. As illustrated in Figure 3-20, leaf moisture content can be assessed by comparing the reflectances near 0.8, 1.6, and 2.2 μm. Even though there are diagnostic water bands at 1.4 and 1.9 μm, it should be kept in mind that these are also regions of high atmospheric absorption due to atmospheric water vapor.

The amount of green biomass also affects the reflectance signature of biologic materials, as illustrated in Figure 3-21 for alfalfa through its growth cycle. The bare-field signature is presented by the zero biomass curve. As the vegetation grows, the spectral signature becomes dominated by the vegetation signature. In principle, the biomass can be measured by comparing the reflectance in the 0.8–1.1 μm region to the reflectance near 0.4 μm.

Figure 3-22 shows another example of the changes that occur in the spectral signature of a beech leaf through its growing cycle, which in turn reflect changes in chlorophyll concentration. Both the position and slope of the rise (called red edge) near 0.7 μm change as the leaf goes from active photosynthesis to total senescence.

High spectral resolution analysis of the red edge allows the detection of geochemical stress resulting from alteration in the availability of soil nutrients. A number of researchers have noted a blue shift, consisting of about 0.01 μm of the red edge or chlorophyll shoulder to slightly shorter wavelengths in plants influenced by geochemical stress (Figs. 3-23 and 3-24). This shift, due to mineral-induced stress, may be related to subtle changes in the cellular environment.

In many situations, geologic surfaces are partially or fully covered by vegetation. Thus, natural spectral signatures will contain a mixture of features that characterize the cover and the underlying material (see Fig. 3-25). The relative contributions depend on the precent of the vegetation cover and the intensity of the indentifying feature (i.e., absorption band or step) being observed.

3-3-3   Depth of Penetration

The reflectivity of surfaces in the visible and near infrared regions is fully governed by the reflectivity of the top few microns. Weathered rocks often show a discrete iron-rich surface layer that could be compositionally different from the underlying rock, as in the case of desert varnish. Thus, it is important to determine the depth of penetration of the sensing radiation. This is usually done by conducting laboratory measurements. Bucking-

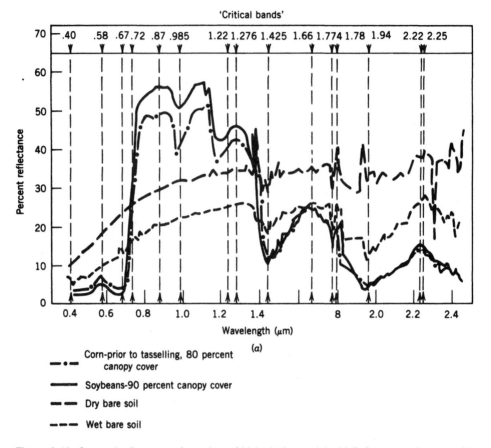

Figure 3-19 legend:

— · —   Corn-prior to tasselling, 80 percent canopy cover

———   Soybeans-90 percent canopy cover

— —   Dry bare soil

— — —   Wet bare soil

Figure 3-19. Spectral reflectance of a variety of biological materials. (*a*) Reflectance of some cultivated vegetation compared to reflectance of bare soil and wet soil.

ham and Sommer (1983) made a series of such measurements using progressively thicker samples. They found that as the sample thickness increases, absorption lines become more apparent (i.e., higher contrast). After a certain critical thickness, an increase in sample thickness does not affect the absorption intensity. This corresponds to the maximum thickness being probed by the radiation.

A typical curve showing the relationship between sample thickness and absorption intensity is shown in Figure 3-26. The penetration depth for 0.9 μm radiation in ferric materials is at most 30 μm, and this thickness decreases as the concentration of ferric material increases. In the case of the 2.2 μm radiation, a 50 μm penetration depth was measured for kaolinite.

The penetration effect can be easily quantified by including the complex wave vector k in the field expression (Equation 2-10). When the medium is absorbing, the index of refraction is given by

$$\sqrt{\varepsilon_r} = n = N_r + iN_i \tag{3-15}$$

$$\rightarrow k = \sqrt{\varepsilon_r}k_0 = N_r k_0 + iN_i k_0 \tag{3-16}$$

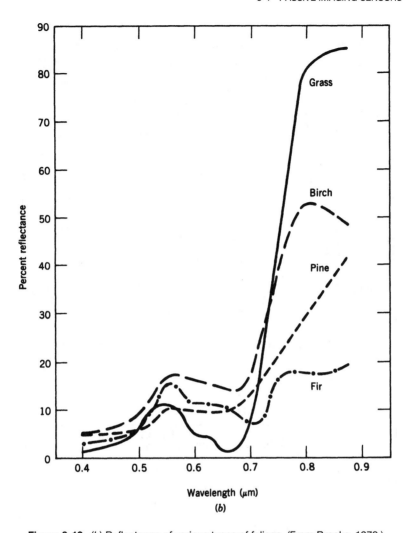

Figure 3-19. (*b*) Reflectance of various types of foliage. (From Brooks, 1972.)

where $k_0 = 2\pi/\lambda$. Thus, the field expression becomes

$$\mathbf{E} = \mathbf{A}e^{i(N_r k_0 + iN_i k_0)r - i\omega t} = \mathbf{E}_0 e^{-N_i k_0 r} \tag{3-17}$$

where $\mathbf{E}_0$ is the field when there is no absorption. Equation 3-17 shows that the field decreases exponentially as its energy is absorbed by the medium. The "skin" depth or penetration depth d is defined as the thickness at which the field energy is reduced by a factor e^{-1}. Thus,

$$d = \frac{1}{N_i k_0} = \frac{\lambda}{4\pi N_i} \tag{3-18}$$

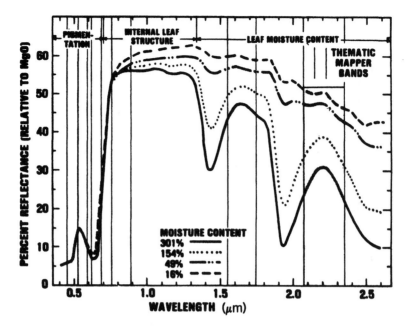

Figure 3-20. Progressive changes in the spectral response of a sycamore leaf with varying moisture content (From Short, 1982.)

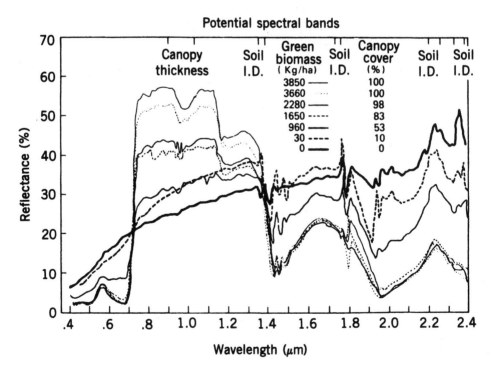

Figure 3-21. Variations in spectral reflectance as functions of amounts of green biomass and percent canopy cover (From Short, 1982.)

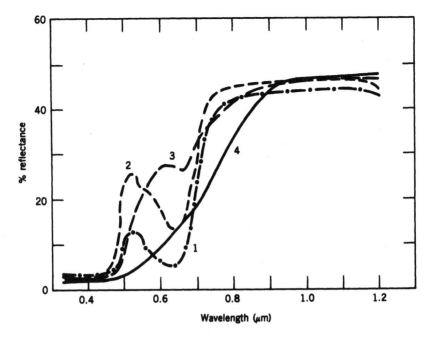

Figure 3-22. Reflectance spectra for a healthy beech leaf (1) and beech leaves in progressive phases of senescence (2–4). (From Knipling, 1969.)

3-4 PASSIVE IMAGING SENSORS

A large number of visible and infrared imaging sensors have been flown in space to study the Earth and planetary surfaces. These include the Gemini, Apollo, and Skylab cameras; the series of Landsat cameras, including the Multispectral Scanner (MSS), Landsat Thematic Mapper (TM), and the Enhanced Landsat Thematic Mapper Plus (ETM+); the series of SPOT Satellite Imagers; and the Advanced Spaceborne Thermal Emission and Reflection Radiometer (ASTER). A number of planetary imaging cameras have also been flown, including the Galileo Jupiter Multispectral Imager and the Mars Orbiter Camera.

3-4-1 Imaging Basics

Passive imaging systems collect information about the surface by studying the spectral characteristics of the electromagnetic energy reflected by the surface, as shown in Figure 3-1. As the source energy propagates through the atmosphere on its way to the surface, the spectral characteristics of the atmosphere are imprinted on the source signal, as shown in Figure 3-2 for the case of the Earth's atmosphere. This incoming energy is then reflected by the surface, and propagates through the atmosphere to the collecting aperture of the imaging sensor.

Let the incoming radiant flux density be F_i. The spectrum of this incoming radiant flux at the surface is given by

$$S_i(\lambda) = S(\lambda,\, T_s) \times \left(\frac{R_s}{d} \right)^2 a(\lambda) \tag{3-19}$$

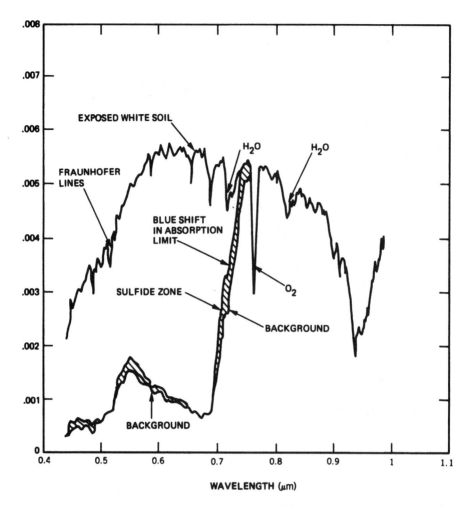

Figure 3-23. Blue shift in the spectrum of conifers induced by a sulfide zone. (From Collins et al., 1983. Reproduced from *Economic Geology,* Vol. 78, p. 728.)

where $S(\lambda, T_s)$ is the energy radiated from the source with temperature T_s, but modified by the absorption spectrum $a(\lambda)$ of the atmosphere. The second term represents the fact that the energy is radiated spherically from the sun surface with radius R_s to the body at distance d.

The fraction of this incoming energy that is reflected by the surface is described by the surface reflectance $\rho(\lambda)$, also known as the surface albedo. The most complete description of the surface reflectance is given by the so-called bidirectional reflectance distribution function (BRDF), which gives the reflectance of the surface as a function of both the illumination and the viewing geometry. In the simpler case of a Lambertian surface, this function is a constant in all directions, which means that the reflected energy is spread uniformly over a hemisphere. If the surface area responsible for the reflection is dS, the radiant flux at the sensor aperture is

$$S_r(\lambda) = S_i(\lambda)\rho(\lambda)dS\,\frac{1}{2\pi r^2}\,a(\lambda) = S(\lambda, T_s)\left(\frac{R_s}{d}\right)^2 a^2(\lambda)\,\frac{\rho(\lambda)dS}{2\pi r^2} \qquad (3\text{-}20)$$

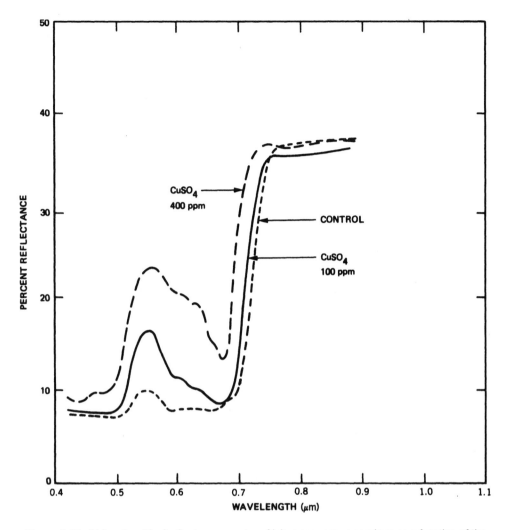

Figure 3-24. Bidirectional leaf reflectance spectra of laboratory grown sorghum as a function of the presence of copper as $CuSO_4$ in the soil. (From Chang and Collins, 1983. Reproduced from *Economic Geology,* Vol. 78, p. 743.)

where r is the distance between the aperture and the surface area reflecting the incoming energy. If the aperture size is denoted by dA, the power captured by the aperture, per unit wavelength, will be

$$P(\lambda) = S_r(\lambda)dA = S(\lambda, T_s)\left(\frac{R_s}{d}\right)^2 a^2(\lambda)\frac{\rho(\lambda)dSdA}{2\pi r^2} \qquad (3-21)$$

The aperture will typically collect the incoming radiation for a short time, known as the dwell time or aperture time. In addition, the sensor will typically collect radiation over a finite bandwidth with an efficiency described by the sensor transfer function $h(\lambda)$. Denoting the dwell time by τ, we find the total energy collected by the sensor from the surface element to be

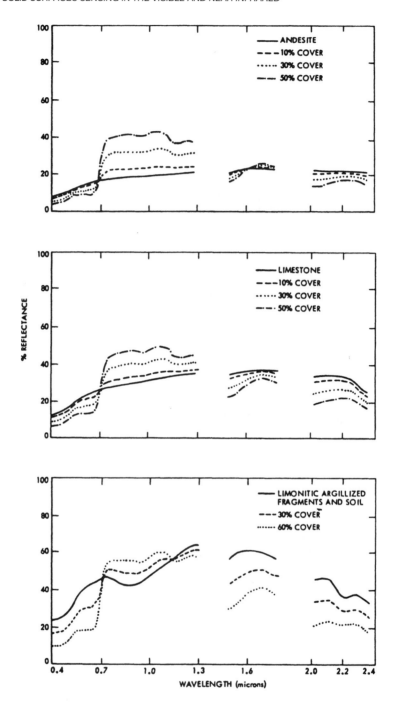

Figure 3-25. Vegetation effects of green grass cover on spectral reflectance of various materials, (From Siegal and Gillespie, © 1980 John Wiley & Sons. Reprinted with permission.)

$$E_r = \int_{\lambda_1}^{\lambda_2} S(\lambda, T_s)\left(\frac{R_s}{d}\right)^2 a^2(\lambda)\frac{\rho(\lambda)dSdA}{2\pi r^2}h(\lambda)\tau d\lambda \qquad (3\text{-}22)$$

If the relative bandwidth is small, this expression can be approximated by

$$E_r \approx S(\lambda_0, T_s)\left(\frac{R_s}{d}\right)^2 a^2(\lambda_0)\frac{\rho(\lambda_0)dSdA}{2\pi r^2}h(\lambda_0)\tau\Delta\lambda \qquad (3\text{-}23)$$

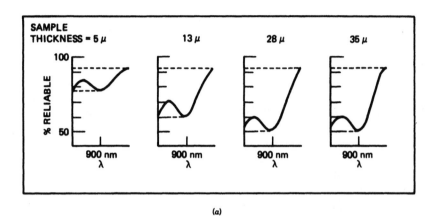

(a)

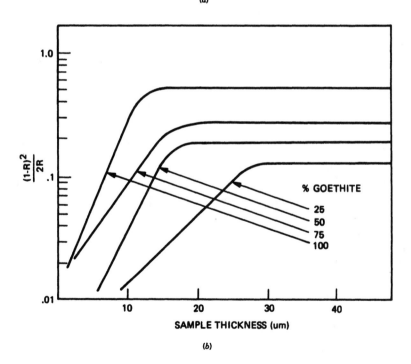

(b)

Figure 3-26. (a) Representation of the effect of increasing sample thickness on the absorption intensity near 0.9 μm in goethite-bearing samples. (b) Relationship between sample thickness and absorption intensity. (From Buckingham and Sommer, 1983. Reproduced from *Economic Geology*, Vol. 78, p. 689.)

where λ_0 is the wavelength at the center of the measurement bandwidth, and $\Delta\lambda$ is the bandwidth. This received energy is compared with the intrinsic noise of the sensor system to determine the image signal-to-noise ratio.

3-4-2   Sensor Elements

The main elements of an imaging system are sketched in Fig. 3-27. The collecting aperture size defines the maximum wave power that is available to the sensor. The collector could be a lens or a reflecting surface such as a plane or curved mirror.

The focusing optics focuses the collected optical energy onto the detecting element or elements. The focusing optics usually consists of numerous optical elements (lenses and/or reflectors) for focusing, shaping, and correcting the wave beam.

A scanning element is used in some imaging systems to allow a wide coverage in the case where few detecting elements are used. When a large array of detecting elements or film is used, the scanning element is usually not necessary.

In order to acquire imagery at different spectral bands, the incident wave is split into its different spectral components. This is achieved by the dispersive element, which could consist of a set of beamsplitters/dichroics, a set of filters on a spinning wheel, or dispersive optics such as a prism or grating (see Fig. 3-28).

The wave is finally focused on the detecting element where its energy is transformed into a chemical imprint in the case of films or into a modulated electrical current in the case of array detectors.

An imaging system is commonly characterized by its response to electromagnetic energy that originates from a fictitious "point source" located infinitely far away from the sensor system.

Electromagnetic waves from such a point source would reach the sensor system as plane waves. Because the sensor system consists of elements that are of finite size, the

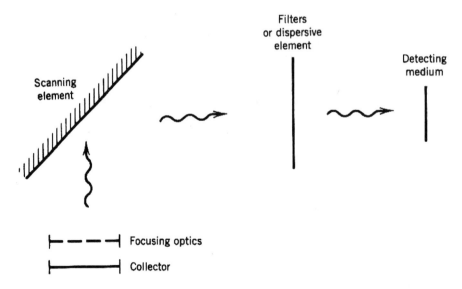

Figure 3-27. Sketch of major elements of an imaging sensor. The elements are not to scale and their order could be different depending on the exact system configuration.

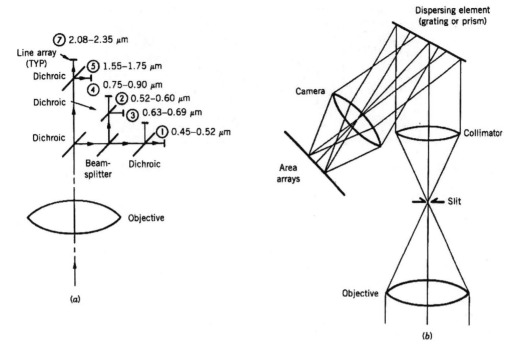

Figure 3-28. Multispectral wave dispersion techniques. (*a*) Beamsplitter used on the Landsat TM. (*b*) Dispersive optics. A third technique is the use of a spinning wheel with a bank of filters.

waves will be diffracted as they propagate through the sensor until they reaches the focal plane. Here, the point source no longer appears as a single point, but the energy will be spread over a finite patch in the focal plane. The descriptive term *point spread function* is commonly used to describe the response of an imaging system to a point source.

The exact shape of the point spread function depends on the design of the optical system of the sensor, and on the physical shapes of the apertures inside the optical system. For example, light from a point source that propagated through a circular aperture with uniform transmission would display an intensity pattern that is described by

$$I(u, v) = I_0 \frac{\pi a^2}{\lambda^2} \left[\frac{2J_1(ka\sqrt{u^2 + v^2})}{ka\sqrt{u^2 + v^2}} \right]^2 \qquad (3\text{-}24)$$

where a is the radius of the circular aperture, λ is the wavelength of the incident radiation, $k = 2\pi/\lambda$ is the wave number, and u and v are two orthogonal angular coordinates in the focal plane. This diffraction pattern, shown in Figure 3-29, is commonly known as the Airy pattern, after Airy, who derived Equation (3-24) in 1835. The first minimum in the Airy pattern occurs at the first root of the Bessel function J_1 at

$$w = \sqrt{u^2 + v^2} = 0.610 \frac{\lambda}{a} = 1.22 \frac{\lambda}{D} \qquad (3\text{-}25)$$

where $D = 2a$ is the diameter of the aperture. The area inside the first minimum of the Airy function is commonly referred to as the size of the Airy disk. The light from the

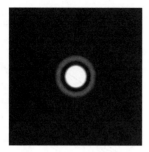 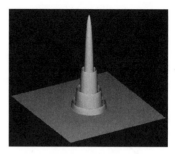

Figure 3-29. Diffraction pattern of a circular aperture with uniform illumination. The image on the left shows the logarithm of the intensity and is scaled to show the first three diffraction rings in addition to the central peak, a total dynamic range of three orders of magnitude. The diffraction pattern is also shown as a three-dimensional figure on the right, displayed using the same logarithmic scale.

point source is not focused to a point; instead, it is spread over the neighborhood of the point in the shape of the Airy function—a process known as diffraction. Looking again at Equation (3-24) or (3-25), we observe that the size of the patch of light that represents the original point source in the instrument focal plane is a function of the electrical size of the aperture as measured in wavelengths. The larger the aperture, the smaller the Airy disk for a given wavelength. On the other hand, if the aperture size is fixed, the Airy disk will be larger at longer wavelengths than at shorter wavelengths. This is the case in any telescope system that operates over a number of wavelengths. The Airy disk at 0.7 μm would be nearly twice as wide as that at 0.4 μm for a fixed aperture size.

The resolution, or resolving power, of an imaging system describes its ability to separate the images of two closely spaced point sources. The most commonly used definition of resolution is that defined by Lord Rayleigh, according to which two images are regarded as resolved when they are of equal magnitude and the principal maximum of the one image coincides with the first minimum of the second image. Therefore, Equation (3-25) describes the smallest angular separation between two point sources of equal brightness if a telescope with a circular aperture with radius a were used to observe them. Figure 3-30 shows three cases in which two point sources are completely resolved (left graph), the two points are just resolved (middle graph), exactly satisfying the Rayleigh criterion, and the two point sources are not resolved (right graph).

It should be mentioned that not all imaging systems employ circular apertures of the type that would result in a point source being displayed as an Airy function in the image plane of the sensor. Figure 3-31 shows the point spread function of a square aperture with equal intensity across the aperture, and also that of a circular aperture in which the intensity is tapered as a Gaussian function from the center with a 50% reduction at the edges of the circular aperture. Notice the reduction in the brightness of the sidelobes of the circular aperture with a Gaussian taper when compared to Figure 3-29. This technique, known as *apodization,* is commonly used to reduce the sidelobes of imaging systems, but comes at the expense of energy throughput coupled with a slight broadening of the central disk of the point spread function.

The resolution definition described by Equation (3-24) assumes that the diffraction pattern of the system is adequately sampled in the focal plane. In modern remote sensing systems, the focal plane of the sensor is commonly populated with electronic detectors. Due to the finite size of these detectors, energy from a small solid angle, corresponding to

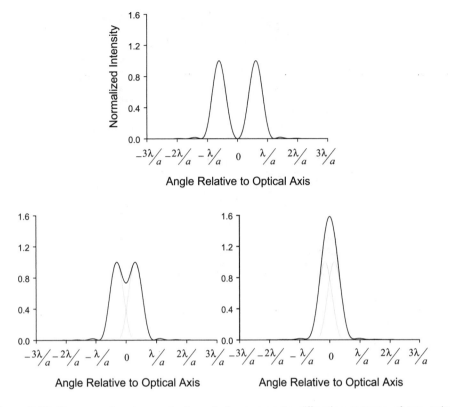

Figure 3-30. These graphs show cuts through the composite diffraction patterns of two point sources. Three cases are shown. The two point sources in the top graph are completely resolved. The two points in the bottom-left graph are just resolved, exactly satisfying the Rayleigh criterion. The two point sources in the bottom-right graph are not resolved. Note that here the intensities are displayed using a linear scale, so the diffraction rings are only barely visible.

the instantaneous field of view (IFOV) of an individual detector, will be integrated and reported as a single value (Figure 3-32). Therefore, if the IFOV of the detector is larger than the size of the Airy disk, it is possible that two point sources can be further apart than the resolution limit of the telescope optics, but the light from these two sources can still end up on a single detector and, hence, they will not be distinguishable. In this case, the resolution of the sensor is driven by the size of detector, and not by the resolving power of the optics. In any case, the intersection of the detector IFOV and the surface being imaged gives the size of the surface element that is imaged by that detector. This area is commonly known as the size of a "pixel" on the ground, referring to the fact that this area will be reported as one picture element in the final image. This area represents the reflecting surface area dS in Equations (3-20)–(3-23).

3-4-3 Detectors

The detector transforms the incoming wave into a form of recordable information. Optical films are one type of detector. Another type is the electrooptic detector, which transforms the wave energy into electrical energy that is usually transmitted to a digital recording

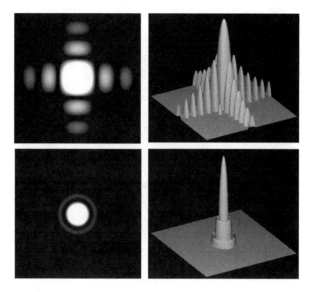

Figure 3-31. Diffraction patterns of a square aperture with uniform illumination (top) and a circular aperture with Gaussian tapered illumination with 50% transmission at the edges of the aperture (bottom). The figures are scaled the same as those in Figure 3-29.

medium. Electrooptic detectors are generally classified on the basis of the physical processes by which the conversion from radiation input to electrical output is made. The two most common ones are thermal detectors and quantum detectors.

Thermal detectors rely on the increase of temperature in heat-sensitive material due to absorption of the incident radiation. The change in temperature leads to change in resistance (in bolometers, usually using a Wheatstone bridge) or voltage (in thermocouplers,

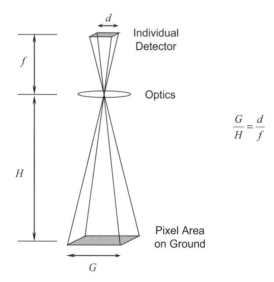

Figure 3-32. Imaging geometry showing the instantaneous field of view of a single detector.

usually using thermoelectric junctions), which can be measured. Typically, thermistor bolometers use carbon or germanium resistors with resistance change of about 4% per degree. Thermal detectors are usually slow, have low sensitivity, and their response is independent of wavelength. They are not commonly used in modern visible and near-infrared remote sensors.

Quantum detectors use the direct interaction of the incident photon with the detector material, which produces free-charge carriers. They usually have high sensitivity and fast response, but they have limited spectral region of response (see Fig. 3-33). The *detectivity* of a quantum detector is defined as

$$D = \frac{1}{NEP} \qquad (3\text{-}26)$$

where *NEP* is the noise equivalent power, which is defined as the incident power on the detector that would generate a detector output equal to the r.m.s. noise output. In other words, the noise equivalent power is that incident power that would generate a signal-to-noise ratio that is equal to 1.

For many detectors, both the detectivity and the *NEP* are functions of the detector area and the signal bandwidth. To take this into account, the performance of quantum detectors is usually characterized by the *normalized detectivity, D**, which is equal to

$$D^* = D \sqrt{A\Delta f} = \frac{\sqrt{A\Delta f}}{NEP} = \frac{(S/N)\sqrt{A\Delta f}}{W} \qquad (3\text{-}27)$$

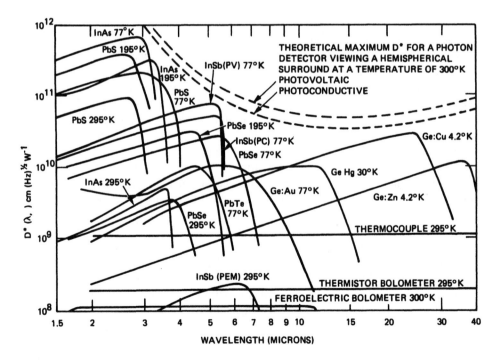

Figure 3-33. Comparison of the *D** of various infrared detectors when operated at the indicated temperature (Hudson, 1969).

where A is the detector area, Δf is the circuit bandwidth, S/N is the signal-to-noise ratio, and W is the radiation incident on the detector.

Quantum detectors are typically classified into three major categories: photoemissive, photoconductive, and photovoltaic.

In photoemissive detectors, the incident radiation leads to electron emission from the photosensitive intercepting surface. The emitted electrons are accelerated and amplified. The resulting anode current is directly proportional to the incident photon flux. These detectors are usually operable at wavelengths shorter than 1.2 μm because the incident photon must have sufficient energy to overcome the binding energy of the electron in the atom of the photosensitive surface. This is expressed by the Einstein photoelectric equation:

$$E = \tfrac{1}{2} m v^2 = hv - \phi \qquad (3\text{-}28)$$

where ϕ = work function = energy to liberate an electron from the surface, m is the electron mass, and v is the velocity of the ejected electron. Thus, the photon energy hv must be greater than ϕ in order to liberate an electron. The critical wavelength of the incident wave is that wavelength for which the photon energy is equal to the work function of the detector material and is given by

$$\phi = hv_c \rightarrow \lambda_c(\mu) = \frac{1.24}{\phi(\text{eV})}$$

The lowest ϕ for photoemissive surfaces is for alkali metals. Cesium has the lowest, $\phi = 1.9$ eV, which gives $\lambda_c = 0.64$ μm. Lower values of ϕ can be achieved with composite surfaces. For instance, for a silver–oxygen–cesium composite, $\phi_c = 0.98$ eV and $\lambda_c = 1.25$ μm.

In a photoconductive detector, incident photons with energy greater than the energy gap in the semiconducting material produce free-charge carriers, which cause the resistance of the photosensitive material to vary in inverse proportion to the number of incident photons. This requires substantially less energy than electron emission, and, consequently, such detectors can operate at long wavelengths in the thermal infrared. The energy gap for silicon, for example, is 1.12 eV, allowing operation of silicon detectors to about 1.1 μm. Indium antimonide has an energy gap of 0.23 eV, giving a cutoff wavelength of 5.9 μm.

In the case of photovoltaic detectors, the light is incident on a p–n junction, modifying its electrical properties, such as the backward bias current.

Quantum detectors can also be built in arrays. This feature allows the acquisition of imaging data without the need for scanning mechanisms that are inefficient and provide short integration time per pixel. For this reason, most of the advanced imaging sensors under development will use detector arrays such as charge-coupled device (CCD) arrays (Fig. 3-34).

Silicon CCD detectors are among the most commonly used detectors in the visible and NIR part of the spectrum. These detectors typically have excellent sensitivity in the wavelength region from 0.4 to 1.1 microns. Fundamentally, a CCD array is made up of a one- or two-dimensional array of metal oxide silicon (MOS) capacitors that collect the charge generated by the free-charge carriers. Each capacitor accumulates the charge created by the incident radiation from a small area, known as a pixel, in the total array. To register the image acquired by the sensor, the charges accumulated on these capacitors must be read from the array. This is typically done using a series of registers into which the charges are transferred sequentially, and the contents of these registers are then sent to a

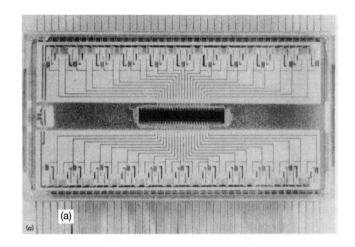

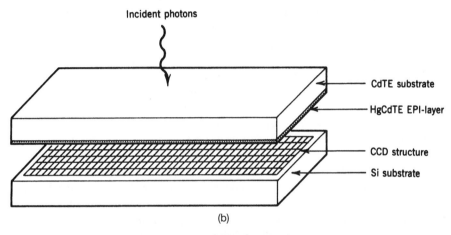

Figure 3-34. Charge-coupled devices (CCD) linear array photograph (*a*) and sketch illustrating the different substrates (*b*). (*c*) a photograph of a two-dimensional CCD array.

common output structure where the charges are converted to a voltage. The voltage is then digitized and registered by the read-out electronics. The speed at which the charges can be read determines the frame rate at which a CCD detector can be used to acquire images. Various different read-out schemes are used to increase this frame rate. The fastest frame rates typically result when two CCD arrays are implemented side by side. One device is illuminated by the incoming radiation and is used to integrate charges, whereas the second device is covered with a metal layer to prevent light from entering the device. Once the charges have accumulated on the imaging CCD, they are quickly transferred to the covered device. These charges can then be read from the second device while the first is again used to integrate charges from the scene being imaged.

Traditional CCD detectors have reduced sensitivity to the shorter wavelengths in the blue portion of the visible spectrum. The polysilicon gate electrodes, used to clock out the charge from the capacitors in the imaging part of the array, strongly absorb blue wavelength light. As pixel geometries get smaller, this problem is exacerbated. The blue response of the sensor decreases rapidly with pixel size. The short wavelength sensitivity can be improved using a technique known as thinning and back-illumination. Essentially, the CCD is mounted upside down on a substrate, and charges are collected through the back surface of the device, away from the gate electrodes. On the longer-wavelength side of the spectrum, the sensitivity of silicon CCDs can be improved by doping the silicon with impurities such as arsenic. These devices, known as extrinsic CCDs, have sensitivities that extend well into the infrared part of the spectrum.

Recently, CMOS detectors have generated much interest. The basic difference between CMOS and CCD detectors is that in the case of the CMOS detectors, the charge-to-voltage conversion takes place in each pixel. This makes it easy to integrate most functions on one chip using standard CMOS processes. This integration into a single structure usually means that signal traces can be shorter, leading to shorter propagation delays and increased read-out speed. CMOS arrays also support random access, making it easy to read only a portion of the array, or even one pixel. The drawback for CMOS detectors is that the charge-to-voltage conversion transistors must be placed in the pixel, and therefore take up area that would otherwise be part of the optically sensitive part of the detector. This leads to a lower fill factor for CMOS devices as compared to CCDs, and generally lower sensitivity.

Indium antimonide (InSb) detectors are commonly used to cover the near-infrared part of the spectrum, as is the case for bands 5 and 7 of the Landsat ETM+ sensor.

In the case of films, black and white, true color, and false color films can be used. The true color film is sensitive to the visible part of the spectrum. The false color film records a small part of the highly reflective portion of the infrared spectrum (0.75–0.9 μm) characteristic of vegetation. With this type of film, a yellow filter is used to remove the blue part so that the blue-, green-, and red-sensitive layers of emulsion in the film are available to record the green, red, and infrared parts of the radiation, respectively. Since healthy vegetation is strongly reflective in the infrared region, it appears bright red in false color infrared images, whereas unhealthy vegetation appears blue to gray.

3-5   TYPES OF IMAGING SYSTEMS

Depending on how the sensor acquires and records the incoming signal, imaging systems can be divided into three general categories: framing cameras, scanning systems, and pushbroom imagers (see Fig. 3-35 and Table 3-3).

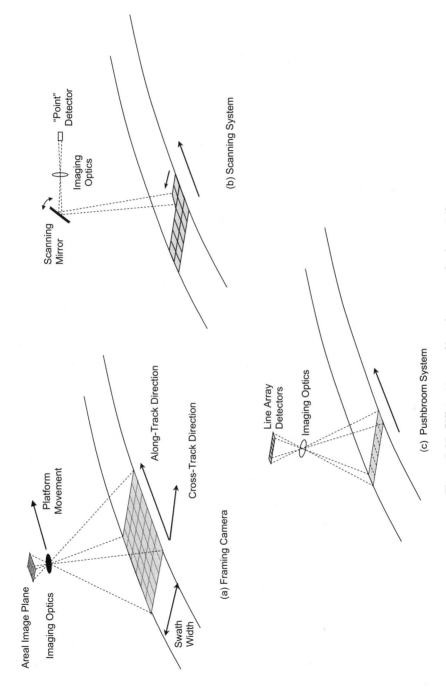

Figure 3-35. Different types of imaging sensor implementations.

89

Table 3-3. Comparison of Different Imaging Systems

Type	Advantage	Disadvantage
Film framing camera	Large image format High information density Cartographic accuracy	Transmission of film Potential image smearing Wide field of view optics
Electronic framing camera	Broad spectral range Data in digital format Simultaneous sampling of image, good geometric fidelity	Difficulty in getting large arrays or sensitive surface Wide field of view optics
Scanning systems	Simple detector Narrow field-of-view optics Wide sweep capability Easy to use with multiple wavelengths	Low detector dwell time Moving parts Difficult to achieve good image, geometric fidelity
Pushbroom imagers	Long dwell time for each detector Across-track geometric fidelity	Wide field-of-view optics

A framing camera takes a snapshot of an area of the surface, which is then projected by the camera optics on a film or a two-dimensional array of detectors located in the camera focal plane. As detector arrays with more pixels become available, framing cameras will become more common. An example of a framing camera is the panchromatic camera used on the two Mars Exploration Rovers that landed on Mars in January 2004. Framing cameras have the major advantage that excellent geometric fidelity can be achieved because the entire image is acquired at once. This comes at the price that the optics system must typically have excellent response over a wide field of view.

Scanning systems use a scanning mirror that projects the image of one surface resolution element on a single detector. To make an image, across-track scanning is used to cover the imaged swath across the track. In some cases, a limited number of detectors are used so that each scan covers a set of across-track lines instead of a single one. In this case, the imaging system is known as a whiskbroom scanner. The platform motion carries the imaged swath along the track. The major disadvantage of such a system is the presence of moving parts and the low detection or dwell time for each pixel. In addition, images acquired with scanning systems typically have poorer geometric fidelity than those acquired with framing cameras. Examples of scanning systems are the Landsat instruments such as the Multispectral Scanner (MSS) and Thematic Mapper (TM) and the Enhanced Thematic Mapper Plus (ETM+).

Pushbroom imagers delete the scanning mechanism and use a linear array of detectors to cover all the pixels in the across-track dimension at the same time. This allows a much longer detector dwell time on each surface pixel, thus allowing much higher sensitivity and a narrower bandwidth of observation. Examples of such systems are the SPOT and the ASTER cameras. A pushbroom system can be thought of as a framing camera with an image frame that is long in the across-track direction and much narrower in the along-track direction. Pushbroom sensors do not require a moving scan mirror in order to acquire an image. As a result, these sensors can be expected to exhibit longer operating life than a scanner. In addition, the fixed geometry afforded by the detector arrays results in high geometric accuracies in the line direction, which will simplify the image reconstruction and processing tasks.

The imaging spectrometer goes one step further. It utilizes a spectrometer section to separate the spectral channels and an area array detector to acquire images simultaneously in a large number of spectral channels. A narrow strip of the surface, one resolution element wide and a swath width long, is imaged through a slit followed by a dispersive element that disperses the energy in the line image into a series of line images of different spectral bands. This dispersed line image is projected onto a two-dimensional detector array. Each array line will detect the image of one spectral image line. Figures 3-36 and 3-37 illustrate one possible concept and design of an imaging spectrometer.

As the remote sensing instrument is carried along the satellite orbit, it typically images a strip on the surface along the orbit track. The width of this strip is known as the *swath* imaged by the camera. In the case of a framing camera, the swath may simply be the width of each image frame in the direction orthogonal to the satellite movement. Successive frames are combined to form an image of which the length is limited by on-board storage or data downlink capability. In the case of a scanning system, the width of the swath is determined by the angle over which the scanning is done. The width of the swath and the altitude of the camera above the surface determine the field of view (FOV) of the imaging instrument, usually expressed as the total angular extent over which an image is acquired. Note that this definition of the FOV means that the *instrument* FOV may be different than the FOV of the optics of the telescope. This will be the case when either the instrument uses a scanning technique for imaging, or when the detector array in the focal plane only covers part of the telescope field of view.

3-6 DESCRIPTION OF SOME VISIBLE/INFRARED IMAGING SENSORS

The complexity of an imaging sensor is usually directly related to the number of spectral channels, the number of detector elements, and the imaging resolution. Figure 3-38 gives a comparison of some of the most advanced imaging systems.

To illustrate the complexity associated with some of the advanced sensors, let us consider the data rate associated with a 100 channel imager with a 25 m resolution and a 100

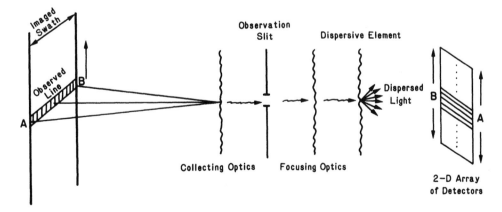

Figure 3-36. Conceptual sketch of an imaging spectrometer. A narrow strip AB is imaged at a specific instant in time. The light from this strip is dispersed by the dispersive element such that each line on the array of detectors will correspond to the image of the strip in a very narrow spectral band.

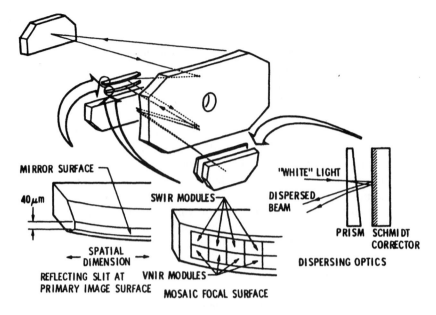

Figure 3-37. One possible design for the optical system of the imaging spectrometer.

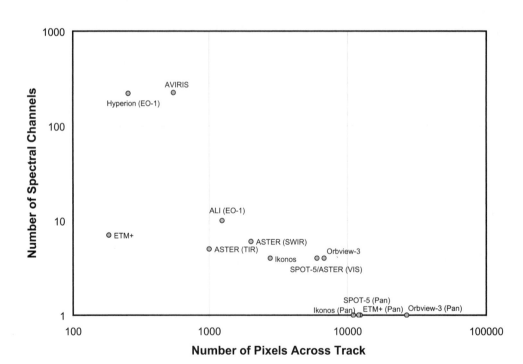

Figure 3-38. Comparison of different imaging systems.

km image swath. As a satellite in earth orbit moves at about 7.5 km/sec, the number of imaged pixels per second is given by

$$n = \frac{100,000}{25} \times \frac{7500}{25} = 1.2 \times 10^6$$

and the total number of pixels is

$$N = 100 \text{ channels} \times n = 1.2 \times 10^8$$

Assuming that each pixel brightness is digitized to 8 bits, the total bit rate is then $8 \times 1.2 \times 10^8$, which is approximately 1 gigabit per second. This is well beyond the capability of present data transmission systems. In fact, it is often the case that the capability of space-borne imaging systems is limited more by the data downlink transmission rate than by the instrument technology. This problem is particularly bad in the case of deep space missions. Thus, some intelligent data-reduction system is required.

3-6-1 Landsat-Enhanced Thematic Mapper Plus (ETM+)

The Landsat Enhanced Thematic Mapper Plus is a multispectral imaging system of the scanning type. It is flown on the Landsat-7 (1999 launch) spacecraft, and its characteristics are summarized in Table 3-4. The sensor is a derivative of the Thematic Mapper (TM) instruments flown on earlier Landsat satellites. The primary changes of the ETM+ over the TM's are the addition of a panchromatic band and improved spatial resolution for the thermal band.

The Landsat-7 spacecraft was put into a 705 km altitude orbit with an inclination of 98.2 degrees. A 185 km swath is imaged (Fig. 3-39). This allows complete coverage utilizing the 233 orbits in 16 days, which is the repeat cycle of the orbit. The satellite orbit is sun synchronous at approximately 10:00 a.m. local time. The data is transmitted to the ground as two streams of 150 Mbps each for a total rate of 300 Mbps using an X-band communications link. The primary receiving station is the U.S. Geological Survey's (USGS) EROS Data Center (EDC) in Sioux Falls, South Dakota. Images can be acquired by EDC through real-time downlink or playback from an on-board, 380 gigabit (100 scenes) solid-state recorder. Since data is split in two streams, one can contain real-time data while the other is playing back data from the recorder, or both

Table 3-4. Enhanced Thematic Mapper Plus Characteristics

Band number	Spectral range (μm)	Spatial resolution (m)	Quantization levels (bits)
1	0.45–0.52	30	8
2	0.53–0.61	30	8
3	0.63–0.69	30	8
4	0.78–0.9	30	8
5	1.55–1.75	30	8
6	10.4–12.5	60	8
7	2.09–2.35	30	8
8	0.52–0.9	15	8

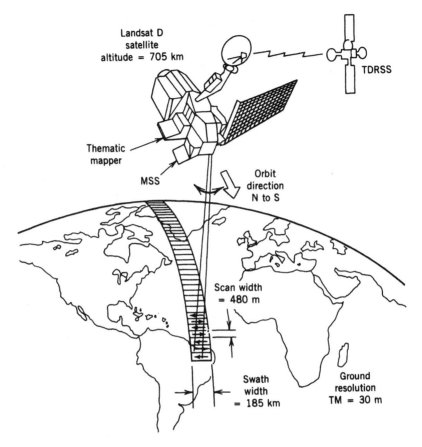

Figure 3-39. Landsat-D mapping geometry. (From Blanchard and Weinstein, © 1980 IEEE.)

streams can be used to play back data from the recorder simultaneously. Images can also be received by a world-wide network of receiving stations either in real time or by direct downlink at X-band.

The ETM+ optics are similar to that of the older Thematic Mapper, shown in Figure 3-40, and its key parameters are given in Table 3-5. The bidirectional scan mirror moves the view of the telescope back and forth across the ground track (Fig. 3-35). The ground track is subdivided into 16 raster lines (32 for the panchromatic channel), which correspond to an array of 16 along-track detectors per spectral channel, and 32 for the panchromatic channel. There are six arrays of 16 detectors, each with an optical filter to define the corresponding spectral band for the visible and near IR regions. The thermal IR channel has a eight-element array. Thus, for each scan, 16 across-track lines are mapped in the visible and near IR regions, 32 are mapped in the panchromatic band, and eight lines are mapped in the thermal IR channel. This corresponds to a strip that has a width of 16 × 30 = 480 m.

The satellite moves at a speed of about 7 km/sec. Thus, we require about 7000/480 = 14.6 scans or a 7.3 Hz back-and-forth scan per second. The scan mirror assembly consists of a flat mirror supported by flex pivots on each side, a torquer, a scan angle monitor, two leaf spring bumpers, and scan mirror electronics. The motion of the mirror in each direc-

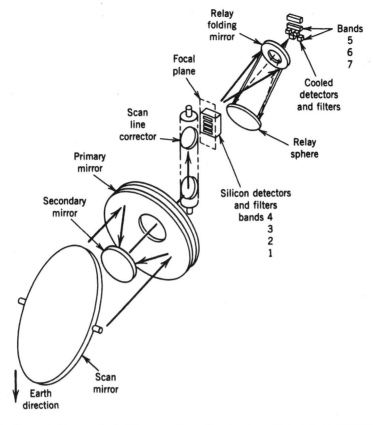

Figure 3-40. Thematic Mapper Optical System. (From Blanchard and Weinstein, © 1980 IEEE.)

Table 3-5. Significant ETM+ Parameters

Orbit	Sun synchronous
	705.3 km altitude
	98.9 min period
	98.2° inclination
	16 day repeat cycle
Scan	185 km swath
	7.0 Hz rate
	85% efficiency
Optics	40.6 cm aperture
	f/6 at prime focus
	42.5 μrad, IFOV, bands 1–4
	f/3 at relay focus
	43.8 μrad IFOV, bands 5, 7
	170 μrad IFOV, band 6
Signal	52 kHz, 3 dB, bands 1–5, 7
	13 kHz, 3 dB, band 6
	1 sample/IFOV
	8 bits/sample
	84.9 Mbps multiplexed output

tion is stopped by the bumper and is boosted by precision torque pulses during the turn-around period. The scan mirror is constructed of beryllium with an egg crate internal structure for maximum stiffness and minimum inertia. It has an elliptical shape and a size of 52 cm × 41 cm.

The telescope is a Ritchey–Chretién design with primary and secondary mirror surfaces of hyperbolic figure to provide for simultaneous correction of spherical aberration and coma. An f/6 design allows a reasonable-size detector with a 40.64 cm aperture; the focal length is 2.438 m. The mirrors are made from ultra-low-expansion (ULE) glass, and have enhanced silver coatings. A detector of 0.1 mm across gives an IFOV of 42.5 μrad, which corresponds to a 30 m spot from 705 km altitude.

A scan line corrector preceding the detectors compensates for the tilt of the array swath due to the spacecraft motion, so the scan lines will be straight and perpendicular to the ground track.

The prime focal plane is located at the focus of the primary telescope and contains the filters and silicon detectors of the first four spectral bands. Two mirrors, a folding mirror and a spherical mirror, are used to also relay the image from the primary focal plane to a second focal plane where the remaining three filters and detectors are located. These detectors require cooling to achieve good performance. Bands 5 and 7 use indium antimonide (InSb) detectors, and band 6 uses a mercury cadmium telluride (HgCdTe) detector. These detectors are kept at temperatures around using radiative cooling.

3-6-2 Advanced Spaceborne Thermal Emission and Reflection Radiometer (ASTER)

ASTER, an advanced multispectral imager operating in the pushbroom configuration, is one of five instruments that were launched on board NASA's Terra spacecraft in December 1999. ASTER is a cooperative effort between NASA and Japan's Ministry of Economy, Trade, and Industry (METI), formerly known as Ministry of International Trade and Industry (MITI).

The Terra satellite is in a sun-synchronous orbit at an altitude of 705 km at the equator, with a descending equator-crossing time of 10:30 a.m. local time. Terra flies in a loose formation with the Landsat 7 satellite, crossing the equator approximately 30 minutes after the Landsat satellite. The orbit repeat period is 16 days.

The complete ASTER system actually consists of three telescopes covering a wide spectral region with 14 bands from the visible to the thermal infrared. The visible and near-infrared system has three bands with a spatial resolution of 15 m, and an additional backward-looking telescope that is used for stereo imaging. Each ASTER scene covers an area of 60 × 60 km. The visible and near-infrared telescopes can be rotated to 24° on either side of the nadir line, providing extensive cross-track pointing capability. Table 3.6 shows the spectral passbands of the ASTER system. Here, we shall describe the visible and near-infrared system; the others will be discussed in more detail in subsequent chapters.

The ASTER visible and near-infrared subsystem consists of two independent telescope assemblies, used for stereo imaging, to minimize image distortion in the backward- and nadir-looking telescopes. The detectors for each of the bands consist of 5000 element silicon charge-coupled detectors (CCDs). Only 4000 of these detectors are used at any one time for the total image swath width of 60 km. The reason for using only 80% of the available detectors has to do with the stereo imaging geometry. In this mode,

Table 3-6. ASTER Characteristics

Subsystem	Band no.	Spectral range (μm)	Spatial resolution (m)	Quantization levels (bits)
VNIR	1	0.52–0.60	15	8
	2	0.63–0.69		
	3N	0.78–0.86		
	3B	0.78–0.86		
SWIR	4	1.60–1.70	30	8
	5	2.145–2.185		
	6	2.185–2.225		
	7	2.235–2.285		
	8	2.295–2.365		
	9	2.360–2.430		
TIR	10	8.125–8.475	90	12
	11	8.475–8.825		
	12	8.925–9.275		
	13	10.25–10.95		
	14	10.95–11.65		

the first image is acquired with the nadir-looking telescope, and the second with the backward-looking telescope. This means that a time lag occurs between the acquisition of the backward image and the nadir image. During this time, earth rotation displaces the relative image centers. based on orbit position information supplied by the Terra platform, the ASTER visible and near-infrared subsystem automatically extracts the appropriate 4000 pixels from each image to ensure that the same ground area is covered in each stereo image.

The ASTER visible and near-infrared optical system is a reflecting–refracting, improved Schmidt design. The backward-looking telescope focal plane contains only a single detector array and uses an interference filter for wavelength discrimination. The focal plane of the nadir-looking telescope contains three line arrays and uses a dichroic prism and interference filters for spectral separation, allowing all three bands to view the same area simultaneously. The telescope and detectors are maintained at 296 ± 3°K using thermal control and cooling from a cold plate. On-board calibration of the two telescopes is accomplished using a halogen lamp as a radiation source. These measures ensure that the absolute radiometric accuracy is ± 4% or better.

The visible and near-infrared subsystem produces the highest data rate of the three ASTER imaging subsystems. With all four bands operating (three nadir-looking and one backward-looking) the data rate including image data, supplemental information, and subsystem engineering data is 62 Mbps.

3-6-3   Mars Orbiter Camera (MOC)

The Mars Orbiter Camera was launched on the Mars Global Surveyor (MGS) spacecraft in November 1996, and arrived at Mars after a 300 day journey in September 1997. The initial orbit insertion of the MGS spacecraft around Mars left the spacecraft (by design) in a highly elliptical orbit with an orbital period of 44.993 hours and altitudes at periapsis and apoapsis of 262.9 km and 54,025.9 km, respectively. The periapsis altitude, however, was low enough to put the spacecraft well within the Martian atmosphere. To place the MGS spacecraft in the appropriate nearly circular and nearly polar orbit for science operations, a technique known as aerobraking was used. Aerobraking essentially uses the drag in the Martian atmosphere to slow the spacecraft down near periapsis, which in turn lowers the apoapsis, slowly circularizing the orbit. The final mapping orbit with altitudes at periapsis and apoapsis of approximately 370 km and 435 km, respectively, and a period of 117 minutes was reached in March 1999.

The MOC is a pushbroom system that incorporates both wide-angle (140 degree) and narrow-angle (0.4 degree) optics for producing global coverage (7.5 km/pixel), selective moderate-resolution images (280 m/pixel), and very selective high-resolution (1.4 m/pixel) images. The narrow-angle camera optics is a 35 cm aperture, f/10 Ritchie–Cretién telescope with a 2048 element CCD detector array (13 micron detector size) operating with a passband of 0.5 to 0.9 microns with a maximum resolution of 1.4 meters per pixel on the surface. At a spacecraft ground track velocity of ~3 km/s, the narrow-angle camera exposure time is approximately 0.44 milliseconds. A 12 MB buffer is used to store images between acquisition and transmission to earth. This camera system has returned spectacular images of the Martian surface at data rates that vary between 700 bps to 29,260 (real-time) bps. The camera system has a mass of 23.6 kg and consumes 6.7 W of power in the standby mode, and 18.7 W when acquiring narrow-angle data. Figure 3.41 shows a narrow-angle MOC image of gullies and sand dunes in a crater on Mars.

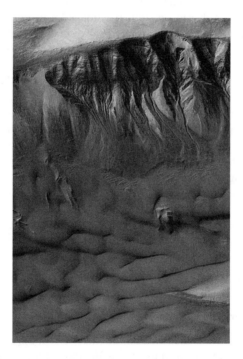

Figure 3-41. The picture shown here was taken by the Mars Orbiter Camera narrow angle (high resolution) camera and "colorized" by applying the colors of Mars obtained by the MOC wide angle cameras. The picture shows gullies in a crater at 42.4°S, 158.2°W, which exhibits patches of wintertime frost on the crater wall, and dark-toned sand dunes on the floor. (Courtesy NASA/JPL/Malin Space Science Systems.)

3-6-4   Mars Exploration Rover Panchromatic Camera (Pancam)

The stereo panchromatic cameras used on the two Mars Exploration Rovers, Spirit and Opportunity, are axamples of framing cameras. The two rovers arrived at Mars in early 2004 at Gusev Crater and Meridiani Planum, respectively, to answer the fundamental question of whether there ever was water present on the surface of Mars.

The Pancam systems are two cameras that combine to form a stereo system. The camera optics for each "eye" consists of identical three-element symmetrical lenses with an effective focal length of 38 mm and a focal ratio of f/20, yielding an IFOV of 0.28 mrad/pixel and a square FOV of 16.8° × 16.8° per eye. The optics and filters are protected from direct exposure to the Martian environment by a sapphire window. A filter wheel is used to capture multispectral images. Each filter wheel has eight positions with narrow-band interference filters covering the 400–1100 nm wavelength region. Two filter pass-bands are common between the left and right cameras, and the left camera has one clear filter. The remaining 13 filter positions (seven on the right camera, and six on the left) have different center wavelengths, allowing spectral measurements at a total of 15 different wavelengths between the two cameras.

The images are captured using a 1024 × 2048 pixel CCD array detector for each "eye" of the stereoscopic system. The arrays are operated in frame transfer mode, with one 1024 × 1024 pixel region constituting the active imaging area and the adjacent 1024 × 1024 region serving as a frame-transfer buffer. The individual detectors are 12 μm in both directions. The arrays are capable of exposure times from 0 msec (to characterize the "readout smear" signal acquired during the ~5 msec required to transfer the image to the frame-transfer buffer) to 30 sec. Analog-to-digital converters provide a digital output with 12 bit encoding and SNR > 200 at all signal levels above 20% of full scale.

Radiometric calibration of both Pancam cameras is performed using a combination of preflight calibration data and inflight images of a Pancam calibration target carried by each rover. The Pancam calibration target is placed within unobstructed view of both camera heads and is illuminated by the sun between 10:00 a.m. and 2:00 p.m. local solar time for nominal rover orientations. The calibration target has three gray regions of variable reflectivity (20%, 40%, and 60%) and four colored regions with peak reflectance in the blue, green, red, and near-IR regions for color calibration.

Figure 3-42 shows a panchromatic image of the site (later named Eagle Crater) where the second Mars Exploration Rover, Opportunity, landed on January 23, 2004. The layered rocks visible in this panoramic image measure only 10 centimeters tall. Further detailed investigation of these rocks by Opportunity's suite of scientific instruments located on its robotic arm showed fine layers that are truncated, discordant, and at angles to each other, indicating that the sediments that formed the rocks were laid down in flowing water.

Figure 3-42. Data from the Mars Exploration Rover Opportunity's panoramic camera's near-infrared, blue, and green filters were combined to create this approximate, true-color image of the rock outcrop near the rover's landing site. (Courtesy NASA/JPL/Cornell.)

3-7   ACTIVE SENSORS

With the advances in laser power and efficiency, laser sources could be used to illuminate surfaces and remotely sense their properties. Laser sources have two characteristics that give them unique aspects: (1) the transmitted energy can be pulsed, and (2) the transmitted energy has a narrow well-known spectral bandwidth.

A pulsed laser sensor can be used to measure surface topography from orbital altitude. An example of such a system is the Mars Orbiter Laser Altimeter (MOLA) instrument, launched on the Mars Gobal Surveyor satellite in November 1996. The transmitter is a Q-switched Nd:YAG laser operating at a wavelength of 1064 nm. Pulses with energy of 48 mJ/pulse are transmitted at a 10 Hz rate and illuminate a spot of approximately 130 m on the surface of Mars. The receiver has a 50 cm mirror and a silicon-avalanche photodiode is used as the detector. The system has a range resolution of 37.5 cm and a vertical accuracy (shot-to-shot) of 37.5 cm. The absolute vertical accuracy is better than 10 m, limited primarily by the accuracy with which the Mars Global Surveyor orbit can be reconstructed. Elevation measurements are made at intervals of 330 m along the spacecraft track. The instrument mass is 25.85 kg, the power consumption is 34.2 W, and it produces a continuous data rate of 618 bps.

The MOLA instrument only measures profiles of elevation directly underneath the Mars Global surveyor satellite. An extension of this system would scan the laser beam across track to acquire surface topography over a wide swath (see Fig. 3-43). This technique is now used routinely from aircraft by several commercial firms worldwide to perform high-accuracy, three-dimensional mapping. Depending on the laser altimeter used and the altitude of operation of the aircraft, ground spot sizes as small as of 20 cm with elevation accuracies of a few centimeters can be achieved. Typical swath widths range from as small as 50 m to about 10 km, depending on the spot sizes used in the mapping. If the laser altimeter is able to measure the intensity of the returning laser pulse in addition to its round-trip time of flight, an image of the surface reflectance

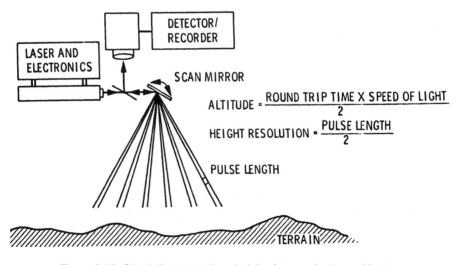

Figure 3-43. Sketch illustrating the principle of a scanning laser altimeter.

at the laser wavelength may be generated. In addition, the narrow spectral spread of a laser source allows the use of the fluorescence technique to identify certain surface materials.

3.8   SURFACE SENSING AT VERY SHORT WAVELENGTHS

All radiation at wavelengths shorter than 0.35 μm is strongly absorbed in planetary atmospheres. Thus, this spectral region can be used for surface sensing only in the case of planets without atmosphere (Moon, Mercury, and asteroids) or, in the case of Earth, from low-flying aircraft.

One of the most useful sensing techniques is γ-ray spectroscopy, which provides a spectral measurement of the particles emitted by a number of radioactive geologic materials. The spectral or energy unit commonly used in this region is the electron volt (1 eV = 1.610^{-19} J). Spectral lines of γ-ray emission are usually at many MeV, which corresponds to wavelengths of a hundred to a thousand angstroms.

3-8-1   Radiation Sources

In the case of the Earth, the main natural sources of γ-radiation are uranium (U), thorium (Th), and potassium-40 (^{40}K).

The radioisotopes of uranium found in nature (^{238}U, which constitutes 99.3% of natural uranium, and ^{235}U, which constitutes 0.7% of natural uranium) are long-lived α emitters and parents of radioactive decay chains. Some of the daughter products with their corresponding energies are given in Table 3-7.

Other major sources of γ-rays in the Earth surface are (1) thorium (^{232}Th), which is a long-lived (1.38×10^{10} years half-life) α emitter and is the parent of a radioactive decay chain that contains, among others, ^{228}Ra, ^{216}Pi, ^{216}Pb, and ^{208}Tl; and (2) potassium-40 (^{40}K), which is widespread.

Table 3-7. Daughter Products of ^{238}U

Symbol	Half-life	Radiation type	Energy (MeV)
^{238}U	4.5×10^9 yr	α	4.18, 4.13
^{234}Th	24.5 days	β	0.19, 0.1
		γ	0.09, 0.06, 0.03
^{234}Pa	1.1 min	β	2.31, 1.45, 0.55
		γ	1.01, 0.77, 0.04
^{234}U	2.5×10^{15} yr	α	4.77, 4.72
		γ	0.05
^{230}Th	8.3×10^4 yr	α	4.68, 4.62
^{226}Ra	1620 yr	α	4.78, 4.59
		γ	0.19
^{214}Bi	19.7 min	β	3.18, and many others
		γ	2.43, 2.2, and many others

Note: α (alpha) particles are positively charged, β (beta) particles are negatively charged, γ (gamma) particles have no charge.

3-8-2 Detection

Gamma-rays are absorbed or scattered by matter with partial or total loss of energy by the photoelectric effect, Compton effect, and pair production (see Fig. 3-44). The ejected electrons dissipate their energy by ionization of surrounding atoms. Two methods are generally used to measure the amount of ionization: (1) measurement of the light emitted, and (2) collection of the created charges.

If the detector absorbing material is optically transparent to the light released by ionization, then a burst of light will accompany each γ-ray interaction. This property is exhibited by a number of plastics, halides, organic liquids, and phosphors. Photodetectors are then used to detect the emitted light. One phosphor commonly used is the inorganic crystal of thallium-activated sodium iodide, NaI (Tl). It has the important property of being able to absorb all of the incident γ-radiation, even the highest-energy rays, due to its high density of 3.67 g/cm³. The spectral resolution of this type of detector is limited by

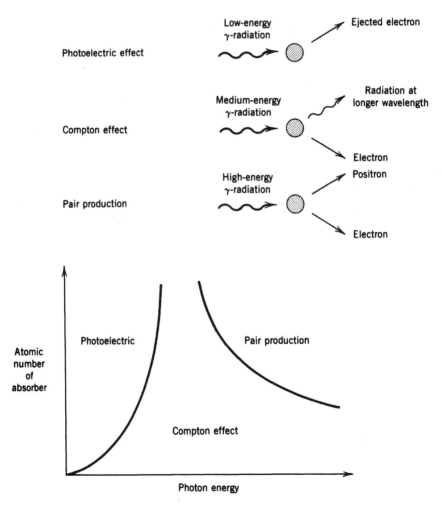

Figure 3-44. Interaction of γ-rays with matter.

the number of light photons released by the ionizing electrons created by γ-ray interactions. The resolution is proportional to the square root of the number of photons released.

Semiconductor detectors have a higher spectral resolution capability. They measure the charge created by ionization within the detector. The resolution in this type of detector is mainly limited by the electronic noise and is somewhat independent of the photon energy.

3-9 IMAGE DATA ANALYSIS

Multispectral remote sensing imagers provide a series of spatial images recorded at a number of different spectral passbands. Each of these images consists of a two-dimensional array of picture elements or pixels; the brightness of each pixel corresponds to the average surface reflectivity for the corresponding surface element. The images from such a multispectral imager can be thought of as a data cube in which horizontal slices represent spatial images for a particular passband, and vertical columns represent the spectral signature of a particular surface picture element.

The spectral signature is the most diagnostic tool in remotely identifying the composition of a surface unit. A trade-off generally exists between the spectral resolution, spectral coverage, radiometric accuracy, and identification accuracy. In the case of hyperspectral imagers, the spectral signature corresponds to high-resolution (spectrally) radiometric measurements over a fairly broad region of the spectrum. In this case, surface units can be separated, classified, and identified based upon some unique characteristic in their reflectivity spectrum, such as a diagnostic absorption band or combination of absorption bands, a diagnostic reflectivity change at a certain wavelength, or ratio of reflectivities in two separate spectral regions. Because hyperspectral imagers record a large number of spectral measurements for every pixel in the image, these instruments produce large data volumes. As an example, a 128 channel image of 15 m pixels and an image size of 1024 × 1024 pixels (covering an area of 15.36 km × 15.36 km), would require storage of 1 Gbit, assuming 8 bits per pixel are recorded. At a nominal speed of 7.5 km/sec, a satellite would have to record this data approximately every 2 seconds, for an approximate data rate of 500 Mbits/sec! For this reason, most visible and near-infrared imagers record significantly fewer channels than this. The challenge then is to extract the useful information from relatively few spectral measurements.

Various different analysis techniques are used to extract qualitative and quantitative information about the surface from the recorded images. The analysis of radiometric and spectral signatures in surface studies can generally be divided into three steps of increasing complexity: (1) detection and delineation, (2) classification, and (3) identification. We shall describe these in more detail in the next few sections using data acquired by the ASTER instrument (see Section 3-6-2 for a description) over the Cuprite Mining District located in Nevada. This area has been used extensively as a testing and verification site for remote sensing instruments. Hydrothermal alteration of Cambrian sedimentary rocks and Cenozoic volcanic rocks by acid–sulfate solutions took place during the Miocene. This occurred at shallow depths, producing silicified rocks containing quartz and minor amounts of alunite and kaolinite, opalized rocks containing opal, alunite and kaolinite, and argillized rocks containing kaolinite and hematite. These units are readily mappable using remote sensing instruments, as we shall show later.

Figure 3-45 shows the nine visible and near infrared channels of an ASTER image of the Cuprite area. The area shown is a subset of a larger image and covers approximately 15 km × 15 km. The bright feature near the middle of the image is Stonewall Playa. The road that cuts through the image from the upper left to middle bottom is Highway 95. Note how similar the individual images are. We shall now analyze this image in more detail in the next few sections, illustrating the most commonly used data analysis techniques.

3-9-1 Detection and Delineation

The first step in the analysis of surface polychromatic or multispectral images is to recognize and delineate areas with different reflectivity characteristics. This can be done manually or with computers by simply delineating areas with image brightness within a certain range of values. Conceptually, this means that the data space is divided into a number of subspaces, and all pixels in the image that have brightness values that fall in a particular

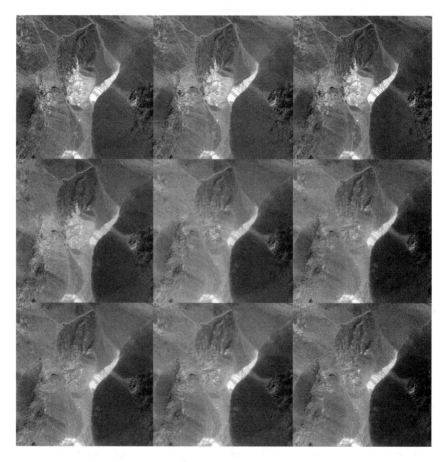

Figure 3-45. Individual spectral channel images for the nine visible and near-infrared channels of the ASTER instrument. The channels progress from left to right and from top to bottom in increasing wavelength, with channel 1 in the upper left and channel 9 in the lower right. The images cover an area of roughly 15 km × 15 km of an area near Cuprite, Nevada.

subspace are assumed to have similar characteristics. In general, change in brightness is associated with changes in surface chemical composition, biological cover, or physical properties (roughness, slope, etc.). Unfortunately, change in brightness can also result from changes in the illumination geometry or atmospheric conditions. If this is not taken into consideration during the data analysis, areas may be misclassified. An obvious example would be areas in the shadow of a cloud misclassified as low-reflectivity areas. Slopes facing the sun will in general appear brighter than those facing away from the sun, even if the two areas have identical characteristics.

The simplest form of image analysis is a simple photo-interpretation of an image. Usually color images are used, displaying three spectral images as the red, green, and blue components of the color image, respectively. Surfaces with similar colors and brightness are delineated. The main difficulty is to decide which spectral images to use in the color composite and which spectral band to assign to each color. Even in the case of the Landsat ETM+ images, where six channels are available (counting all the channels with 30 m resolution), this leads to 120 possible combinations. Experience has shown, however, that not all these combinations are that useful, and therefore, only a few combinations are commonly used. For example, geologic interpretation in semiarid areas is typically done using bands 1, 4, and 7 of the TM (or ETM+) instrument as blue, green, and red displays. Sometimes TM bands 2, 4, and 7 (ASTER bands 1, 3, and 6) are used to reduce the effects of atmospheric scattering, which is more severe in the shorter-wavelength TM band 1. Atmospheric scattering is particularly troublesome in humid, tropical areas, where the usefulness of the shorter-wavelength bands is reduced dramatically. In these situations, the optimum combination for geologic interpretation is to display bands 4, 5, and 7 as blue, green, and red, respectively.

Figure 3-46 shows two ASTER color combinations of the Cuprite scene, bands 1, 2, and 3 (the three visible channels displayed as blue, green, and red) on the left, and bands 1, 3, and 6 (roughly equivalent to TM bands 2, 4, and 7) on the right. Because of the high correlation between the reflectances in the individual bands, especially in the visible part

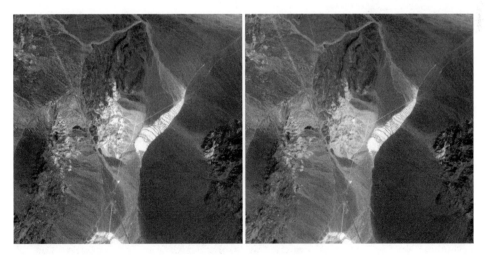

Figure 3-46. Two color combination displays for the Cuprite scene shown in Figure 3-45. On the left are ASTER channels 1, 2, and 3 displayed as blue, green, and red, respectively, giving a pseudonatural color image. On the right are channels 1, 3, and 6 displayed as blue, green, and red. See color section.

of the spectrum, there is little color information in the image on the left. The image on the right shows more detail in the colors, and differentiates well between the opalized rocks with kaolinite and/or alunite (shown in cyan) and the unaltered volcanics, which are shown in reddish colors.

Several techniques are used to increase the separation of surfaces in correlated images. One such technique transforms the image from the red, green, and blue (RGB) representation of color to the intensity, hue, and saturation (IHS) representation. (Most image analysis software packages include this capability.) The color information is contained in the hue of the transform, and the purity of the color is contained in the saturation. After transformation, the saturation image layer is then stretched to increase the color separation. The modified image layers are then transformed back to the red, green, and blue representation for display. Figure 3-47 shows the same two images as Figure 3-46 after enhancing the colors using the IHS transform. Note the dramatic increase in color separation between the two sets of images. Note in particular the area to the left of the road that runs from the top left to the center bottom. This area, shown in gold tones on the left image, and light green tones in the right image, is not easily distinguished in either of the original images in Figure 3-46. We shall show below that this area contains higher concentrations of alunite and kaolinite.

In addition to increasing color separation, this color transformation is often used to enhance images in the following way. One would start with a medium-resolution multispectral image and transform the three image layers of the RGB representation to the IHS representation. After stretching the hue and possibly the saturation channels, the intensity channel is replaced with a higher-resolution image such as the panchromatic band of a SPOT image that has been registered with the original data. The new IHS layers are then transformed back to the RGB representation. The resulting color image appears to have much higher resolution than the original one because of the enhanced resolution of the intensity layer. Sometimes, an image from a completely different type of sensor, such as an imaging radar, may be used as the new intensity layer. Since radar images typically provide excellent separation of surfaces with different geometrical properties such as roughness,

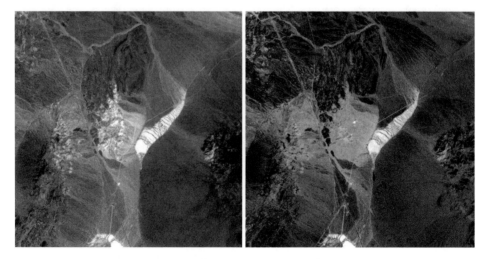

Figure 3-47. The same images shown in Figure 3-46 after performing a color stretch using the IHS transform. Note the dramatic increase in color separation evident in both images. See color section.

this combination better delineates surfaces based on both spectral information (contained in the color of the image) and geometrical properties contained in the intensity of the image.

The RGB/IHS transformation uses three channels of data as input. As the number of data channels increases, the potential number of color combinations increases rapidly. As an example, a six-channel system has 120 potential color combinations, whereas a 200-channel system would have 7,880,400 such three-band combinations! A more powerful method of data analysis, known as *principal components* analysis, offers one solution to the problem of which channels to use in the color display. Mathematically, principal component analysis simply transforms an *n*-dimensional dataset into its eigenvectors. If the original data are real numbers, all eigenvectors are orthogonal. The eigenvector image corresponding to the largest eigenvalue, usually called the first principal component (PC1), is the image with the largest variation in brightness, whereas the PCN image, corresponding to the smallest eigenvalue, contains the least variation. For most scenes, the first three principal components will account for more than 95% of the variation in the data. Figure 3-48 shows the nine principal component images for the Cuprite scene. The

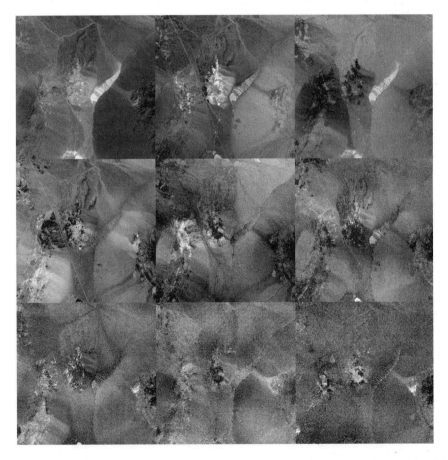

Figure 3-48. Principal component images for the nine visible and near-infrared channels of the Cuprite scene. The principal components progress from left to right and from top to bottom, with PC1 in the upper left and PC9 in the lower right.

first principal component image contains 91% of the variation in the data. The next three principal components contain another 8.6% of the variation, with the remaining 0.4% of the variation contained in the last five principal components. The last three principal component images are quite noisy. All the images have been stretched to show the same dynamic range for display purposes.

In practice, it is found that the first eigenvector spectrum typically is closely related to the average scene radiance, which is the incoming solar radiance convolved with the average scene reflectance and the atmospheric attenuation. The PC1 image is typically dominated by topographic effects, strongly highlighting slopes and shadows as seen in the upper-left image of Figure 3-48. For this reason, the PC1 layer is not commonly used in the further analysis of the data. It is more common to use the next three principal component images, corresponding to PC2, PC3, and PC4, in a color analysis of the image. Figure 3-49 shows the color image displaying these three principal components as the blue, green, and red channels, respectively. This image shows very little variation in brightness, showing that little information about the absolute albedo remains. Spectral differences, as manifested in the optimum stretch of the colors, are highlighted excellently. Notice how well the alteration zone to the left of the road is identified in this image as compared to the original visible color composite shown on the left in Figure 3-46.

In some cases, principal component analysis is performed separately on subsets of the total number of bands. For example, one might elect to perform the analysis separately on the visible channels of an ASTER image, and then again separately on the near-infrared channels. In this way, it may be possible to extract optimally the difference in reflectance

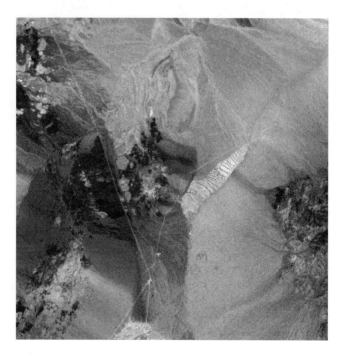

Figure 3-49. The principal components PC2, PC3, and PC4 are displayed as blue, green, and red, respectively in this color combination of the Cuprite scene. Notice the excellent color separation compared to Figures 3-46 and 3-47. See color section.

properties of each subset of the spectrum. One drawback of principal component-based display techniques is that spectral features that occur in only a small number of pixels will be lost in the averaging process and will, therefore, not be visible in the final product.

In the case of multispectral images, the delineation process should take albedo variation in any one of the spectral channels into consideration. In many situations, more accurate delineation of surface units can be achieved by using ratios of reflectivity in two different spectral bands. This would allow the minimization of nonrelevant effects such as slope changes. To illustrate, let us assume that the reflectivity of a certain surface unit as a function of wavelength λ and incidence angle θ is given by

$$R(\lambda, \theta) = g(\lambda)f(\theta) \qquad (3\text{-}29)$$

where $g(\lambda)$ denotes the "pure" spectral response of the surface, and $f(\theta)$ denotes the modification of the response by the imaging geometry. If we consider two neighboring areas A and B of identical composition but having different slope aspects, then their reflectivity will be different at each and every spectral band. However, if we consider the ratio r at two separate bands,

$$r = \frac{R(\lambda_1, \theta)}{R(\lambda_2, \theta)} = \frac{g(\lambda_1)}{g(\lambda_2)} \qquad (3\text{-}30)$$

the effect of the slope change is eliminated and only the change in the composition is delineated.

Ratio images are also used to highlight specific spectral differences. For example, if a particular surface has a spectral signature that shows high reflectance at λ_1 and low reflectance at λ_2, the ratio image will enhance this difference and delineate these surfaces more clearly. Clay minerals, for example show relatively strong absorption in band 7 of TM images (due to the absorption by the hydroxyl group at 2.2–2.3 microns—see Figure 3-11), and little absorption in band 5. The ratio image of bands 5/7 is therefore commonly used to delineate clays that are typically associated with hydrothermal alteration. Similarly, TM band ratios 3/1 or 5/4 are commonly used to delineate the presence of iron oxide. Another example is the strong reflectance shown by vegetation in the near infrared region, compared to the low reflectance in the red part of the visible spectrum, which would lead to large values in a TM band (4/2).

Figure 3-50 shows the spectra of some minerals commonly associated with hydrothermal alteration, resampled to show the expected spectra for the ASTER visible and near-infrared passbands. Also shown are the ASTER passbands in the bottom of the figure. Most of the clays show strong absorption in band 6 at the fundamental wavelength of the Al–OH bending mode. These same minerals show little absorption in ASTER band 4, however. An ASTER band 4/7 ratio image would, therefore, be expected to highlight the presence of these minerals. The iron oxide minerals in Figure 3-50 show strong spectral ratios when comparing ASTER bands 4/3 or 3/1. Figure 3-51 shows a color image of these three ratios displayed as red (4/7), green (3/1) and blue (4/3). The alteration zones, especially the one to the right of Highway 95, are highlighted well in this image.

Ratio images have the advantage that slope and illumination effects are reduced significantly. On the other hand, they have the disadvantage that noise and instrument artifacts in the images are typically amplified. In addition, those surfaces that have spectral fea-

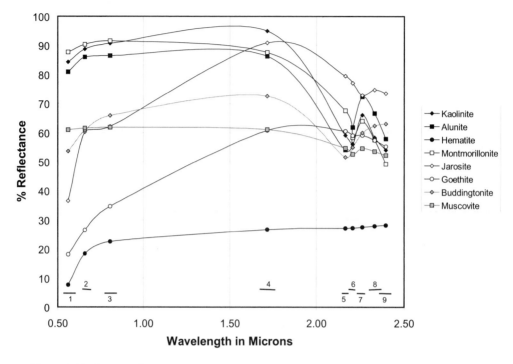

Figure 3-50. Spectra of some minerals commonly associated with hydrothermal alteration, resampled to the ASTER visible and near-infrared passbands. Also shown are the ASTER passbands in the bottom of the figure.

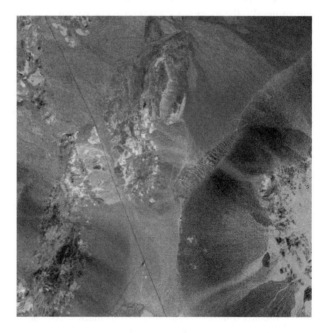

Figure 3-51. Spectral ratio image of the Cuprite scene. The ratios are 4/7 (red), 3/1 (green), and 4/3 (blue). See color section.

tures that are similar in the different bands, that is, for which the spectral features are correlated, will show little variation in the ratio images.

3-9-2 Classification

The next step after delineation is to classify units based on a set of criteria. Classifications extend not only to individual images, but also to a number of images taken at different times of the same area or of different areas.

Classification of images involves using a set of rules to decide whether different pixels in an image have similar characteristics. These rules in effect divide the total data space into subsets separated by so-called decision boundaries. All pixels that fall within a volume surrounded by such decision boundaries are then labeled as belonging to a single class. The classification criteria range from the most simple, such as all areas with identical reflectivity in a certain spectral band being put into the same class, to more sophisticated criteria, such as comparing the measured spectra over a large wavelength range. Some intermediate criteria include albedo (simple and composite), specific spectral absorption bands, spectral response slope in specific spectral regions, or the presence of specific spectral features.

Two major approaches are used in classifying images: supervised and unsupervised classifications. In the case of supervised classification, a user will specify so-called *feature vectors* to be used in the comparison process. These vectors can be thought of as defining the centroids of the decision volumes that are separated by the decision boundaries. These feature vectors can be extracted from the image to be classified or could come from a library of spectral signatures either measured in the laboratory or in the field. In the case of unsupervised classification, the computer is allowed to find the feature vectors without help from an image analyst. In the simplest form, known as the *K-means algorithm, K* feature vectors are typically selected at random from the data space.

Once the feature vectors are identified, classification rules are used to assign pixels in the image to one of the feature vectors. Many different classification rules are used, ranging from the simple nearest-neighbor distance classifier, to neural network schemes, to sophisticated schemes that take into account the expected statistical distributions of the data. To apply these rules during the classification process, a so-called *distance measure* is typically defined. The nearest-neighbor scheme, for example, simply calculates the Euclidian distance between a pixel and each of the feature vectors as if each spectral measurement represented an orthogonal axis in the data space. In other cases, the distance definition includes some measure of the probability that a pixel may be similar to a particular feature vector. During the classification process, the distance between a pixel and each of the feature vectors is computed. The pixel is then labeled the same as the feature vector for which this distance is the smallest. If this smallest distance is larger than some threshold specified by the analyst, the pixel will not be classified.

Figure 3-52 shows the results of an unsupervised classification of the Cuprite scene. The classification was arbitrarily asked to produce six classes and the initial feature vectors were selected randomly from the data space. The classification was then performed iteratively, updating the feature centroids after each iteration, until fewer than 0.01% of the pixels were changing between iterations. We note that the two classes colored dark blue and green are found mostly in the altered zones, especially the one to the right of Highway 95. All we know at this point is that we found six stable "classes" of terrain from the data itself. To attach any significance to the classes, we need to compare the

Figure 3-52. Results of an unsupervised classification of the Cuprite scene. The classification was initialized using randomly chosen features, and then iterated. The number of classes was arbitrarily set to six. See color section.

spectra of the feature centroids to some library of spectra. These could be either laboratory spectra or spectra measured in the field. Since it is unlikely that such large areas will be covered with a homogeneous layer of a single mineral type, field spectra may be more useful at this stage. Nevertheless, the spectra of the classes colored dark blue in Figure 3-52 show large band 4/5 ratios, consistent with the spectra of the alteration minerals shown in Figure 3-50.

The results of any classification depend strongly on the selection of the feature vectors. In selecting these in the case of supervised classification, the analyst may be guided by the results of previous analysis such as those described in the previous section. For example, feature vectors could be selected based on unusual spectral ratios observed in a ratio analysis, or they may be selected based on specific colors found in a principal component analysis. Finally, the analyst may have some field experience with the scene, and may be picking feature vectors based on known characteristics of certain areas in the scene.

We used the principal component image in Figure 3-49 as a guide to select areas to use as the classification feature vectors in a supervised classification. We selected three areas from the image corresponding to the pink and dark blue tones in the alteration zones, and the green alluvial areas near the top of the image, which represent the unaltered terrain. We added to these areas a feature selected from one of the areas from the green class in the alteration zone to the right of Highway 95 from the unsupervised classification. The classification result is shown in Figure 3-53. The alteration zones are clearly highlighted in this result.

Figure 3-53. Results of a supervised classification of the Cuprite scene. The classification was initialized using features selected from a combination of the principal component results shown in Figure 3-49 and the unsupervised classification results in Figure 3-52. The number of classes was limited to four. See color section.

In the classification schemes discussed above, it was assumed that each pixel would be assigned to only one feature vector class. In reality, pixels represent areas on the ground that are rarely homogenously covered by only one type of surface. Instead, these surfaces are typically a mixture of different surface covers, each with a different spectral response. The spectral signature measured by the remote sensing instrument is made up of the weighted sum of the individual surface element spectra; the weights in the summation depend on the relative abundance of that type of cover in the area represented by the pixel. Subpixel classification schemes attempt to identify these relative abundances for each pixel. The fundamental assumption for the subpixel classification scheme is that the measured spectrum $S_{\text{total}}(\lambda)$ is a linear mixture of individual spectra:

$$S_{\text{total}}(\lambda) = \sum_{i=1}^{N} a_i S_{ei}(\lambda) + n \qquad (3\text{-}31)$$

where a_i represents the relative fraction of the measured spectrum contributed by the *endmember* spectrum $S_{ei}(\lambda)$, and n denotes additive noise. Under the assumption that we have identified an exhaustive set of endmember spectra to choose from, and that all fractions must be positive, Equation (3-31) is constrained by

$$\sum_{i=1}^{N} a_i = 1; \qquad a_i \geq 0 \qquad (3\text{-}32)$$

If we describe the measured and endmember spectra as M-dimensional vectors, where M is the number of spectral channels, we can rewrite Equation (3-31) as

$$\vec{S} = E\vec{A} + \vec{N} \tag{3-33}$$

The endmember matrix is an $M \times N$ matrix, with each of the N columns representing one endmember spectrum, and is the same for the entire image. Usually, the number of spectral channels is larger than the number of endmember spectra. In that case, the *unmixing* solution is found as

$$\vec{A} = (E^T E)^{-1} E^T \cdot \vec{S} \tag{3-34}$$

Linear unmixing for large dimensional datasets is computationally expensive. The matrix multiplications shown in Equation (3-34) must be performed for each pixel. On the other hand, linear unmixing provides more quantitative information than a simple classification.

Identifying the endmember spectra to use in the analysis is a major challenge, since the results are strongly influenced by the choice of the endmember spectra. As in the case of simple classification, several techniques are used to select these. One could use laboratory or field spectra. In that case, the remote sensing dataset must be calibrated. If the analyst has some prior knowledge about the scene, endmember spectra could be selected from the scene itself. Alternatively, multidimensional scatter plots could be used to identify the extreme values in the multidimensional histogram of the spectra in the image. Since all the pixels in the image fall within the volume enclosed by these extreme endpoints, they form an exhaustive set of endmembers.

As an illustration, we used the four features used in the supervised classification shown in Figure 3-53 as our "endmembers" and performed an unmixing of the Cuprite scene. There is no guarantee that these spectra are indeed endmembers in this image, so care should be taken when interpreting the results, as negative abundances may result. The results are shown in Figure 3-54, where we display only the abundances associated with the three spectra selected from the alteration zone. The color assignment is the same as that used in Figure 3-53. As expected, the features that were originally selected from the alteration zone show high abundances in these areas, and relatively low amounts in the rest of the image. When comparing the results to published mineral maps of the Cuprite area, it is found that the reddish areas in Figure 3-54 show high abundance of hematite and goethite.

3-9-3 Identification

The last step in the spectral analysis of imaging data is the unique identification of the classified elements. This requires a detailed knowledge of the spectral signatures of the materials being sought, as well as of all the other materials in the scene, and the development of a spectral signature library of all expected natural materials. In the ideal case, if a certain material, or family of materials, is the only one that has a certain spectral feature, such as an absorption line at a certain wavelength, the identification becomes simple. The identification feature could be a single absorption line or an association of lines.

Figure 3-54. This image shows the relative abundances of different materials after linear unmixing. The spectra used in the supervised classification were used as endmembers. See color section.

If spectral signatures from a reference library were used in the classification to begin with, the identification is automatic. However, if the classification was done using areas selected from the image, or data-driven endmembers, the final identification step involves comparing the spectra of the endmembers or feature vectors to those in a reference library. If a reference spectral library is not available, field checking in one area of each class will allow identification of the constituents in the whole scene.

One example is the case of identifying the presence of sodium on the surface of Io, the first satellite of Jupiter, by ground spectral observation. Figure 3-55 shows the reflectance of Io in the visible and near infrared regions and compares it to the reflectance spectra of a number of materials, including sulfur dioxide frost. Two specific features led to the identification of Io surface material as consisting mainly of sulfur dioxide frost:

1. The sharp drop of the reflectance for wavelengths lower than 0.45 αm. This wavelength corresponds to the energy of the forbidden gap in sulfur (see Fig. 3-15).
2. The deep absorption band at 4.08 μm, which corresponds to the absorption of SO_2 frost.

Another, more recent, example is the identification of the mineral jarosite on the Meridiani plain on Mars by the Mössbauer spectrometer on the Mars Exploration Rover Opportunity. Figure 3-56 shows the measured spectrum of one of the rocks dubbed "El Capitan" within the rock outcrop inside the crater where the rover Opportunity landed (see

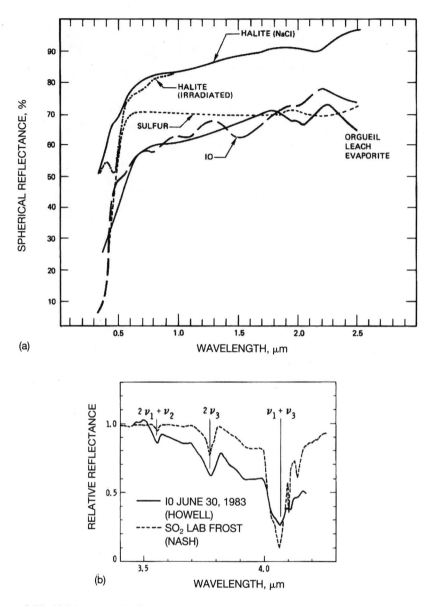

Figure 3-55. (a) Io's spectral reflectance showing the step drop near 0.45 μm (from Fanale et al., 1974). (b) Io's spectral reflectance showing the deep absorption at 4.08 μm associated with SO_2 frost compared to laboratory reflectance spectrum of SO_2 frost (courtesy of D. Nash, JPL).

Fig. 3-42). The Mössbauer spectrometer graph shows the presence of an iron-bearing mineral called jarosite. The pair of yellow peaks in the graph indicates a jarosite phase, which contains water in the form of hydroxyl as a part of its structure. These data give evidence of water-driven processes that have existed on Mars. Three other phases are also identified in this spectrum: a magnetic phase (blue), attributed to an iron oxide mineral; a silicate phase (green), indicative of minerals containing double-ionized iron (Fe 2+); and a third phase (red) indicative of minerals with triple-ionized iron (Fe 3+).

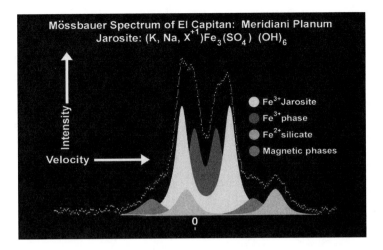

Figure 3-56. This graph shows a spectrum, taken by the Mars Exploration Rover Opportunity's Mössbauer spectrometer. The Mössbauer spectrometer graph shows the presence of an iron-bearing mineral called jarosite. See color section. (Courtesy NASA/JPL/University of Mainz.)

EXERCISES

3-1. Consider a planet of radius R with its rotation axis inclined at an angle ψ relative to the normal to the plane of its orbit around the sun. Assume that the solar rays reaching the planet are all parallel to the orbital plane.

(a) Calculate the sun angle (angle between the local vertical and the sun direction) at the equator as a function of longitude δ and the location of the planet in its orbit (i.e., the angle α in Figure 3-57). Consider $\delta = 0$ to be the point closest to the sun. Plot the sun angle for $\psi = 0°$ and $27°$ and for $\delta = 0$.

(b) Calculate the sun angle at the poles. Plot for $\psi = 0°$ and $27°$.

(c) Calculate the sun angle as a function of latitude γ along zero longitude as a function of α. Plot for $\psi = 0°$ and $27°$ and for $\gamma = 45°$.

(d) Calculate the sun angle as a function of longitude δ for $\gamma = 45°$, $\psi = 27°$ and $\alpha = 0°$, $90°$, and $180°$

3-2. A narrowband filter with a variable center frequency is used to observe a spectral region where the emitted intensity is given by

$$W(v) = 1 - \alpha e^{-(v-v_0)^2/v_s^2}$$

The filter transmission function is given by

$$F(v) = e^{-(v-v_c)^2/v_f^2}$$

The intensity measured by a detector behind the filter is given by

$$I = \int T(v)dv = \int_{-\infty}^{+\infty} W(v)F(v)dv$$

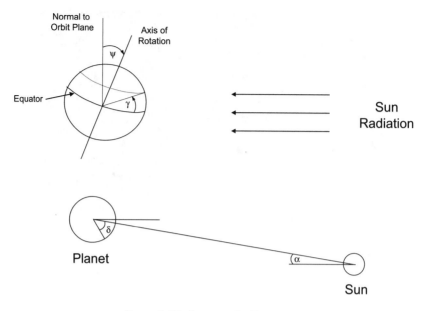

Figure 3-57. Geometry for Exercise 3-1.

    (a) Derive the expression for I as a function of v_s, v_f, v_c, and α.

    (b) Plot I as a function of v_c/v_s for $v_f/v_s = 0.1$, 1, and 10, and for $\alpha = 0.1$ and 1.

    (c) Repeat the calculations for

$$F(v) = \begin{cases} 1 \text{ for } -1 \le \dfrac{v - v_c}{v_f} \le 1 \\ 0 \text{ otherwise} \end{cases}$$

3-3.   Figure 3-58 shows the energy levels for three different materials.

    (a) Silicon has a bandgap of 1.12 eV. Calculate the cutoff wavelength for this material and sketch the expected absorption spectrum.

    (b) Now suppose silicon is doped with arsenic, which places a layer of electrons at 1.07 eV. Calculate the cutoff wavelength for this material and sketch the absorption spectrum.

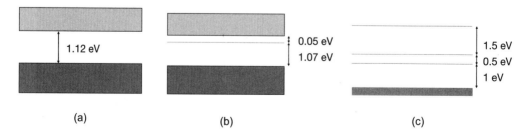

Figure 3-58. Energy levels of three different materials.

(c) This material has three discrete energy levels at 1 eV, 1.5 eV, and 3. eV. Calculate the positions of the primary absorption lines, and sketch the absorption spectrum of this material.

3-4. A class of hypothetical granular materials have energy levels that depend on the crystalline field E as shown in Figure 3-59. Describe the behavior of the spectral reflectance as a function of E. Only the transitions shown in Figure 3-59 are allowed.

3-5. Calculate and plot the reflection coefficient as a function of the angle of incidence for a half space with index of refraction equal to $n = 1.7$ and for $n = 9$. In both cases, the upper half space has $n = 1$. Consider both horizontal and vertical polarizations.

3-6. Consider a half space with index of refraction $n = n_r + in_i$ and $n_i = \alpha e^{-(v-v_0)^2/v_s^2}$. Assume that $\alpha \ll n_r$. Calculate the reflection coefficient for normal incidence and plot its behavior as a function of v/v_0 for $n_r = 3$, $\alpha = 0.1$, and $v/v_s = 0.05$.

3-7. A telescope is orbiting the Earth at 705 km altitude. The telescope lens diameter is 40 cm, the focal length is 120 cm, and the square focal plane is 20.916 cm on a side.

(a) Calculate the power density at the lens if the Earth albedo is 50%. (Assume that the solar energy density at the surface of the Earth is 1.37 kW/m².)

(b) Calculate the total energy intercepted from a 30 m × 30 m ground pixel if the shutter stays open for 0.1 millisecond.

(c) Calculate the field of view and the swath width of the telescope.

(d) If the instrument is a framing camera, calculate the number of detectors required to acquire a square image with 30 m × 30 m pixels on the ground. Also, calculate the dwell time if we assume that we use a scanning mirror to stare at the same area with a 75% duty cycle.

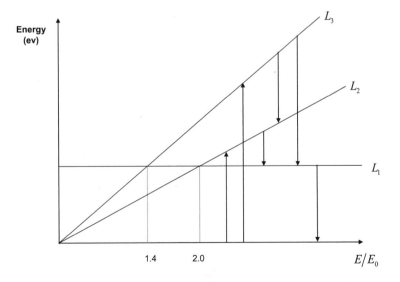

Figure 3-59. Energy levels and allowable transitions for a hypothetical material.

(e) In the pushbroom configuration with a single line of detectors, calculate the number of detectors required for a pixel size of 30 m × 30 m on the ground. Also calculate the dwell time and total energy per pixel assuming that smearing is to be kept to less than 1/3 of a pixel.

(f) For a scanning mirror system, calculate the mirror scan rate, the dwell time per pixel, and the total energy intercepted from each pixel, assuming that we use an array of 16 detectors arranged in the along-track direction.

(g) Now assume that we place a filter in the focal plane with a 1% transmission bandwidth centered at 0.5 microns. Calculate the number of photons received from a pixel for each of the three cases above.

(h) Calculate the data rate for each of the three cases, assuming that we digitize the data from each pixel to 8 bits.

3-8. A telescope is orbiting the Earth at 705 km altitude. The telescope lens diameter is 40 cm, the focal length is 120 cm, and the square focal plane is 10 cm on a side. The pixels in the detector are 10 microns on a side. Calculate the swath width, size of the pixels on the ground, and the number of pixels across the swath assuming a pushbroom design. Also calculate the maximum dwell time per pixel and the resulting data rate.

The instrument is now changed into a spectrometer. The incoming light is dispersed using a prism such that the different colors are separated spatially in a direction orthogonal to the pushbroom line array. To measure the different spectral components, the prism mechanism is scanned so that different colors sweep over the line array. If we are to measure 64 spectral channels, calculate the scan rate of the dispersion mechanism. What is the dwell time per spectral channel now? What is the data rate?

Assume that the wavelength region 0.4 microns to 2.4 microns is dispersed over 640 microns in the focal plane. Now assume that we stack 64 line arrays next to each other to cover the dispersed spectrum. Calculate the bandpass of each channel, the dwell time per spectral channel, and the resulting data rate.

3-9. A scanning laser altimeter is in a 250 km altitude orbit. The laser sends a series of 20 nanosecond pulses with a peak power of 10 kW and a beam divergence of 0.5 milliradian. What is the beam footprint on the surface and the height measurement accuracy? Let us assume that the surface has a 10% albedo and scatters in an isotropic fashion in the upper hemisphere. What is the power collected by a receiving collector of 50 cm diameter? What number of photons does this energy correspond to assuming that $\lambda = 0.5$ μm? Assuming that the detector has an area of 10^{-5} cm^2 and a bandwidth $\Delta f = 1/\tau$ with $\tau = 20$ nanoseconds, calculate the detector minimum D^*. In order to eliminate the signal due to the solar background, a 1% (i.e., 0.0005 μm bandwidth) filter is used before the detector. Assuming that the solar energy density at the surface is 2 Kw/m^2/μm, what is the signal to background ratio? How can this ratio be increased? Do you think this system can be operated during the day? Assuming that a mapping swath of 40 km is required and the surface has to be sampled every 200 m, how many pulses does the laser need to transmit per second? Assume that the satellite velocity is 7 km/sec.

3-10. Consider a planet with radius 1900 km, located at a distance of 1.4×10^9 km from the sun and 1.25×10^9 km from the Earth. The average surface temperature of the planet is 70°K. The surface reflectivity is 0.3 across the visible and infrared spectrum. Plot the received energy by a 1 m^2 aperture telescope in Earth orbit as a function of wavelength. Assume that the sun is a blackbody with a temperature of 6000°K and a radius of 7×10^5 km. Now let us assume that an area of size 100 km × 100 km at the equator consists of an active volcanic caldera with a temperature of 700°K. Can the earth-orbiting telescope uniquely detect the presence of the caldera? Can it measure its size and temperature? Explain.

REFERENCES AND FURTHER READING

Abrams, M. J., R. P. Ashley, L. C. Rowan, A. F. H. Goetz, and A. B. Kahle. Mapping of hydrothermal alteration in the Cuprite Mining District, Nevada, using aircraft scanner imagery for the 0.46–2.36 μm spectral region. *Geology,* **5,** 713, 1977.

Abshire, J. B., X. Sun, and R. S. Afzal. Mars Orbiter Laser Altimeter: Receiver model and performance analysis. *Applied Optics,* **39,** 2440–2460, 2000.

Anderson, F. S., A. F. C. Haldemann, N. T. Bridges, M. P. Golombek, T. J. Parker, and G. Neumann. Analysis of MOLA data for the Mars Exploration Rover landing sites. *Journal of Geophysical Research,* **108(E12),** 8084, doi:10.1029/2003JE002125, 2003.

Bateson, C.A., G. P. Asner, and C. A. Wessman. Endmember bundles: A new approach to incorporating endmember variability into spectral mixture analysis. *IEEE Transactions on Geoscience and Remote Sensing,* **GE-38,** 1083–1094, 2000.

Bell, J. F., III, et al. Mars Exploration Rover Athena Panoramic Camera (Pancam) investigation. *Journal of Geophysical Research,* **108(E12),** 8063, doi:10.1029/2003JE002070, 2003.

Blanchard, L. E., and O. Weinstein. Design challenges of the thematic mapper. *IEEE Transactions on Geoscience and Remote Sensing,* **GE-18,** 146–160, 1980.

Brown, M., H. G. Lewis, and S. R. Gunn. Linear spectral mixture models and support vector machines for remote sensing. *IEEE Transactions on Geoscience and Remote Sensing,* **GE-38,** 2346–2360, 2000.

Bryant, R., M. S. Moran, S. A. McElroy, C. Holifield, K. J. Thome, T. Miura, and S. F. Biggar. Data continuity of Earth Observing 1 (EO-1) Advanced Land I satellite imager (ALI) and Landsat TM and ETM+. *IEEE Transactions on Geoscience and Remote Sensing,* **GE-41,** 1204–1214, 2003.

Buckingham, W. F., and S. E. Sommer. Minerological characteristics of rock surfaces formed by hydrothermal alteration and weathering: Application to remote sensing. *Economic Geology,* **78,** 664–674, 1983.

Chahine, M., et al. Interaction mechanisms within the atmosphere. In R. N. Colwell (Ed.), *Manual of Remote Sensing.* American Society of Photogrammetry, Falls Church, VA, 1983.

Chang, S. H., and W. Collins. Confirmation of airborne biogeophysical mineral exploration technique using laboratory methods. *Economic Geology,* **78,** 723–736, 1983.

Collins, W., et al. Airborne biogeophysical mapping of hidden mineral deposits. *Economic Geology,* **78,** 737–749, 1983.

Crisp, J. A., M. Adler, J. R. Matijevic, S. W. Squyres, R. E. Arvidson, and D. M. Kass. Mars Exploration Rover mission. *Journal of Geophysical Research,* **108(E12),** 8061, doi:10.1029/2002JE002038, 2003.

Crosta, A. P., C. R. De Souza Filho, F. Azevedo, and C. Brodie. Targeting key alteration minerals in epithermal deposits in Patagonia, Argentina, using ASTER imagery and principal component analysis. *International Journal of Remote Sensing,* **24,** 4233–4240, 2003.

Dozier, J., and A. Strahler. Ground investigations in support of remote sensing. Chapter 23 in R. N. Colwell (Ed.), *Manual of Remote Sensing.* American Society of Photogrammetry, Falls Church, VA, 1983.

Edgett, K. S. Low-albedo surfaces and eolian sediment: Mars Orbiter Camera views of western Arabia Terra craters and wind streaks. *Journal of Geophysical Research,* **107(E6),** 10.1029/2001JE001587, 2002.

Fanale F., et al. Io: a surface evaporite deposit. *Science,* **186,** 922–925, 1974.

Fanale F., et al. Io's surface. In D. Morrison (Ed.), *Sattelites of Jupiter.* University of Arizona Press, 1982.

Goetz, A. F., B. N. Rock, and L. C. Rowan. Remote sensing for exploration: An overview. *Economic Geology,* **78,** 573–590, 1983.

Goetz, A., and L. Rowan. Geologic remote sensing. *Science,* **211,** 781–791, 1981.

Goetz, A., G. Vane, J. Solomon, and B. Rock. Imaging spectrometry for Earth remote sensing. *Science,* **228,** 1147–1153, 1985.

Hubbard, B. E., J. K. Crowley, and D. R. Zimbelman. Comparative alteration mineral mapping using visible to shortwave infrared (0.4–2.4 μm) Hyperion, ALI, and ASTER imagery. *IEEE Transactions on Geoscience and Remote Sensing,* **GE-41,** 1401–1410, 2003.

Hudson, R. J. *Infrared Systems Engineering.* Wiley, New York, 1969.

Hunt, G. R., Spectral signatures of particulate minerals in the visible and near infrared. *Geophysics,* **42,** 501–513, 1977.

Hunt, G. R., and R. P. Ashley. Spectra of altered rocks in the visible and near infrared. *Economic Geology,* **74,** 1613, 1974.

Hunt, G. R., and J. W. Salisbury. Mid-infrared spectral behavior of igneous rocks. U.S. Air Force Cambridge Research Laboratories Report AFCRL-TR-74-0625, Environmental Research Papers, no. 496, 1974.

Hunt, G. R., and J. W. Salisbury. Mid-infrared spectral behavior of sedimentary rocks. U.S. Air Force Cambridge Research Laboratories Report AFCRL-TR-75-0356, Environmental Research Papers, no. 520, 1975.

Hunt, G. R., and J. W. Salisbury. Mid-infrared spectral behavior of metamorphic rocks. U.S. Air Force Cambridge Research Laboratories Report AFCRL-TR-76-0003, Environmental Research Papers, no. 543, 1976.

Liang, S., H. Fang, and M. Chen, Atmospheric correction of Landsat ETM+ land surface imagery. I. Methods. *IEEE Transactions on Geoscience and Remote Sensing,* **GE-39,** 2490–2498, 2001.

Liang, S., H. Fang, J. T. Morisette, M. Chen. C. J. Shuey, C. L. Walthall, and C. S. T. Daughtry. Atmospheric correction of Landsat ETM+ land surface imagery. II. Validation and applications. *IEEE Transactions on Geoscience and Remote Sensing,* **GE-40,** 2736–2746, 2002.

Malin, M. C., and K. S. Edgett. Mars Global Surveyor Mars Orbiter Camera: Interplanetary cruise through primary mission. *Journal of Geophysical Research,* **106,** 23,429–23,570, 2001.

Maselli, F., C. Conese, T. De Filippis, and S. Norcini. Estimation of forest parameters through fuzzy classification of TM data. *IEEE Transactions on Geoscience and Remote Sensing,* **GE-33,** 77–84, 1995.

Milliken, R. E., J. F. Mustard, and D. L. Goldsby. Viscous flow features on the surface of Mars: Observations from high-resolution Mars Orbiter Camera (MOC) images. *Journal of Geophysical Research,* **108(E6),** 5057, doi:10.1029/2002JE002005, 2003.

Minnaert, M. The reciprocity principle in Lunar photometry. *Astrophysical Journal,* **93,** 403–410, 1941.

Nassen, K. The cause of color. *Scientific American,* December 1980.

Nishii, R., S. Kusanobu, and S. Tanaka. Enhancement of low spatial resolution image based on high-resolution bands. *IEEE Transactions on Geoscience and Remote Sensing,* **GE-34,** 1151–1158, 1996.

Rogalski, A. *Infrared Detectors.* Gordon and Breach Science Publishers, Amsterdam, 2000.

Rowan, L. C., A. F. H. Goetz, and R. P. Ashley. Discrimination of hydrothermally altered and unaltered rocks in visible and near-infrared multispectral images. *Geophysics,* **42,** 522, 1977.

Sabins, F. F. *Remote Sensing: Principles and Interpretation,* 3rd ed., W. H. Freeman, New York, 1996.

Schowengerdt, R. A. *Remote Sensing: Models and Methods for Image Processing,* 2nd ed., Academic Press, 1997.

Short, N. *The Landsat Tutorial Workbook: Basics of Satellite Remote Sensing.* NASA Ref. Publ. 1078, U.S. Government Printing Office, Washington, DC, 1982.

Siegel, B., and A. Gillespie. *Remote Sensing in Geology.* Wiley, New York, 1980.

Solaiman, B., R. K. Koffi, M.-C. Mouchot, and A. Hillion. An information fusion method for multispectral image classification postprocessing. *IEEE Transactions on Geoscience and Remote Sensing,* **GE-36,** 395–406, 1998.

Song, C., and C. E. Woodcock. Monitoring forest succession with multitemporal Landsat images: Factors of uncertainty. *IEEE Transactions on Geoscience and Remote Sensing,* **GE-41,** 2557–2567, 2003.

Swain, P. H., and S. M. Davis. *Remote Sensing: The Quantitative Approach.* McGraw-Hill, New York, 1978.

Thornbury, W. D. *Principles of Geomorphology.* Wiley, New York, 1969.

Ton, J., J. Sticklen, and A. K. Jain. Knowledge-based segmentation of Landsat images. *IEEE Transactions on Geoscience and Remote Sensing,* **GE-29,** 222–232, 1991.

Whitney, G., M. J. Abrams, and A. F. H. Goetz. Mineral discrimination using a portable ratio determining radiometer. *Economic Geology,* **78,** 688–698, 1983.

Wilson, W. H. Solar ephemeris algorithm. Vis. Lab. Rep. SIO 80-13, Univ. of Calif., Scripps Inst. of Oceanography, 1980.

Yamaguchi, Y., A. B. Kahle, H. Tsu, T. Kawakami, and M. Pniel. Overview of Advanced Spaceborne Thermal Emission and Reflection Radiometer (ASTER). *IEEE Transactions on Geoscience and Remote Sensing,* **GE-36,** 1062–1071, 1998.

Yamaguchi, Y., and C. Naito. Spectral indices for lithologic discrimination and mapping by using the ASTER SWIR bands. *International Journal of Remote Sensing,* **24,** 4311–4323, 2003.

Zhukov, B., D. Oertel, F. Lanzl, and G. Reinhackel. Unmixing-based multisensor multiresolution image fusion. *IEEE Transactions on Geoscience and Remote Sensing,* **GE-37,** 1212–1226, 1999.

Zuber, M. T., D. E. Smith, S. C. Solomon, D. O. Muhleman, J. W. Head, J. B. Garvin, J. B. Abshire, and J. L. Bufton. The Mars Observer Laser Altimeter investigation, *Journal of Geophysical Research,* **97,** 7781–7797, 1992.

4

SOLID-SURFACE SENSING: THERMAL INFRARED

Any object that is at a physical temperature that is different from absolute zero emits electromagnetic radiation. This radiation is described mathematically by Planck's radiation law. Planck's results were announced in 1900, and followed research on the topic by Rayleigh, Jeans, Wien, Stefan, and Boltzmann, who all studied different aspects of the problem.

As will be shown later on, Planck's radiation law describes radiation that occurs at all wavelengths. The radiation peaks at a wavelength that is inversely proportional to the temperature. For most natural bodies, the peak thermal emission occurs in the infrared region. In the case of the sun, numerous stars, and high-temperature radiators, their high temperature leads to a peak emission in the visible and UV regions of the spectrum.

Electromagnetic radiation is only one of three ways in which heat energy is transferred from one place to another. *Conduction* describes the transport of heat from one object to another through physical contact, such as the case in which a pot on a stove is heated by being in physical contact with the heating element. The liquid in the bottom of the pot that is in physical contact with the metal is also heated by conduction. Gasses and liquids can also be heated by *convection*. In this case, the heat energy is transported by the convective movement of the heated material. The same principle also heats the air in the atmosphere of a planet using the heat energy of the surface of the planet. It is important to realize that of these three mechanisms, only electromagnetic radiation can transport the energy through a vacuum. It is this form of transport that allows the heat of the sun to reach the earth.

The heat conduction property of the surface layer is a key factor in the response of the surface to the periodic heat input from the sun. This conduction allows the surface to absorb the heat input from the sun, changing the physical temperature of the surface in the process. The fact that the surface is not at absolute zero temperature means that the surface itself will again radiate electromagnetic energy according to Planck's law. It is this electromagnetic radiation that is measured by remote sensing instruments, allowing us to study the thermal properties of the surface layer.

Introduction to the Physics and Techniques of Remote Sensing. By C. Elachi and J. van Zyl
Copyright © 2006 John Wiley & Sons, Inc.

4-1 THERMAL RADIATION LAWS

Planck's law describes the spectral distribution of the radiation from a blackbody as

$$S(\lambda, T) = \frac{2\pi hc^2}{\lambda^5} \frac{1}{e^{ch/\lambda kT} - 1} \tag{4-1}$$

which is usually written as

$$S(\lambda, T) = \frac{c_1}{\lambda^5} \frac{1}{e^{c_2/\lambda T} - 1} \tag{4-2}$$

where
$S(\lambda, T)$ = spectral radiant emittance in W/m^3 (Watts per unit area per unit wavelength)
λ = wavelength of the radiation
h = Planck's constant = 6.626×10^{-34} Wsec^2
T = absolute temperature of the radiatior in K
c = velocity of light = 1.9979×10^8 m/sec
k = Boltzmann's constant = 1.38×10^{-23} Wsec/K
$c_1 = 2\pi hc^2 = 3.74 \times 10^{-16}$ Wm^2
$c_2 = ch/k = 0.0144$ mK

We note that the temperature in these expressions is the physical or kinetic temperature of the body. This temperature is the one that would be measured if we were to place a thermometer in physical contact with the body.

Integrating the emittance over the whole spectrum gives an expression for the total flux emitted by a blackbody of unit area. This is known as the Stefan–Boltzmann law:

$$S = \int_0^\infty S(\lambda, T)d\lambda = \frac{2\pi^5 k^4}{15c^2 h^3} T^4 = \sigma T^4 \tag{4-3}$$

where $\sigma = 5.669 \times 10^{-8}$ W/m^{3}K^4. Differentiating $S(\lambda, T)$ with respect to the wavelength and solving for the maximum gives Wien's law, which is the expression of the wavelength of maximum emittance:

$$\lambda_m = \frac{a}{T} \tag{4-4}$$

where $a = 2898$ μmK. Thus the sun, with a temperature of 6000 K, will have maximum emittance at $\lambda_m = 0.48$ μm. A surface at temperature 300 K will have maximum emittance at $\lambda_m = 9.66$ μm, that is, in the infrared region.

Another useful expression is the value of the emittance at $\lambda = \lambda_m$:

$$S(\lambda_m, T) = bT^5 \tag{4-5}$$

where $b = 1.29 \times 10^{-5}$ W/m^3/K^5. To illustrate, for $T = 300$ K and a spectral band of 0.1 μm around the peak, the emitted power is about 3.3 W/m^2.

The radiation laws can also be written in terms of the number of photons emitted. This form is useful in discussing the performance of photon detectors. Dividing the expression of $S(\lambda, T)$ by the energy associated with one photon, hc/λ, the result is the spectral radiant photon emittance (in photon/sec/m^3):

$$Q(\lambda, T) = \frac{2\pi c}{\lambda^4} \frac{1}{e^{ch/\lambda kT} - 1} \tag{4-6}$$

The corresponding form of the Stefan–Boltzmann law becomes

$$Q = \sigma' T^3$$

where $\sigma' = 1.52 \times 10^{15}$ m^{-2} sec^{-1} T^{-3}. Therefore, the rate at which photons are emitted from a blackbody varies as the third power of its absolute temperature. To illustrate, for $T = 300$, the total number of emitted photons is 4×10^{22} photons/m^2/sec.

4-1-1   Emissivity of Natural Terrain

The formulas in the previous section describe the radiation from a blackbody. These bodies have the highest efficiency in transforming heat energy into electromagnetic energy. All natural terrains have a lower efficiency, which is expressed by the spectral emissivity factor $\varepsilon(\lambda)$:

$$\varepsilon(\lambda) = \frac{S'(\lambda, T)}{S(\lambda, T)} \tag{4-7}$$

which is the ratio of the radiant emittance of the terrain to the radiant emittance of a blackbody at the same temperature. The mean emissivity factor ε is given by

$$\varepsilon = \frac{\displaystyle\int_0^\infty \varepsilon(\lambda) S(\lambda, T) d\lambda}{\displaystyle\int_0^\infty S(\lambda, T) d\lambda} = \frac{1}{\sigma T^4} \int_0^\infty \varepsilon(\lambda) S(\lambda, T) d\lambda \tag{4-8}$$

Three types of sources can be distinguished by the way that the spectral emissivity varies (see Fig. 4-1):

1. Blackbody, where $\varepsilon(\lambda) = \varepsilon = 1$
2. Graybody, where $\varepsilon(\lambda) = \varepsilon =$ constant less than 1
3. Selective radiator, where $\varepsilon(\lambda)$ varies with wavelength

When radiant energy is incident on the surface of a thick medium, a fraction ρ is reflected by the interface and a fraction τ is transmitted through the interface. To satisfy conservation of energy, these fractions are related by

$$\rho + \tau = 1 \tag{4-9}$$

A blackbody absorbs all of the incident radiant energy. In this case $\tau = 1$ and $\rho = 0$ (i.e., all the energy is transmitted through the interface and then absorbed).

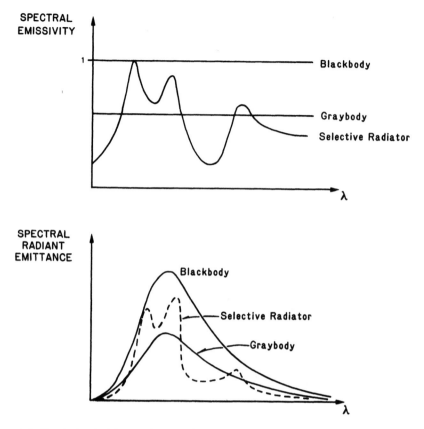

Figure 4-1. Spectral emissivity ε and spectral radiant emittance $S(\lambda, T)$ as a function of wavelength.

Based on Kirchoff's law, the absorptance α of a body is equal to its emittance ε at the same temperature. In other terms, the transmittance through the interface is the same from both directions. Thus we have

$$\alpha = \varepsilon = \tau \qquad (4\text{-}10)$$

and from Equation 4-9,

$$\varepsilon = \tau = 1 - \rho \qquad (4\text{-}11)$$

The reflectivity ρ is also called the surface albedo A.

The emissivity of a material is also a function of the direction of emission. The previous expressions give the hemispherical emissivity, which is the emissivity of a source radiating into a hemisphere. Directional emissivity $\varepsilon(\theta)$ is the emissivity at angle θ relative to the normal to the surface. Normal emissivity corresponds to $\theta = 0$. In the most general case, natural surfaces have an emissivity $\varepsilon(\lambda, \theta)$, which is a function of both wavelength and direction of emission.

For metals, the emissivity is low (a few percent), particularly when the metal is polished (ρ is high, ε is low). It increases with temperature and drastically increases with the formation of an oxide layer. For nonmetals, emissivity is high, usually more than 0.8, and decreases with increasing temperature.

The infrared radiation from a natural opaque surface originates within a fraction of a millimeter of the surface. Thus, emissivity is a strong function of the surface state or cover. For instance, a thin layer of snow or vegetation will drastically change the emissivity of a soil surface.

Care should be taken in trying to guess the emissivity of a material on the basis of its visual appearance, which is its reflectance in the visible region. A good illustration is snow. In the visible region, snow is an excellent diffuse reflector and from Kirchoff's law we might think that its emissivity is low. However, at 273 K most of the spectral radiant emittance occurs between 3 and 70 μm (maximum at 10.5 μm). Hence, a visual estimate is meaningless. In actuality, snow is an excellent emitter in the infrared region and it has low reflectance.

For most natural bodies, the spectral radiant emittance is not a monotonous curve but contains spectral lines that characterize the body constituents. In the thermal infrared spectral region, there are a number of absorption lines associated with fundamental and overtone vibrational energy bands. The use of a spectral signature in the thermal infrared region for unit identification will be discussed later in this chapter.

4-1-2 Emissivity from the Sun and Planetary Surfaces

The emission from the sun corresponds approximately to a blackbody with a 6000 K surface temperature. This gives a peak emission at 0.48 μm.

In the case of planetary surfaces, the temperature ranges from about 40 K on Pluto to about 700 K on Mercury. This corresponds to peak emission at wavelengths between 72 and 4.1 μm, respectively. In the case of Earth, the surface temperature ranges from about 240 K to about 315 K, with a most common temperature of about 300 K. This corresponds to peak emissions of 12.4, 9.18, and 9.66 μm, respectively. It should be pointed out that in the case of cloud-covered planets, most of the surface thermal emission is blocked by the atmosphere.

One interesting parameter is the relative variation of the emittance (i.e., $\Delta S/S$) as a function of temperature changes when the observation wavelength is fixed at the peak emission wavelength for a temperature T_p. From Equation 4-5 we have

$$\left.\frac{\Delta S}{S}\right|_{T_p, \lambda_m} = 5\frac{\Delta T}{T_p} \tag{4-12}$$

Thus, a 1% change in the surface temperature would result in a 5% emittance change.

It is particularly interesting to compare the spectral distribution of the emitted radiation to the spectral distribution of the reflected radiation for the different bodies in the solar system. Figure 4-2 shows these spectral distributions for Mercury, Venus, Mars, Jupiter, and Saturn. The wavelength at which the emitted and reflected energy are equal is usually between 2 μm (for Mercury) and 11 μm (for Saturn). The exact value depends on the surface emissivity $\varepsilon = 1 - A$, where A is the albedo or reflectivity ($A = \rho$).

The reflected energy spectrum is given by (see Chapter 3)

$$S_r(\lambda) = S(\lambda, T_s) \times \left(\frac{R_s}{d}\right)^2 \times A \tag{4-13}$$

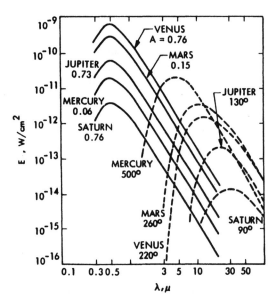

Figure 4-2. Reflected (continuous line) and emitted (dashed line) energy spectra for Mercury, Venus, Mars, Jupiter, and Saturn for the indicated temperatures and surface albedo. The albedo is assumed constant across the spectrum.

where the first term is the energy radiated from the sun with temperature T_s, the second term represents the fact that the energy is radiated spherically from the sun surface with radius R_s to the body at distance d, and the third term is the body average reflectivity.

4-2   HEAT CONDUCTION THEORY

Remote sensing instruments measure the electromagnetic radiation that is emitted from objects that are at a temperature above absolute zero. To understand better how to interpret what these instruments measure, it is necessary that we understand how materials respond to a change in temperature. The heat conduction equation is

$$\nabla^2 T - \frac{1}{\kappa} \frac{\partial T}{\partial t} = 0 \tag{4-14}$$

where T is the temperature and κ is the diffusivity in m²/sec. κ is related to the material thermal conductivity K (in Cal/m/sec/degree), density ρ (in kg/m³), and specific heat or thermal capacity C (in Cal/kg/degree) by

$$\kappa = \frac{K}{\rho C}$$

The heat conduction equation results from expressing the conservation of heat in an infinitesimal volume, dV. The heat flux across a surface (in Cal/m²/sec) is given by

$$\mathbf{f} = -K\nabla T \tag{4-15}$$

Consider an infinitesimal rectangular parallelepiped centered at point P, with edges parallel to the coordinate axes and of lengths $2dx$, $2dy$, and $2dz$ (see Fig. 4-3). The rate at which heat flows into the parallelepiped through the face $ABCD$ is equal to

$$4\left(f_x - \frac{\partial f_x}{\partial x} dx\right) dy\, dz$$

Similarly, the rate at which heat flows out through the face is given by

$$4\left(f_x + \frac{\partial f_x}{\partial x} dx\right) dy\, dz$$

Thus, the rate of gain of heat from flow across these two faces is equal to

$$-8\frac{\partial f_x}{\partial x} dx\, dy\, dz$$

Repeating the same for the other directions, the total rate of gain of heat in the volume from flow across its surface is

$$-8\left(\frac{\partial f_x}{\partial x} + \frac{\partial f_y}{\partial y} + \frac{\partial f_z}{\partial z}\right) dx\, dy\, dz = -8\nabla \cdot \mathbf{f}\, dx\, dy\, dz = -\nabla \cdot \mathbf{f}\, dV \qquad (4\text{-}16)$$

The rate of gain of heat associated with temperature change in the volume is also given by

$$\rho C\, \frac{\partial T}{\partial t} dV \qquad (4\text{-}17)$$

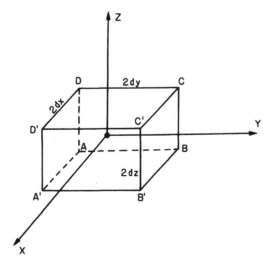

Figure 4-3. Geometry for derivation of the heat equation.

Equating the above two equations gives

$$\nabla \cdot \mathbf{f} + \rho C \frac{\partial T}{\partial t} = 0 \Rightarrow -\nabla \cdot (K\nabla T) + \rho C \frac{\partial T}{\partial t} = 0$$

or

$$\nabla^2 T - \frac{\rho C}{K} \frac{\partial T}{\partial t} = 0 \Rightarrow \nabla^2 T - \frac{1}{\kappa} \frac{\partial T}{\partial t} = 0$$

If heat is produced in the solid and is supplied at the rate $A(x, y, z, t)$ per unit time and unit volume, the heat conduction equation, for K constant, becomes

$$\nabla^2 T - \frac{1}{\kappa} \frac{\partial T}{\partial t} = -\frac{A(x, y, z, t)}{K} \tag{4-18}$$

This situation occurs when there is a geothermal or radioactive heat source.

Another common situation is when K is not a constant; then

$$\rho C \frac{\partial T}{\partial t} = \nabla \cdot (K\nabla T) + A \tag{4-19}$$

If K varies only as a function of position, the above equation can be solved with no great difficulty. However, if K is also a function of T, then the equation becomes nonlinear and usually the solution can be acquired only by numerical techniques.

The complete solution of the heat conduction equation requires knowledge of the initial and boundary conditions. Usually, the temperature throughout the body is given at the instant $t = 0$, and the solution should be equal to this initial function when $t \to 0$. In some situations, the temperature is continuously periodic. In this case, the initial condition is not relevant after a certain amount of time has passed.

The boundary condition can be of different nature:

1. Prescribed surface temperature. This is the easiest case, but not very common.
2. No flux across the surface (i.e., insulated surface). $\partial T / \partial n = 0$ at all points of the surface.
3. Prescribed flux across the surface.
4. Linear heat transfer at the surface. This occurs when the flux across the surface is proportional to the temperature difference between the surface and the surroundings:

$$K \frac{\partial T}{\partial n} = -H(T - T_0) \tag{4-20}$$

If in addition there is a constant flux F into the surface, then

$$K \frac{\partial T}{\partial n} = -H(T - T_0) + F \tag{4-21}$$

5. Nonlinear heat transfer. In most practical cases, the flux of heat from the surface is a nonlinear function of the temperature difference. For instance, in the case of a graybody, it is equal to $\varepsilon\sigma(T^4 - T_0^4)$, where T and T_0 are the temperatures on each side of the boundary. However, if $(T - T_0)/T_0 \ll 1$, then the flux can be approximated by

$$F = \varepsilon\sigma(T^2 - T_0^2)(T^2 + T_0^2)$$

$$= \varepsilon\sigma(T - T_0)(T + T_0)(T^2 + T_0^2) \tag{4-22}$$

$$\Rightarrow F \simeq 4\,\varepsilon\sigma T_0^3(T - T_0)$$

which is a linear heat transfer case.

Solutions for different cases and situations can be found in the book by Carslaw and Jaeger, *Conduction of Heat in Solids.*

4-3 EFFECT OF PERIODIC HEATING

The case of a semiinfinite solid with a surface temperature that is a harmonic function of time is of particular interest in remote sensing because of the periodic surface heating from the sun. Let us assume that the surface temperature for the semiinfinite solid $x > 0$ is given by

$$T(0, t) = A\cos(\omega t - \phi) \tag{4-23}$$

Later on, we shall relate this temperature to the forcing function provided by the heat input from the sun. We seek a solution to the heat conduction equation of the type

$$T(x, t) = u(x)e^{i(\omega t - \phi)} \tag{4-24}$$

Substituting in the heat conduction equation (4-16), we get

$$\frac{d^2u}{dx^2} - i\frac{\omega}{\kappa}u = 0 \tag{4-25}$$

which has a solution

$$u = Be^{\pm x\sqrt{i(\omega/\kappa)}} = Be^{\pm x(1+i)\sqrt{\omega/2\kappa}} = Be^{\pm kx}Be^{\pm ikx} \tag{4-26}$$

where $k = \sqrt{\omega/2\kappa}$. Only the solution with the decaying exponential is realistic. Thus, the solution to Equation (4-24) is given by

$$T(x, t) = Be^{-kx}e^{i(\omega t - \phi - kx)}$$

$$\Rightarrow T(x, t) = B_1 e^{-kx}\cos(\omega t - \phi - kx) + B_2 e^{-kx}\sin(\omega t - \phi - kx)$$

and the one that satisfies the boundary condition at the interface is

$$T(x, t) = Ae^{-kx}\cos(\omega t - \phi - kx) \tag{4-27}$$

This represents a temperature wave decaying as a function of x (see Fig. 4-4) with a wave number k and a wavelength $\lambda = 2\pi/k = 2\pi\sqrt{2\kappa/\omega}$.

For typical rock materials with $\kappa = 10^{-6}$ m²/sec, the wavelength is about 1 m for a frequency of 1 cycle/day and 19 m for 1 cycle/year. For a metallic conductor with $\kappa = 10^{-4}$ m²/sec, the wavelength is 10 m for 1 cycle/day and 190 m for 1 cycle/year. Table 4-1 gives the thermal properties for a number of geologic materials. It is clear from the values of κ that diurnal heating of the surface only affects approximately the top meter and annual heating affects the top 10 to 20 meters.

From the above solution, and Equation (4-15), the flux of heat F at the surface is

$$F = -K\frac{\partial T}{\partial x}\bigg|_{x=0} = kKA[\cos(\omega t - \phi) - \sin(\omega t - \phi)]$$

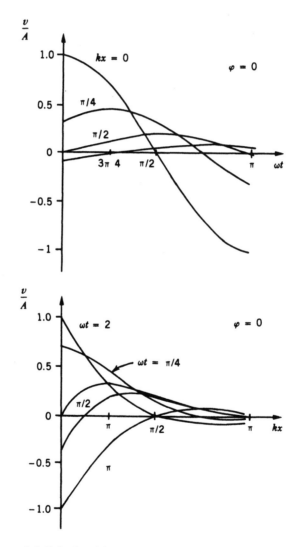

Figure 4-4. Behavior of the temperature wave as a function of depth.

TABLE 4-1. Thermal Properties of Some Materials

Material	K (Cal/msec °C)	ρ (kg/m³)	C (Cal/kg/°C)	κ (m²/sec)	P (Cal/m² sec$^{1/2}$ °C)
Water	0.13	1000	1010	1.3×10^{-7}	370
Basalt	0.5	2800	200	9×10^{-7}	530
Clay (moist)	0.3	1700	350	5×10^{-7}	420
Granite	0.7	2600	160	16×10^{-7}	520
Gravel	0.3	2000	180	8×10^{-7}	320
Limestone	0.48	2500	170	1.1×10^{-7}	450
Dolomite	1.20	2600	180	26×10^{-7}	750
Sandy soil	0.14	1800	240	3×10^{-7}	240
Sandy gravel	0.60	2100	200	14×10^{-7}	500
Shale	0.35	2300	170	8×10^{-7}	340
Tuff	0.28	1800	200	8×10^{-7}	320
Marble	0.55	2700	210	10×10^{-7}	560
Obsidian	0.30	2400	170	7×10^{-7}	350
Pumice, loose	0.06	1000	160	4×10^{-7}	90

Source: From Janza, 1975.

which can be written as

$$F = \sqrt{2}kKA \cos\left(\omega t - \phi + \frac{\pi}{4} \right) \tag{4-28}$$

Equations 4-27 and 4-28 form a coupled pair. Thus, if we have a semiinfinite half-space that is heated by a periodic flux $F_0 \cos(\omega t - \phi)$, the steady-state temperature is given by

$$T(x, t) = \frac{F_0 e^{-kx}}{kK\sqrt{2}} \cos\left(\omega t - kx - \phi - \frac{\pi}{4} \right) \tag{4-29}$$

This involves the term $kK = \sqrt{(\omega/2)K\rho C}$. The term $\sqrt{K\rho c}$ is called thermal inertia P (units of Cal/m² deg sec$^{1/2}$). The steady state temperature is then expressed as

$$T(x, t) = \frac{F_0 e^{-kx}}{P\sqrt{\omega}} \cos\left(\omega t - kx - \phi - \frac{\pi}{4} \right) \tag{4-30}$$

This expression shows that there is a phase difference, that is, a time lag, between the peak flux input to the surface, and the peak temperature response of the surface to that input. In terms of the soil temperature on the earth, it is well known that whereas the peak solar flux occurs at noon when the sun is at its zenith, the soil temperature peaks later in the afternoon, typically around 2 p.m. Also, it is important to remember that remote sensing instruments will record the radiation from the surface that is driven primarily by the temperature of the surface. Therefore, whereas the *reflected* solar radiation will peak at the same time the incoming solar flux that heats the surface would peak, the *radiated* emission from the surface due to its own temperature will peak at a later time.

Of course, the solar flux, though a periodic function, does not vary as a single-frequency sinusoid as in the simple example described above. We shall discuss this case in more detail in the next section. In a more general case where the surface temperature is a periodic function that can be expressed as a Fourier series, we have the following:

$$T(0, t) = A_0 + \sum_{n=1}^{\infty} A_n \cos(n\omega t - \phi_n) \tag{4-31}$$

The total solution for the temperature is then

$$T(x, t) = A_0 + \sum_{n=1}^{\infty} A_n e^{-k_n x} \cos(n\omega t - \phi_n - k_n x) \tag{4-32}$$

where $k_n = \sqrt{n\omega/2\kappa} = k\sqrt{n}$.

4-4   USE OF THERMAL EMISSION IN SURFACE REMOTE SENSING

The fact that the surface temperature, and, therefore, the thermal emission, is dependent on the surface thermal inertia, which in turn is a function of the surface material, leads to the possibility of using the surface thermal emission to infer some of the surface thermal properties. The variations in the surface temperature are driven by the periodic solar illumination, which provides a periodic flux.

4-4-1   Surface Heating by the Sun

In the case of interest to us (i.e., solar heating of the Earth surface), the boundary condition is given by

$$-K\frac{\partial T}{\partial x}\bigg|_{x=0} = (1 - \rho)\sigma T_s^4 - \varepsilon\sigma T^4 + I(t) = -\varepsilon\sigma(T^4 - T_s^4) + I(t) \tag{4-33}$$

where T_s is the effective sky radiance temperature (at long wavelength) and I is the incoming solar radiation modulated by the atmospheric transmission. The first and third terms to the right are the radiation fluxes input from the sky and the sun, respectively. The second term is the radiation flux output from the surface. This condition neglects other heat transmission mechanisms, such as atmospheric conduction and convection and latent heat effects. Let us assume that $T - T_s \ll T_s$, then the boundary condition can be simplified to

$$-K\frac{\partial T}{\partial x}\bigg|_{x=0} = -4\varepsilon\sigma T_s^3(T - T_s) + I(t) \tag{4-34}$$

The solution of the heat conduction equation subject to the above condition can be obtained by using the following substitution:

$$h(x, t) = T - \frac{1}{q}\frac{\partial T}{\partial x} \tag{4-35}$$

where $q = 4\varepsilon\sigma T_s^3/K$. Then,

$$\frac{\partial h}{\partial x} = \frac{\partial T}{\partial x} - \frac{1}{q}\frac{\partial^2 T}{\partial x^2} = \frac{\partial T}{\partial x} - \frac{1}{q\kappa}\frac{\partial T}{\partial t}$$

$$\frac{\partial^2 h}{\partial x^2} = \frac{\partial^2 T}{\partial x^2} - \frac{1}{q\kappa}\frac{\partial^2 T}{\partial x\partial t} = \frac{1}{\kappa}\frac{\partial T}{\partial t} - \frac{1}{q\kappa}\frac{\partial^2 T}{\partial x\partial t} = \frac{1}{\kappa}\frac{\partial}{\partial t}\left(T - \frac{1}{q}\frac{\partial T}{\partial x}\right) = \frac{1}{\kappa}\frac{\partial h}{\partial t}$$

Thus,

$$\frac{\partial^2 h}{\partial x^2} - \frac{1}{\kappa}\frac{\partial h}{\partial t} = 0 \qquad (4\text{-}36)$$

and the boundary condition becomes

$$h(0, t) = T(0, t) - \frac{1}{q}\frac{\partial T}{\partial x}\bigg|_{x=0} = T(0, t) - T(0, t) + T_s + \frac{I(t)}{qK} = T_s + \frac{I(t)}{qK} \qquad (4\text{-}37)$$

which is simpler than the boundary condition in Equation 4-34. The solar illumination I is a function of the ground reflectivity A in the solar spectral region (i.e., mostly visible and near infrared), the solar declination δ, the site latitude γ, and the local slope. It can be written as

$$I(t) = (1 - A)S_0 CH(t) \qquad (4\text{-}38)$$

where S_0 is the solar constant, C is a factor to account for the reduction of solar flux due to cloud cover and

$$H(t) = \begin{cases} M[Z(t)]\cos Z'(t) & \text{during the day} \\ 0 & \text{during the night} \end{cases}$$

where $Z'(t)$ is the local zenith angle for inclined surfaces, $Z(t)$ is the zenith angle, and M is the atmospheric attenuation, which is a function of the zenith angle Z. The function $H(t)$ is periodic relative to t and can be written as

$$H(t) = \sum_{n=0}^{\infty} A_n \cos(n\omega t - \phi_n) \qquad (4\text{-}39)$$

leading to

$$h(0, t) = T_s + \frac{(1 - A)S_0 C}{qK}\sum_{n=0}^{\infty} A_n \cos(n\omega t - \phi_n)$$

This leads to a solution for $h(x, t)$ (from Equation 4-32):

$$h(x, t) = T_s + \frac{(1 - A)S_0 C}{qK}\sum_{n=0}^{\infty} A_n e^{-k\sqrt{n}x}\cos(n\omega t - \phi_n - k\sqrt{n}x) \qquad (4\text{-}40)$$

The solution for $T(x, t)$ is then derived from Equation 4-35 as

$$T(x, t) = Be^{qx} - qe^{qx} \int_x^\infty h(\xi, t)e^{-q\xi}\, d\zeta \tag{4-41}$$

Putting $\eta = -x + \xi$ and requiring that T stays finite as $x \to \infty$, then

$$T(x, t) = -q \int_0^\infty h(x + \eta, t)e^{-q\eta}\, d\eta \tag{4-42}$$

and the solution for $T(0, t)$ is

$$T(0, t) = -q \int_0^\infty h(\eta, t)e^{-q\eta}\, d\eta$$

$$= T_s + \frac{(1 - A)S_0 C}{K} \sum_{n=0}^\infty \frac{A_n \cos(n\omega t - \phi_n - \delta_n)}{\sqrt{(q + k\sqrt{n})^2 + nk^2}} \tag{4-43}$$

where $\tan \delta_n = (q + 1)/qk\sqrt{n}$.

The effect of geothermal heat flux Q at the surface can now be introduced by adding a second solution $T = Qx/K + Q/qK$, which satisfies the heat conduction equation and the boundary condition. Thus, we need to add the term Q/qK to the expression of $T(0, t)$.

Figure 4-5 illustrates the solution for the diurnal temperature behavior as a function of thermal inertia, albedo in the visible region A, geothermal flux, and emissivity in the infrared region ε.

The mean diurnal temperature $\overline{T}$ is derived by integrating $T(0, t)$ over the diurnal cycle T_d:

$$\overline{T} = \frac{1}{T_d} \int_0^{T_d} T(0, t)\,dt = T_s + \frac{Q}{s} + (1 - A)\frac{S_0 C A_0}{s} \cos(\phi_0) \tag{4-44}$$

where

$$A_0 \cos(\phi_0) = \frac{1}{T_d} \int_0^{T_d} H(t)\,dt$$

and $s = qK = 4\varepsilon\sigma T_s^3$. Note that $\overline{T}$ is independent of the surface inertia and that it can be used in conjunction with albedo measurements and topographic information to derive Q.

By analyzing the curves in Figure 4-5, it is apparent that the day-night temperature difference ΔT is strongly dependent on the thermal inertia P, weakly dependent on the albedo A, or the emissivity. This fact has been used to map thermal inertia variations.

4-4-2 Effect of Surface Cover

The diurnal temperature variation is limited to the top meter or less of a semiinfinite solid. The thickness of this region is proportional to the square root of the diffusivity κ. Thus, if the surface is covered by a thin layer of material with low κ (e.g., low K or large ρC), the covered material will barely be affected by the surface temperature variation.

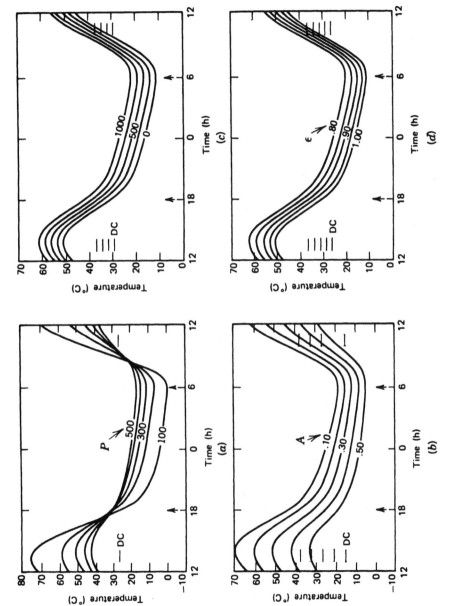

Figure 4-5. Diurnal temperature curves for varying (a) thermal inertia in cal/m2 sec1/2, (b) albedo, (c) geothermal heat flux in HFU, and (d) emissivity. The mean diurnal temperature DC is indicated by horizontal lines. Fixed parameters used are inertia 300 cal/m2 sec1/2, albedo 0.3, emissivity 1.0, latitude 30°, solar declination 0°, sky radiant temperature 260° and cloud cover 0.2 (Watson, 1973).

The solution for this problem was given by Watson (1973), and some of the results are illustrated in Figure 4-6. This figure shows the effect of a layer of dry, sandy soil ($\kappa = 2 \times 10^{-7}$ m^2/sec, $P = 150$ Cal/m^2/sec$^{1/2}$) and a layer of dry lichen and moss ($\kappa = 1.4 \times 10^{-7}$ m^2/sec, $P = 40$ Cal/m^2/sec$^{1/2}$) superimposed on a half-space with thermal inertia $P = 400$ Cal/m^2/sec$^{1/2}$, which is similar to that of common rocks. It is clear that a layer of about 10 cm could almost completely insulate the subsurface from diurnal temperature variation.

4-4-3 Separation of Surface Units Based on Their Thermal Signature

The curves in Figure 4-5 show a number of interesting behaviors that can be used in deriving surface properties. First, a single measurement of surface-emitted heat cannot be used by itself to derive the surface thermal properties because it depends on a number of independent parameters (see Eq. 4-43). However, if we measure the difference in emitted heat at two times of the day, the following can be observed:

1. The emissivity effect is negligible (Fig. 4-5d) and, therefore, can be neglected.
2. The albedo effect is significant; thus, an albedo measurement in the visible and near infrared region should be made.
3. The thermal inertia effect is significant if the times of observation are selected appropriately.

Considering that satellite observations on a global and regular basis can be made at best every 12 hours, on ascending and descending orbits, the best times for observation would be at around 2:00 a.m. and 2:00 p.m. Thus, if a broadband thermal imager is used in conjunction with a broadband visible/near infrared imager (to derive A), the pair of thermal observations would allow the derivation of the thermal inertia P on a pixel-by-pixel basis. Thermal inertia can be thought of as a measure of the resistance of the material to change of temperature. As indicated in Table 4-1, P varies significantly between different materials. For instance, it is more than quadruple between loose pumice and sandy gravel or basalt or limestone. Thus a thermal inertia map can be used to classify surface units.

4-4-4 Example of Application in Geology

The observation of the surface in the reflective visible and near-infrared spectral regions can lead to ambiguous results because different materials can develop similar weathering stains on the surface, and, conversely, similar materials may weather (chemical weathering) in different ways. Thus, it is desirable to augment the surface reflectance observation with data that can give information about the body properties of the near-surface material to some significant depth. Thermal remote sensing plays such a role.

Figures 4-7a, b, c, and d show the day visible, day IR, night IR, and resulting thermal inertia image of a portion of Death Valley. In Figure 4-7d, the bright areas correspond to high thermal inertia, and the dark areas correspond to low thermal inertia. The brightest features in the scene correspond to areas underlain by dolomite, limestone, quartzite, or granite. These rock types stand out distinctly on the thermal inertia image and are quite easy to map. The darkest features correspond to areas of young alluvium in the mountain valleys, whereas the floor of Death Valley is bright to medium gray.

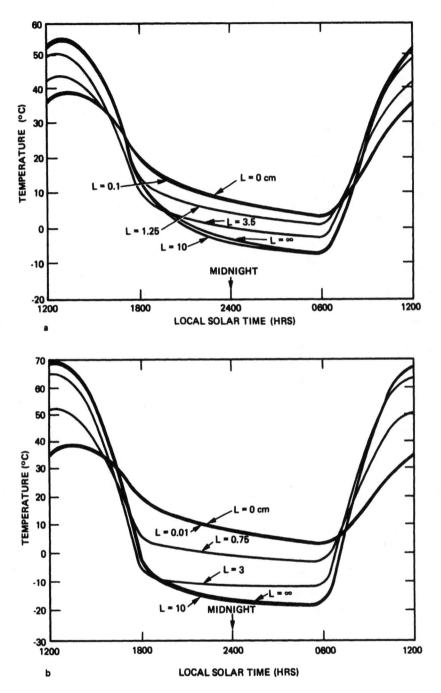

Figure 4-6. Plots of diurnal surface temperature versus local time for two different layers of thickness L over a half-space. (a) Layer with $P = 150$ Cal/m^2 sec$^{1/2}$ and $\kappa = 2 \cdot 10^{-7}$ m^2/sec. (b) Layer with $P = 40$ Cal/m^2 sec$^{1/2}$ and $\kappa = 1.4 \cdot 10^{-7}$ m^2/sec. The half-spaces have $P = 400$ Cal/ m^2 sec$^{1/2}$ (Watson, 1973).

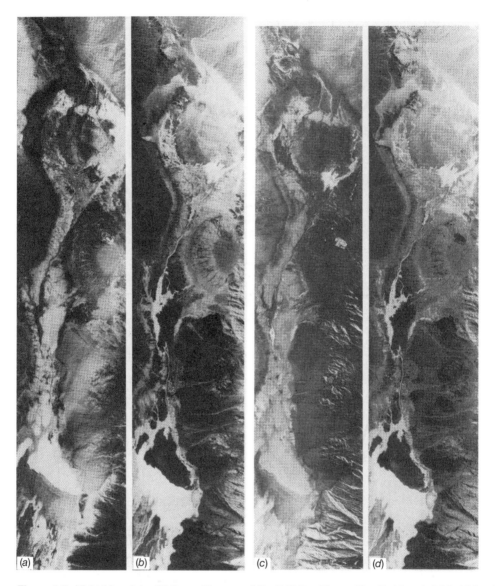

Figure 4-7. Night (*a*) and day (*b*) thermal images of Death Valley. Thermal inertia (*c*) and visible (.54 to .58 μm) (*d*) images of Death Valley. (Courtesy of A. Kahle, Jet Propulsion Laboratory.)

Thermal inertia images also allow accurate delineation of the bedrock–alluvium contact. This differentiation is often difficult to make on visible or reflected IR images, particularly when moderate to low spatial resolution prevents the use of textural information.

4-4-5 Effects of Clouds on Thermal Infrared Sensing

The presence of clouds affects thermal infrared sensing in two ways. First, the clouds reduce the local amount of incoming radiation, which results in spatially varying heating of the surface. This is manifested in thermal infrared images as patchy warm and cool patterns, where the cooler areas are typically those in the cloud shadows. Second, heavy overcasts tend to reduce the contrast in thermal images because of the reradiation of energy between the surface and cloud layer. This additional energy adds a background noise signal that reduces the overall image contrast.

Figure 4.8 shows images acquired in the visible and the thermal infrared region with the ASTER instrument on the Terra spacecraft of an area northeast of Death Valley in Nevada. Note the bright clouds in the upper-right part of the visible image. The thermal infrared image shows the typical warm and cool patchy response in the same area.

4-5   USE OF THERMAL INFRARED SPECTRAL SIGNATURES IN SENSING

In general, the spectral emissivity in the thermal and far infrared region (8 to 100 μm) is not a constant or monotonous function, but contains a number of bands and lines that are characteristic of the surface composition. The opacity of the Earth's atmosphere above 25 μm (see Fig. 1-13) limits the useful thermal regions for Earth-surface remote sensing to 8 to 25 μm.

Spectral signatures of most solids in the thermal infrared region are mainly the result of vibrational processes, and the most intense features result from the fundamental modes that occur in the thermal infrared region. In order to better understand the features in the emissivity spectra of inorganic solids, we will first describe briefly the restrahlen effect and the Christiansen frequencies.

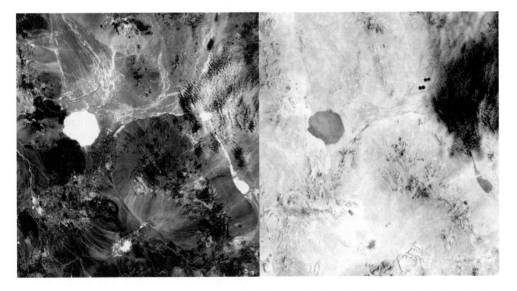

Figure 4-8. Visible (left) and thermal infrared (right) images showing the effects of clouds. The clouds are in the upper-right portion of the image and appear bright in the visible image, with some cloud shadows visible. The infrared image shows a patchy warm and cold appearance because of differential heating between the areas under the clouds and those not covered by the clouds.

As indicated in Equation 3-6, the reflection coefficient of a plane interface at normal incidence is

$$|R^2| = \frac{(N_r - 1)^2 + N_i^2}{(N_r + 1)^2 + N_i^2} \tag{4-45}$$

If we are near the center of a strong absorption band, then $N_i \gg N_r$ and

$$|R^2| = 1$$

which corresponds to total reflection or no emissivity. In other terms, the surface boundary can be described as completely opaque. This is called the restrahlen effect and related to it are the minima in the thermal infrared emissivity spectra.

If we are in a situation where $N_r \approx 1$, the reflection coefficient becomes

$$|R^2| \approx \frac{N_i^2}{4 + N_i^2} \tag{4-46}$$

and if we are in a low absorption case (i.e., away from a spectral line), then R is very small, which corresponds to a strong emissivity. Because of the spectral dispersion of many materials, there are specific spectral regions in the thermal infrared in which the index of refraction is equal or close to unity. These spectral regions correspond to the Christiansen frequencies. Figure 4-9a shows the refraction index of quartz and illustrates the fact that $n(\lambda)$ is equal or close to unity at 7.4, 9.3, and 12 μm. Figure 4-9b shows the transmission of a thin film of quartz powder. A maximum is clearly observed at 7.4 μm and 12 μm (when the particles are small). The lack of a peak at 9.3 μm is attributed to the fact that quartz is strongly absorbing at this frequency, thus leading to the situation in which N_i is large and $T = 1 - R$ is small (from Eq. 4-46).

A number of geologic materials exhibit vibrational spectral features in the thermal infrared region (see Figs. 4-10 and 4-11). Silicates (quartzite, quartz monzonite, monzonite, latite, and olivine basalt) have emittance minima as a result of the restrahlen effect associated with the fundamental Si–O vibration near 10 μm. Multiple bands occur in that region due to the asymmetric stretching motions of O–Si–O, Si–O–Si, O⁻–Si–O⁻, and the symmetric O⁻–Si–O⁻. The SiO stretching modes and the H–O–Al bending mode near 11 μm are responsible for the absorption bands in clays (kaolinite and montmorillonite). In Figures 4-10 and 4-11, these bands show as a single wide absorption region mainly because of the experimental measurement technique, in which the grinding of the samples to fine particles reduces the periodicity of the lattice, and also as a result of lattice disorder and impurities.

The emittance maxima in silicates that are associated with the Christiansen frequency occur near 8 μm (Fig. 4-11), just before the onset of the intense absorption due to the SiO vibration. The locations of the Christiansen peak and of the minimum band migrate fairly systematically to larger wavelengths as the material moves from felsic to mafic and ultramafic (Fig. 4-11).

At longer wavelengths, stretching and bending modes lead to additional spectral features. These are summarized in Figure 4-12.

In the case of carbonates (limestone and dolomite), the spectrum is dominated by the features resulting from the vibrations of the CO_3^{2-} group, which are an asymmetric stretch near 7 μm and a planar bend near 11.5 μm and near 14.5 μm (Fig. 4-10).

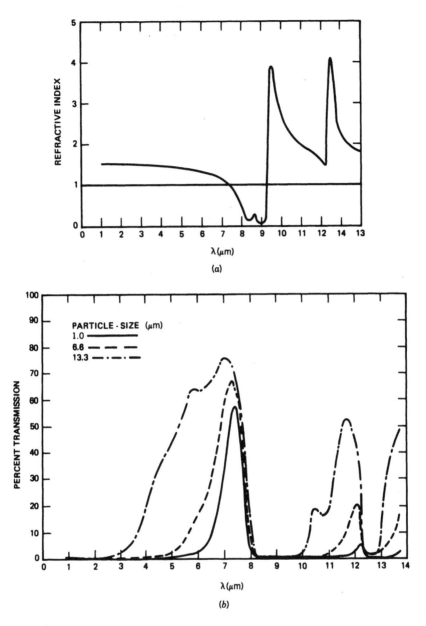

Figure 4-9. (a) Dispersion of quartz. (b) Transmission through a 12.8 μm thick layer of quartz powder with particle sizes of 1, 6.6, and 15.5 μm. (From Henry, 1948.)

4-6 THERMAL INFRARED SENSORS

Thermal imagers operate in the same fashion as visible and near infrared sensors. The major differences are that the signal is usually significantly weaker (see Fig. 4-2) and presently available detectors are typically less sensitive (see Fig. 3-48). This leads to sensors with lower resolution and longer dwell time to increase the energy that is integrated.

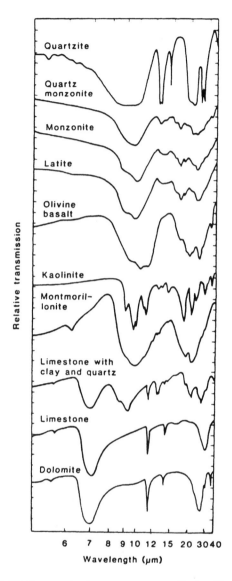

Figure 4-10. Infrared transmission spectra for some common silicates. Regions of low transmission are associated with restrahlen bands and correspond to regions of low emittance (Hunt and Salisbury, 1974, 1975, and 1976).

For example, the thermal infrared channel on the ETM+ instrument has a resolution that is a factor of three worse than the visible and near infrared channels (see Table 3-6) and the ASTER thermal infrared channels have resolutions that are a factor of six worse than the visible channels. Another major difference between thermal infrared imagers and visible imagers is the need for cooling the detectors, and sometimes even the optics at longer wavelengths. This must be done to ensure that the radiation contributed by the emission from the imaging system itself is small compared to the thermal energy collected from the scene.

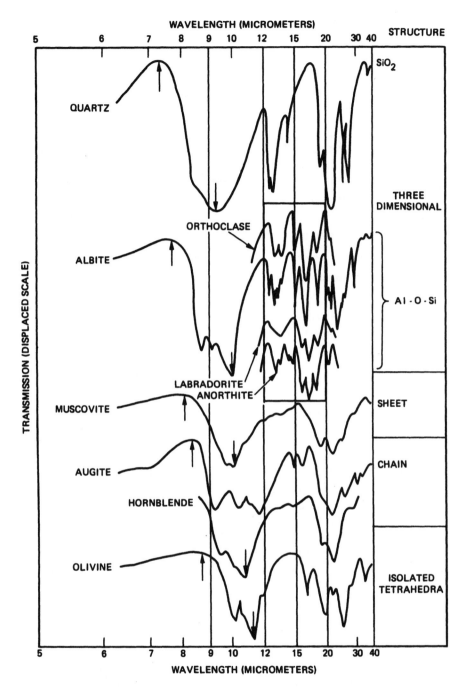

Figure 4-11. Transmission spectra of minerals of different composition and structure. The up arrows show the Christiansen peak. The down arrows show the absorption band minimum. (From Hunt in Siegel and Gillespie, © 1980 John Wiley & Sons. Reprinted with permission.)

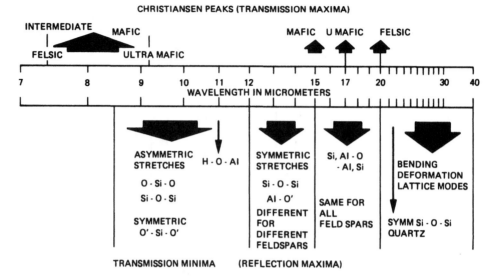

Figure 4-12. Diagram illustrating the location of features and the type of vibrational processes that produce the spectral signature of silicates in the thermal infrared region. (From Hunt in Siegel and Gillespie, © 1980 John Wiley & Sons. Reprinted with permission.)

Three main methods are used to cool detectors and optics in space. Passive radiators are the most reliable and efficient cooling systems used in space. Passive cooling is based on the fact that all objects radiate an amount of energy that is proportional to the their surface area, their emissivity, and T^4, where T is the temperature difference between the radiator and space. Often, radiators are combined with louvers, which allow the spacecraft to regulate the amount of heat transfer. Thermal control can further be enhanced by using highly insulated sun shields to reduce heating of the spacecraft by the incoming solar radiation. Radiators can be used to cool down to about 60 K under ideal conditions, although it is more common to use active cooling significantly below about 100 K.

For temperatures below about 100 K, active cooling is required. Stored cryogen coolers use a reservoir of coolant, such as superfluid helium or solid hydrogen. These systems rely on the boiling or sublimation of the low-temperature fluid or solid to absorb the waste heat and eject it overboard in the form of a vented gas. Detector temperatures as low as 1.5 K can be reached with such open-cycle cooling systems. The disadvantage of this cooling method is that cooling only happens as long as there is sufficient coolant available. Since the coolant has to evaporate in order to cool the detectors, this means that these cooling systems typically have a finite lifetime and a significant amount of extra mass must be launched into space. Also, to carry as much coolant as possible into orbit, the cooling fluid is typically topped off just prior to launch. This requires special equipment and procedures at the launch site, and typically limits the time between launch attempts.

For applications requiring substantial cooling over an extended period of time, mechanical coolers, or cryo coolers, are often the preferred solution. Mechanical coolers work much like refrigerators in the sense that a coolant is compressed and then allowed to expand. The coolant is not lost in this type of process, and the lifetime is limited only by the mechanical functions of the cooler. A potential drawback of mechanical cooling is vibration. Unless the coolers are balanced very carefully, they can introduce vibrations into the structure, which, in turn, can affect sensor pointing, or violate jitter requirements. Chemical sorption coolers do not suffer from the vibration problems associated with mechanical coolers. These sorption coolers consist of a number of sorption beds, which contain a metal hydride powder that absorbs a gas such as hydrogen. Heating such a bed pressurizes the hydrogen gas, while cooling a bed allows the hydride to absorb the gas. The operation of a sorption cryo cooler is, therefore, based on alternately heating and cooling beds of the sorbent material to pressurize, circulate, and absorb the gas in a closed Joule–Thomson refrigeration cycle.

Mercury–cadmium–telluride (HgCdTe) is one of the most widely used semiconductors for infrared photodetectors, which are currently available in arrays of 1024 × 1024 pixels and can be used to wavelengths up to about 12 microns. For wavelengths beyond 12 microns, extrinsic silicon and germanium devices are most often used. Arsenic-doped silicon (Si:As) operate to about 25 micron wavelengths, and stressed gallium-doped germanium (Ge:Ga) detectors operate to about 200 microns. Arsenic-doped silicon detectors are available in 1024 × 1024 arrays, whereas the Multiband Imaging Photometer instrument on the Spitzer Space Telescope flies a 32 × 32 gallium-doped germanium array for its 70 micron band, and a 2 × 20 Ge:Ga array, mechanically stressed to extend its photoconductive response to 160 microns. In addition to arsenic-doped silicon detectors, antimonide-doped silicon (Si:Sb) detectors are also used for thermal infrared imaging. These have been shown to have good responses to about 40 microns.

The performance of a thermal infrared imaging system is typically quoted as a *noise equivalent delta T*, or NEDT. The NEDT is that temperature fluctuation that would result in a signal-to-noise ratio of 1. Modern spaceborne instruments have NEDTs that are less than 0.1 K. We shall briefly describe a few thermal infrared sensors below.

4-6-1 Heat Capacity Mapping Radiometer

In 1978, a Heat Capacity Mapping Mission (HCMM) satellite was put into orbit with the objective of mapping surface thermal inertia using day–night thermal imaging as discussed in Section 4-4-3. The sensor (heat capacity mapping radiometer, or HCMR) had two channels. One spectral channel covered the reflectance band from 0.5 to 1.1 μm, which provided measurements of the surface albedo in the sun-illumination region. The second spectral channel viewed the thermal infrared band between 10.5 and 12.5 μm. The thermal channel had a measurement accuracy of 0.3 K (noise equivalent temperature difference). The satellite was put in a 620 km altitude orbit, which allowed the imaging of the surface twice every day, at 2:00 a.m. and 2:00 p.m. in equatorial regions, and 1:30 a.m. and 1:30 p.m. at northern mid-latitudes.

From the nominal altitude of 620 km, the spatial resolution of the infrared and visible channels was 600 m and 500 m at nadir, respectively. The data-coverage swath was approximately 716 km and was achieved by using a scanning mirror placed in front of the sensor optics. Table 4-2 gives a summary of the major characteristics of the HCMR. Figure 4-13 gives a block diagram of the optics.

The optics consisted of a flat elliptical scanning mirror that had a scan rate of 14 revolutions per second and scanned over a 60° angle. The collector was a catadioptric with an afocal reflecting telescope. The telescope had a modified Dall–Kirkham configuration that reduced the optical beam from 20.32 cm to 2.54 cm in diameter. Spectral separation was provided by a dichroic beam splitter positioned in the collimated beam from the sec-

Table 4-2. Heat Capacity Mapping Radiometer Summary Data Sheet

Orbital altitude = 620 km
Angular resolution = 0.83 mrad
Resolution = 0.6 km × 0.6 km at nadir (infrared)
0.5 km × 0.5 km at nadir (visible)
Scan angle = 60° (full angle)
Scan rate = 14 revolutions/sec
Swath width = 716 km
Information bandwidth = 53 kHz/channel
Thermal channel = 10.5 to 12.5 μm; NEDT = 0.4 K at 280 K
Usable range = 260 to 340 K
Visible channel = 0.55 to 1.1 μm; SNR = 10 at ~1% albedo
Dynamic range = 0 to 100% albedo
Nominal telescope optics diameter = 20 cm
Calibration = Infrared: view of space, seven-steep staircase electronic calibration, and blackbody
calibration once each scan
Visible: preflight calibration

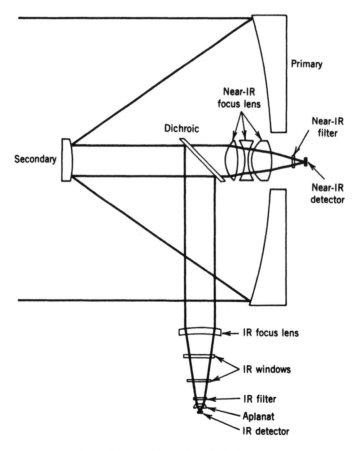

Figure 4-13. HCMR optical block diagram.

ondary mirror, which acted as a folding mirror for the 10.5 to 12.5 μm band and transmitted energy at shorter wavelengths.

The reflectance channel optics consisted of a long wavelength (greater than 0.55 μm) pass interference filter, focusing optics, and an uncooled silicon photodiode. The long wavelength cutoff of the silicon detector limited the band pass to wavelengths of less than 1.1 μm. The sensitive area of the detector was approximately 0.15 mm square.

The thermal infrared beam was focused onto the mercury–cadmium–telluride detector using a germanium lens. Final focusing and spectral trimming were accomplished by a germanium aplanat located at the detector.

The detectors produced an electrical signal that was proportional to the difference in radiant energy between the scene and space. A space-clamping technique was used to establish a zero reference level once every rotation when the mirror was looking toward space. Calibration signals consisting of a six-step staircase wave form were inserted on every scan line at the signal amplifier input to provide constant calibration of the amplifier. The data from the two channels were then digitized and multiplexed together for transmission to Earth.

4-6-2 Thermal Infrared Multispectral Scanner

In order to fully utilize the thermal infrared spectral emissivity of surface materials for mapping and identification, a six-channel thermal infrared multispectral scanner (TIMS) was developed in 1982 for airborne investigations. The spectral bands covered are: 8.2–8.6 μm, 8.6–9.0 μm, 9.0–9.4 μm, 9.4–10.2 μm, 10.2–11.2 μm, and 11.2–12.2 μm. The TIMS consists of a 19 cm diameter newtonian reflector telescope mounted behind an object plane 45° flat scanning mirror followed by a spectrometer. In the focal plane, the entrance slit to the spectrometer acts as the field stop and defines the instantaneous field of view of 2.5 mrad.

The total field of view scanned is 80°. The detector array consists of six mercury–cadmium–telluride elements cooled by liquid nitrogen. The sensitivity ranges between 0.1° and 0.3°C noise-equivalent temperature change at a mean temperature of 300 K. This is comparable to a noise-equivalent change in spectral emissivity of 0.002 to 0.006. This sensor allows the acquisition of spectral images covering the SiO_2 vibrational bands and the CO_3^{2-} vibrational bands. Even though the spectral resolution is low, a significant amount of material classification is possible. Figure 4-14 shows an example of the data acquired with TIMS, which illustrates its mapping and units-discrimination capability.

4-6-3 ASTER Thermal Infrared Imager

The ASTER thermal infrared telescope has five bands with a spatial resolution of 90 m; see Table 3-6. The telescope is a Newtonian catadioptric system with an aspheric primary mirror and lenses for aberration correction. Each band uses 10 mercury–cadmium–telluride (HgCdTe) detectors. Spectral discrimination is achieved by optical band-pass filters that are placed over each detector element. The ASTER thermal infrared system uses a mechanical cooler for maintaining the detectors at 80 K.

A scanning mirror functions both for scanning and pointing. In the scanning mode, the mirror oscillates at about 7 Hz. For calibration, the scanning mirror rotates 180° from the nadir position to view an internal blackbody that can be heated or cooled. The ASTER re-

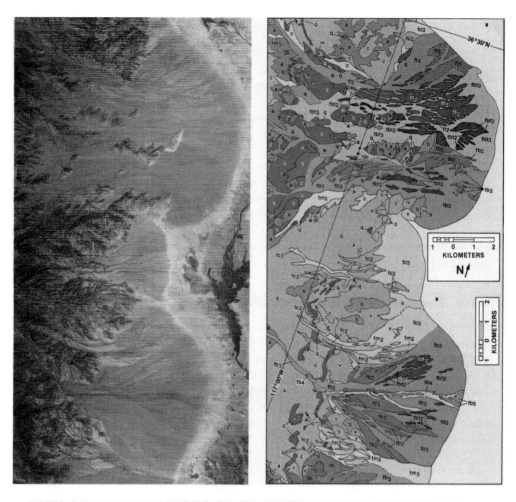

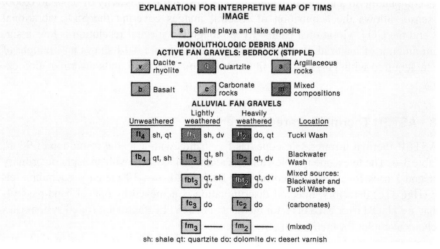

EXPLANATION FOR INTERPRETIVE MAP OF TIMS IMAGE

s Saline playa and lake deposits

MONOLITHOLOGIC DEBRIS AND ACTIVE FAN GRAVELS: BEDROCK (STIPPLE)

v Dacite – rhyolite   **q** Quartzite   **a** Argillaceous rocks

b Basalt   **c** Carbonate rocks   **m** Mixed compositions

ALLUVIAL FAN GRAVELS

Unweathered	Lightly weathered	Heavily weathered	Location
ft_4 sh, qt	ft_3 sh, dv	ft_2 do, qt	Tucki Wash
fb_4 qt, sh	fb_3 qt, sh dv	fb_2 qt, dv	Blackwater Wash
	fbt_3 qt, sh dv	fbt_2 qt, dv do	Mixed sources: Blackwater and Tucki Washes
	fc_3 do	fc_2 do	(carbonates)
	fm_3 ——	fm_2 ——	(mixed)

sh: shale qt: quartzite do: dolomite dv: desert varnish

Figure 4-14. Thermal infrared image of northern Death Valley acquired by TIMS. See color section. (From Kahle and Goetz, 1983.)

quirement for the NEDT was less than 0.3 K for all bands with a design goal of less than 0.2 K. The total data rate for the ASTER thermal infrared subsystem, including supplementary telemetry and engineering telemetry, is 4.2 Mbps. Because the TIR subsystem can return useful data both day and night, the duty cycle for this subsystem is set at 16%. The mechanical cryo cooler, like that of the SWIR subsystem, operates with a 100% duty cycle.

Figure 4-15 show images of the five ASTER thermal infrared channels for the Cuprite, Nevada scene discussed in the previous chapter. Also shown is the image for visible band 1 for comparison. Note the significant difference in resolution between the visible image and the thermal images. Also, the thermal images appear to have much less contrast than the visible image. Note the significant topographic effects shown in the thermal images near the right-hand edge of the images. The light and dark contrasts resemble shadowing in visible images. In this case, however, the effects are not shadowing; they are due to differential heating of the slopes. Those slopes facing the sun are heated more efficiently than those facing away from the sun. The result is that slopes facing the sun appear brighter in the thermal images, whereas those facing away from the sun appear darker.

Figure 4-16 shows the visible image, plus the five thermal infrared principal component images. The first principal component image contains 99.73% of the variation, followed by PC2 with 2.04%. The last three principal component images are progressively more dominated by scan line noise. Figure 4-17 shows a color image displaying the first three principal components as red, green, and blue, respectively. The image was sharp-

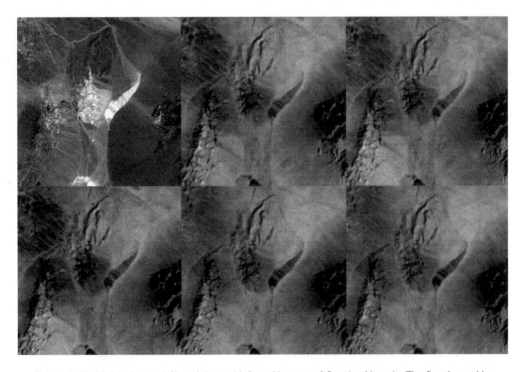

Figure 4-15. Visible (upper left) and thermal infrared images of Cuprite, Nevada. The five thermal infrared images follow in sequence from band 10 to band 14, starting in the middle of the upper row, and increasing from left to right.

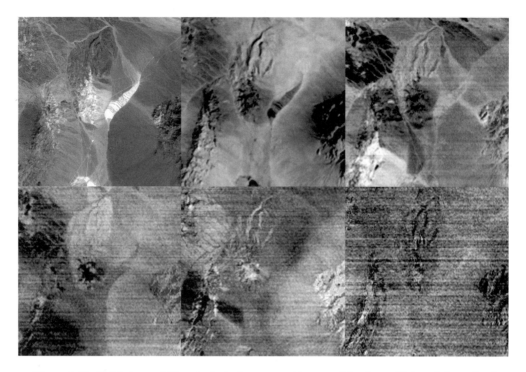

Figure 4-16. Visible (upper left) and principal component images of the thermal infrared channels of the Cuprite, Nevada scene. The principal components are displayed in decreasing order starting in the middle of the upper row, and decreasing from left to right.

ened using an HSV transform as described in Chapter 3, and the brightness channel was replaced by the visible image shown in Figure 4-16. Note the excellent detail in the alteration area to the right of Highway 95.

Color combination images are often used to enhance the interpretation of ASTER thermal infrared images. One common combination displays bands 13, 12, and 10 as red, green, and blue, respectively. In this combination, variations in quartz content appear as more or less red, carbonate rocks are green, and mafic volcanic rocks are purple.

4-6-4 Spitzer Space Telescope

The Spitzer Space Telescope [previously known as the Space Infrared Telescope Facility (SIRTF)], the fourth and last of NASA's so-called Great Observatories, was launched from Cape Canaveral, Florida into an Earth-trailing heliocentric orbit on 25 August 2003. Even though this observatory strictly speaking is not used for solid surface sensing, which is the topic of this chapter, we include a brief discussion here to illustrate some of the points mentioned earlier regarding cooling and detectors for infrared imaging.

The Spitzer observatory introduced a new paradigm for cryogenic missions in which the telescope was launched at ambient temperature; only the focal plane instruments were cooled to cryogenic temperatures. This "warm launch" architecture, coupled with the heliocentric orbit and careful design of the cryostat, allow a very long cryogenic lifetime (predicted to be longer than 5 years) with a relatively small amount of cryogen (about 360

Figure 4-17. Sharpened color thermal infrared principal component image of the Cuprite, Nevada scene. The colors are PC1 (red), PC2 (green), and PC3 (blue). The image was sharpened using the HSV transform in which the visible image was used as the brightness layer.

liters of superfluid helium). Following the launch, the telescope gradually cooled to approximately 6 K over a period of about 45 days. A telescope temperature near 6 K is required to bring the telescope thermal background down low enough for observing at the longest Spitzer wavelengths. The helium bath temperature is maintained at 1.24 K, and the outer shell temperature is approximately 34 K.

The Spitzer telescope is a Ritchey–Chrétien design with an 85 cm beryllium primary mirror, with diffraction limited performance at 5.5 microns achieved on orbit. The rest of the telescope is built entirely of beryllium, to produce a lightweight telescope capable of stable operation at cryogenic temperatures. The focal plane is populated by three instruments. The InfraRed Array Camera (IRAC) provides images at 3.6, 4.5, 5.8, and 8.0 microns. All four detector arrays are 256×256 square pixels using indium antimonide (InSB) detectors at 3.6 and 4.5 microns, and arsenic doped silicon (Si:As) detectors for the remaining two bands. The InfraRed Spectrograph (IRS) performs both low- and high-resolution spectroscopy. Low-resolution spectra can be obtained from 5.2 to 38.0 microns. High-resolution spectra can be obtained from 9.9 to 37.2 microns. The spectrograph consists of four modules, each of which is built around a 128×128 pixel array. The detectors are arsenic-doped silicon (Si:As) at the lower half of the bandwidth for each of the low- and high-resolution spectrometers, and antimonide-doped silicon (Si:Sb) at the upper half of the bands for both. The Multiband Imaging Photometer for Spitzer (MIPS) is designed to provide photometry and superresolution imaging as well as efficient mapping capabilities in three wavelength bands centered near 24, 70, and 160 microns. The MIPS 24 micron detector is a 128×128 pixel arsenic-doped silicon (Si:As) array, identical to those used by the IRS instrument. MIPS uses a 32×32 gallium-doped germanium (Ge:Ga) array for 70 microns and a 2×20 Ge:Ga array, mechanically stressed to extend its photoconductive response to 160 microns, On-board calibrators are provided for each array. Additionally, MIPS has a scan mirror to provide mapping with very efficient use of telescope time.

Cryogenic missions like Spitzer have a finite lifetime. Therefore, it is crucial to plan observations such that both the science return and mission lifetime are maximized. In the case of the Spitzer telescope, this is achieved by operating the three instruments in the following sequence: IRAC for about 9 days, followed by MIPS for about 8 days, and, finally, IRS for about 6 days. Since MIPS makes observations at the longest wavelengths, it requires the telescope to be the coldest. The telescope is, therefore, cooled down just prior to the MIPS campaign. Once the MIPS campaign is over, the telescope temperature is allowed to drift upward slowly, reducing the amount of cryogen needed. Since the IRAC instrument operates at the shortest wavelengths, it can tolerate the increased telescope temperatures best. Using this scheme, it is expected that the Spitzer telescope will be operating for longer than 5 years before the superfluid helium will be exhausted. Once this happens, the telescope temperature will naturally increase, and observations at the longer wavelengths will no longer be possible. It is expected, however, that the shorter wavelength bands of the IRAC instrument will still be able to acquire useful images, even at the elevated telescope temperatures.

4-6-5 2001 Mars Odyssey Thermal Emission Imaging System (THEMIS)

THEMIS is one of five instruments on the Mars Odyssey spacecraft that was launched on April 7, 2001, and reached Mars on October 23, 2001. After three months of using the Martian atmosphere to aerobrake, it finally reached an nearly circular orbit of approximately 400 km altitude.

The THEMIS instrument consists of two pushbroom spectrometers: one that operates in the visible (5 bands), and one that operates in the thermal infrared (9 distinct bands) part of the spectrum. The telescope is an all-reflective, three-mirror f/1.7 anastigmatic design with an effective aperture of 12 cm and an effective focal length of 20 cm. A dichroic beam splitter separates the visible and infrared radiation. The infrared detector is a 320×240 silicon microbolometer array with a field of view of 4.6° crosstrack by 3.5° downtrack. The thermal infrared bands are centered at 6.78, 7.93, 8.56, 9.35, 10.21, 11.04, 11.79, 12.57, and 14.88 microns. The resolution from orbit is about 100 m/pixel, and the images cover a swath that is 32 km wide. The instrument mass is about 12 kg, and it consumes an average of 12 W of power. The compressed data rate from the infrared instrument to the spacecraft is 0.6 Mbits/sec.

Figure 4-18 shows a day/night pair of THEMIS images of crater ejecta in the Terra Meridiani region on Mars. The daytime image brightness is affected by both the morphological and physical properties of the surface. Morphological details are enhanced by the fact the slopes facing the sun are warmer than those that face away from the sun. This difference in temperature mimics the shadowing seen in visible images. In these images, the physical properties are dominated by the density of the materials. The dust-covered areas heat up faster than the exposed rocks. Therefore, the brighter areas in the daytime images are typically dust-covered, whereas the darker areas are rocks. Infrared images taken during the nighttime exhibit only the thermo-physical properties of the surface. The effect of differential heating of sun-facing versus non-sun-facing areas dissipates quickly at night. The result is that physical properties dominate as different surfaces cool at different rates through the nighttime hours. Rocks typically have higher thermal inertia, cool slowly, and are, therefore, relatively bright at night. Dust and other fine-grained materials have low thermal inertia, cool very quickly, and are dark in nighttime infrared images.

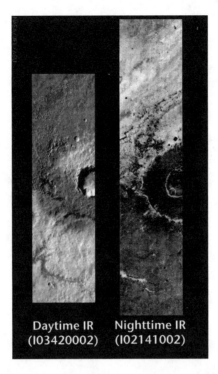

Figure 4-18. Day and night thermal infrared images of crater ejecta in the Terra Meridiani region on Mars, acquired by the THEMIS instrument on the Mars Odyssey spacecraft. (Courtesy NASA/JPL/ Arizona State University.)

Figure 4-19 shows a false-color THEMIS infrared image that was acquired over the region of Ophir and Candor Chasma in Valles Marineris on Mars. The image was constructed using infrared filters centered at 6.3, 7.4, and 8.7 microns. The color differences in this image represent compositional differences in the rocks, sediments, and dust that occur in this region of Mars.

4-6-6 Advanced Very High Resolution Radiometer (AVHRR)

The AVHRR instrument is a cross-track scanning radiometer that measures emitted radiation in four (for the AVHRR/1) to six (for the latest AVHRR/3) bands ranging from the blue visible band to the thermal infrared. Table 4-3 shows the imaging bands and their main application for the AVHRR/3 instrument.

The AVHRR instruments are carried on the polar orbiting satellites operated by the United States National Oceanic and Atmospheric Administration (NOAA). The satellites orbit at an altitude of 833 km, with an orbit inclination of 98.9°, and complete 14 orbits each day. The AVHRR telescope is an 8 inch afocal, all-reflective Cassegrain system. Cross-track scanning is provided by an elliptical beryllium mirror rotating at 360 rpm (six times per second) about an axis parallel to the Earth. Each scan spans an angle of $\pm 55.4°$ from the nadir, for a total swath width of 2399 km. The instantaneous field-of-view (IFOV) of each sensor is approximately 1.4 milliradians, giving a resolution of about 1.1 km at the nadir point.

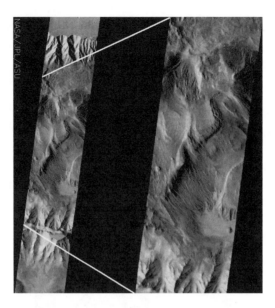

Figure 4-19. False-color THEMIS infrared image of the Ophir and Candor Chasma region of Valles Marineris on Mars. See color section. (Courtesy NASA/JPL/Arizona State University.)

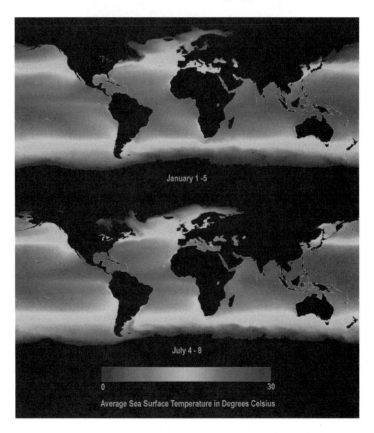

Figure 4-20. Mean sea surface temperature for the period 1987–1999. The top panel is for the period January 1–5, and the bottom panel is for July 4–8. (Courtesy NASA/JPL-Caltech.) See color section.

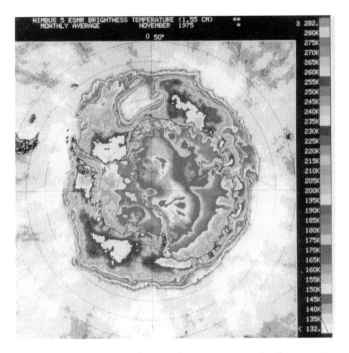

Figure 1-6. Passive microwave image of Antarctic ice cover acquired with a spaceborne radiometer. The color chart corresponds to the surface brightness temperature.

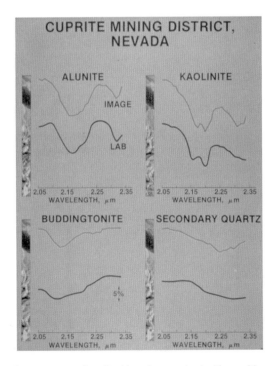

Figure 1-10. Images of an area near Cuprite, Nevada, acquired with an airborne imaging spectrometer. The image is shown to the left. The spectral curves derived from the image data are compared to the spectral curves measured in the laboratory using samples from the same area. (Courtesy of JPL.)

Introduction to the Physics and Techniques of Remote Sensing. By C. Elachi and J. van Zyl
Copyright © 2006 John Wiley & Sons, Inc.

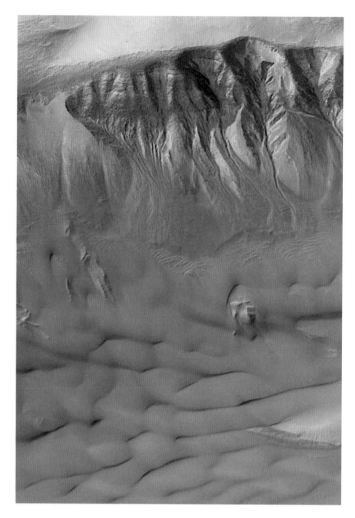

Figure 3-41. The picture shown here was taken by the Mars Orbiter Camera narrow angle (high resolution) camera and "colorized" by applying the colors of Mars obtained by the MOC wide angle cameras. The picture shows gullies in a crater at 42.4°S, 158.2°W, which exhibits patches of wintertime frost on the crater wall, and dark-toned sand dunes on the floor. (Courtesy NASA/JPL/Malin Space Science Systems.)

Figure 3-42. Data from the Mars Exploration Rover Opportunity's panoramic camera's near-infrared, blue, and green filters were combined to create this approximate, true-color image of the rock outcrop near the rover's landing site. (Courtesy NASA/JPL/Cornell.)

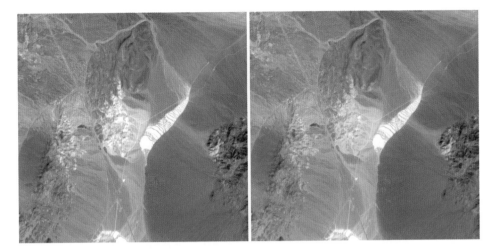

Figure 3-46. Two color combination displays for the Cuprite scene shown in Figure 3-45. On the left are ASTER channels 1, 2, and3 displayed as blue, green, and red, respectively, giving a pseudonatural color image. On the right are channels 1, 3, and 6 displayed as blue, green, and red.

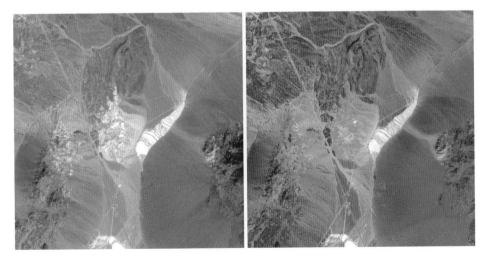

Figure 3-47. The same images shown in Figure 3-46 after performing a color stretch using the IHS transform. Note the dramatic increase in color separation evident in both images.

Figure 3-49. The principal components PC2, PC3, and PC4 are displayed as blue, green, and red, respectively in this color combination of the Cuprite scene. Notice the excellent color separation compared to Figures 3-46 and 3-47.

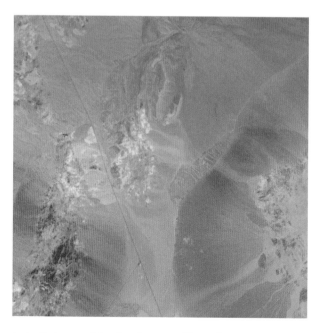

Figure 3-51. Spectral ratio image of the Cuprite scene. The ratios are 4/7 (red), 3/1 (green), and 4/3 (blue).

Figure 3-52. Results of an unsupervised classification of the Cuprite scene. The classification was initialized using randomly chosen features, and then iterated. The number of classes was arbitrarily set to six.

Figure 3-53. Results of a supervised classification of the Cuprite scene. The classification was initialized using features selected from a combination of the principal component results shown in Figure 3-49 and the unsupervised classification results in Figure 3-52. The number of classes was limited to four.

Figure 3-54. This image shows the relative abundances of different materials after linear unmixing. The spectra used in the supervised classification were used as endmembers.

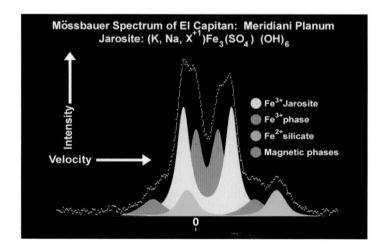

Figure 3-56. This graph shows a spectrum, taken by the Mars Exploration Rover Opportunity's Mössbauer spectrometer. The Mössbauer spectrometer graph shows the presence of an iron-bearing mineral called jarosite. (Courtesy NASA/JPL/University of Mainz.)

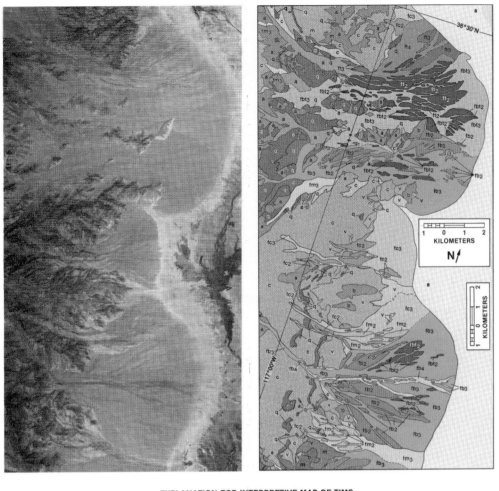

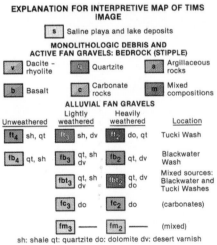

EXPLANATION FOR INTERPRETIVE MAP OF TIMS IMAGE

| s | Saline playa and lake deposits |

MONOLITHOLOGIC DEBRIS AND ACTIVE FAN GRAVELS: BEDROCK (STIPPLE)

| v | Dacite – rhyolite | q | Quartzite | a | Argillaceous rocks |
| b | Basalt | c | Carbonate rocks | m | Mixed compositions |

ALLUVIAL FAN GRAVELS

Unweathered	Lightly weathered	Heavily weathered	Location
ft_4 sh, qt	ft_3 sh, dv	ft_2 do, qt	Tucki Wash
fb_4 qt, sh	fb_3 qt, sh dv	fb_2 qt, dv	Blackwater Wash
	fbt_3 qt, sh dv	fbt_2 qt, dv do	Mixed sources: Blackwater and Tucki Washes
	fc_3 do	fc_2 do	(carbonates)
	fm_3 ——	fm_2 ——	(mixed)

sh: shale qt: quartzite do: dolomite dv: desert varnish

Figure 4-14. Thermal infrared image of northern Death Valley acquired by TIMS. (From Kahle and Goetz, 1983.)

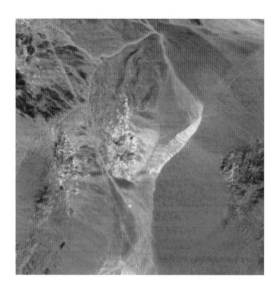

Figure 4-17. Sharpened color thermal infrared principal component image of the Cuprite, Nevada scene. The colors are PC1 (red), PC2 (green), and PC3 (blue). The image was sharpened using the HSV transform in which the visible image was used as the brightness layer.

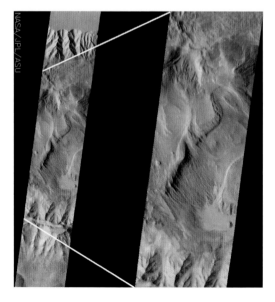

Figure 4-19. False-color THEMIS infrared image of the Ophir and Candor Chasma region of Valles Marineris on Mars. (Courtesy NASA/JPL/Arizona State University.)

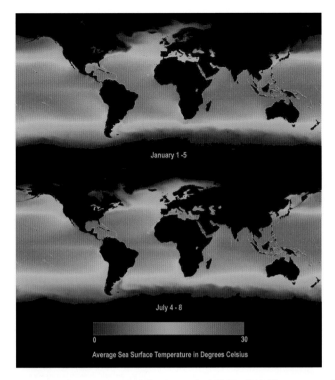

Figure 4-20. Mean sea surface temperature for the period 1987–1999. The top panel is for the period January 1–5, and the bottom panel is for July 4–8. (Courtesy NASA/JPL-Caltech.)

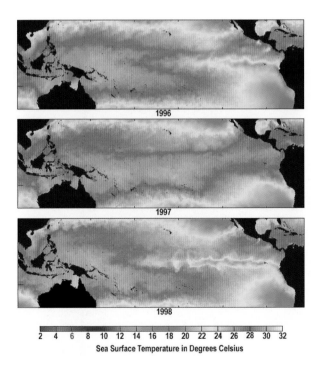

Figure 4-21. Weekly averages of the sea surface temperature for a portion of the Pacific Ocean for the last week in December for the years 1996, 1997, and 1998. The anomalously high sea surface temperature in 1997 is associated with the El Niño weather pattern. (Courtesy NASA/JPL-Caltech.)

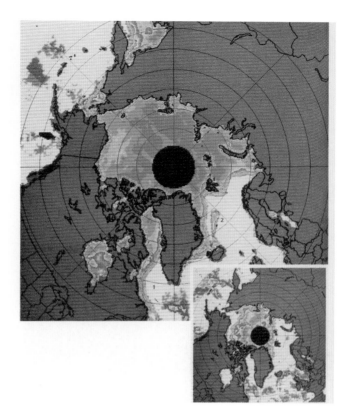

Figure 5-4. Microwave images of the north polar region. The main image corresponds to the maximum extent of ice in the winter. The inset corresponds to its minimum extent in the summer. See color section.

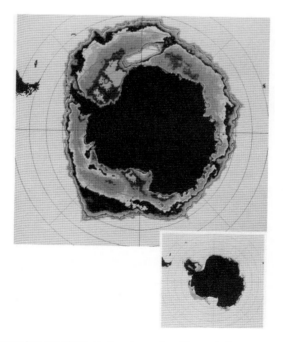

Figure 5-5. Microwave images of the south polar region. The main image corresponds to the maximum extent of ice in the local winter. The inset corresponds to its minimum extent in the local summer.

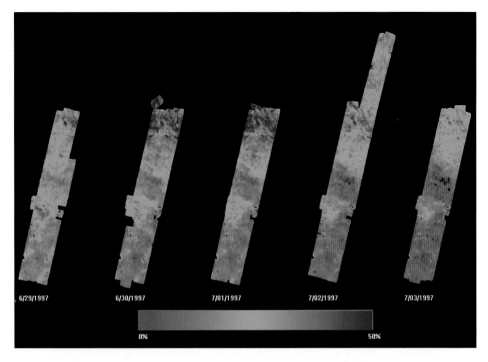

Figure 5-6. Soil moisture distribution measured with the ESTAR radiometer as part of the Southern Great Plains Experiment in Oklahoma in 1997. (Courtesy NASA/JPL-Caltech.)

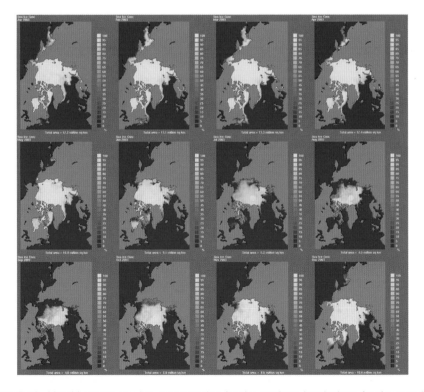

Figure 5-18. Monthly mean sea ice concentration for the northern hemisphere for the year 2003 (Fetterer and Knowles, 2002).

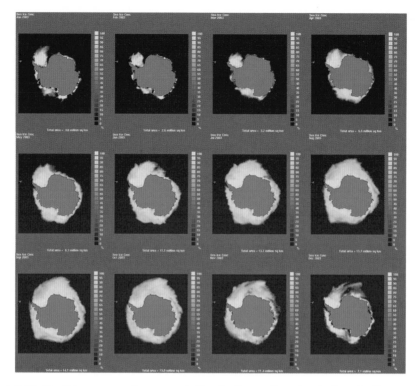

Figure 5-19. Monthly mean sea ice concentration for the southern hemisphere for the year 2003 (Fetterer and Knowles, 2002).

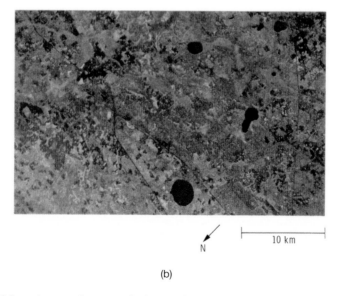

(b)

Figure 6-8. (b) Spaceborne radar composite image of the same area in Florida acquired at three different angles. The color was generated by assigning blue, green, and red to the images acquired at 28, 45, and 58° incidence angles, respectively.

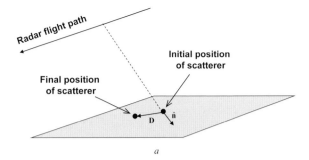

b

Figure 6-50. (a) Along-track interferometry imaging geometry. (b) Interferogram acquired over the Straits of Juan de Fuca with the NASA/JPL AIRSAR system in 1999. The interferogram shown was acquired with the L-band system, with a baseline of 20 m in the along-track direction. Given the normal flight parameters of the NASA DC-8 aircraft, this image has a velocity ambiguity of 2.4 m/s. (Courtesy NASA/JPL-Caltech.)

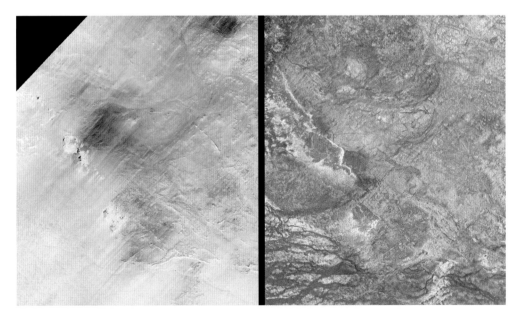

Figure 6-56. A comparison of images of the Safsaf Oasis area in south-central Egypt. On the left is a LandSat Thematic Mapper image showing bands 7, 4, and 1 displayed as red, green, and blue. The image on the right is from the SIR-C/X-SAR system, displaying L-band HH, C-band HH, and X-band VV as red, green, and blue, respectively. Each image represents an area of approximately 30 km by 25 km. (Courtesy NASA/JPL-Caltech.)

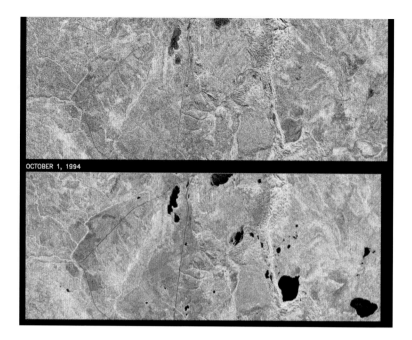

OCTOBER 1, 1994

Figure 6-57. Seasonal images of the Price Albert area in Canada. Both images were acquired with the SIR-C/X-SAR system and display the L-band return in red, the C-band return in green, and the X-band return in blue. The image on the top was acquired on April 10, 2004, and the one on the bottom on October 1, 1994. (Courtesy NASA/JPL-Caltech.)

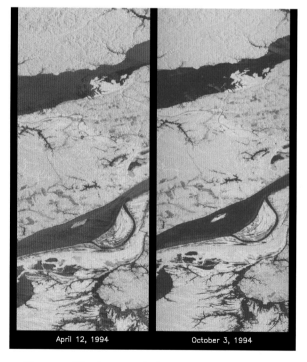

April 12, 1994   October 3, 1994

Figure 6-58. Seasonal images of the Amazon rain forest near Manaus, Brazil. Both images were acquired with the SIR-C/X-SAR system, and display the L-band HH return in red, the L-band HV return in green, and the inverse of the L-band VV return in blue. The image on the left was acquired on April 12, 2004, and the one on the right on October 3, 1994. The images are about 8 km wide, and 25 km long. The yellow and red areas represent flooding under the forest canopy. (Courtesy NASA/JPL-Caltech.)

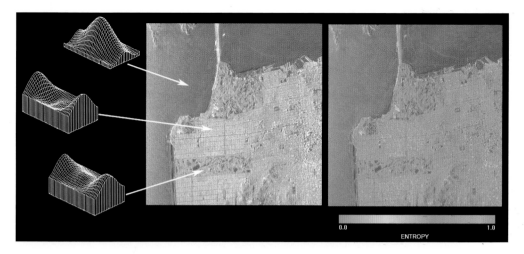

Figure 6-59. The polarization responses for three different areas in the San Francisco L-band image (acquired with the NASA/JPL AIRSAR system) as indicated by the arrows are shown here for comparison. The figure on the right shows the entropy of the covariance matrix for each multilook pixel. The entropy is a measure of the randomness contained in the covariance matrix.

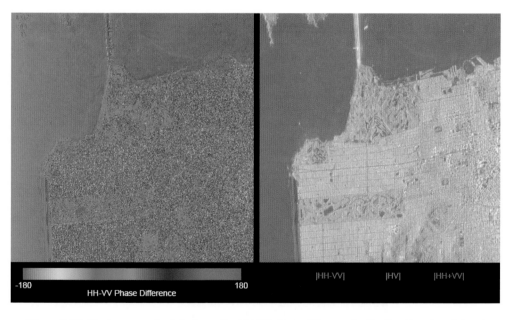

Figure 6-60. The image on the left shows the HH-VV phase difference for the San Francisco L-band image. The display on the right highlights areas with phase differences near 0° in blue, those with phase differences near 180° in red, and areas with random phases in green.

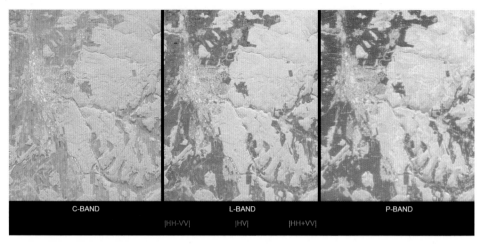

C-BAND   L-BAND   P-BAND

|HH-VV| |HV| |HH+VV|

Figure 6-61. Three frequency images of a portion of the Black Forest in Germany (acquired with the NASA/JPL AIRSAR system). The town of Villingen is shown as the red areas in the left half of the images. The dark areas surrounding the town are agricultural fields, and the brighter areas in the right half of the images are coniferous forests. Radar illumination is from the top.

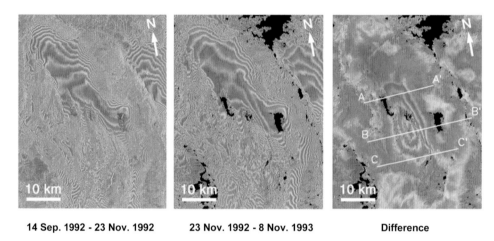

14 Sep. 1992 - 23 Nov. 1992   **23 Nov. 1992 - 8 Nov. 1993**   **Difference**

Figure 6-62. Deformation signals measured at C-band following the M 6.1 Eureka Valley earthquake in California. The left two images are the individual interferograms constructed from three acquisitions. The earthquake occurred between the second and third acquisitions. (Reprinted from *Science,* Peltzer and Rosen, 1995.)

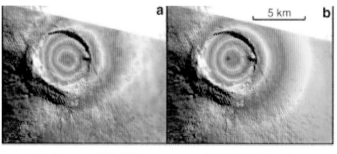

Interferogram 1992-1998   **Predicted deformation**

Figure 6-63. Observed and predicted deformation signals on Darwin volcano in the Galapagos Islands. The prediction assumes a Mogi point source at 3 km depth. (Reprinted from *Nature.* Amelung et al., 1995.)

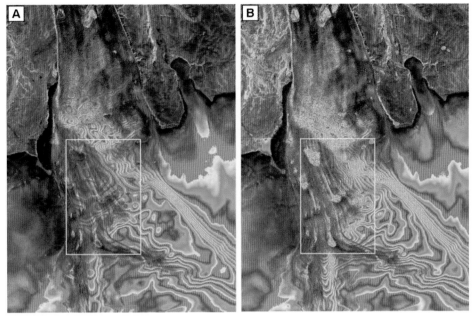

21-22 September 1995　　　　**26-27 October 1995**

Figure 6-64. Two interferograms over the Ryder Glacier in Greenland acquired with the ERS-1 radar system. The interferogram on the right shows a dramatic increase in the speed of the glacier, as evidenced by the closer spacing in the interference fringes in this image as compared to the one on the left. (Reprinted with permission from Joughin et al., 1996b. Copyright 1996 AAAS.)

 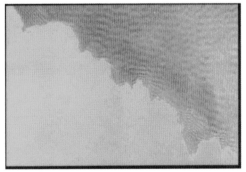

Radar Backscatter　　　　**Along-Track Interferometric Phase**

Figure 6-67. Radar backscatter (left) and along-track interferometric phase (right) of a section of the Hawaiian coastline. Red colors mean the surface is moving away from the radar; illumination is from the top. The complicated wave patterns are only visible in the phase image. (Courtesy NASA/JPL-Caltech.)

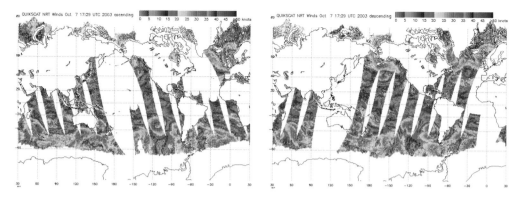

Figure 6-74. Image swaths acquired by the SeaWinds instrument over a 24 hour period ending on October 7, 2003. The left image shows the ascending swaths; the right image shows the descending swaths. (Courtesy NASA/JPL-Caltech.)

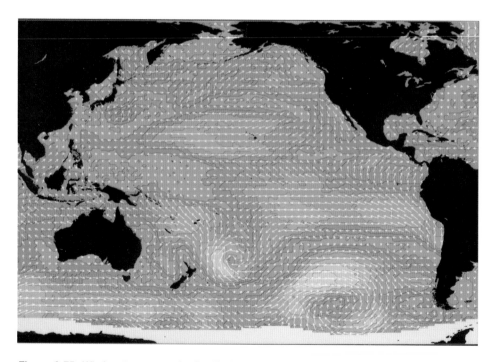

Figure 6-75. Wind patterns over the Pacific for July 1978 derived from the SASS measurement. (Courtesy of P. Woicesyhn, Jet Propulsion Laboratory.)

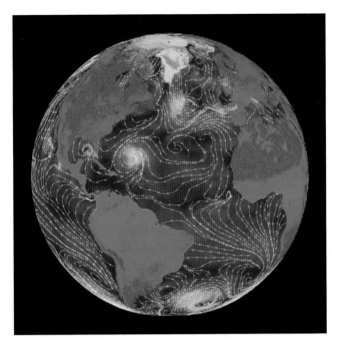

Figure 6-77. Global ocean winds as measured by QuikScat on September 20, 1999. Colors over the ocean indicate wind speed, with orange as the fastest wind speeds and blue as the slowest. White streamlines indicate the wind direction. (Courtesy NASA/JPL-Caltech.)

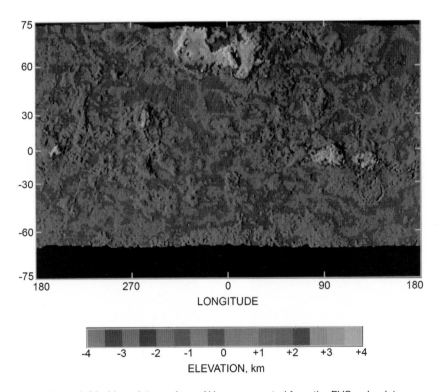

Figure 6-86. Map of the surface of Venus generated from the PVO radar data.

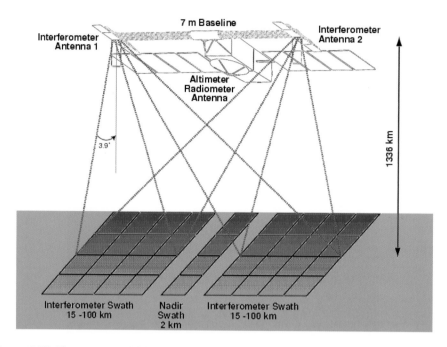

Figure 6-87. The proposed WSOA instrument concept integrated with the Jason altimeter on the Proteus satellite bus. (Courtesy NASA/JPL-Caltech.)

Conventional Altimeter Data

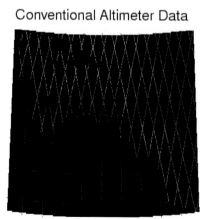

Wide-Swath Altimeter Data

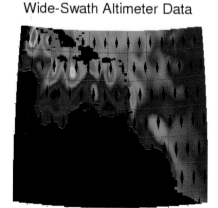

Figure 6-88. A comparison of the coverage between the TOPEX/POSEIDON conventional altimeter (left) and the proposed WSOA instrument (right) in the same orbit. (Courtesy NASA/JPL-Caltech.)

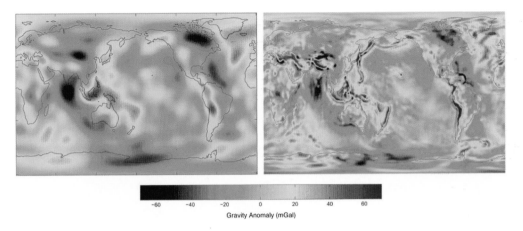

Figure 7-7. Comparison of gravity anomaly models before the GRACE mission (left) and after 363 days of GRACE data (right). The increase in resolution is clearly seen in the image on the right. The unit mGal corresponds to 0.00001 m/s^2. (Courtesy University of Texas Center for Space Research.)

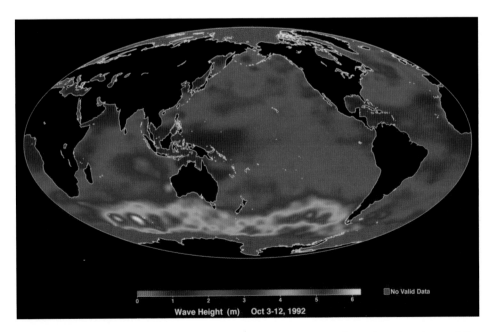

Figure 7-10. Global measurement of surface wave height derived from the echo shape of the TOPEX/POSEIDON altimeter. (Courtesy NASA/JPL-Caltech.)

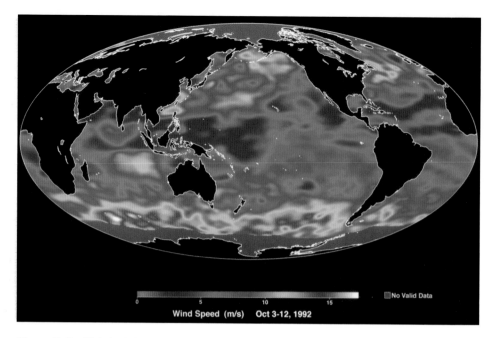

Figure 7-12. Global wind-speed measurement derived from the TOPEX/Poseidon altimeter measurements. (Courtesy NASA/JPL-Caltech.)

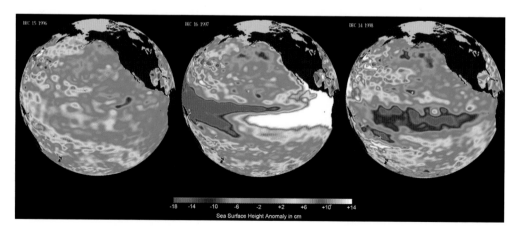

Figure 7-13. This figure shows sea surface height anomalies measured by the TOPEX/Poseidon mission in December 1996 (left), December 1997 (middle), and December 1998 (right). Sea surface height anomalies are the difference between the dynamic and static ocean topography. The left image shows a "normal" year, the middle an El Niño event, and the image on the right a La Niña pattern. (Courtesy NASA/JPL-Caltech.)

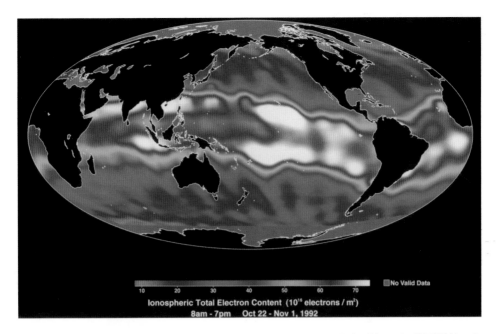

Figure 7-14. Global ionospheric total electron content measurement derived from the TOPEX/Poseidon altimeter measurements. (Courtesy NASA/JPL-Caltech.)

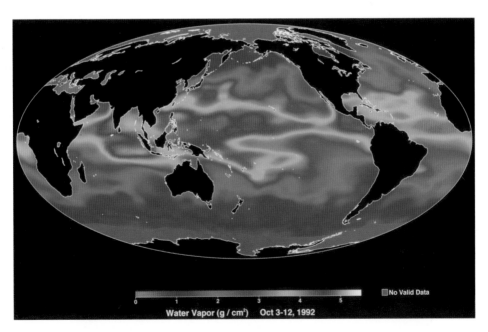

Figure 7-15. Global atmospheric water vapor distribution derived from the TOPEX/Poseidon radiometer measurements. (Courtesy NASA/JPL-Caltech.)

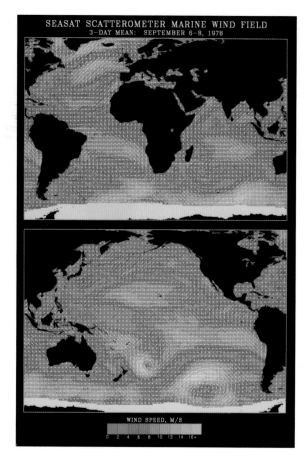

Figure 7-18. Average global winds derived from the Seasat scatterometer for September 6–8, 1978. See color section. (Courtesy of P. Woceishyn, Jet Propulsion Laboratory.)

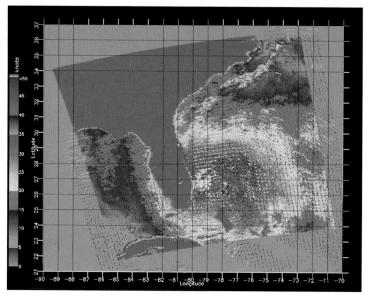

Figure 7-19. SeaWinds measurements of the winds associated with Hurricane Frances approaching the coast of Florida on September 4, 2004. (Courtesy NASA/JPL-Caltech.)

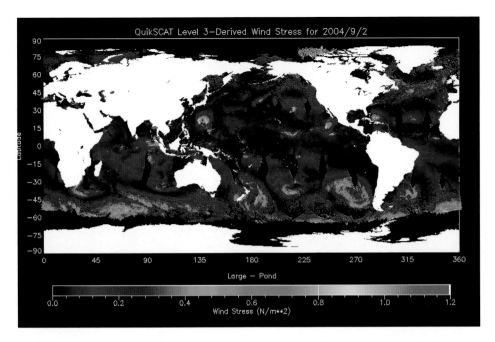

Figure 7-20. SeaWinds measurements of the global wind stress for September 2, 2004. Hurricane Frances is visible near Florida, while Typhoon Songda is approaching Japan. (Courtesy NASA/JPL-Caltech.)

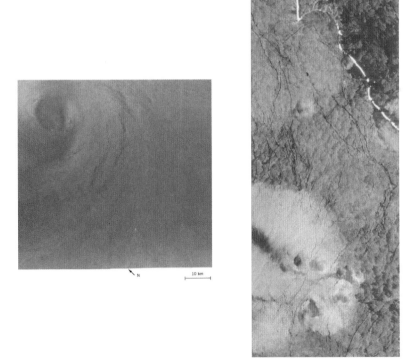

Figure 7-27. Shown on the left is a small-scale, well-organized tropical storm near the mouth of the Gulf of California. Except for its small size, the storm has all the characteristics of a hurricane, for example, the cyclonic spiral arms and a well-defined center of low winds. The image on the right shows an intense rain storm in the western Pacific imaged by the SIR-C radar. The storm is the yellowish area near the bottom of the image. See text for more discussion. (Courtesy NASA/JPL-Caltech.)

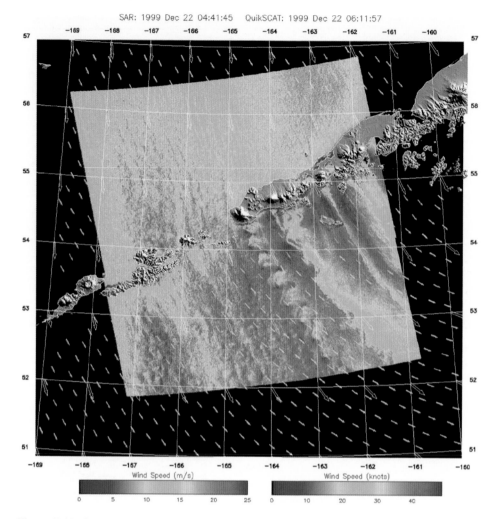

Figure 7-29. Combined QuikSCAT scatterometer and RADARSAT SAR wind-speed products reported by Monaldo et al. (2004). The small arrows represent the color-coded scatterometer wind vectors. The background image is the SAR-retrieved wind-speed field. The large arrows represent the NOGAPS model wind directions used to initialize the SAR wind-speed retrievals. When the QuikSCAT arrows blend into the SAR image, the wind speeds agree. This wind field covers a portion of the Aleutian Islands. (From Monaldo et al. © 2004 IEEE.)

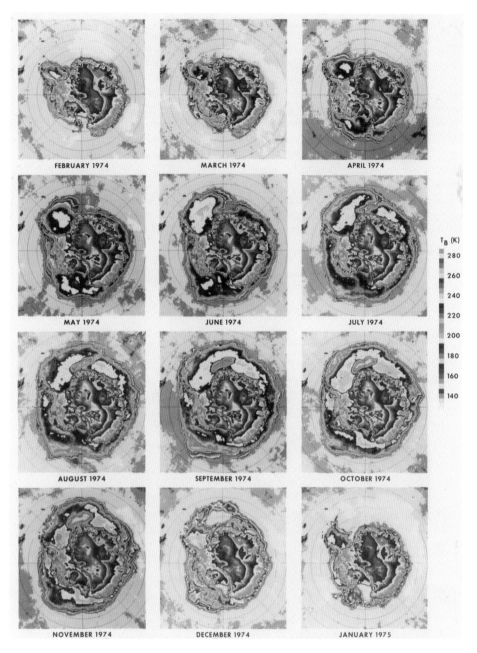

Figure 7-32. Passive microwave imagery of floating ice acquired by a spaceborne sensor. The twelve images show the thermally emitted radiance temperature (color code on the right) at a microwave frequency of 19.35 GHz. The images show the change of sea ice coverage from a minimum extent in January/February to a maximum extent in September.

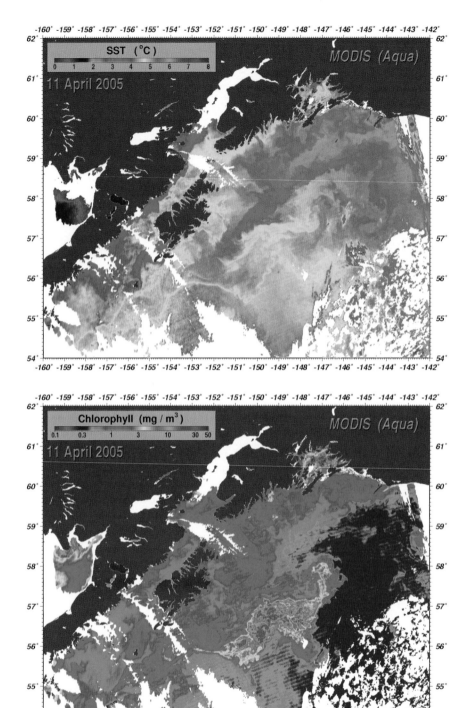

Figure 7-34. Measurements of sea surface temperature (top) and chlorophyll concentration (bottom) off the coast of Alaska made with the MODIS instrument on the NASA Aqua spacecraft. High concentrations of chlorophyll are shown in red. The images were acquired on 11 April 2005. (Courtesy NASA/Goddard Space Flight Center.)

ClO and O3 (column abundances) from UARS MLS

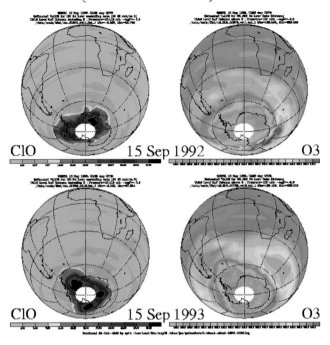

Figure 9-13. Examples of ClO and O_3 measurements made with the MLS instrument on the UARS satellite. (Courtesy NASA/JPL-Caltech.)

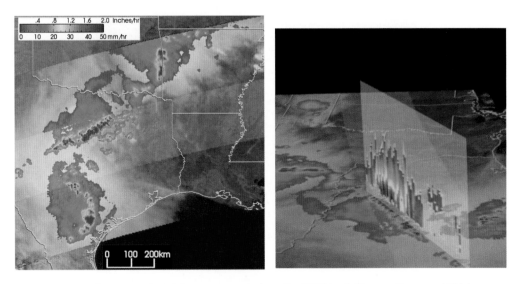

Figure 9-21. Examples of rainfall measurements from the TRMM satellite over Texas and Oklahma. The image on the left shows the horizontal distribution of rainfall, whereas the image on the right shows the vertical structure in the rainfall for one of the three storms visible in the left image. (Courtesy NASA/Goddard Space Flight Center.)

Figure 11-24. MISR measurements of cloud heights associated with hurricanes Frances and Ivan. The panel on the left in each case is a natural color image from the nadir-looking camera. (Courtesy NASA/JPL-Caltech.)

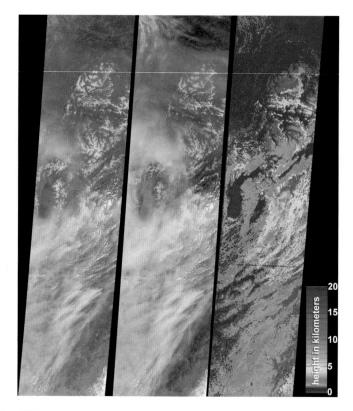

Figure 11-25. MISR measurements of the extent and height of smoke from numerous fires in the Lake Baikal region on June 11, 2003 are shown in the panel on the right. Areas where heights could not be retrieved are shown as dark gray. See text for detailed discussion. (Courtesy NASA/JPL-Caltech.)

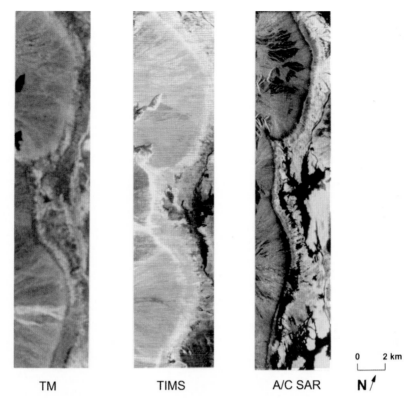

Figure A-1. Images of Death Valley, California acquired in the three major regions of the electromagnetic spectrum: (*a*) Landsat image acquired in visible/NIR, (*b*) airborne TIMS image acquired in the thermal IR, and (*c*) airborne SAR image acquired in the microwave.

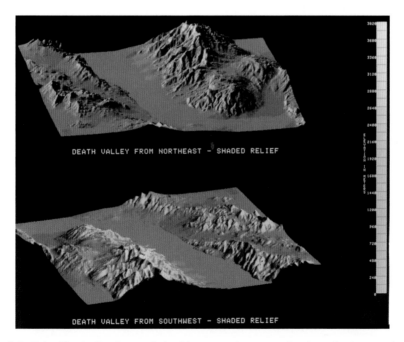

Figure A-2. False illumination image derived by computer processing from the topography database.

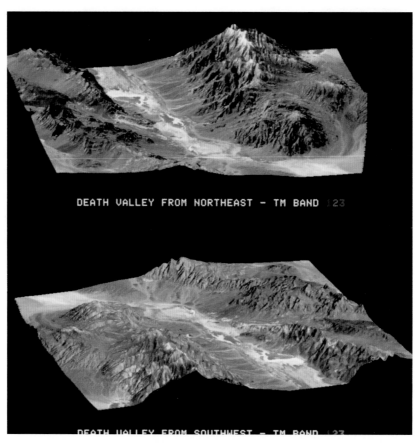

DEATH VALLEY FROM NORTHEAST — TM BAND 123

DEATH VALLEY FROM SOUTHWEST — TM BAND 123

Figure A-3. Two perspective views generated from a combination of the Landsat image and the digital topography base. (Courtesy of M. Kobrick, JPL.)

COTTONBALL BASIN AND TELESCOPE PEAK

LANDSAT TM BAND 123
2X VERTICAL EXAGGERATION
VIEW FROM NORTHEAST

Figure A-4. Surface perspective view generated from a combination of the Landsat image and the digital topography base. (Courtesy of M. Kobrick, JPL.)

The thermal infrared channels of the AVHRR imagers are used to produce maps of sea surface temperature (McClain et al., 1985, Walton, 1988). Figure 4-20 shows the mean sea surface temperature for the years 1987–1999 for January 1–5, and July 4–8, respectively. Clear seasonal differences are evident in these images, with warmer water extending further north in July, and further south in January. Figure 4-21 shows the sea surface temperature for the last week in December for the years 1996, 1997, and 1998. The anomalously high sea surface temperatures associated with the El Niño weather pattern in 1997 are clearly shown in the middle panel.

Table 4-3. AVHRR Instrument Parameters

Band number	Wavelength in microns	IFOV in milliradians	Typical use
1	0.58–0.68	1.39	Daytime cloud and surface mapping
2	0.725–1.1	1.41	Land–water boundaries
3(A)	1.58–1.64	1.30	Snow and ice detection
3(B)	3.55–3.93	1.51	Night cloud mapping; sea surface temperature
4	10.30–11.30	1.41	Night cloud mapping; sea surface temperature
5	11.50–12.50	1.30	Sea surface temperature

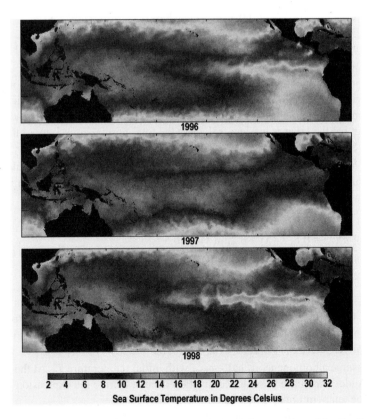

Figure 4-21. Weekly averages of the sea surface temperature for a portion of the Pacific Ocean for the last week in December for the years 1996, 1997, and 1998. The anomalously high sea surface temperature in 1997 is associated with the El Niño weather pattern. (Courtesy NASA/JPL-Caltech.) See color section.

EXERCISES

4-1. Plot the ratio of the thermal emission from two blackbodies with temperatures of 300 K and 6000 K. Cover only the spectral region from 5 μm to 25 μm.

4-2. Compare the reflected and emitted energy from the Earth's surface over the 5 μm to 25 μm spectral region. Consider the case of surface material with reflectivity equal to 0.1, 0.5, and 0.9. Assume that the sun is a blackbody with a temperature of 6000 K, and that the total energy density at the Earth's surface is 1.1. kW/m². The earth is a graybody with a temperature of 300 K. Neglect the effects of the atmosphere.

4-3. Repeat the previous exercise for the case of Mercury, Mars, and Europa (one of the Jovian satellites), assuming their surface temperatures to be 640 K, 200 K, and 120 K, respectively.

4-4. Derive the expression for the wavelength of maximum emittance λ_m given in Equation 4-4. What is the corresponding wavelength for maximum photon emittance?

4-5. Calculate the total emissivity factor for a 300 K body characterized by the following spectral emissivity:

$$\varepsilon = \begin{cases} 0.1 & \text{for } -\Delta\lambda \leq \lambda \leq \Delta\lambda \\ 0.5 & \text{elsewhere} \end{cases}$$

Consider the cases of $\lambda_0 = 8$ μm, 10 μm, and 20 μm, and $\Delta\lambda = 0.1$ μm and 1 μm.

4-6. Consider a medium that consists of a half-space characterized by a normal emissivity of $\varepsilon_1(\lambda)$ when exposed to free space, overlaid by a layer of thickness L of a material with normal emissivity of $\varepsilon_2(\lambda)$ when exposed to free space. Assuming that the layer is lossless, calculate the total normal emissivity of the medium as a function of wavelength and L.

Now assume that the layer has a loss factor of $\alpha(\lambda)$. What would the total emissivity be in this case? Plot the total emissivity as a function of λ for the following cases:

(a) $\varepsilon_1(\lambda) = 0.2$, $\varepsilon_2(\lambda) = 0.6$, $\alpha(\lambda)L = 0.5$
(b) $\varepsilon_1(\lambda) = 0.6$, $\varepsilon_2(\lambda) = 0.2$, $\alpha(\lambda)L = 0.5$
(c) $\varepsilon_1(\lambda) = 0.6$, $\varepsilon_2(\lambda) = 0.2$, $\alpha(\lambda)L = 2.0$
(d) $\varepsilon_1(\lambda) = 0.6$, $\varepsilon_2(\lambda) = 0.2$, $\alpha(\lambda)L = 0.1$

4-7. Using Figure 4-5a, plot the difference $\Delta T = T(t) - T(t + 12 \text{ hours})$ as a function of times for different values of thermal inertia. What is the best time pair to achieve the best discrimination? Plot the corresponding ΔT as a function of thermal inertia P.

4-8. Assume a homogeneous half-space with initial temperature T_0. At the time $t = 0$, a sudden increase in solar illumination occurs. Derive the expression of the surface temperature as a function of time.

4-9. Consider a surfaces of the following materials: limestone, montmorillonite, kaolinite, olivine basalt, and quartz monzonite. Using Figure 4-10 select a set of observation wavelengths that would allow the discrimination and identification of these five materials. State the identification criteria.

4-10. Consider an imaging thermal sensor in orbit at altitude h. The filter/detector has a sensitive bandwidth $\Delta\lambda$ around the central response wavelength λ_0. The optics has a collector area A and is such that the detector covers a surface area S on the surface of the Earth. The surface emissivity is ε and its temperature is T.

(a) Derive the expression giving the total number of emitted photons N which are collected by the sensor.

(b) Plot N as a function of T for the following case: $\lambda_0 = 10$ μm, $\Delta\lambda = 1$ μm, $A = 400$ cm^2, $S = 0.25$ km^2, $h = 600$ km, and $\varepsilon = 0.8$.

(c) How sensitive should the detector be in order to detect changes of less than 1 K over the range of T from 270 K to 315 K.

(d) Repeat (b) and (c) for the case $S = 0.1$ km^2 and $\Delta\lambda = 0.25$ μm.

REFERENCES AND FURTHER READING

Abrams, M. The Advanced Spaceborne Thermal Emission and Reflection Radiometer (ASTER): Data products for the high spatial resolution imager on NASA's Terra platform. *International Journal of Remote Sensing,* **21,** 847–859, 2000.

Barton, I. J. Satellite-derived sea surface temperatures: Current status. *Journal of Geophysical Research,* **100,** 8777–8790, 1995.

Carslaw, H. S., and J. C. Jaeger. *Conduction of Heat in Solids.* University Press, Oxford, 1967.

Christensen, P. R., J. L. Bandfield, M. D. Smith, V. E. Hamilton, and R. N. Clark. Identification of a basaltic component on the Martian surface from Thermal Emission Spectrometer data. *Journal of Geophysical Research,* **105,** 9609–9621, 2000.

Christensen, P. R., et al., The Mars Global Surveyor Thermal Emission Spectrometer experiment: Investigation description and surface science results. *Journal of Geophysical Research,* **106,** 23, 823–871, 2001.

Christensen, P. R., R. V. Morris, M. D. Lane, J. L. Bandfield, and M. C. Malin. Global mapping of Martian hematite mineral deposits: Remnants of water-driven processes on Mars. *Journal of Geophysical Research,* **106,** 23, 873–885, 2001.

Christensen, P. R., et al. The Thermal Emission Imaging System (THEMIS) for the Mars 2001 Odyssey Mission. *Space Science Reviews,* **110,** 85–130, 2004.

Coll, C., E. Valor, V. Caselles, and R. Niclos. Adjusted Normalized Emissivity Method for surface temperature and emissivity retrieval from optical and thermal infrared remote sensing data. *Journal of Geophysical Research,* **108(D23),** 4739, doi:10.1029/2003JD003688, 2003.

Cooper, B. L., J. W. Salisbury, R. M. Killen, and A. E. Potter. Midinfrared spectral features of rocks and their powders. *Journal of Geophysical Research,* **107,** 5017, 10.1029/2000JE001462, 2002.

Engelbracht, C.W., E. T. Young, G. H. Rieke, G. R. Iis, J. W. Beeman, and E. E. Haller. Observing and calibration strategies for FIR imaging with SIRTF. *Experimental Astronomy,* **10,** 403–413, 2000.

Fazio, G. G., P. Eisenhardt, and J-S. Huang. The Space Infrared Facility (SIRTF): A new probe for study of the birth and evolution of galaxies in the early universe. *Astrophysics and Space Science,* **269,** 541–548, 1999.

Gillespie, A. R., and A. B. Kahle. The construction and interpretation of a digital thermal inertia image. *Photogrammetry Engineering Remote Sensing,* **43,** 983, 1977.

Gillespie, A., S. Rokugawa, T. Matsunaga, J. S. Cothern, S. Hook, and A. B. Kahle. A temperature and emissivity separation algorithm for Advanced Spaceborne Thermal Emission and Reflection Radiometer (ASTER) images. *IEEE Transactions on Geoscience and Remote Sensing,* **GE-36,** 1113–1126, 1998.

Goetz, A. F. H., and L. Rowan. Geologic remote sensing. *Science,* **211,** 781–791, 1981.

Henry, R. L. The transmission of powder films in the infrared. *Journal of Optical Society of America,* **38,** 775–789, 1948.

Hovis, W. A. Infrared spectral reflectance of some common minerals. *Applied Optics,* **5,** 245, 1966.

Hunt, G. R., and J. W. Salisbury. Mid-infrared spectral behavior of igneous rocks. U.S.A.F. Cambridge Res. Lab. Tech. Report AFCRL-TR-74-0625, 1974. Also report # TR-75-0356 and TR-76-0003 for the cases of sedimentary rocks and metamorphic rocks, respectively.

Janza, F. K. (Ed.). *Manual of Remote Sensing,* Vol. I. American Society of Photogrammetry. Falls Church, VA, 1975.

Kahle, A. B. A simple thermal model of the earth's surface for geological mapping by remote sensing. *Journal of Geophysical Research,* **82,** 1673, 1977.

Kahle, A. B., D. P. Madura, and J. M. Soha. Middle infrared multispectral aircraft scanner data: Analysis for geological applications. *Applied Optics,* **19,** 2279, 1980.

Kahle, A. B., A. R. Gillespie, and A. F. H. Goetz. Thermal inertia imaging: A new geologic mapping tool. *Geophysics Research Letters,* **3,** 26, 1976.

Kahle, A. B., and A. Goetz. Mineralogical information from a new airborne thermal infrared multispectral scanner. *Science,* **222,** 24–27, 1983.

Kahle, A. B., and L. C. Rowan. Evaluation of multispectral middle infrared aircraft images for lithologic mapping in the East Tintic Mountains, Utah. *Geology,* **8,** 234–239, 1980.

Lane, M. D., P. R. Christensen, and W. K. Hartmann. Utilization of the THEMIS visible and infrared imaging data for crater population studies of the Meridiani Planum landing site, *Geophysical Research Letters,* **30,** 1770, doi:10.1029/2003GL017183, 2003.

Liang, S. An optimization algorithm for separating land surface temperature and emissivity from multispectral thermal infrared imagery. *IEEE Transactions on Geoscience and Remote Sensing,* **GE-39,** 264–274, 2001.

Lyon, R. J. P. Analysis of rocks by spectral emission (8 to 25 microns). *Economic Geology,* **78,** 618–632, 1965.

McClain E. P., W. G. Pichel, and C. C. Walton, Comparative performance of AVHRR-based multichannel sea surface temperatures. *Journal of Geophysical Research,* **90,** 11, 587, 1985.

Milam, K. A., K. R. Stockstill, J. E. Moersch, H. Y. McSween Jr., L. L. Tornabene, A. Ghosh, M. B. Wyatt, and P. R. Christensen. THEMIS characterization of the MER Gusev crater landing site. *Journal of Geophysical Research,* **108(E12),** 8078, doi:10.1029/2002JE002023, 2003.

Moroz, V. I. *Physics of the Planets.* Nauka Press, Moscow, 1976.

Nedeltchev, N. M., Thermal microwave emission depth and soil moisture remote sensing. *International Journal of Remote Sensing,* **20,** 2183–2194, 1999.

Ninomiya, Y., T. Matsunaga, Y. Yamaguchi, K. Ogawa, S. Rokugawa, K. Uchida, H. Muraoka, and M. Kaku. A comparison of thermal infrared emissivity spectra measured in situ, in the laboratory, and derived from thermal infrared multispectral scanner (TIMS) data in Cuprite, Nevada, U.S.A. *International Journal of Remote Sensing,* **18,** 1571–1581, 1997.

Quattrochi, D.A., and J.C. Luvall. Thermal infrared remote sensing for analysis of landscape ecological processes: Methods and applications, *Landscape Ecology,* **14,** 577–598, 1999.

Sabins, F. *Remote Sensing: Principles and Interpretation.* Freeman, San Francisco, 1978.

Siegal B., and A. Gillespie. *Remote Sensing in Geology.* Wiley, New York, 1980.

Smith, J. A., N. S. Chauhan, T. J. Schmugge, and J. R. Ballard, Jr. Remote sensing of land surface temperature: The directional viewing effect. *IEEE Transactions on Geoscience and Remote Sensing,* **GE-35,** 972–974, 1997.

Walton, C. C. Nonlinear multichannel algorithms for estimating sea surface temperature with AVHRR satellite data. *Journal of Applied Meteorology, 27,* 115, 1988.

Watson, K. Periodic Heating of a layer over a semi-infinite solid. *Journal of Geophysical Research* **83,** 5904–5910, 1973.

Watson, K. Geologic applications of infrared images. *Proceedings of IEEE,* **63,** 128–137, 1975.

Werner, M. W. SIRTF and NASA's Origins Program. *Advances in Space Research,* **30,** 2149–2150, 2002.

Zhengming, W., and J. Dozier. Land-surface temperature measurement from space: Physical principles and inverse modeling. *IEEE Transactions on Geoscience and Remote Sensing,* **GE-27,** 268–278, 1989.

5

SOLID-SURFACE SENSING: MICROWAVE EMISSION

Thermal radiation from natural surfaces occurs mainly in the far infrared region; however, it extends throughout the electromagnetic spectrum into the submillimeter and microwave region. In this latter region, the radiant emittance is given by the Rayleigh–Jeans approximation of Planck's law, which corresponds to the case of $ch/\lambda \ll kT$. In this limit, the spectral radiant emittance is given by

$$S(\lambda) = \frac{2\pi ckT}{\lambda^4} \qquad (5\text{-}1)$$

where $S(\lambda)$ is in W/m^3. Usually, in microwave radiometry, $S(\lambda)$ is expressed in terms of energy per unit frequency. The transformation is given by

$$v = \frac{c}{\lambda} \rightarrow dv = -\frac{c}{\lambda^2}d\lambda$$

and

$$|S(v)dv| = |S(\lambda)d\lambda| \rightarrow S(v) = \frac{\lambda^2}{c}S(\lambda) \qquad (5\text{-}2)$$

Thus,

$$S(v) = \frac{2\pi kT}{\lambda^2} = \frac{2\pi kT}{c^2}v^2 \qquad (5\text{-}3)$$

where $S(v)$ is in W/m^2 Hz. The surface radiance or brightness $B(\theta, v)$ is related to $S(v)$ by

$$S(v) = \int_{\Omega} B(\theta, v)\cos\theta \, d\Omega' = \int_0^{2\pi}\int_0^{\pi/2} B(\theta, v)\cos\theta\sin\theta \, d\theta \, d\phi$$

Introduction to the Physics and Techniques of Remote Sensing. By C. Elachi and J. van Zyl
Copyright © 2006 John Wiley & Sons, Inc.

that is, the surface brightness is the spectral radiant emittance per unit solid angle. If the brightness is independent of θ, the surface is called Lambertian and

$$S(v) = \pi B(v)$$

and the surface brightness (units of W/m^2 Hz steradians) is given by

$$B(v) = \frac{2kT}{\lambda^2} = \frac{2kT}{c^2} v^2 \tag{5-4}$$

The Rayleigh–Jeans approximation is very useful in the microwave region. It is mathematically simpler than Planck's law, and it can be shown that the difference between this approximation and Planck's law is less than 1% if $v/T < 3.9 \times 10^8$ Hz K. For example, for a black body at 300 K, the Rayleigh–Jeans approximation holds as long as $v < 117$ GHz

5-1 POWER–TEMPERATURE CORRESPONDENCE

In the case of a graybody with emissivity $\varepsilon(\theta)$, the radiant power per unit bandwidth emitted from a surface element ds in a solid angle $d\Omega'$ is equal to

$$P(v) = \frac{2kT}{\lambda^2} \varepsilon(\theta) \, ds \, d\Omega' \tag{5-5}$$

An aperture with effective area A at a distance r from the surface element would represent a solid angle of $d\Omega' = A/r^2$. Therefore, if we have a linearly polarized antenna at distance r of effective area A and normalized pattern (i.e., equal to unity at the boresight) $G(\theta, \phi)$ intercepting the emitted field, the power collected in a spectral band dv is (replacing $d\Omega'$ by A/r^2):

$$P(v) = \frac{2kT}{\lambda^2} \varepsilon(\theta) \frac{G(\theta, \phi)}{2} \frac{dsA \, dv}{4\pi r^2} = \frac{AkT}{\lambda^2} \varepsilon(\theta) G(\theta, \phi) \, d\Omega \, dv \tag{5-6}$$

where $d\Omega = ds/r^2$ is the solid angle of the emitting elementary area as viewed from the antenna. The factor $\frac{1}{2}$ is due to the fact that the emitted radiation is unpolarized and the polarized antenna will detect only half the total incident power (see Chapter 2). The total received power in a spectral band Δv can then be expressed as

$$P_r = AkT \int_{\Delta v} \int_{\Omega} \frac{\varepsilon(\theta) G(\theta, \phi) \, d\Omega \, dv}{\lambda^2} \tag{5-7}$$

Usually, $\Delta v \ll v$, then

$$P_r = \frac{AkT \, \Delta v}{\lambda^2} \int_{\Omega} \varepsilon(\theta) G(\theta, \phi) \, d\Omega \tag{5-8}$$

This expression can be written as

$$P_r = kT_{eq} \Delta v \tag{5-9}$$

with T_{eq} given by the following equivalent expressions:

$$T_{eq} = \frac{TA}{\lambda^2} \int \varepsilon(\theta) G(\theta, \phi) \, d\Omega$$

$$T_{eq} = T \frac{\int \varepsilon(\theta) G(\theta, \phi) \, d\Omega}{\Omega_0}$$

$$T_{eq} = \frac{T}{4\pi} \int \varepsilon(\theta) g(\theta, \phi) \, d\Omega$$

where

$$\Omega_0 = \frac{\lambda^2}{A} = \text{antenna beam solid angle}$$

$$g(\theta, \phi) = \frac{4\pi A}{\lambda^2} G(\theta, \phi) = \text{antenna gain}$$

Thus, the effective temperature observed by the receiver is equal to the surface temperature multiplied by a factor that depends on the surface angular emissivity and the receiving antenna pattern. In order to derive accurately the surface temperature from microwave radiometer measurements, it is necessary to account for the antenna pattern. In the case of planets with atmospheres, the effects of atmospheric emission and absorption are usually significant, as will be discussed in a later chapter.

5-2 SIMPLE MICROWAVE RADIOMETRY MODELS

The derivation of the equivalent temperature in the previous section assumed that the graybody is the only source of radiation. Figure 5.1 shows the more realistic case in which a graybody surface at a temperature T_g is radiating in the presence of the sky, which is at an equivalent temperature T_s. If the planet has no atmosphere, T_s is a constant value. On the other hand, if the planet has an atmosphere, T_s is larger when looking toward the horizon because of the thicker atmosphere as measured along the line of

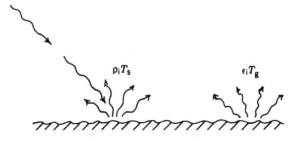

Figure 5-1. The total energy radiated from the surface consists of the energy emitted by the surface plus the energy reflected by the surface.

sight. The two contributions to the radiated energy in this case are the emitted energy from the surface, which is the same as that discussed in the previous section, and the energy originally radiated by the sky and subsequently reflected by the surface. Putting these two contributions in Equation (5-5), and following the same derivation as in the previous section, one finds that the equivalent microwave temperature of this surface can be expressed as

$$T_i(\theta) = \rho_i(\theta)T_s + \varepsilon_i(\theta)T_g \tag{5-10}$$

where i indicates the polarization. The above simple expression assumes clear atmosphere (no cloud emissions) and neglects atmospheric absorption. The emissivity ε_i and reflectivity ρ_i are related by $\varepsilon_i = 1 - \rho_i$; therefore,

$$T_i(\theta) = T_g + \rho_i(\theta)(T_s - T_g) \tag{5-11}$$

or

$$T_i(\theta) = T_s + \varepsilon_i(\theta)(T_g - T_s) \tag{5-12}$$

To illustrate, let us consider the case of a sandy surface of dielectric constant equal to 3.2 (i.e., $n = \sqrt{3.2}$). Figure 5-2a shows the surface reflection coefficient and the corresponding surface contribution to the total radiance temperature for both horizontal and vertical polarization. The reflectivity ρ_v goes to zero at the Brewster angle $\theta_b = 60.8°$. Figures 5-2b and c show the contribution of the sky temperature for the case of a planet without atmosphere $[T_s(\theta) = T_s = \text{constant}]$ and the case of a planet with atmosphere $[T_s(\theta)$ is larger toward the horizon]. Figures 5-2d and e show the resulting total microwave brightness temperature. It is clear that the measured value of T is strongly dependent on the observation angle, the polarization, and the model for the atmospheric temperature. In general, the following is usually true:

$T_v = T_g$ at the Brewster angle

$T_v - T_h$ is maximum near the Brewster angle

$T_v = T_h = T_g$ for surfaces with very low dielectric constant, and $T_v = T_h = T_s$ for surfaces with very high dielectric constant

The subscripts v and h correspond to vertical and horizontal polarization, respectively.

5-2-1 Effects of Polarization

The microwave brightness temperature depends on three independent factors: the surface temperature T_g, the sky temperature T_s, and the surface dielectric constant or index of refraction n. The fact that the brightness temperature at a certain angle is a function of the polarization allows us to derive two of these parameters if the third one is known.

Let us assume that the sky temperature T_s is known. In the case of planets with no atmosphere or if the atmosphere is transparent at the frequency of observation, T_s is basically the temperature of space. This can be directly measured with the sensor by looking away from the planet. In the case in which the atmospheric contribution is significant, a model for T_s must be used.

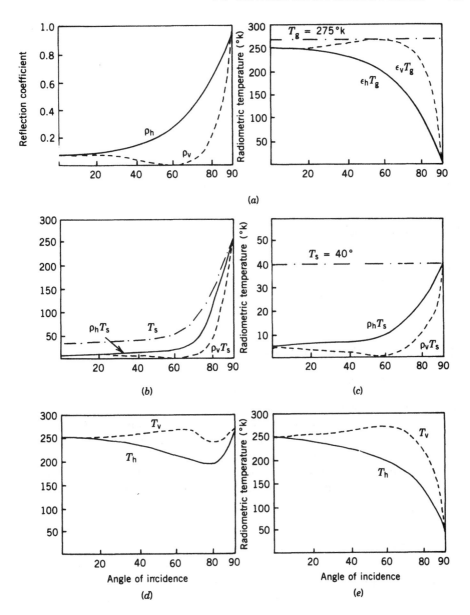

Figure 5-2. Observed radiometric temperature of a half-space with dielectric constant $\varepsilon = 3.2$ and temperature $T_g = 275$ K. The graphs in (b) and (d), and (c) and (e) correspond, respectively, to a planet with atmosphere or without atmosphere.

From Equation 5-12, we can then derive the following expression for the emissivity:

$$\frac{\varepsilon_v(\theta)}{\varepsilon_h(\theta)} = \frac{T_v(\theta) - T_s(\theta)}{T_h(\theta) - T_s(\theta)} \tag{5-13}$$

The measured ratio on the right-hand side of Equation (5-13) plus the theoretical expression for $\varepsilon_v(\theta)/\varepsilon_h(\theta)$ would then allow the derivation of the surface dielectric constant.

Once the dielectric constant of the surface is known, the emissivity at each of the two polarizations can be calculated. Using this information, the surface temperature can then be derived from

$$T_g = T_s + \frac{T_i - T_s}{\varepsilon_i} \qquad (5\text{-}14)$$

5-2-2 Effects of the Observation Angle

The fact that T_h and T_v have different dependence on the observation angle does allow the derivation of all three unknown natural parameters T_s, T_g, and dielectric constant. One possible approach is as follows.

By looking at $T_v(\theta)$ and $T_v(\theta) - T_h(\theta)$, a peak should be observed near the Brewster angle (see Fig. 5-2). Knowing this angle will allow the derivation of the surface dielectric constant. This in turn will provide $\rho_h(\theta)$ and $\rho_v(\theta)$. Then Equation (5-12), T_g and $T_s(\theta)$ can be derived.

In reality, the above approach most likely will be iterative. For instance, if it is found that the derived T_g varies with θ, then a mean value can be taken and another iteration made.

5-2-3 Effects of the Atmosphere

The Earth's atmosphere absorption is relatively small at frequencies lower than 10 GHz. Clouds are also transparent at these frequencies. At higher frequencies, the absorption increases appreciably mainly due to the presence of water vapor and oxygen. The water vapor absorption increases from 10^{-3} dB/km at 10 GHz to 1 dB/km at 400 GHz for 1 g/m^3 density and 1 bar pressure. In addition, sharp absorption lines are present at 22.2 GHz and near 180 GHz. Oxygen has strong absorption lines at 60 GHz and 118.8 GHz. The atmospheric absorption in the high-frequency region plays a major role in the behavior of T_s and in the transmission of the surface radiation. This effect is discussed in detail in Chapter 9.

5-2-4 Effects of Surface Roughness

In a large number of situations, natural surfaces have a rough interface and contain near-surface inhomogeneities. Thus, the surface reflectivity and emissivity are strongly dependent on the surface roughness and subsurface inhomogeneities, and their expressions are fairly complex.

Assuming that the observation frequency is fairly high so that the subsurface penetration is negligible, the sky contribution to the microwave brightness can be expressed as

$$B_s(\nu) = \frac{2k\nu^2}{c^2} \int_\Omega T_s(\theta')\sigma(\theta, \theta', \phi) \sin\theta'\, d\theta'\, d\phi'] \qquad (5\text{-}15)$$

where $\sigma(\theta, \theta', \phi')$ is the scattering coefficient of a wave coming from direction (θ', ϕ') into the direction of observation θ, the integral being over the hemisphere covering the sky. This expresses the fact that incoming sky radiation from all directions can be partially scattered into the direction of observation θ (see Fig. 5-3).

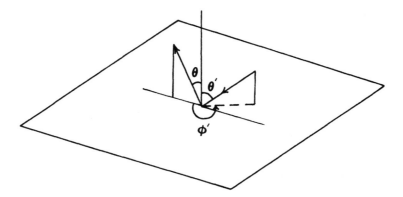

Figure 5-3. Geometry of wave scattering from a rough surface.

Similarly, the contribution from the surface can be expressed as

$$B_g(v) = \frac{2kv^2}{c^2} T_g \left[1 - \int_\Omega \sigma(\theta, \theta', \phi') \sin \theta' \, d\theta' \, d\phi' \right] \tag{5-16}$$

Thus, the observed microwave temperature is:

$$T = T_g \left[1 - \int_\Omega \sigma(\theta, \theta', \phi') \sin \theta' \, d\theta' \, d\phi' \right] + \int_\Omega T_s(\theta') \sigma(\theta, \theta', \phi') \sin \theta' \, d\theta' \, d\phi' \tag{5-17}$$

If the sky temperature is independent of θ', then

$$T = T_g + (T_s - T_g) \int_\Omega \sigma(\theta, \theta', \phi') \sin \theta' \, d\theta' \, d\phi' \tag{5-18}$$

which is identical to Equation 5-12 except that the surface reflectivity $\rho(\theta)$ has been re-placed by its equivalent in the case of a rough surface, that is, the integral of the backscat-ter cross section.

5-3 APPLICATIONS AND USE IN SURFACE SENSING

In the case of a graybody surface, the radiant power emitted is a function of the surface temperature T and its emissivity ε (Equation 5-5). The emissivity in turn is a function of the surface composition and roughness.

The range of variation of $P(v)$ as a result of variation in surface temperature is usually fairly limited. In the case of the Earth, variations in temperature at one location as a func-tion of time would rarely exceed 60 K, which gives a relative variation of about 20% (60 K relative to 300 K).

Variations due to change in surface composition or roughness are much larger. To il-lustrate, if we consider the case of nadir observation of three types of smooth materials such as water ($n = 9$ at low microwave frequencies), solid rocks ($n = 3$), and sand ($n = 1.8$) at a thermodynamic temperature of 300 K, the variations in the equivalent surface mi-crowave temperature can easily double (see Table 5-1).

TABLE 5-1. Microwave Temperature of Three Representative Types of Material with $T_g = 300$ K and $T_s = 40$ K

Type of surface material	Index of refraction	Dielectric constant	Normal reflectivity ρ	Microwave temperature (K)
Water	9	81	0.64	134
Solid rock	3	9	0.25	235
Sand	1.8	3.2	0.08	280

5-3-1 Application in Polar Ice Mapping

One of the most useful applications of spaceborne microwave radiometry for surface studies is in the mapping of polar ice cover and monitoring its temporal changes. The large emissivity difference between ice and open water (their dielectric constants are approximately 3 and 80, respectively) leads to a strong contrast in the received radiation, thus allowing easy delineation of the ice cover. The key advantage of the microwave imaging radiometer, relative to a visible or near-infrared imager, is the fact that it acquires data all the time, even during the long dark winter season and during times of haze or cloud cover.

To illustrate, let us consider the case of normal incidence and specular reflection. From Equation 5-11, the difference between the radiometric temperatures of two areas a and b with the same thermodynamic temperature is

$$\Delta T = T_a - T_b = (\rho_a - \rho_b)(T_s - T_g) = \Delta\rho(T_s - T_g)$$

where

$$\rho = \left(\frac{\sqrt{\varepsilon}-1}{\sqrt{\varepsilon}+1}\right)^2 = \left(\frac{n-1}{n+1}\right)^2 \tag{5-19}$$

In the case of ice ($\varepsilon = 3$) and water ($\varepsilon = 80$) we have

$$\rho \text{ (ice)} = 0.07$$

$$\rho \text{ (water)} = 0.64$$

$$\rightarrow \Delta\rho = 0.57$$

Usually, $T_s = 50$ K and the surface temperature of the ice is $T_g = 272$ K; then $\Delta T = 127$ K, which is significant. Because of its higher reflectivity, the water surface will look significantly cooler than the ice surface.

If the nature of the ice changes (in salinity, ice age, etc.), a slight change in ε results. This in turn, causes the reflectivity of the surface to change slightly. This change can be written as

$$\Delta\rho = \rho\frac{2\Delta\varepsilon}{\sqrt{\varepsilon}(\varepsilon-1)} \tag{5-20}$$

Thus, if $\varepsilon = 3$ and $\Delta\varepsilon = 0.6$ (20% change), then

$$\frac{\Delta\rho}{\rho} = 0.34 \rightarrow \Delta\rho = 0.34 \times 0.07 = 0.024$$

and

$$\Delta T = 5.4 \text{ K}$$

which is well above the sensitivity of modern orbiting radiometers.

Figures 5-4 and 5-5 show examples of microwave radiometer data covering the northern and southern polar regions at different times of the year. The changes in the ice cover

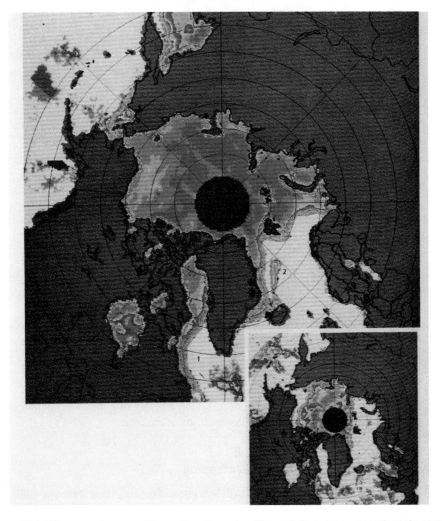

Figure 5-4. Microwave images of the north polar region. The main image corresponds to the maximum extent of ice in the winter. The inset corresponds to its minimum extent in the summer. See color section.

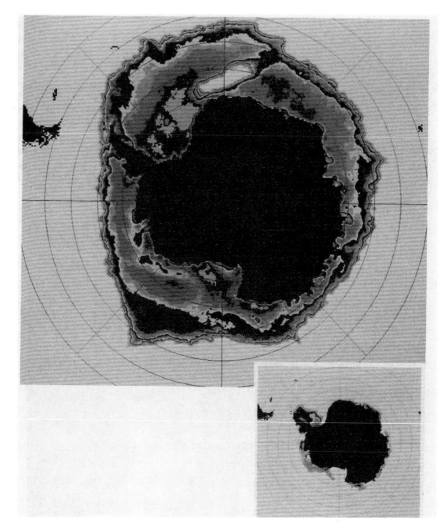

Figure 5-5. Microwave images of the south polar region. The main image corresponds to the maximum extent of ice in the local winter. The inset corresponds to its minimum extent in the local summer. See color section.

are clearly apparent. In the ice-covered area, the changes in brightness temperature are mainly due to changes in the nature and composition of the ice.

5-3-2 Application in Soil Moisture Mapping

The fact that water has a high microwave dielectric constant and, therefore, low emissivity in comparison to most natural surfaces leads to the possibility of mapping variations in soil moisture. The variation of the surface dielectric constant as a function of soil moisture has been measured by a number of researchers for different soil types. Figure 5-7 shows such an example and the corresponding variations in emissivity and brightness temperature for an illustrative surface and sky temperature. It is apparent that the brightness temperature variations are significant. For unfrozen soils, the surface microwave brightness temperature at the L band (1.4 GHz) decreases by as much as 70 K or more (depending on the vegetation cover) as the soil moisture increases from dry to saturated. By inverting this relationship, soil moisture can be measured radiometrically with an accuracy of better than 0.04 g cm^{-3} under vegetation cover of water content up to about 5 kg m^{-2} (typical of mature crops and shrubland). Soil moisture estimation will thus be feasible over about 65% of the land surface. Above this threshhold, the soil moisture estimation accuracy degrades significantly as the attenuation by vegetation masks the soil emission. These retrieval accuracies have been demonstrated by airborne L-band radiometers over a range of natural and agricultural terrains. As an example, Figure 5-6 shows results from the ESTAR airborne instrument obtained during the 1997 Southern Great Plains Experiment in Oklahoma.

To retrieve the soil moisture from the measured microwave temperature, models typically assume the radiation to come from a horizontal surface at temperature T_g covered by a uniform canopy of vegetation that is at a temperature T_c. The radiation from the underlying soil surface is attenuated by the vegetation canopy. This attenuation is characterized by an optical depth τ (see Chapter 9) that relates to the vegetation water content according to

$$\tau = bW_c \tag{5-21}$$

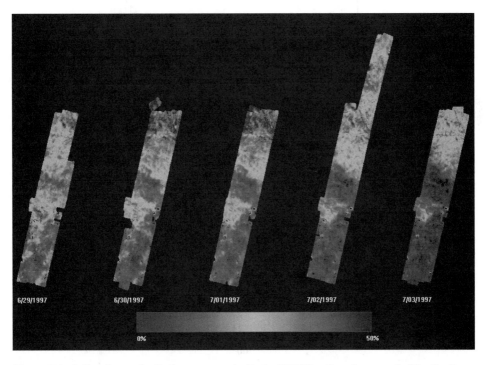

Figure 5-6. Soil moisture distribution measured with the ESTAR radiometer as part of the Southern Great Plains Experiment in Oklahoma in 1997. [The data were acquired as part of NASA's Earth-Sun System Division and archived and distributed by the Goddard Earth Sciences (GES) Data and Information Services Center (DISC) Distributed Active Archive Center (DAAC).] See color section.

The constant b is a vegetation opacity coefficient that is determined experimentally and has a value of approximately 0.1 at the L band. Following the same derivation as for the bare surface in the previous section, it is possible to write the microwave temperature of this combination as

$$T = (1 - \rho_i)T_g e^{-\tau} + T_c(1 - e^{-\tau}) + \rho_i T_c(1 - e^{-\tau})e^{-\tau} \qquad (5\text{-}22)$$

The subscript i refers to the polarization (either horizontal or vertical). The first term represents the radiation from the underlying soil as attenuated by the vegetation. The second term is the upward radiation from the vegetation layer, and the third term is the downward radiation from the vegetation layer that is reflected by the underlying soil surface. Note that both the optical depth τ of the vegetation and the soil reflectivity ρ_i are functions of the incidence angle θ.

The soil reflectivity is related to the soil moisture content through the dielectric constant (see Figure 5-7) and is modified by surface roughness. Therefore, to estimate the soil moisture, we need to infer the soil reflectivity from Equation (5-22). The problem can be simplified by recognizing that for dawn (6 a.m. local time) orbit overpasses, the soil

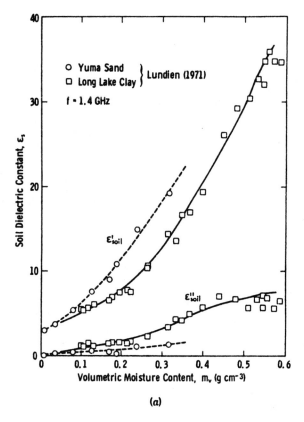

(a)

Figure 5-7. (a) Relative dielectric constants of sandy and high-clay soils as a function of volumetric moisture content at 1.4 GHz (Ulaby et al., 1982).

and vegetation temperatures are approximately equal to the surface air temperature, and the brightness temperature can be written as

$$T = T_a(1 - \rho_i e^{-2\tau}) \tag{5-23}$$

where T_a is the surface air temperature.

The basic soil moisture retrieval approach is to invert Equation (5-23) for the surface reflectivity using a single radiometer channel (L band, horizontal polarization) and ancillary information to estimate and correct for T_a and W_c (Jackson and le Vine, 1996). Once the surface reflectivity is known, the surface dielectric constant is calculated using the Fresnel equations. Finally, the surface soil moisture is estimated from its dielectric constant using a dielectric mixing model. This approach has been used in several studies (e.g., Jackson et al., 2000) with excellent results. Ancillary information on landscape, topography, and soil characteristics can be used to improve the soil moisture retrieval by

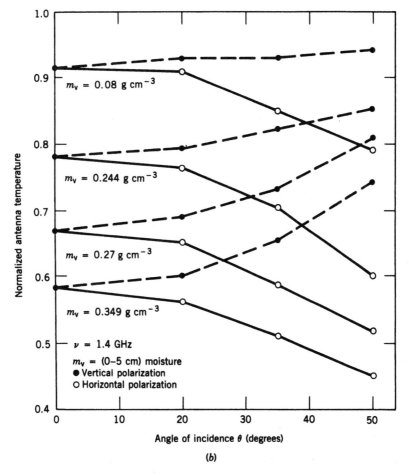

Figure 5-7. (*b*) Measured normalized (relative to surface temperature) antenna temperature of a smooth surface function of angle for various soil moistures (Newton and Rouse, 1980).

providing a surface roughness correction and soil texture information for use with the dielectric mixing model.

If both vertical and horizontal polarizations are used in the measurement, a two-channel retrieval approach can be implemented for estimating soil moisture. Multichannel retrieval approaches are based on the premise that additional channels can be used to provide information on the vegetation and/or temperature characteristics of the scene, so that reliance on ancillary data can be reduced. Such approaches have been evaluated using both simulations and experimental data (Wigneron et al., 2000.). The retrieval method involves adjusting the scene parameters until a best fit is achieved between the computed and observed brightness temperatures.

5-3-3 Measurement Ambiguity

From Equation 5-9, it is apparent that the equivalent microwave temperature depends on the nature of the antenna pattern and on the behavior of $\varepsilon(\theta)$ as a function of θ. As illustrated in Figure 5-8, this can lead to errors due to radiation collected via the sidelobes and misinterpreted as radiation in the main lobe.

If the surface temperature T varies as a function of x, as indicated in Figure 5-8, the measured temperature is equal to (considering a one-dimensional surface)

$$T_{eq} = \frac{1}{\Omega_0} \int T(\theta)\varepsilon(\theta)G(\theta)d\Omega \qquad (5\text{-}24)$$

where $\tan \theta = x/h$ and h is the sensor altitude. As the antenna moves over the surface, the measured brightness temperature is a convolution integral between the normalized antenna gain $G(\theta)$ and the effective surface brightness εT. A similar expression can be derived in the more real case of a two-dimensional surface.

The extent of the measurement ambiguity can be characterized by the fraction of the total measured temperature contributed by the sidelobes. This fraction is

$$F = \frac{1 - \int_{\Omega_0} G(\theta)d\Omega}{\int_{2\pi} G(\theta)d\Omega} \qquad (5\text{-}25)$$

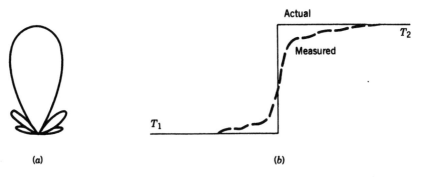

(a)   (b)

Figure 5-8. An antenna with the pattern shown in (a) will measure a temperature profile slightly different than the actual profile at a step discontinuity (b) due to radiation received through the sidelobes.

where Ω_0 is the main lobe solid angle. In the one-dimensional case, this is equal to

$$F = 1 - \frac{\displaystyle\int_{-\theta_0}^{+\theta_0} G(\theta)d\theta}{\displaystyle\int_{-\pi}^{+\pi} G(\theta)d\theta} \tag{5-26}$$

This fraction depends on the type of antenna and on any tapering used to reduce the side-lobes.

5-4   DESCRIPTION OF MICROWAVE RADIOMETERS

In almost all surface applications, imaging microwave radiometers are used instead of nadir line scan systems. In real-aperture radiometers, the imaging is achieved by mechanically or electronically scanning the receiving antenna beam across the swath. The satellite motion allows the imaging along the swath. Synthetic aperture radiometers utilize interferometry principles originally developed in radio astronomy to measure the visibility function of the scene, from which the actual radiometer image can then be constructed.

An imaging radiometer consists of three basic elements: (1) an antenna and its associated scanning mechanism or series of correlators, which collects incoming radiation from specified beam-pointing directions; (2) a receiver, which detects and amplifies the collected radiation within a specified frequency band; and (3) a data handling system, which performs digitizing, multiplexing, and formatting functions on the received data as well as other calibration and housekeeping data. After transmission to the ground, the sensor data are then (1) converted to units of antenna temperature using calibrated references, (2) corrected for antenna pattern effects and atmospheric effects to derive surface microwave temperature, and (3) interpreted to derive geophysical parameters such as surface temperature and soil moisture. For example, in the case of polar ice mapping, some of the desired geophysical data are the extent, percent coverage, and motion of the ice cover.

5-4-1   Antenna and Scanning Configuration for Real-Aperture Radiometers

The antenna size L and the operating wavelength λ define the angular resolution of the sensor as

$$\theta_r = \frac{\lambda}{L} \tag{5-27}$$

If we assume that the antenna is circular, the antenna beam spot directly underneath the satellite will also be circular, corresponding to a surface resolution of

$$r = h\theta_r = \frac{\lambda h}{L} \tag{5-28}$$

where h is the altitude of the satellite. If the antenna points away from the nadir direction, as is the case with scanning systems, the beam spot is no longer circular, and becomes both bigger and more elongated in the direction along the line of sight. In this case, the spot size in the two directions can be approximated by

$$r_{\perp} = \frac{\lambda h}{L \cos \theta}; \qquad r_{\parallel} = \frac{\lambda h}{L \cos^2 \theta} \qquad (5\text{-}29)$$

where θ is the angle with respect to nadir at the center of the beam. The subscripts $\perp$ and $\parallel$ refer to the directions perpendicular and parallel to the line of sight of the antenna. These expressions assume uniform illumination of the aperture. If weighted illumination (apodization) is used, the resolutions will be slightly larger; see Chapter 3.

A wide variety of antennas have been used with imaging radiometers. They are mainly of the reflector type or waveguide array type. In order to get wide coverage, the beam is scanned either mechanically, which is commonly done with reflector antennas, or electrically, which is commonly done with phased-array antennas. One particular scan geometry that has become popular for both reflector antennas and phased arrays is the conical scan, in which the beam is offset at a fixed angle from the nadir and scanned about the vertical (nadir) axis (see Fig. 5-9). The beam thus sweeps the surface of a cone. If the full 360° of a scan is utilized, double coverage fore and aft of the spacecraft is obtained. The main advantage of this type of scan is that the angle of incidence of the antenna beam at the Earth's surface is constant, independent of scan position. In addition, the line of sight through the atmosphere is also of constant length. This significantly increases the accura-

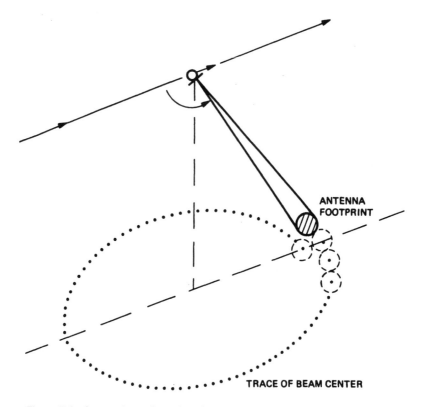

Figure 5-9. Geometric configurations for a conically scanned imaging radiometer.

cy with which the brightness temperature data can be interpreted in terms of changes in the surface parameters. The constant path through the atmosphere relaxes the need to know accurately the effects of the atmosphere on the propagating waves.

5-4-2 Synthetic-Aperture Radiometers

One of the drawbacks of real-aperture radiometers is the relatively poor resolution of the images. For example, a radiometer operating at 1.4 GHz using a 6 meter diameter antenna in a conical scan at 45° from an orbit altitude of 500 km will have a resolution of about 37 km, not taking into account the spherical shape of the earth. This is not really sufficient for many applications that require higher-resolution local imaging.

Synthetic-aperture radiometers promise much improved spatial resolution from space. They are implemented using an interferometric configuration in which two (or more) antennas are used to measure the radiation from the scene simultaneously. The separation between any pair of antennas is known as the baseline of the interferometer. As we shall show below, the measurement from each baseline represents a point in the Fourier transform of the scene. The image of the scene is produced by inverting this measured Fourier transform.

To illustrate the principle of synthetic-aperture radiometry, consider the idealized single-baseline interferometer shown in Figure 5-10. The two antennas are assumed to be identical, and are physically separated by a distance B. These antennas are receiving the radiation from a point source that is far enough away from the antennas that we can assume the waves to travel along parallel "rays" as shown in the figure. The waves from the point source arrive at an angle θ with respect to the direction perpendicular to the interferometer baseline, from a direction aligned with the axis of the baseline. The voltage from each antenna is amplified, the two voltages are cross-correlated, and the results integrated.

Assuming that the "voltage" antenna pattern is represented by $A(\theta)$, we can write the received signals as

$$V_1(\theta) = A(\theta)E_0 e^{i(kd+\varphi)}; \qquad V_2(\theta) = A(\theta)E_0 e^{i(\varphi)} \qquad (5\text{-}30)$$

where E_0 and φ are the amplitude and phase of the incoming radiation, respectively. The difference in the path length to the two antennas, d, is a function of the baseline length, and is given by

$$d = B \sin \theta \qquad (5\text{-}31)$$

After these signals are amplified, they are cross-correlated and integrated. The result is the complex visibility function of the interferometer, given by

$$V(\theta) = V_1(\theta)V_2^*(\theta) = |A(\theta)|^2 |E_0|^2 e^{ikB \sin\theta} \qquad (5\text{-}32)$$

Figure 5-11 shows the real part of the theoretical visibility function for two antennas, each with a diameter of 0.5 m, separated by a 10 m baseline. The wavelength is 21 cm. The dotted line is the antenna pattern $|A(\theta)|^2$ for the two individual antennas.

Now consider these two antennas to be mounted on an aircraft or spacecraft such that they point to the nadir and the baseline is oriented horizontally, with the antenna separation in the cross-track direction. This represents a one-dimensional implementation of

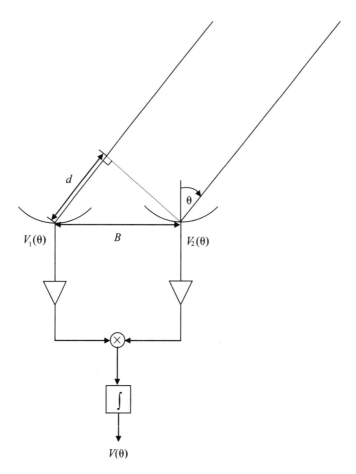

Figure 5-10. Ideal single-baseline interferometer showing the signal arriving from an off-axis source.

aperture synthesis. To simplify the discussion, we shall assume that the antennas have beams that are very narrow in the along-track direction but wide in the cross-track direction. The surface being imaged is an extended source of microwave thermal radiation. The received signal in this case would be the integral over all solid angles "seen" by the antennas in a narrow strip across track:

$$V_B = \int_{-\pi/2}^{\pi/2} |A(\theta)|^2 |E_0|^2 e^{ikB \sin\theta} d\theta \tag{5-33}$$

Here θ represents the angle in the cross-track direction, measured relative to the nadir. This expression is appropriate for the so-called "quasicoherent" case, in which the product of the system bandwidth and the integration time is much smaller than unity. If this is not the case, Equation (5-33) should be modified to take into account the spatial decorrelation of the signals (see Thompson et al., 1986 for more details). Using the Rayleigh–Jeans approximation for the incoming radiation, and writing the visibility in terms of an equivalent temperature, we can show that the measured equivalent surface temperature is

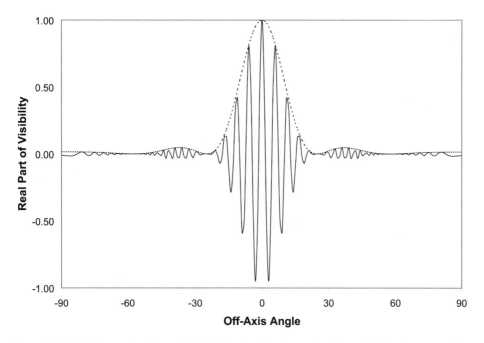

Figure 5-11. Real part of a theoretical visibility function for a single-baseline interferometer. The baseline is 10 m and the wavelength is 21 cm. The dotted line is the antenna pattern of an individual antenna, assumed to be 0.5 m in diameter.

$$T_{\text{eq } B} = \int_{-\pi/2}^{\pi/2} |A(\theta)|^2 \ \varepsilon(\theta) T_g(\theta) e^{ikB \sin\theta} d\theta \tag{5-34}$$

where $T_g(\theta)$ is the actual surface temperature at each angle, $\varepsilon(\theta)$ is the emissivity of the surface at that angle, and the subscript B refers to the baseline used in the measurement. Such a single-baseline measurement represents only a single measurement in the frequency domain. It is not possible to reconstruct the one-dimensional profile of surface temperatures and emissivities from such a single measurement. Ideally, Equation (5-34) should be repeated by varying the baseline length continuously from zero to infinity to cover the entire Fourier domain. In practice, these measurements are made using a number of discrete baseline values that are determined by the physical arrangement of the antennas. For example, the ESTAR (Le Vine et al., 1994) airborne system uses five antennas spaced as shown in Figure 5-12. By combining these antennas in pairs, it is then possible to obtain all baselines with spacings of $n\lambda/2$ with $0 \leq n \leq 7$ Therefore, if we denote each distinct baseline by $n\lambda/2$, we can rewrite (5-34) as

$$T_{\text{eq}}(n) = \int_{-\pi/2}^{\pi/2} |A(\theta)|^2 \ \varepsilon(\theta) T_g(\theta) e^{in\pi \sin\theta} d\theta = \int_{-\pi/2}^{\pi/2} T(\theta) |A(\theta)|^2 e^{in\pi \sin\theta} d\theta \tag{5-35}$$

This expression is a weighted Fourier transform of the scene brightness temperature. To find the actual profile of surface emissivity multiplied by surface temperature, Equation (5-35) must be inverted using the all the individual visibility measurements. However, in order to invert these measurements, the Fourier plane must be sampled adequately.

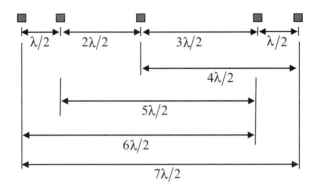

Figure 5-12. The ESTAR antenna spacings.

As we shall show below, combining the visibility measurements can be thought of as synthesizing an antenna array. Any array synthesized from discrete elements has an antenna pattern that shows maxima, known as grating lobes, at angles where the signal path length differences as given by Equation (5-31) are multiples of 2π. To ensure that we can invert the measurements to find the scene brightness temperature profile, the individual antenna spacings must be such that the grating lobes fall outside the angles over which we want to perform the brightness temperature measurement. If the individual antennas have very wide beams, such as that of a half-wave dipole, the spacing has to be multiples of $\lambda/2$ (Ruff et al., 1988). If the individual antennas have narrower beams, the antennas can be moved further apart. For our discussion, we shall assume that the antennas are spaced such that the individual baselines are

$$B_n = n\frac{\lambda}{2}; \qquad n = 0, \pm 1, \ldots, \pm N \tag{5-36}$$

In this case, the one-dimensional image reconstruction is (see Ruff et al., 1988)

$$\hat{T}(\theta') = \sum_{n=-N}^{N} T_{\text{eq}}(n) e^{-in\pi \sin\theta'} \tag{5-37}$$

To show how this measurement improves the spatial resolution of the radiometer, note that we can combine Equations (5-35) and (5-36):

$$\hat{T}(\theta') = \int_{-\pi/2}^{\pi/2} T(\theta) G(\theta, \theta') d\theta; \qquad G(\theta, \theta') = \sum_{n=-N}^{N} |A(\theta)|^2 e^{in\pi(\sin\theta - \sin\theta')} \tag{5-38}$$

The quantity $G(\theta, \theta')$ in Equation (5-38) is the gain of the "synthesized" antenna, which is a function of the physical positions of the antennas that form the baselines, weighted by the pattern of the (identical) individual antennas. If these individual antennas have gain patterns that are nearly constant over the range of angles that are of interest, $G(\theta, \theta')$ reduces to the well-known array factor in antenna array theory. In that case, we can show that the synthesized antenna pattern is (see Ruff et al., 1988)

$$G(\theta, \theta') = \frac{\sin\left[\dfrac{\pi}{2}(2N+1)(\sin\theta - \sin\theta')\right]}{\sin\left[\dfrac{\pi}{2}(\sin\theta - \sin\theta')\right]} \tag{5-39}$$

Figure 5-13 shows the antenna patterns that would be synthesized in different directions if $N = 4$ baselines are used. Using the Rayleigh criterion for angular resolution (see Chapter 3), the angular resolution of the synthesized antenna in the nadir direction is

$$\Delta\theta = \sin^{-1}\left(\frac{2}{2N+1}\right) \approx \frac{\lambda}{B_{\max}} \tag{5-40}$$

The approximation holds if $N \gg 1$, and $B_{\max} = N\lambda$ is the maximum separation between any two antennas in the array. This expressions shows that the angular resolution of the synthetic-aperture radiometer is the same as that of an antenna of size equal to the maximum baseline.

The analysis above assumed that the antennas are physically spaced every $\lambda/2$. This means that there are many baselines that are synthesized more than once. Sparse or thinned arrays, such as the ESTAR array, use minimum redundancy to synthesize all the baselines up to the maximum spacing. This is most important in spaceborne applications, where the mass of the system must be minimized. Ruff and coworkers (1988) report case

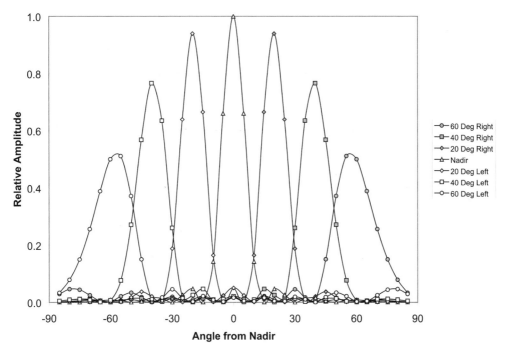

Figure 5-13. Theoretical antenna patterns synthesized in different pointing directions if $N = 4$ baselines are used.

studies of several thinned-array configurations, the largest of which would synthesize up to $B_{max} = 1032 \times \lambda/2$ with only 63 elements.

In practice, the inversion of the visibilities is done by approximating the integral in Equation (5-35) by a sum:

$$T'_{eq} = \sum_{m=1}^{N} G_{nm}T_m; \qquad G_{nm} = |A(\theta_m)|^2 \, e^{ikB_n \sin\theta_m}\Delta\theta \qquad (5\text{-}41)$$

There are a total of N such expressions corresponding to each of the N baselines. This can be written in matrix form as

$$T'_{eq} = GT \qquad (5\text{-}42)$$

with T_{eq} a vector with n elements, $T = \varepsilon T_g$ a vector with m elements, and G a $n \times m$ matrix. In practice this will generally be an underdetermined set of equations, that is, $m > n$. In that case, the least squares solution for Equation (5-42) is

$$\hat{T} = \{G^T(GG^T)^{-1}\}T'_{eq} \qquad (5\text{-}43)$$

The superscript T refers to the transpose of the matrix. The matrix $G^T(GG^T)^{-1}$ is known as the pseudoinverse of G. This is the type of inversion that has been used for the ESTAR system (Le Vine et al., 1994).

The discussion so far has been limited to improving the resolution in the cross-track direction, such as in the case of the ESTAR instrument. In this case, antennas that are long in the along-track direction are used to provide adequate resolution in that direction. The synthetic-aperture concept can easily be modified to the two-dimensional case. The proposed European Soil Moisture and Ocean Salinity Mission (SMOS) will use antennas placed periodically on a Y-shaped structure to provide two-dimensional aperture synthesis. With 23 antennas along each arm of the Y spaced at distances of , an angular resolution of 1.43° can be realized (Kerr et al., 2000; Bará et al., 1998). Figure 5-14 shows the theoretical antenna patterns for a Y-shaped configuration as proposed for SMOS with identical weights assigned to all baselines in the inversion. The hexagonal symmetry is clearly seen in the contour plot. The sidelobes are relatively high, and likely unacceptable for real imaging applications. These can be reduced by applying a tapered weighting to the baselines, similar to the apodization described in Chapter 3. Bará et al. show antenna patterns for different types of weighting.

5-4-3 Receiver Subsystems

The performance of a radiometer system is commonly reported as the smallest change in the scene temperature ΔT that can be measured in the radiometer output. This performance is obviously strongly influenced by the exact implementation of the radiometer receiver and processing system. Radiometer receivers typically are of the superheterodyne type. The input signal to the radiometer receiver is mixed with a local oscillator signal to produce an intermediate-frequency (IF) signal that is then amplified, integrated, and detected (see Chapter 6). The input radiometer signal is a random signal that is proportional to the temperature of the scene being imaged. This signal is band-limited by the bandwidth B of the receiver system, and is typically referred to as the antenna temperature T_a.

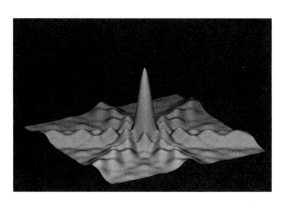

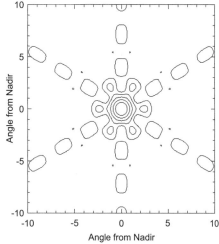

Figure 5-14. Theoretical pattern synthesized by a Y-shaped antenna configuration with 23 antennas per arm spaced at distances of 0.89 λ. No weighting is applied to the individual baselines.

The receiver electronics also generate thermal noise that is added to this incoming signal, and is likewise characterized by an equivalent receiver temperature T_{rec}. If this combined signal is integrated for a time period τ, it can be shown that the ratio of the fluctuating part to the average is $1/\sqrt{B\tau}$. If we can assume that the system parameters (gain, etc.) are constant, it can be shown that such a total power radiometer would have the ideal sensitivity

$$\Delta T_{\text{ideal}} = \frac{T_a + T_{\text{rec}}}{\sqrt{B\tau}} \tag{5-44}$$

Unfortunately, in practice it is not possible to completely remove temporal variations in system parameters. If the receiver gain fluctuates by an amount ΔG around its nominal value G_s, an additional fluctuation in the radiometer output is generated. In this case, because the gain fluctuations are statistically independent from the random fluctuations of the incoming signal and the signal generated inside the receiver, the sensitivity of a total power radiometer will be reduced to

$$\Delta T_{TP} = (T_a + T_{\text{rec}}) \sqrt{\frac{1}{B\tau} + \left(\frac{\Delta G_s}{G_s}\right)^2} \tag{5-45}$$

To reduce the effects of gain fluctuations, most spaceborne radiometers are of the Dicke-switched superheterodyne type. An example of such a system is shown in Figure 5-15. Following the antenna, a switch determines what is to be fed into the receiver—the antenna signal or a calibration signal. Measurements of the calibration targets are used later in the data processing to calibrate the signal data. The Dicke switch switches periodically at a high rate between the incoming signal and a known reference source. A synchronous detector at the other end of the receiver demodulates the signal. By these high-frequency comparisons of the signal with a known reference source, the effect of low-frequency gain variations in the amplifiers and other active devices is minimized. In this system, the criti-

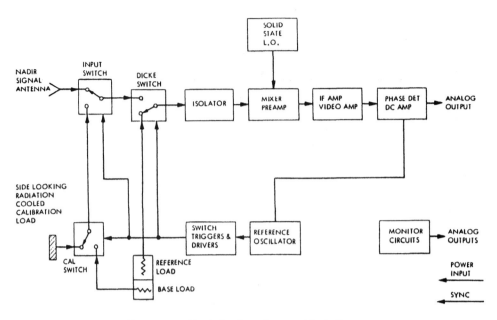

Figure 5-15. Example of a radiometer block diagram.

cal noise component is the mixer, and considerable effort is made to keep it as low noise as possible to improve the receiver sensitivity. If the equivalent temperature of the reference source is T_{ref}, the sensitivity of the Dicke radiometer can be written as

$$\Delta T_{\text{Dicke}} = \sqrt{\frac{2(T_a + T_{\text{rec}})^2 + 2(T_{\text{ref}} + T_{\text{rec}})^2}{B\tau} + \left(\frac{\Delta G_s}{G_s}\right)^2 (T_a + T_{\text{ref}})^2} \qquad (5\text{-}46)$$

In the special case where $T_{\text{ref}} = T_a$, the radiometer is known as a balanced Dicke radiometer and the sensitivity becomes

$$\Delta T_{\text{balanced}} = 2\frac{T_a + T_{\text{rec}}}{\sqrt{B\tau}} = 2\Delta T_{\text{ideal}} \qquad (5\text{-}47)$$

Various different techniques are used to balance Dicke radiometers. Details are provided by Ulaby et al., 1981 (Chapter 6).

5-4-4 Data Processing

The data are first calibrated by converting the radiometer output voltage levels to the antenna temperature, referenced to the antenna input ports. The data calibration process uses prelaunch calibration tests as well as the radiometric data from the calibration targets of known microwave temperature.

The next step is to derive surface microwave brightness temperature. The antenna receives radiation from regions of space defined by the antenna pattern. The antenna pattern is usually strongly peaked along the beam axis, and the spatial resolution is defined by the angular region over which the antenna power pattern is less than 3 dB down from its value at beam center. However, all antennas have sidelobes and some of the received energy comes from outside the main 3 dB area. Thus, some ambiguities occur, as discussed earlier.

In the case of synthetic-aperture radiometers, the data processing involves inverting the visibility measurements as described earlier. A weighting is typically applied to the individual visibilities during the inversion. For more details, see Le Vine et al., 1994, or Bará et al., 1998.

5-5 EXAMPLES OF DEVELOPED RADIOMETERS

A number of microwave radiometers have been flown on Earth-orbiting and planetary missions. For illustration, this section gives brief descriptions of a number of recent ones, as well as some proposed missions.

5-5-1 Scanning Multichannel Microwave Radiometer (SMMR)

SMMR was launched on Seasat and Nimbus 7 satellites in 1978, and is a five-frequency (6.6, 10.7, 18, 21, and 32 GHz), dual-polarized imaging radiometer. Its characteristics are given in Table 5-2, and a view of the sensor is shown in Figure 5-16. It consists of six independent Dicke-type superheterodyne radiometers fed by a single antenna. Figure 5-17 shows a block diagram of the SMMR sensor. At 37 GHz, two channels simultaneously measure the horizontal and vertical components of the received signal. At the other four frequencies, the channels alternate between two polarizations during successive scans. In this manner, 10 data channels, corresponding to five dual-polarized signals, are provided by the instrument. The ferrite switches, isolator, and reference and ambient loads are packaged as a single unit for low-loss and isothermal operation. The mixers are Shottky-barrier diode, balanced, double-sideband mixers with integral IF preamplifiers having a 10 to 110 MHz bandwidth. The local oscillators are fundamental-frequency, cavity-stabilized Gunn diodes.

The SMMR has a scanning antenna system consisting of an offset parabolic reflector with a 79 cm diameter collecting aperture and a multifrequency feed assembly. The antenna reflector is mechanically scanned about a vertical axis, with a sinusoidally varying velocity, over a ±25% azimuth angle range. The antenna beam is offset 42° from nadir; thus,

TABLE 5-2. SMMR Instrument Characteristics (Nominal)

Characteristics	1	2	3	4	5
Frequency (GHz)	6.6	10.69	18	21	37
RF bandwidth (MHz)	220	220	220	220	220
Integration time (msec)	126	62	62	62	30
Sensitivity, $\Delta T_{rms}(K)$	0.9	0.9	1.2	1.5	1.5
Dynamic range (K)	10–330	10–330	10–330	10–330	10–330
Absolute accuracy (K) (long-term)	2	2	2	2	2
IF frequency range (MHz)	10–110	10–110	10–110	10–110	10–110
Antenna beamwidth (deg)	4.2	2.6	1.6	1.4	0.8
Antenna beam efficiency (%)	87	87	87	87	87

Figure 5-16. SMMR instrument in its handling fixture.

the beam sweeps out the surface of a cone and provides a constant incidence angle at the Earth's surface. Calibration is achieved by alternately switching in a "cold horn" viewing deep space and a "calibration load" at instrument ambient temperature at the scan extremes. The multifrequency feed horn is a ring-loaded corrugated conical horn with a sequence of waveguide tapers, resonators, and orthomode transducers at the throat, to which are coupled the 10 output ports. The cold calibration horns are similar in design to the feed horn. One horn serves the 6.6 and 10.7 GHz channels, another the 18 and 21 GHz channels, and a third the 37 GHz channels. These horns are scaled to provide equal beamwidths of 15°.

A data-programmer unit in the electronics assembly provides the timing, multiplexing, and synchronization signals; contains A/D converters, multiplexers, and shift registers; and provides formatting and buffering functions between the instrument and the spacecraft digital systems.

5-5-2   Special Sensor Microwave Imager (SSM/I)

The first SSM/I was launched in 1987 as part of the Defense Meteorological Satellite Program (DMSP). Since then, there has been at least one SSM/I orbiting the Earth at any

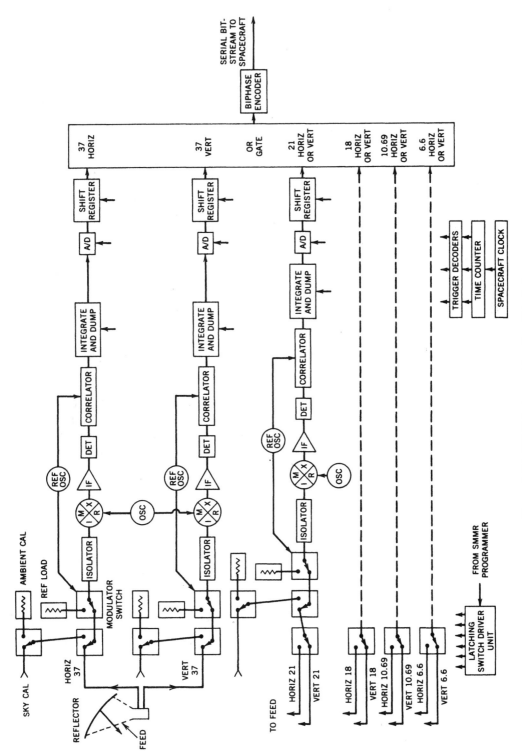

Figure 5-17. SMMR instrument functional block diagram.

time. The SSM/I instruments are carried on sun-synchronous polar orbiting satellites with an average inclination of 101° at an altitude of 860 km with a swath width of 1392 km. The SSM/I is a conically scanning radiometer that images the surface at a constant incidence angle of 53°. SSM/I operates at four frequencies: 19, 22, 37, and 85 GHz. Dual-polarization measurements are made at 19, 37, and 85 GHz for a total of seven radiometer channels.

The SSM/I instrument utilizes an offset parabolic reflector of dimensions 60 × 66 cm, fed by a corrugated, broad-band, seven-port horn antenna. The resolutions are 70 km × 45 km at 19 GHz, 60 km × 40 km at 22 GHz, 38 km × 30 km at 37 GHz, and 16 km × 14 km at 85 GHz. The reflector and feed are mounted on a drum that contains the radiometers, digital data subsystem, mechanical scanning subsystem, and power subsystem. The reflector–feed–drum assembly is rotated about the axis of the drum by a coaxially mounted bearing and power transfer assembly. All data, commands, timing and telemetry signals, and power pass through slip-ring connectors to the rotating assembly. The SSM/I rotates continuously at 31.6 rpm and measures the scene brightness temperatures over an angular sector of ±51.2° about the forward or aft directions. (Some of the satellites scan around the forward direction, and some around the aft; they are not all identical.) The spin rate provides a period of 1.9 sec during which the spacecraft sub-satellite point travels 12.5 km. Each scan of 128 discrete, uniformly spaced radiometric samples is taken at the two 85 GHz channels and, on alternate scans, 64 discrete samples are taken at the remaining five lower-frequency channels. The antenna beam intersects the Earth's surface at an incidence angle of 53.1° (as measured from the local Earth normal). This, combined with the conical scan geometry, results in a swath width of approximately 1400 km. The SSM/I sensor weighs 48.6 kg and consumes 45 watts. The data rate is 3276 bps.

One example of the use of SSM/I data is the routine mapping of sea ice concentrations in the Arctic and Antarctic regions. The algorithm used to identify sea ice is based on the observation that the microwave emission from sea ice and open water have very different polarization and frequency responses (Cavalieri et al., 1984, 1991). In general, open water shows brightness temperatures measured at vertical polarization significantly higher than those measured at horizontal polarization. The microwave emission from sea ice, however, shows little difference between brightness temperatures measured at the two different polarizations. Open water shows higher emission at 37 GHz than at 19 GHz, whereas the opposite is true for multiyear sea ice. First-year sea ice shows little difference in emission at 19 and 37 GHz. Figures 5-18 and 5-19 show the average sea ice concentrations for the Arctic and the Antarctic for 2003. The dynamic changes in sea ice concentration are clearly visible in these images. Data like this can be found on the Internet site of the National Snow and Ice Distribution Center (NSIDC).

5-5-3 Tropical Rainfall Mapping Mission Microwave Imager (TMI)

TMI is a total power microwave radiometer on the Tropical Rainfall Mapping Mission (TRMM) satellite that is used to provide quantitative information about precipitation in a wide swath under the satellite. TMI is based on the SSM/I design, with the addition of a dual-polarized 10.7 GHz channel for a total of five frequencies (10.7, 19.4, 21.3, 37, and 85.5 GHz).

The TMI antenna is an offset parabola, with an aperture size of 61 cm projected along the propagation direction. The antenna beam views the earth surface with an angle relative to the nadir of 49°, which results in an incident angle of 52.8° at the surface. The TMI

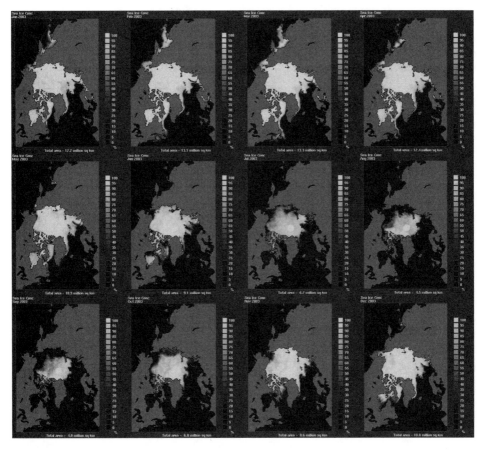

Figure 5-18. Monthly mean sea ice concentration for the northern hemisphere for the year 2003 (Fetterer and Knowles, 2002). See color section.

antenna rotates in a conical scan about the nadir axis at a constant speed of 31.6 rpm. Only the forward 130° of the scan circle is used for taking data. The rest is used for calibrations and other instrument housekeeping purposes. From the TRMM orbit of 350 km altitude, the 130° sector scanned yields a swath width of 758.5 km. During each complete revolution of the antenna, the TRMM subsatellite point advances by a distance of 13.9 km. Since the smallest footprint (85.5 GHz channels) size is only 6.9 km (down-track direction) by 4.6 km (cross-track direction), there is a gap of 7.0 km between successive scans. However, this is the only frequency at which there is a small gap. For all other frequency channels, footprints from successive scans overlap the previous scans.

5-5-4 Advanced Microwave Scanning Radiometer for EOS (AMSR-E)

The AMSR-E instrument is the latest addition to the family of spaceborne microwave radiometers, and was launched on NASA's Aqua satellite in April 2003. It is a six-frequency, conically scanning radiometer operating at frequencies 6.925, 10.65, 18.7, 23.8, 36.5, and 89 GHz. Table 5-3 summarizes some of the parameters of the AMSR-E instrument.

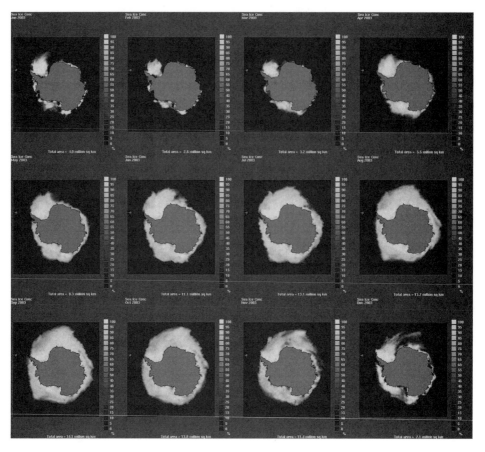

Figure 5-19. Monthly mean sea ice concentration for the southern hemisphere for the year 2003 (Fetterer and Knowles, 2002). See color section.

TABLE 5-3. AMSR-E Instrument Characteristics

Center frequency (GHz)	6.925	10.65	18.7	23.8	36.5	89
Ground spot size (km × km)	74 × 43	51 × 30	27 × 16	31 × 18	14 × 8	6 × 4
Sensitivity (K)	0.3	0.6	0.6	0.6	0.6	1.1

AMSR-E uses an offset parabolic antenna with a 1.6 m diameter that points at a fixed angle of 47.4° from the nadir, which means that it measures the radiation from a local surface angle of 55°. The instrument records energy over an arc of ±61° about the nadir track. From the orbital altitude of 705 km, this provides a swath width of 1445 km.

EXERCISES

5.1. Consider the case of a half-space with index of refraction n_1 covered by a thin layer of thickness L and index of refraction n_2. Derive the brightness temperature T of this medium, for both horizontal and vertical polarization. Plot the following:

(a) T as a function of L/λ for nadir looking

(b) T as a function of θ for $L/\lambda = 0.1$, 1, and 10

In both cases, assume $T_g = 300$ K and $T_g = 50$ K = constant. Consider both the cases of $n_1 = 3$, $n_2 = 8$, and $n_1 = 9$, $n_2 = 1.8$.

5.2. Calculate the brightness temperature of a rough surface characterized by the following scatter functions:

$$\sigma(\theta, \theta', \phi') = \rho_0/2\pi, \text{ where } \rho_0 = \text{constant}$$

$$\sigma(\theta, \theta', \phi') = \rho_0 \cos^2(\theta + \theta')/2\pi$$

5.3. Consider the case of a homogeneous half-space. Calculate and plot as a function of the look angle θ the value of $\Delta T/\Delta n$, which represents the change in T due to change in surface index of refraction n. Do the calculation for both horizontal and vertical polarization. Assume $T_g = 300$ K and examine both the cases where $T_s = 40$ K and $T_s = 40/\cos\theta$ K. Consider both the cases of $n = 2$ and $n = 9$

5.4. A radiometer is being used to map and locate the edge of the ocean ice cover where the surface microwave brightness temperature has a sudden jump from T_I to T_s. The antenna used has a pattern given by

$$G(\theta) = \begin{cases} G_0 & \text{for } -10° \leq \theta \leq 10° \\ 0 & \text{otherwise} \end{cases}$$

Plot the observed change in the temperature profile for the case $T_I = 295$ K and $T_s = 270$ K. Assume that the sensor altitude is 600 km and that it moves at a speed of 7 km/sec.

5.5. Repeat Exercise 5-4 for the case in which the antenna pattern is

$$G(\theta) = G_0\left(\frac{\sin\alpha}{\alpha}\right)^2$$

with $\alpha = \pi\theta/\theta_0$ and $\theta_0 = 10°$

5.6. Repeat the previous two exercises for the case in which the surface temperature is equal to T_I everywhere except for $-X \leq x \leq X$, where it is equal to T_s. Consider the two cases for which $X = 5$ km and $X = 20$ km.

5.7. An imaging radiometer uses a conical scanning approach to achieve a swath width of 1400 km from a 600 km altitude. The antenna is 2 m in diameter and the instrument operates at 10 GHz. The system has a bandwidth of 100 MHz and a system noise temperature of $T_{rec} = 700$ K. The gain fluctuations are controlled to $\Delta G_s/G_s = 10^{-3}$ Calculate:

(a) The surface resolution and antenna pointing angle (ignore the curvature of the earth)

(b) The scanning rate in revolutions per minute, assuming that the satellite moves at 7 km/sec

(c) The maximum dwell time for each pixel

(d) The sensitivity of the radiometer operated as a total power radiometer

(e) The sensitivity of the radiometer operated as an unbalanced Dicke radiometer if $T_a - T_{ref} = 10$ K

(f) The sensitivity of the radiometer operated as a balanced Dicke radiometer

Assume the antenna temperature is 300 K. In parts (e)–(f), assume that the integration time is equal to half the time it takes for the antenna footprint to move the distance equal to a footprint size. Repeat the calculations for an operating frequency of 36 GHz.

5.8. Repeat the previous exercise for a system that uses a planar scanning approach.

5.9. Plot the real part of the complex visibility function of the antenna configuration shown in Figure 5-A1. Assume that the individual antenna patterns are

$$A(\theta) = \cos \theta \quad -\frac{\pi}{2} \leq \theta \leq \frac{\pi}{2}$$

Assume that $x = \lambda/2$.

5.10. The microwave temperature of a soil surface covered by a vegetation canopy can be written as

$$T = (1-\rho_i)T_g e^{-\tau} + T_c(1-e^{-\tau}) + \rho_i T_c(1-e^{-\tau})e^{-\tau}$$

where T_g and T_c are the temperatures of the soil and the canopy, respectively. The subscript i refers to the polarization (either horizontal or vertical). The optical depth τ of the canopy is a function of the vegetation water content according to

$$\tau = bW_c$$

The constant b is a vegetation opacity coefficient that is determined experimentally and has a value of approximately 0.1 at the L band.

For dawn (6 a.m. local time) orbit overpasses, the soil and vegetation temperatures are approximately equal to the surface air temperature, and the brightness temperature can be written as

$$T = T_a(1 - \rho_i e^{-2\tau})$$

where T_a is the surface air temperature. Assume that the soil has a dielectric constant of 15. Calculate the expected change in microwave temperature for a 5%

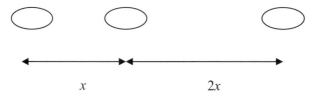

Figure 5-A1. Geometry for Problem 5-9.

change in soil dielectric constant, and plot this as a function of the vegetation water content.

REFERENCES AND FURTHER READING

Allison, J. J., E. B. Rodgers, T. T. Wilheit, and R. W. Fett. Tropical cyclone rainfall as measured by the Nimbus 5 electrically scanning microwave radiometer. *Bulletin of American Meterological Society,* **55**(9), 1974.

Bará, J., A. Camps, F. Torres, and I. Corbella. Angular resolution of two-dimensional, hexagonally sampled interferometric radiometers. *Radio Science,* **33**(5), 1459–1473, 1998.

Bayle, F., J.-P. Wigneron, Y. H. Kerr, P. Waldteufel, E. Anterrieu, J.-C. Orlhac, A. Chanzy, O. Marloie, M. Bernardini, S. Sobjaerg, J.-C. Calvet, J.-M. Goutoule, and N. Skou. Two-dimensional synthetic aperture images over a land surface scene. *IEEE Transactions on Geoscience and Remote Sensing,* **GE-40,** 710–714, 2002.

Blume, H. C., B. M. Kendall, and J. C. Fedors. Measurements of ocean temperature and salinity via microwave radiometry. *Boundary Layer Meteorology,* **13,** 295, 1977.

Campbell, W. J., P. Gloersen, W. J. Webster, T. T. Wilheit, and R. O. Ramseier. Beaufort sea ice zones as delineated by microwave imagery. *Journal of Geophysical Research,* **81**(6), 1976.

Cavalieri, D. J., P. Gloersen, and W. J. Campbell. Determination of sea ice parameters with the NIMBUS-7 SMMR. *Journal of Geophysical Research,* **89 (D4),** 5355–5369, 1984.

Cavalieri, D. J., J. Crawford, M.R. Drinkwater, D. Eppler, L.D. Farmer, R.R. Jentz, and C.C. Wackerman. Aircraft active and passive microwave validation of sea ice concentration from the DMSP SSM/I. *Journal of Geophysical Research,* **96(C12),** 21,989–22,009, 1991.

Eppler, D. T., L. D. Farmer, A. W. Lohanick, M. A. Anderson, D. Cavalieri, J. Comiso, P. Gloersen, C. Garrity, T. C. Grenfell, M. Hallikainen, J. A. Maslanik, C. Mätzler, R. A. Melloh, I. Rubinstein, and C. T. Swift. Passive microwave signatures of sea ice. In *Microwave Remote Sensing of Sea Ice,* F. D. Carsey (Ed.), pp. 47–71, American Geophysical Union, Washington, D.C., 1992.

Fetterer, F., and K. Knowles. *Sea Ice Index.* National Snow and Ice Data Center, Boulder, CO. Digital media. 2002. Updated 2004.

Grody, N. C., and A. N. Basist. Global identification of snowcover using SSM/I measurements. *IEEE Transactions on Geoscience and Remote Sensing,* **GE-34,** 237–249, 1996.

Hofer, R., E. G. Njoku, and J. W. Waters. Microwave radiometric measurements of sea surface temperature from the Seasat satellite: First Results. *Science,* **212,** 1385, 1981.

Hollinger, J. P., J. L. Peirce, and G. A. Poe. SSM/I instrument evaluation. *IEEE Transactions on Geoscience and Remote Sensing,* **GE-28,** 781–790, 1990.

Jackson, T. J., and T. J. Schmugge. Vegetation effects on the microwave emission of soils. *Remote Sensing of the Environment,* **36,** 203–212, 1991.

Jackson, T. J., and D. M. Le Vine. Mapping surface soil moisture using an aircraft-based passive microwave instrument: algorithm and example. *Journal of Hydrology,* **184,** 85–99, 1996.

Jackson, T. J., D. M. Le Vine, A. Y. Hsu, A. Oldak, P. J. Starks, C. T. Swift, J. Isham, and M. Haken. Soil moisture mapping at regional scales using microwave radiometry: the Southern Great Plains hydrology experiment. *IEEE Transactions on Geoscience and Remote Sensing,* **GE-37,** 2136–2151, 1999.

Jackson, T. J. Multiple resolution analysis of L-band brightness temperature for soil moisture. *IEEE Transactions on Geoscience and Remote Sensing,* **GE-39,** 151–164, 2001.

Jackson, T. J., and A. Y. Hsu. Soil moisture and TRMM microwave imager relationships in the Southern Great Plains 1999 (SGP99) experiment. *IEEE Transactions on Geoscience and Remote Sensing,* **GE-39,** 1632–1642, 2001.

Judge, J., J. F. Galantowicz, A. W. England, and P. Dahl. Freeze/thaw classification for prairie soils using SSM/I radiobrightnesses. *IEEE Transactions on Geoscience and Remote Sensing,* **GE-35,** 827–832, 1997.

Judge, J., J. F. Galantowicz, and A. W. England. A comparison of ground-based and satellite-borne microwave radiometric observations in the Great Plains. *IEEE Transactions on Geoscience and Remote Sensing,* **GE-39,** 1686–1696, 2001.

Kawanishi, T., T. Sezai, Y. Ito, K. Imaoka, T. Takeshima, Y. Ishido, A. Shibata, M. Miura, H. Inahata, and R. W. Spencer. The Advanced Microwave Scanning Radiometer for the Earth Observing System (AMSR-E), NASDA's contribution to the EOS for global energy and water cycle studies. *IEEE Transactions on Geoscience and Remote Sensing,* **GE-41,** 184–194, 2003.

Kerr, Y., J. Font, P. Waldteufel, A. Camps, J. Bará, I. Corbella, F. Torres, N. Duffo, M. Vall.llosera, and G. Caudal. Next generation radiometers: SMOS a dual pol L-band 2D aperture synthesis radiometer. In *IEEE 2000 Aerospace Conference Proceedings,* Big Sky, MT, March 18–25, 2000.

Kerr, Y. H., P. Waldteufel, J.-P. Wigneron, J. Martinuzzi, J. Font, and M. Berger. Soil moisture retrieval from space: The Soil Moisture and Ocean Salinity (SMOS) mission. *IEEE Transactions on Geoscience and Remote Sensing,* **GE-39,** 1729–1735, 2001.

Fetterer, F., and K. Knowles. *Sea Ice Index.* Boulder, CO: National Snow and Ice Data Center. Digital media, 2002.

Kunzi, K. F., A. D. Fisher, D. H. Staelin, and J. W. Waters. Snow and ice surfaces measured by the Nimbus 5 microwave spectrometer. *Journal of Geophysical Research,* **81**(27), 1976.

Le Vine, D. M., A. J. Griffis, C. T. Swift, and T. J. Jackson. ESTAR: A synthetic aperture radiometer for remote sensing applications. *Proceedings of the IEEE,* **82-12,** 1787–1801, 1994.

Le Vine, D. M., C. T. Swift, and M. Haken. Development of the synthetic aperture microwave radiometer, ESTAR. *IEEE Transactions on Geoscience and Remote Sensing,* **GE-39,** 199–202, 2001.

Le Vine, D.M., T. J. Jackson, C. T. Swift, M. Haken, and S. W. Bidwell. ESTAR measurements during the Southern Great Plains experiment (SGP99). *IEEE Transactions on Geoscience and Remote Sensing,* **GE-39,** 1680–1685, 2001.

Lipes, R. G., R. L. Bernstein, V. J. Cardone, D. G. Katsaros, E. G. Njoku, A. L. Riley, D. B. Ross, C. T. Swift, and F. J. Wentz. Seasat Scanning Multi-Channel Microwave Radiometer: Results of the Gulf of Alaska Workshop. *Science,* **204**(4400), 1979.

Newton, R. W., and J. Rouse. Microwave radiometer measurements of soil moisture content. *IEEE Trans. Ant. Prop.,* **AP-28,** 680–686, 1980.

Njoku, E. G., J. M. Stacey, and F. T. Barath. The Seasat Scanning Multi-Channel Microwave Radiometer (SMMR): Instrument description and performance. *IEEE Journal of Oceanic Engineering,* **OE-5,** 100, 1980.

Njoku, E. G., E. J. Christensen, and R. E. Cofield. The Seasat Scanning Multi-Channel Microwave Radiometer (SMMR): Antenna pattern corrections development and implementation. *IEEE Journal of Oceanic Engineering,* **OE-5,** 125, 1980.

Njoku, E. G. Passive microwave remote sensing of the earth from space—A review. *Proceedings of the IEEE,* **70**(7), 728–750, 1982.

Njoku, E. G., T. J. Jackson, V. Lakshmi, T. K. Chan, and S. V. Nghiem. Soil moisture retrieval from AMSR-E. *IEEE Transactions on Geoscience and Remote Sensing,* **GE-41,** 215–229, 2003.

Paloscia, S., G. Macelloni, E. Santi, and T. Koike. A multifrequency algorithm for the retrieval of soil moisture on a large scale using microwave data from SMMR and SSM/I satellites. *IEEE Transactions on Geoscience and Remote Sensing,* **GE-39,** 1655–1661, 2001.

Parkinson, C. L., D. J. Cavalieri, P. Gloersen, H. J. Zwally, and J. C. Comiso. Arctic sea ice extents, areas, and trends, 1978-1996. *Journal of Geophysical Research,* **104**(C9), 20,837–20,856, 1999.

Partington, K. C. A data fusion algorithm for mapping sea-ice concentrations from Special Sensor Microwave/Imager data. *IEEE Transactions on Geoscience and Remote Sensing,* **GE-38,** 1947–1958, 2000.

Ruff, C. S., C. T. Swift, A. B. Tanner, and D. M. Le Vine. Interferometric synthetic aperture microwave radiometry for the remote sensing of the Earth. *IEEE Transactions on Geoscience and Remote Sensing,* **GE-26**(5), 597–611, 1988.

Schmugge, T. J. Remote sensing of soil moisture: Recent advances. *IEEE Transactions on Geoscience and Remote Sensing,* **GE-21**(3), 336–344, 1983.

Skou, N. *Microwave Radiometer Systems—Design and Analysis.* Artech House, Dedham, MA, 1989.

Staelin, D. H. Passive remote sensing at microwave wavelengths. *Proceedings of the IEEE,* **57**(4), 427–439, 1969.

Steffen, K., J. Key, D. Cavalieri, J. Comiso, P. Gloersen, K. S. Germain, and I. Rubinstein. The estimation of geophysical parameters using passive microwave algorithms. In *Microwave Remote Sensing of Sea Ice,* F. D. Carsey (Ed.), pp. 201–231, American Geophysical Union, Washington, DC, 1992.

Thompson, A. R., J. M. Moran, and G. W. Swenson. *Interferometry and Synthesis in Radio Astronomy.* Wiley, New York, 1986.

Ulaby, F. T., and K. R. Carver. Passive microwave remote sensing. In *Manual of Remote Sensing,* Vol. 1, pp. 475–516, American Society of Photogrammetry, Falls Church, VA, 1983.

Ulaby, F. T., R. K. Moore, and A. K. Fung. *Microwave Remote Sensing,* Vol. 1. Addison-Wesley, Reading, MA, 1981.

Ulaby, F. T., R. K. Moore, and A. K. Fung. *Microwave Remote Sensing,* Vol. 2. Addison-Wesley, Reading, MA, 1982.

Ulaby, F. T., R. K. Moore, and A. K. Fung. *Microwave Remote Sensing,* Vol. 3. Artech House, Dedham, MA, 1985.

Wigneron, J. P., P. Waldteufel, A. Chanzy, J.-C. Calvet, and Y. Kerr. Two-dimensional microwave interferometer retrieval capabilities over land surface (SMOS mission). *Remote Sensing of the Enviroment,* **73,** 270–282, 2000.

Zwally, H. J., and P. Gloersen. Passive microwave images of the polar regions and research applications. *Polar Research,* **18**(116), 1976.

6

SOLID-SURFACE SENSING: MICROWAVE AND RADIO FREQUENCIES

Active microwave sensors are playing a large role in remote sensing of planetary surfaces. Spaceborne imaging radars, scatterometers, and altimeters were developed in the late 1970s and early 1980s. Advanced radar implementations, such as polarimetric and interferometric systems, were developed in the 1980s, and were flown in space in the mid to late 1990s. They provide information about the surface physical (topography, morphology, and roughness) and dielectric properties. In some cases, they provide information about the subsurface properties. These sensors are attractive because they operate independently of sun illumination and usually are not sensitive to weather conditions or cloud cover. Thus, they are particularly suitable to monitoring dynamic phenomena with short time constants in which repetitive observation is required regardless of "optical visibility" conditions.

6-1 SURFACE INTERACTION MECHANISM

Any interface separating two media with different electric or magnetic properties will affect an electromagnetic wave incident on it. Let us assume that a plane electromagnetic wave is incident on an interface separating a vacuum half-space (Earth's atmosphere acts almost like a vacuum at the frequencies used in spaceborne radars) and a dielectric half-space of dielectric constant ε (see Fig. 6-1a). The electromagnetic wave will interact with the atoms in the dielectric. These would then become small electromagnetic oscillators and radiate waves in all directions. Thus, some energy will be radiated toward the upper as well as the lower half-space (Fig. 6-1a).

If the surface is perfectly flat, the incident wave will excite the atomic oscillators in the dielectric medium at a relative phase such that the reradiated field consists of two plane

Introduction to the Physics and Techniques of Remote Sensing. By C. Elachi and J. van Zyl
Copyright © 2006 John Wiley & Sons, Inc.

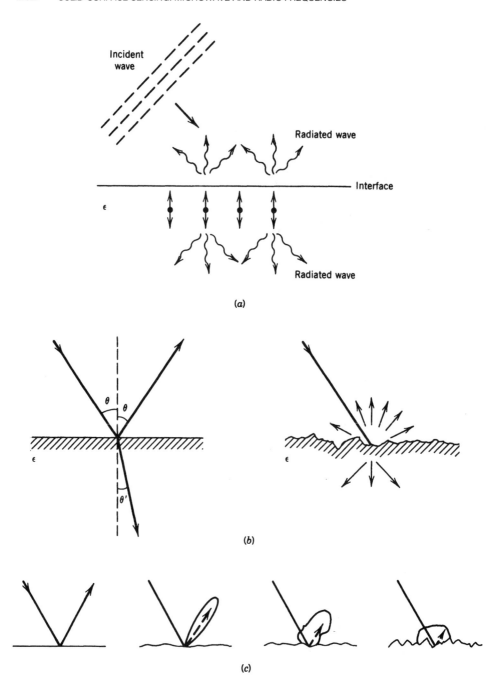

Figure 6-1. (a) An incident wave on a dielectric half-space will excite the dielectric atoms, which become small oscillating dipoles. These dipoles reradiate waves in both half-spaces. (b) In the case of a flat interface (left), the reradiated waves are in two specific directions: reflection and refraction. In the case of a rough surface (right), waves are reradiated in all directions. (c) Pattern of reradiated waves in the upper half-space for increasingly rough interfaces.

waves (see Fig. 6-1*b*): one in the upper medium at an angle equal to the angle of incidence—this is the reflected wave—and one in the lower medium at an angle θ' equal to

$$\theta' = \arcsin\left(\frac{\sin\theta}{\sqrt{\varepsilon}}\right) \tag{6-1}$$

where θ is the incidence angle. This is the refracted or transmitted wave.

If the surface is rough, then some of the energy is reradiated in all directions. This is the scattered field (Fig. 6-1*b*). The amount of energy scattered in all directions other than the Fresnel reflection direction is dependent on the magnitude of the surface roughness relative to the wavelength of the incident wave (see Fig. 6-1*c*). In the extreme case in which the surface is very rough, the energy is scattered equally in all directions. Of particular interest for spaceborne imaging radars is the amount of energy scattered back toward the sensor. This is characterized by the surface backscatter cross section $\sigma(\theta)$.

The backscatter cross section is defined as the ratio of the energy received by the sensor over the energy that the sensor would have received if the surface had scattered the energy incident on it in an isotropic fashion. The backscatter cross section is usually expressed in dB (decibels), and is given by

$$\sigma = 10\log(\text{energy ratio})$$

The backscatter cross section expressed in dB can be a positive number (focusing of energy in the back direction) or a negative number (focusing of energy away from the back direction).

6-1-1 Surface Scattering Models

Radar scattering from a surface is strongly affected by the surface geometrical properties. The surface small-scale geometric shape (also called roughness) can be statistically characterized by its standard deviation relative to a mean flat surface. The surface standard deviation (commonly referred to as the surface roughness) is the root mean square of the actual surface deviation from this average surface. However, knowing the surface standard deviation does not provide a complete description of the surface geometrical properties. It is also important to know how the local surface deviation from the mean surface is related to the deviation from the mean surface at other points on the surface. This is mathematically described by the surface height autocorrelation function. The surface correlation length is the separation after which the deviation from the mean surface for two points are statistically independent. Mathematically, it is the length after which the autocorrelation function is less than $1/e$.

Consider first the case of a perfectly smooth surface of infinite extent that is uniformly illuminated by a plane wave. This surface will reflect the incident wave into the specular direction with scattering amplitudes equal to the well-known Fresnel reflection coefficients as described in the previous section. In this case, no scattered energy will be received in any other direction. If the surface is now made finite in extent, or if the infinite surface is illuminated by a finite-extent uniform plane wave, the situation changes. In this case, the far-field power will decrease proportionally to the distance squared (the well-known R-squared law). The maximum amount of reflected power still appears in the specular direction, but a lobe structure, similar to an "antenna pattern," appears around the specular direction. The exact shape of the lobe structure depends on the size and the shape of the finite illuminated area, and the pattern is adequately predicted by using physical optics calculations. This component of the scattered power is often referred to as the *coherent component* of the scattered field. For angles far away from the specular direction, there will be very little scattered power in the coherent component.

The next step is to add some roughness to the finite surface such that the root mean square (r.m.s.) height of the surface is still much less than the wavelength of the illuminating source. The first effect is that some of the incident energy will now be scattered in directions other than the specular direction. The net effect of this scattered energy is to fill the nulls in the "antenna pattern" of the coherent field described above. The component of the scattered power that is the result of the presence of surface roughness is referred to as the *incoherent component* of the scattered field. At angles significantly away from the specular direction, such as the backscatter direction at larger incidence angles, the incoherent part of the scattered field usually dominates.

As the roughness of the surface increases, less power is contained in the coherent component, and more in the incoherent component. In the limit in which the r.m.s. height becomes significantly larger than the wavelength, the coherent component is typically no longer distinguishable, and the incoherent power dominates in all directions. In this limit, the strength of the scattering in any given direction is related to the number of surface facets that are oriented such that they reflect specularly in that direction. This is the same phenomenon that causes the shimmering reflection of the moon on a roughened water surface.

Several different criteria exist to decide if a surface is "smooth" or "rough." The most commonly used one is the so-called Rayleigh criterion that classifies a surface as rough if the r.m.s. height satisfies $h > \lambda/8 \cos \theta$. Here θ is the angle at which the radar wave is incident on the surface. A more accurate approximation of surface roughness was introduced by Peake and Oliver (1971). According to this approximation, a surface is considered smooth if $h < \lambda/25 \cos \theta$ and is considered rough if $h > \lambda/4 \cos \theta$. Any surface that falls in between these two values is considered to have intermediate roughness.

Depending on the angle of incidence, two different approaches are used to model radar scattering from rough natural surfaces. For small angles of incidence, scattering is dominated by reflections from appropriately oriented facets on the surface. In this regime, physical optics principles are used to derive the scattering equations. As a rule of thumb, facet scattering dominates for angles of incidence less than 20–30°. For the larger angles of incidence, scattering from the small-scale roughness dominates. The best known model for describing this type of scattering is the small-perturbation model, often referred to as the Bragg model. This model, as its name suggests, treats the surface roughness as a small perturbation from a flat surface. More recently, Fung et al. (1992) proposed a model based on an integral equation solution to the scattering problem that seems to describe the scattering adequately in both limits.

First-Order Small-Perturbation Model. The use of the first-order small-perturbation model to describe scattering from slightly rough surfaces dates back to Rice (1951). Rice used a perturbation technique to show that to the first order, scattering cross sections of a slightly rough surface can be written as

$$\sigma_{xx} = 4\pi k^4 h^2 \cos^4 \theta |\alpha_{xx}|^2 \, W(2k \sin \theta, 0); \qquad xx = hh \text{ or } vv$$
$$\sigma_{hv} = 0$$

$$(6\text{-}2)$$

where $k = 2\pi/\lambda$ is the wavenumber, λ is the wavelength, and θ is the local incidence angle at which the radar waves impinge on the surface. The roughness characteristics of the surface are described by two parameters: h is the surface r.m.s. height and $W(\xi_x, \xi_y)$ is the two-dimensional normalized surface roughness spectrum, which is the Fourier transform of the two-dimensional normalized surface autocorrelation function. Note that the surface r.m.s. height should be calculated *after* local slopes have been removed from the surface profile; the slope of the surface changes the radar cross section because of the change in the local incidence angle. Local slopes that tilt toward or away from the radar do not change the surface roughness; instead, they affect the local incidence angle. This is a frequent source of error in laboratory and field experiments.

The two most commonly used roughness spectrum functions are those for surfaces that are described by Gaussian and exponential correlation functions. These two roughness spectrum functions are

$$W_g(k_x, k_y) = \frac{l^2}{\pi} \exp\left[\frac{-(k_x^2 + k_y^2)l^2}{4}\right] \tag{6-3}$$

and

$$W_e(k_x, k_y) = \frac{2l^2}{\pi[1 + (k_x^2 + k_y^2)l^2]^{3/2}} \tag{6-4}$$

The surface electrical properties are contained in the variable α_{xx}, which is given by

$$\alpha_{hh} = \frac{(\varepsilon - 1)}{(\cos\theta + \sqrt{\varepsilon - \sin^2\theta})^2} \tag{6-5}$$

$$\alpha_{vv} = \frac{(\varepsilon - 1)[(\varepsilon - 1)\sin^2\theta + \varepsilon]}{(\varepsilon\cos\theta + \sqrt{\varepsilon - \sin^2\theta})^2} \tag{6-6}$$

Here ε is the dielectric constant, or relative permittivity, of the soil. We note that the small-perturbation model as described here is applicable only to relatively smooth surfaces. The usual assumptions are that the roughness is small compared to the wavelength ($kh < 0.3$) and that the r.m.s. slope satisfies $s < 0.3$. Note that the first-order small-perturbation model predicts no depolarization, that is, no power is observed in the polarization orthogonal to that used for transmission. If the calculations are extended to the second order, a cross-polarized component is predicted (Valenzuela, 1967).

Several researchers have measured profiles of microtopography in order to better describe the roughness characteristics of natural surfaces. The profiles of microtopography are measured using various different approaches. The simplest approach utilizes a large board with a grid painted on it. The board is then pushed into the surface to the lowest point on the surface, and a photograph is taken of the board covered with the surface profile. The profile is then later digitized from the photograph. The advantage of this approach is that it is easy to make the measurement, and the equipment is relatively cheap and easily operated in the field. Disadvantages include the fact that only relatively short profiles can be measured (typically a meter or two at best), and that the soil has to be soft enough to push the board into.

A second approach utilizes a horizontal bar with an array of vertical rods of equal length that are dropped to the surface. The heights of the top of the rods above a known level surface are then measured and recorded. While relatively easier to operate than the boards described above, especially over rocky or hard surfaces, the disadvantage of this method is still the limited length of the profiles than can be measured with one instrument, leading to a large amount of time required to make measurements of reasonably large areas, especially in regions with difficult access.

Laser profilers are also sometimes used to measure microtopography. In this case, a laser is mounted on a frame that allows the laser to be translated as a raster pattern. Measurements are typically take every centimeter or so along a particular profile. These instruments obviously require power to operate, limiting their utility to easily accessible areas. An additional drawback is that the size of the frame usually limits the area that can be measured to a meter or so square. Lasers are also sometimes operated from low-flying aircraft, but in this case, the measurement density is inadequate for microtopography studies.

Stereo photography, either at close range or from specially equipped helicopters, seems to provide the best balance between accuracy and coverage. The photographs are digitized and then correlated against each other to reconstruct the microtopography using the same kind of software developed to construct large-scale digital elevation models from stereo cameras flown on either aircraft or satellites. Although more expensive to acquire, the full three-dimensional surface can be reconstructed over a relatively large area, leading to improved statistics.

Using stereo photographs acquired from a helicopter, Farr (1992) studied the roughness characteristics of several lava flows in the Mojave Desert in southern California. He found that the power spectra of these natural surfaces exhibit the general form

$$W(k) = bk^m \tag{6-7}$$

with the value of the exponent m between -2 and -3, consistent with known behavior of topography at larger scales. His measurements showed values closer to -2, and that the magnitude of m first seems to increase with lava-flow age, but then decreases for surfaces between 0.14 and 0.85 million years old. For surfaces older than 0.85 million years, the magnitude of m seems to increase again. The change in surface roughness is consistent with a model of erosion of the surface in arid terrain. At first, the sharper edges of the lava flows are eroded and wind-blown material tends to fill in the lower spots in the surface, gradually reducing the surface roughness. As the surface ages, however, water erosion starts cutting channels into the surface, roughening the profile again.

Shi et al. (1997) report a different approach to their analysis of surface roughness characteristics. Using 117 roughness profiles measured over various fields in the Washita watershed, they fitted the correlation function of the measured profiles with a general correlation function of the form

$$\rho(r) = \exp[-(r/l)^n] \tag{6-8}$$

Values of $n = 1$ correspond to an exponential correlation function, whereas $n = 2$ corresponds to a Gaussian function. Their results indicate that 76% of the surfaces could be fitted with values of $n \leq 1.4$, leading to the conclusion that the exponential correlation function is the more appropriate description of the surface correlation function.

We note that for values of $kl \gg 1$ the roughness spectrum of the exponential correlation function behaves like Equation (6-7) with an exponent of -3. The results from the Shi, et al. (1997) study seem to indicate that agriculture and pasture fields have roughness spectra that contain more energy at the longer spatial scales than the natural lava flow surfaces studied by Farr.

The electrical properties of a rough surface is described by the complex dielectric constant, or relative permittivity, of the soil, which is a strong function of the soil moisture. This is the result of the fact that the dielectric resonance of both pure and saline water lies in the microwave portion of the electromagnetic spectrum. Dry soil surfaces have dielectric constants on the order of 2 to 3, whereas water has a dielectric constant of approximately 80 at microwave frequencies. Therefore, adding a relatively small amount of water to the soil drastically changes the value of the dielectric constant.

A wet, bare soil consists of a mixture of soil particles, air, and liquid water. Usually, the water contained in the soil is further divided into two parts, so-called *bound* and *free* water. Due to the influence of matric and osmotic forces, the water molecules contained within the first few molecular layers surrounding the soil particles are tightly held by the soil particles, hence the term bound water. The amount of bound water is directly proportional to the surface area of the soil particles, which, in turn, is a function of the soil texture and mineralogy. Because of the relatively strong forces acting on it, bound water exhibits an electromagnetic spectrum that is different from that of regular liquid water. Because the matric forces acting on a water molecule decrease rapidly with distance away from the soil particle, water molecules located more than a few molecular layers away from the soil particles are able to move throughout the soil with relative ease; this is known as free water for this reason. The complex dielectric constant of both bound and free water is a function of frequency, temperature, and salinity of the soil.

Dobson et al. (1985) derived a semiempirical relationship between the dielectric constant and the volumetric soil moisture, m_v, of the form [see also Peplinski et al. (1995)]

$$\varepsilon = \left[1 + \frac{\rho_b}{\rho_s} \, \varepsilon_s^\alpha + m_v^\beta \varepsilon_{fw}^\alpha - m_v \right]^{1/\alpha} \qquad (6\text{-}9)$$

where $\alpha = 0.65$, $\rho_s = 2.66 \ g/cm^3$, ε_s is the dielectric constant of the solid soil (typical value ≈ 4.7), ρ_b is the bulk density of the soil (on the order of 1.1 g/cm^3 for sandy loam soils), and β is a parameter that is a function of the soil texture:

$$\beta = 1.2748 - 0.00519S - 0.00152C \qquad (6\text{-}10)$$

S and C are the percentage of sand and clay in the soil, respectively. The dielectric constant of free water is a function of temperature and frequency, and is given by

$$\varepsilon_{fw} = \varepsilon_{w\infty} + \frac{\varepsilon_{w0} - \varepsilon_{w\infty}}{1 + (2\pi f \tau_w)^2} \qquad (6\text{-}11)$$

In Equation (6-11), τ_w is the relaxation time for water, ε_{w0} is the static dielectric constant for water, $\varepsilon_{w\infty} = 4.9$ is the high-frequency limit of the real part of the dielectric constant for water, and f is the frequency. Both τ_w and ε_{w0} are functions of temperature (Ulaby et al., 1986). At 20°C, the values are $2\pi\tau_w = 0.58 \times 10^{-10}$ seconds and $\varepsilon_{w0} = 80.1$.

Equation (6-9) describes a nonlinear relationship between the dielectric constant and the volumetric soil moisture of a surface, and has been used to explain the observed nonlinear relationship between these two quantities, as shown in Figure 5-7.

The Integral Equation Model. Fung et al. (1992) showed that the expressions for the tangential surface fields on a rough dielectric surface can be written as a pair of integral equations. The scattered fields, in turn, are written in terms of these tangential surface fields. Using this formulation and standard approximations they showed that the scattered field can be interpreted as a single scattering term and a multiple scattering term. When the surface is smooth enough, the single scattering term reduces to the well-known small-perturbation model described above, and the cross-polarized terms reduce numerically to the second-order small-perturbation result. Their results also show that in the high-frequency limit, only the well-known Kirchoff term described by the physical optics model remains significant for surfaces with small r.m.s. slopes. When the surface r.m.s. slopes are large, however, the multiple scattering terms are important.

Fung et al. (1992) showed that the single scattering backscatter cross sections can be written as

$$\sigma_{xy} = \frac{k^2}{2} \exp(-2k^2 h^2 \cos^2 \theta) \sum_{n=1}^{\infty} h^{2n} |I_{xy}^n|^2 \frac{W^n(-2k \sin \theta, 0)}{n!} \tag{6-12}$$

with

$$I_{xy}^n = (2k \cos \theta)^n f_{xy} \exp(-k^2 h^2 \cos^2 \theta) + \frac{k^n \cos^n \theta [F_{xy}(-k \sin \theta, 0) + F_{xy}(k \sin \theta, 0)]}{2} \tag{6-13}$$

The term W^n is the Fourier transform of the nth power of the surface correlation function, which can be calculated using

$$W^n(k) = \frac{2}{\pi} \int_0^{\infty} r \rho^n(r) J_0(kr) dr \tag{6-14}$$

where $J_0(x)$ is the Bessel function of the first kind and order zero. Also,

$$f_{hh} = \frac{-2R_h}{\cos \theta}; \quad f_{vv} = \frac{2R_v}{\cos \theta}; \quad f_{hv} = f_{vh} = 0 \tag{6-15}$$

where R_h and R_v the well-known Fresnel reflection coefficients for horizontal and vertical polarization, respectively. Finally,

$$F_{hh}(-k \sin \theta, 0) + F_{hh}(k \sin \theta, 0) =$$
$$\frac{-2 \sin^2 \theta (1 + R_h)^2}{\cos \theta} \left[\left(1 - \frac{1}{\mu}\right) + \frac{\mu \varepsilon - \sin^2 \theta - \mu \cos^2 \theta}{\mu^2 \cos^2 \theta} \right] \tag{6-16}$$

$$F_{vv}(-k \sin \theta, 0) + F_{vv}(k \sin \theta, 0) =$$
$$\frac{2 \sin^2 \theta (1 + R_v)^2}{\cos \theta} \left[\left(1 - \frac{1}{\varepsilon}\right) + \frac{\mu \varepsilon - \sin^2 \theta - \varepsilon \cos^2 \theta}{\varepsilon^2 \cos^2 \theta} \right] \tag{6-17}$$

$$F_{hv}(-k \sin \theta, 0) + F_{hv}(k \sin \theta, 0) = 0 \tag{6-18}$$

where μ is the relative permeability of the surface, and ε is the relative permittivity, or dielectric constant. Note again that the single scattering term does not predict any depolarization. The cross-polarized return is predicted by the multiple scattering term. The expressions are quite complicated and are given in Fung et al. (1992). Figure 6-2 shows the predicted backscatter cross section as a function of incidence angle for different surface roughness values and different dielectric constants. The plot on the left shows that increasing the surface roughness generally causes an increase in the radar cross sections for all polarization combinations. Notice how the difference between the HH and VV cross-sections becomes smaller as the surface gets rougher. The plot on the right shows that increasing the dielectric constant (or soil moisture) also increases the radar cross sections for all polarizations. In this case, however, increasing the dielectric constant also increases the difference between the HH and VV cross sections.

6-1-2   Absorption Losses and Volume Scattering

All natural materials have a complex dielectric constant ε:

$$\varepsilon = \varepsilon' + i\varepsilon'' \tag{6-19}$$

where the imaginary part corresponds to the ability of the medium to absorb the wave and transform its energy into another type of energy (heat, chemical, etc.). If we consider a wave propagating in a homogeneous medium, then

$$E = E_0 e^{i\sqrt{\varepsilon}kx}$$

If we assume that $\varepsilon'' \ll \varepsilon'$, then

$$\sqrt{\varepsilon} = \sqrt{\varepsilon' + i\varepsilon''} = \sqrt{\varepsilon'} + \frac{i\varepsilon''}{2\sqrt{\varepsilon'}} \tag{6-20}$$

and

$$E = E_0 e^{-\alpha_a x} e^{i\sqrt{\varepsilon'}kx} \tag{6-21}$$

where

$$\alpha_a = \frac{\varepsilon'' k}{2\sqrt{\varepsilon'}} = \frac{\pi\varepsilon''}{\lambda\sqrt{\varepsilon'}}$$

and the power of the wave as a function of x can be written as

$$P(x) = P(0)e^{-2\alpha_a x} \tag{6-22}$$

If α_a is a function of x, then the above equation will become

$$P(x) = P(0) \exp\left(-2\int_0^x \alpha_a(\xi)d\xi\right) \tag{6-23}$$

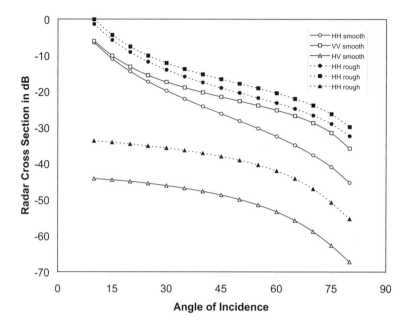

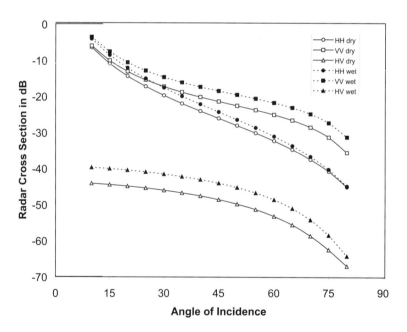

Figure 6-2. The predicted radar cross sections for a slightly rough surface assuming an exponential correlation function. The top figure shows the effect of changing surface roughness and constant dielectric constant; the bottom right shows the effect of changing dielectric constant for constant roughness.

The penetration depth L_p is defined as the depth at which the power decreases to $P(0)e^{-1}$ (i.e., 4.3 dB loss). Thus,

$$L_{\mathrm{p}} = \frac{1}{2\alpha_a} = \frac{\lambda\sqrt{\varepsilon'}}{2\pi\varepsilon''} \tag{6-24}$$

This can also be expressed as a function of the medium loss tangent ($\tan\delta = \varepsilon''/\varepsilon'$):

$$L_{\mathrm{p}} = \frac{\lambda}{2\pi\sqrt{\varepsilon'}\tan\delta} \tag{6-25}$$

Typically, dry low-loss soils have $\varepsilon' = 3$ and $\tan\delta = 10^{-2}$, which give $L_p = 9.2\,\lambda$.

At this depth, the incident power is 4.3 dB weaker than at the surface. In the case of radar observation, if the interface is covered by a layer of thickness L_p, the attenuation due to absorption alone will be $2 \times 4.3 = 8.6$ dB for a normally incident wave. The factor 2 represents the fact that absorption affects both the incident and the scattered waves.

The loss tangents of natural surfaces vary over a wide range. For pure ice, dry soil, and permafrost they are usually less than 10^{-2}. For wet soil, sea ice, and vegetation, they are usually around 10^{-1}. The loss tangent always increases with the percentage of liquid water present in the propagation medium.

In the case of an inhomogeneous medium such as a vegetation canopy, the propagating wave loses part of its energy by scattering in all directions. Part of the scattered energy is in the back direction. An inhomogeneous medium can be modeled as a collection of uniformly distributed identical scatterers of backscatter cross section σ_i and extinction cross section α_i. The extinction cross section includes both absorption and scattering losses. If the effect of multiple scattering is neglected, the backscatter cross section of a layer of scatterers of thickness H can be written as

$$\sigma_\rho = \int_0^H \sigma_\nu e^{-2\alpha z/\cos\theta}\,dz = \frac{\sigma_\nu\cos\theta}{2\alpha}[1 - e^{-2\alpha H\sec\theta}] \tag{6-26}$$

where

$$\sigma_\nu = N\sigma_i$$

$$\alpha = N\alpha_i$$

N is the number of scatterers per unit volume. If the layer covers a surface with backscatter cross section σ_s, the total backscatter coefficient can be written as

$$\sigma = \sigma_\rho + \sigma_s e^{-2\alpha H\sec\theta} + \sigma_{sv} \tag{6-27}$$

The third term represents energy that interacted with the surface and the volume scatterers, and is commonly referred to as the double-bounce term. All the variables in Equations (6-26) and (6-27) could be different for different polarization combinations if the individual scatterers are not randomly oriented.

Dielectric cylinders are commonly used to model branches and trunks of trees. Depending on what type of tree is being modeled, different statistical orientations are assumed for the branches. Grasslands and crops like wheat are also modeled using thin dielectric cylinders as the individual scatterers. In this case, the scatterers are assumed to be oriented with a narrow statistical distribution around the vertical. If the plants being modeled have large leaves, they are typically modeled using dielectric disks.

6-1-3 Effects of Polarization

Electromagnetic wave propagation is a vector phenomenon, that is, all electromagnetic waves can be expressed as complex vectors. Plane electromagnetic waves can be represented by two-dimensional complex vectors. This is also the case for spherical waves when the observation point is sufficiently far removed from the source of the spherical wave. Therefore, if one observes a wave transmitted by a radar antenna when the wave is a large distance from the antenna (in the far field of the antenna), the radiated electromagnetic wave can be adequately described by a two-dimensional complex vector. If this radiated wave is now scattered by an object, and one observes this wave in the far field of the scatterer, the scattered wave can again be adequately described by a two-dimensional vector. In this abstract way, one can consider the scatterer as a mathematical operator that takes one two-dimensional complex vector (the wave impinging upon the object) and changes it into another two-dimensional vector (the scattered wave). Mathematically, therefore, a scatterer can be characterized by a complex 2×2 scattering matrix:

$$\mathbf{E}^{SC} = [\mathbf{S}]\mathbf{E}^{tr} \tag{6-28}$$

where $\mathbf{E}^{tr}$ is the electric field vector that was transmitted by the radar antenna, $[\mathbf{S}]$ is the 2×2 complex scattering matrix that describes how the scatterer modified the incident electric field vector, and $\mathbf{E}^{SC}$ is the electric field vector that is incident on the radar receiving antenna. This scattering matrix is a function of the radar frequency and the viewing geometry.

The voltage measured by the radar system is the scalar product of the radar antenna polarization and the incident wave electric field:

$$V = \mathbf{p}^{rec} \cdot [\mathbf{S}]\mathbf{p}^{tr} \tag{6-29}$$

Here, $\mathbf{p}^{tr}$ and $\mathbf{p}^{rec}$ are the normalized polarization vectors describing the transmitting and receiving radar antennas. The power received by the radar is the magnitude of the voltage squared:

$$P = VV^* = |\mathbf{p}^{rec} \cdot [\mathbf{S}]\mathbf{p}^{rad}|^2 \tag{6-30}$$

Once the complete scattering matrix is known and calibrated, one can synthesize the radar cross-section for any arbitrary combination of transmit and receive polarizations using Equation (6-30). This expression forms the basis of radar polarimetry, which we shall describe in more detail later in this chapter. Figure 6-3 shows a number of such synthesized images for the San Francisco Bay area in California. The data were acquired with the NASA/JPL AIRSAR system. The Golden Gate Bridge is the linear feature at the top middle of the image. Golden Gate Park is the large rectangle about one-third from the bottom of the image. Note the strong variation in the relative return from the urban areas in the

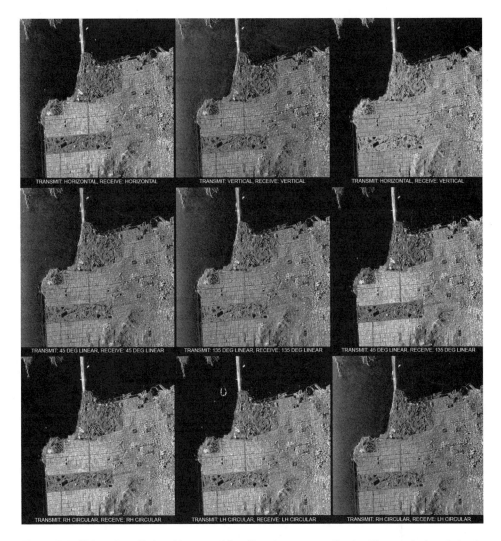

Figure 6-3. This series of L-band images of San Francisco was synthesized from a single polarimetric image acquired by the NASA/JPL AIRSAR system at the L band. The nine images show the copolarized (transmitted and received polarizations are the same) and the cross-polarized (transmitted and received polarizations are orthogonal) images for the three axes of the Poincaré sphere. Note the relative change in brightness between the city of San Francisco, the ocean, and Golden Gate Park, which is the large rectangle about a third from the bottom of the images. The radar illumination is from the left.

top left of the image. These are related to the orientation of the buildings and the streets relative to the radar look direction (Zebker et al., 1987). The contrast between the urban areas and the vegetated areas such as Golden Gate Park is maximized using the 45° linear polarization for transmission, and 135° linear polarization for reception (see Chapter 7 of Ulaby and Elachi, 1990 for a discussion).

To further illustrate the effect of polarization, let us consider a simplified model of a vegetation canopy consisting of short, vertically oriented linear scatterers over a rough surface. If we assume that the scatterers can be modeled as short vertical dipoles, and we ignore the higher-order term that contains the interaction between the dipoles and the underlying surface, then a horizontally polarized incident wave will not interact with the canopy (i.e., $\sigma_\nu = 0$, $\alpha = 0$) and will scatter from the surface, leading to a backscatter cross section:

$$\sigma_{\mathrm{HH}} = \sigma_{\mathrm{sHH}}$$

On the other hand, a vertically polarized wave will interact strongly with the dipoles, leading to an expression for σ_{VV} equal to the first two terms given in Equation 6-27. If we take an intermediate situation in which the polarization vector is rotated by an angle ϕ relative to the horizontal component, then only the vertical component will interact with the dipoles, leading to an expression for σ equal to

$$\sigma_{\phi\phi} = \sigma_{\mathrm{sHH}} \cos^2\phi + \sigma_{\mathrm{sVV}} \sin^2\phi \, e^{-2\alpha H \sec\theta} + \frac{\sigma_\nu \sin^2\phi \cos\theta}{2\alpha}[1 - e^{-2\alpha H \sec\theta}] \qquad (6\text{-}31)$$

Thus, as the polarization vector is rotated from the horizontal state to the vertical one, the effect of the canopy begins playing an increasingly larger role. Figure 6-4 shows a sketch that illustrates this effect.

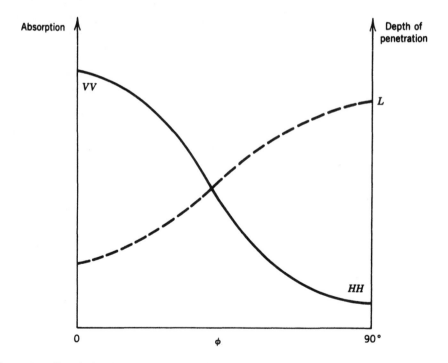

Figure 6-4. Sketch showing total absorption (continuous line) and the mean depth of penetration (dashed line) in a layer consisting of vertically oriented scatterers as a function of the direction ϕ of the polarization vector relative to the incidence plane.

6-1-4 Effects of the Frequency

The frequency of the incident wave plays a major role in the interaction with the surface. It is a key factor in the penetration depth (Equation 6-13), scattering from a rough surface (Equations 6-2 and 6-12), and scattering from finite scatterers.

The penetration depth is directly proportional to the ratio $\lambda/\tan \delta$. The loss tangent does vary with the wavelength. However, for most material, the penetration depth varies linearly with λ in the radar sensor's spectral region. An L-band (20 cm wavelength) signal will penetrate about 10 times deeper than a K_u-band (2 cm wavelength) signal, thus providing access to a significant volumetric layer near the surface.

Figure 6-5 shows the penetration depth for a variety of natural materials. It is clear that the frequency plays a significant role in what will be sensed with an imaging radar. A surface covered with 2 m of pure snow will not be visible at frequencies above 10 GHz, whereas the snow cover is effectively transparent at frequencies of 1.2 GHz or lower.

Moisture is a key factor affecting the penetration depth. Figure 6-5 shows that at L-band frequencies soils with very low moisture have penetration depth of 1 m or more. This situation is encountered in many arid and hyperarid regions of the world such as most of northern Africa, north-central China, southwestern United States, and the Arabian peninsula. However, in most temperate regions where the moisture exceeds a few percentage points, the penetration depth becomes less than a few centimeters, particularly at the high end of the spectrum.

In addition, the scattering from a rough surface is strongly dependent on the frequency (Equation 6-2). In the case of a constant-roughness spectrum, the backscatter cross section increases as the fourth power of the frequency. Even if the surface-roughness spectrum decreases as the square or the cube of the spectral frequency (i.e., $W \sim k^{-2}$ or k^{-3}), the backscatter cross section will still strongly increase as a function of frequency. Figure 6-6 shows the behavior of the backscatter as a function of frequency for a variety of surface types.

Figure 6-7 shows dramatically the effect of frequency in the case of scattering from a forest canopy in the Black Forest in Germany. The images were acquired with the NASA/JPL AIRSAR system in 1991. The low-frequency P-band (68 cm wavelength) HH image shows variations in brightness that are correlated with the topography of the terrain under the trees. The areas that have the higher returns are where the local topography is nearly flat and is dominated by radar signals reflected off the ground, followed by a reflection off the tree trunks, before returning to the radar. When the ground slopes away from the radar or toward it, this term decreases rapidly, leading to a reduction in the observed return near the streams (van Zyl, 1993). The L-band (24 cm wavelength) HH image shows little variation in brightness because the penetration length is shorter than that at P band [see Equation (6-23)], which means that the scattering is dominated by returns from the branches in the canopy.

6-1-5 Effects of the Incidence Angle

As shown in Figure 6-2, the backscatter return is strongly dependent on the angle of incidence. At small angles, the backscattered return varies significantly with the angle of incidence, and it provides information on the surface-slope distribution for topography at a scale significantly larger than the wavelength. At large angles, the return signal provides information about the small-scale structure. Thus, different surfaces could be discriminat-

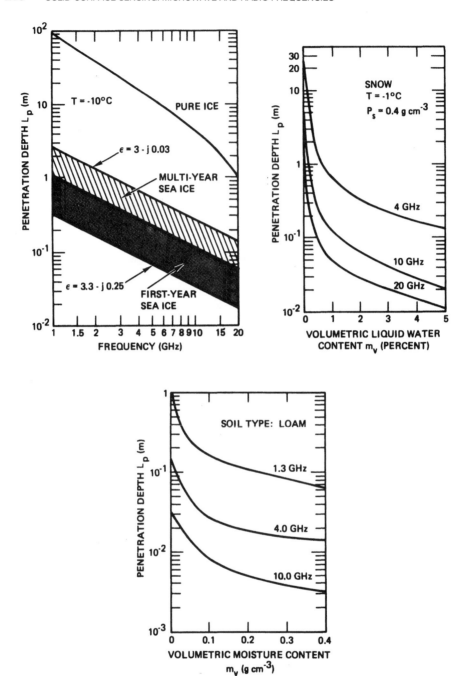

Figure 6-5. Penetration depth as a function of frequency for a variety of materials. Reprinted, by permission, from Fawwaz T. Ulaby et al., *Remote Sensing and Surface Scattering and Emission Theory,* Figures 11-63, 11-65, and 11-66. © 1982 Artech House, Inc.

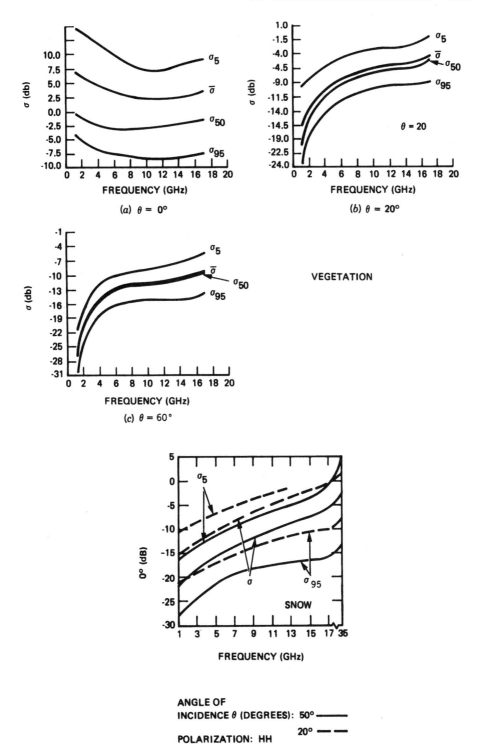

Figure 6-6. Examples of backscatter cross section as a function of frequency for a variety of natural surfaces. (From Ulaby et al., 1982.)

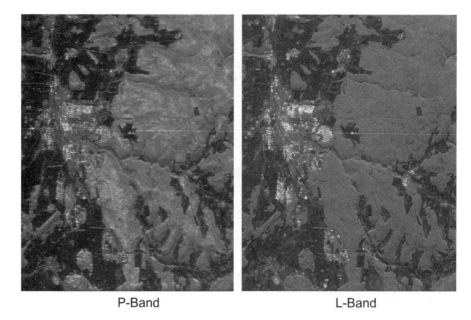

P-Band   L-Band

Figure 6-7. Dual frequency image of the Black Forest in Germany acquired by the NASA/JPL AIR-SAR system. See text for a discussion of the differences.

ed and classified based on their "angular signature" in a similar fashion as "spectral signature" and "polarization signature" can be used for classification. Figure 6-8a illustrates the general behavior of $\sigma(\theta)$ for general categories of surfaces. Figure 6-8b shows spaceborne radar imagery of the same area acquired at three different angles, clearly demonstrating the variations of the angular signature of different types of terrain.

6-1-6   Scattering from Natural Terrain

Models of electromagnetic wave interactions with surfaces and inhomogeneous layers provide information about the general behavior of wave scattering from natural terrain. However, the complexity and wide variety of natural surfaces and vegetation covers make it extremely hard to exactly model these surfaces and their corresponding backscatter behavior as a function of polarization, frequency, and illumination geometry. Nevertheless, many models, both theoretical and empirical, have been constructed to predict scattering from vegetated areas, and this remains an area of active research (see the references for numerous examples).

A good knowledge of the expected range of backscatter cross sections is a critical input to the design of a spaceborne imaging radar. This can be simply illustrated by the fact that a 3 dB refinement in the range of backscattered cross sections to be mapped has an impact of a factor of two on the needed radiated power. Considering that spaceborne radar sensors usually push spacecraft capabilities, such as power, to their present technological limits, a factor of two in the radiated power is significant.

Theoretical models are significantly enhanced with a broad and extensive field-acquired database. This is usually done with truck-mounted or airborne calibrated scat-

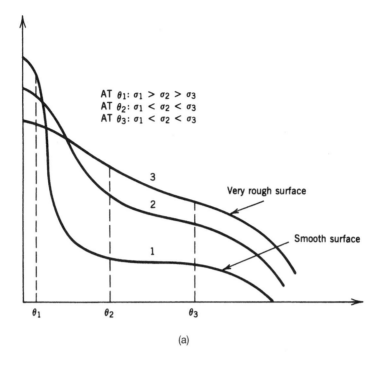

AT θ_1: $\sigma_1 > \sigma_2 > \sigma_3$
AT θ_2: $\sigma_1 < \sigma_2 < \sigma_3$
AT θ_3: $\sigma_1 < \sigma_2 < \sigma_3$

(a)

(b)

Figure 6-8. (a) Illustration showing how different surfaces can be separated by measuring their backscatter return at a limited number of incidence angles. (b) Spaceborne radar composite image of the same area in Florida acquired at three different angles. The color was generated by assigning blue, green, and red to the images acquired at 28, 45, and 58° incidence angles, respectively. See color section.

terometers. Well-controlled experiments have allowed the development of a good data-base for vegetation, which is illustrated in Figure 6-9 (Ulaby et al., 1982). Calibrated measurements were acquired from space in 1973 using the Skylab scatterometer. A histogram of the backscatter cross-section measurement over North America is shown in Figure 6-10. It shows that at 13.9 GHz and 33° incidence angle the mean backscatter cross section of land surfaces is −9.9 dB, with a standard deviation of 3 dB. Figure 6-11 shows the angular pattern for the Skylab data. An interesting factor is that 80% of the measurements (excluding the lowest 10% and highest 10%) are within about 2.3 dB from the mean value, except at very small angles.

6-2 BASIC PRINCIPLES OF RADAR SENSORS

Discrimination between signals received from different parts of the illuminated scene may be accomplished by different techniques involving angular, temporal, or frequency (Doppler) separation. Even though there is a very wide variety of radar systems custom designed for specific applications, most of the basic governing principles of radar sensors are similar. These are briefly discussed in this section. For further details, see the suggested reading list at the end of the chapter.

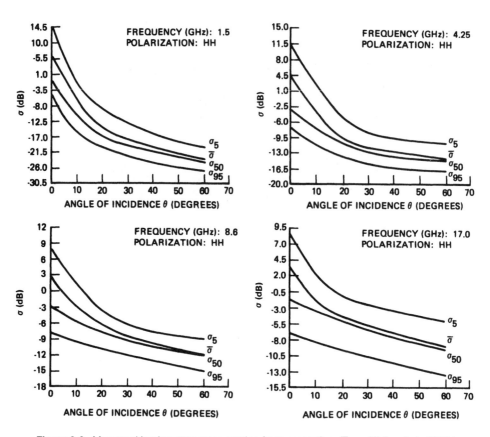

Figure 6-9. Measured backscatter cross section from vegetation. (From Ulaby et al., 1982.)

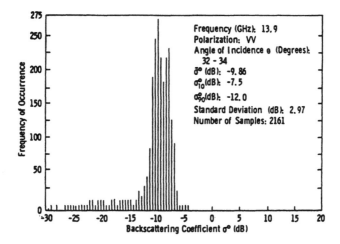

Figure 6-10. Histogram of backscatter data over North America at 13.9 GHz and 33° incidence angle acquired with the Skylab scatterometer. (From Ulaby et al., 1982.)

Imaging radars generate surface images that look very similar to passive visible and infrared images. However, the principle behind the image generation is fundamentally different in the two cases. Passive visible and infrared sensors use a lens or mirror system to project the radiation from the scene on a two-dimensional array of detectors, which could be an electronic array or a film using chemical processes. The two-dimensionality can also be achieved by using scanning systems. This imaging approach conserves the angular relationships between two targets and their images.

Imaging radars, on the other hand, use the time delay between the echoes that are backscattered from different surface elements to separate them in the range (cross-track)

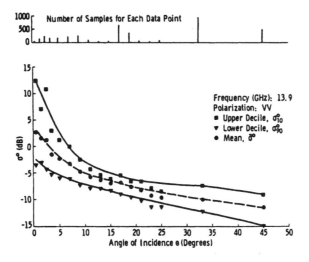

Figure 6-11. Behavior of the backscatter cross section as a function of angle for all the data acquired with the Skylab scatterometer. (From Ulaby et al., 1982.)

dimension , and the angular size (in the case of the real-aperture radar) of the antenna pattern, or the Doppler history (in the case of the synthetic-aperture radar) to separate surface pixels in the azimuth (along-track) dimension.

6-2-1   Antenna Beam Characteristics

Let us consider first the case of the linear array of N radiators (Fig. 6-12). The contribution of the nth radiator to the total far field in direction θ can be written as

$$E_n \sim (a_n e^{i\phi n}) e^{-ikd_n \sin\theta} \tag{6-32}$$

and the total field is proportional to

$$E(\theta) \sim \sum_n a_n e^{i\phi_n - ikd_n \sin\theta} \tag{6-33}$$

where a_n is the relative amplitude of the signal radiated by the nth element and ϕ_n is its phase. If all the radiators are identical in amplitude and phase, and equally spaced, then the total far field for the array is

$$E(\theta) \sim ae^{i\phi} \sum_{n=1}^{N} e^{-inkd \sin\theta} \tag{6-34}$$

This is the vectorial sum of N equal vectors separated by a phase $\psi = kd \sin \theta$ (Fig. 6-13). As seen in Figure 6-13, the sum is strongly dependent on the value of ψ. For $\theta = 0$ and $\psi = 0$, all the vectors add coherently. As ψ increases, the vectors are out of phase relative to each other, leading to a decrease in the total sum. When ψ is such that $N\psi = 2\pi$, the sum is then equal to zero. This corresponds to

$$Nkd \sin \theta = 2\pi \tag{6-35}$$

or

$$\theta = \sin^{-1} \frac{2\pi}{Nkd} \tag{6-36}$$

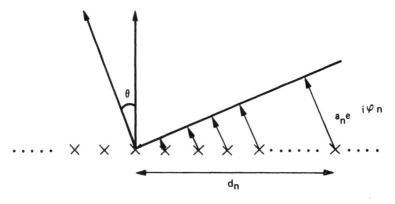

Figure 6-12. Geometry for a linear array.

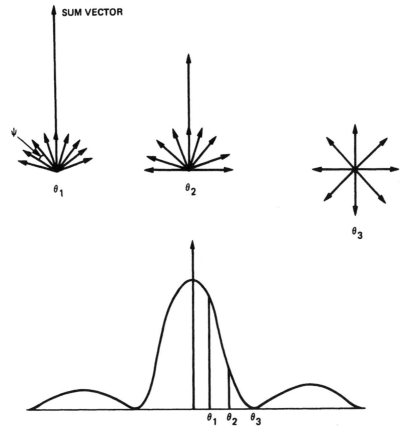

Figure 6-13. Radiation pattern of a linear array.

At this angle, there is a null in the radiated total field. The same will occur for $N\psi = 2m\pi$, where m is an integer, leading to a series of nulls at

$$\theta = \sin^{-1} \frac{2m\pi}{Nkd} \tag{6-37}$$

In between these nulls, there are peaks that correspond to

$$N\psi = 2m\pi + \pi \tag{6-38}$$

$$\Rightarrow \theta = \sin^{-1} \frac{(2m+1)\pi}{Nkd} \tag{6-39}$$

If the antenna, instead of being an array of discrete elements, is continuous, then the summation in Equation 6-33 is replaced by an integral:

$$E \sim \int_{-D/2}^{D/2} a(x)\, e^{-ikx \sin\theta}\, dx \tag{6-40}$$

If $a(x)$ is uniform across the aperture, then

$$E \sim \int_{-D/2}^{D/2} e^{-ikx\sin\theta} \, dx = D \frac{\sin(kD \sin \theta/2)}{kD \sin \theta/2} \tag{6-41}$$

This familiar $\sin \alpha/\alpha$ pattern occurs whenever there is uniform rectangular distribution. For this pattern, the nulls occur at

$$\frac{kD \sin \theta}{2} = m\pi \tag{6-42}$$

or

$$\theta = \sin^{-1} \frac{2m\pi}{kD} = \sin^{-1} \frac{m\lambda}{D} \tag{6-43}$$

which is the same as Equation 6-36 with Nd replaced by D. The first null occurs at

$$\theta_0 = \sin^{-1} \frac{\lambda}{D} \tag{6-44}$$

and if $\lambda \ll D$, then

$$\theta_0 = \frac{\lambda}{D} \tag{6-45}$$

The width of the main beam from null to null is then equal to $2\lambda/D$.

Of particular interest is the angle β between the half-power points on opposite sides of the main beam. This can be determined by solving the equation

$$\left[\frac{\sin(kD \sin(\beta/2)/2)}{kD \sin(\beta/2)/2} \right]^2 = 0.5 \tag{6-46}$$

This transcendental equation can be solved numerically, leading to the solution for β as

$$\beta = 0.88 \frac{\lambda}{D} \tag{6-47}$$

For the uniform illumination, the first sidelobe is about 13 dB (factor of 20) weaker than the main lobe. This sidelobe can be reduced by tapering the antenna illumination. However, this leads to a broadening of the central beam. The following table shows the beam width-sidelobe tradeoff:

Weighting	Half-power beam	First sidelobe
Uniform	0.88 λ/D	−13.2 dB
Linear taper	1.28 λ/D	−26.4 dB
$(\cos)^2$ taper	1.45 λ/D	−32.0 dB

In general, a good approximation for the half-power beamwidth is

$$\beta = \frac{\lambda}{D} \tag{6-48}$$

It should be noted that in the case of radar systems, the antenna pattern comes into play both at transmission and at reception. In this case, the sidelobe level is squared (i.e., doubled when expressed in decibels) and the angle β corresponds to the point at which the power is one quarter (i.e., -6 dB) of the power at the beam center.

Let us reconsider the case of the array in Figure 6-13 with equally spaced antennas, and let us assume it is being used to observe a target P. Depending on the mechanism of combining the signal from each of the antennas, the array can be focused or unfocused. In the most common configuration, the received signal at element A_n is carried via a transmission line of electrical length l_n to a central receiver that combines the signals from all the elements (Fig. 6-14a). If the array transmission lines are selected such that

$$r_1 + l_1 = r_2 + l_2 = \cdots = r_N + l_N \tag{6-49}$$

then all the received signals will add in phase in the receiver and the array is said to be focused at the target point P. If the array is a radar system, the transmitted signal originates at the same point at which the transmission lines are combined, and if the relationship in Equation 6-48 is satisfied, the echoes will still add in phase.

If the transmission lines are such that all the l_ns are equal, the array becomes unfocused, except for a target at infinity.

If each array element has its own receiver and recorder, and the summation is carried on at a later time in a processor, the same array focusing can be achieved if all the receivers have a common reference signal, and the received echoes are combined after an appropriate phase shift is added to them (see Fig. 6-14b). In this case, the signal V_n from element A_n is shifted by a phase $\phi_n = 2kr_n$ and the total output signal is given by

$$V = \sum_n V_n e^{-2ikr_n} \tag{6-50}$$

This scheme is commonly used in so-called digital beam-forming systems. In this case, the signal from each antenna element is sampled and recorded separately. The overall antenna pattern is then synthesized later, as described in Equation (6-50). This allows one to digitally scan the antenna in any direction by choosing the appropriate phases used in Equation (6-50).

Similarly, if a single antenna is moved along the array line and from each location x_n a signal is transmitted and an echo received and coherently recorded, the echoes can then be summed later in a processor to generate the equivalent of a focused (or unfocused) array (see Fig. 6-14c). In this way, the array is synthesized by using a single moving antenna.

Imaging radar sensors typically use an antenna that illuminates the surface to one side of the flight track. Usually, the antenna has a fan beam that illuminates a highly elongated elliptically shaped area on the surface, as shown in Fig. 6-15. The illuminated area across track defines the image *swath*. Within the illumination beam, the radar sensor transmits a very short effective pulse of electromagnetic energy. Echoes from surface points farther away along the cross-track coordinate will be received at proportionally later times (Fig.

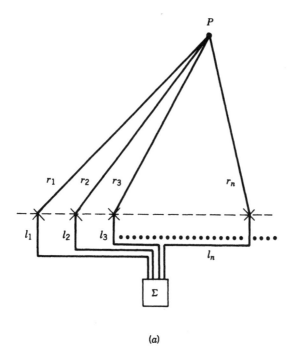

(a)

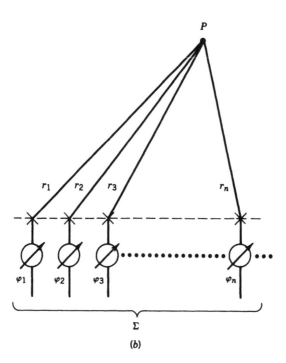

(b)

Figure 6-14. Antenna array focused at point P by adding an appropriate phase shift to each element. The phase shift can be introduced by having transmission lines of appropriate length (a), by having a set of phase shifters, or by having a set of receivers with phase shifters (b). The same can be achieved with a single moving antenna/phase shifter/receiver (c).

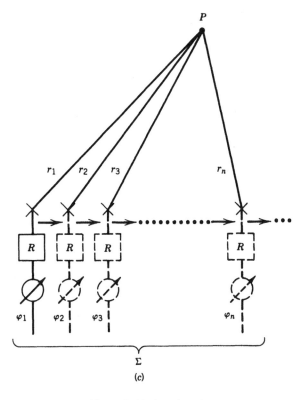

Figure 6-14. (*continued*)

6-15). Thus, by dividing the receive time in increments of equal time bins, the surface can be subdivided into a series of *range bins*. The width in the along-track direction of each range bin is equal to the antenna footprint along the track x_a. As the platform moves, the sets of range bins are covered sequentially, thus allowing strip mapping of the surface line by line. This is comparable to strip mapping with a pushbroom imaging system using a line array in the visible and infrared part of the electromagnetic spectrum. The brightness associated with each image pixel in the radar image is proportional to the echo power contained within the corresponding time bin. As we will see later, the different types of imaging radars differ in the way in which the azimuth resolution is achieved.

6-2-2 Signal Properties: Spectrum

Radar sensors are of two general types: continuous wave (CW) and pulsed wave (Fig. 6-16). In the case of pulsed systems, the number of pulses radiated by the radar per second is called the pulse repetition frequency (PRF).

Let us consider a signal $A(t)$ that is a single pulse of length τ and carrier frequency f_0 (Fig. 6-16):

$$A(t) = A \cos \omega_0 t \quad -\frac{\tau}{2} < t \le \frac{\tau}{2} \tag{6-51}$$

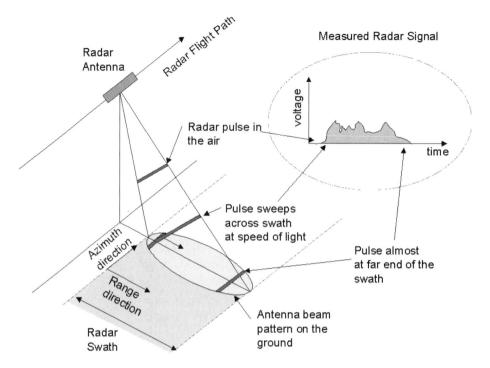

Figure 6-15. Imaging radars typically use antennas that have elongated gain patterns that are pointed to the side of the radar flight track. The pulse sweeps across the antenna beam spot, creating an echo as shown in this figure.

where $\omega_0 = 2\pi f_0$. The corresponding frequency spectrum is

$$F(\omega) = \int_{-\infty}^{+\infty} A(t)e^{-i\omega t}\, dt = A\int_{-\tau/2}^{+\tau/2} \cos(\omega_0 t)e^{-i\omega t}\, dt = A\left[\frac{\sin(\omega_0 - \omega)\tau/2}{\omega_0 - \omega} + \frac{\sin(\omega_0 + \omega)\tau/2}{\omega_0 + \omega}\right]$$

which is shown in Figure 6-16b (only the positive frequencies are shown). The spectrum is centered at the angular frequency ω_0 and the first null occurs at

$$\omega = \omega_0 \pm \frac{2\pi}{\tau}$$

or

$$f = f_0 \pm \frac{1}{\tau}$$

So the null-to-null bandwidth is

$$B' = \frac{2}{\tau}$$

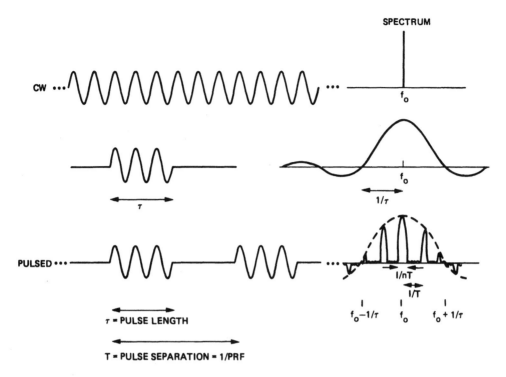

Figure 6-16. Temporal and spectral characteristics of continuous wave (CW) and pulsed wave signals. *n* is the total number of pulses in a pulse train.

Also of interest is the half-amplitude bandwidth, which is equal to

$$B \simeq \frac{1.2}{\tau}$$

In general, the bandwidth of a pulse is defined as

$$B = \frac{1}{\tau} \qquad (6\text{-}52)$$

Thus, short monochromatic pulses have a wide bandwidth, and long pulses have a narrow bandwidth.

The bandwidth, and corresponding time length, of a pulse is of key importance in the capability of the pulse to discriminate neighboring targets (i.e., sensor range resolution). If there are two point targets separated by a distance Δr, the two received echoes will be separated by a time $\Delta t = 2\Delta r/c$. If the sensing pulse has a length τ, it is necessary that $\tau < 2\Delta r/c$ in order for the two echoes not to overlap (Fig 6-17). Otherwise it would be perceived that there is only one dispersed target. The shortest separation Δr that can be measured is given by (see Section 6-2-4)

$$\Delta r = \frac{c\tau}{2} = \frac{c}{2B} \qquad (6\text{-}53)$$

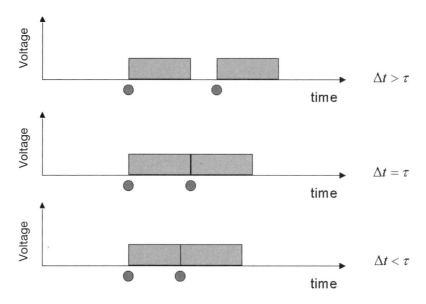

Figure 6-17. The range resolution of a radar is determined by the pulse length. If the leading edge of the echo from the second target arrives at the radar before the trailing edge of the echo from the first target, their returns cannot be separated, as shown in the bottom illustration. In the top illustration, the two targets are far enough apart that return pulses are recognized as two separate returns.

This is called the range resolution. Thus, in order to get a high range resolution (e.g., small Δr), a short pulse or a wide-bandwidth pulse is needed.

The energy in a pulse is equal to

$$E = P\tau \qquad (6\text{-}54)$$

where P is the instantaneous peak power. The energy in a pulse characterizes the capability of the pulse to detect a target, and a high pulse energy is usually desired. This can be achieved by increasing the peak power P. However, the maximum power is severely limited by the sensor hardware, particularly in the case of spaceborne sensors. The other possibility is to increase τ. But, as discussed earlier, a long pulse corresponds to a narrow bandwidth B and a resulting poor range resolution. Thus, in order to have a high detection ability (large E or τ) and a high resolution (large B), a pulse with the seemingly incompatible characteristics of large τ and large B is needed. This is made possible by modulating the pulse (see next section).

In the case of a periodic signal (Fig. 6-16c) consisting of n pulses, the spectrum will consist of a series of lobes of width $1/nT$ modulated by an envelope that corresponds to the spectrum of one pulse. The separation between the lobes is $1/T = \text{PRF}$.

6-2-3 Signal Properties: Modulation

In designing the signal pattern for a radar sensor, we are faced with the dilemma of wanting a long pulse with high energy and a wide bandwidth, which normally implies a short pulse. This dilemma is resolved by using pulse modulation. Two types of modulation are discussed here for illustration: linear frequency modulation, commonly called FM chirp, and binary phase modulation.

In the case of FM chirp, the frequency f_0 is not kept constant throughout the pulse, but it is linearly changed from f_0 to $f_0 + \Delta f$ (see Figure 6-18a). Δf can be positive (up chirp) or negative (down chirp). Intuitively, we can say that the bandwidth is equal to

$$B = |(f_0 + \Delta f) - f_0| = |\Delta f|$$

which is independent of the pulse length τ. Thus, a pulse with large τ and large B can be constructed.

The signal expression for a chirp is mathematically given by

$$A(t) = A \cos\left(\omega_0 t + \frac{\Delta \omega}{2\tau'} t^2 \right) \quad \text{for } 0 \leq t \leq \tau'$$
$$= 0 \qquad\qquad\qquad \text{otherwise}$$

(6-55)

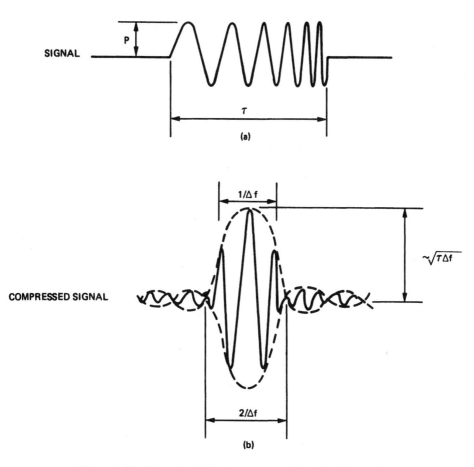

Figure 6-18. Chirp signal (a) and compressed (dechirped) signal (b).

The instantaneous angular frequency of such a signal is

$$\omega(t) = \omega_0 + \frac{\Delta\omega}{\tau'}t \tag{6-56}$$

The reason that a chirp signal of long duration (i.e., high pulse energy) can have a high range resolution (i.e., equivalent to a short duration pulse) is that at reception the echo can be compressed using a matched filter such that at the output the total energy is compressed into a much shorter pulse (see Fig. 6-18b and Appendix D). This output pulse will still have the same total energy of the input pulse but will be significantly shorter.

The length of the output pulse τ is limited by the bandwidth B to (see Appendix D)

$$\tau = \frac{1}{B} = \frac{2\pi}{\Delta\omega} = \frac{1}{\Delta f} \tag{6-57}$$

The compression ratio is then equal to

$$\frac{\tau'}{\tau} = \tau'\frac{\Delta\omega}{2\pi} = \tau'\Delta f \tag{6-58}$$

which is known as the *time–bandwidth product* of the modulated pulse. Before compression, even though the echoes from two neighboring targets will overlap, the instantaneous frequency from each echo at a specific time in the overlap region is different. This allows the separation of the echoes.

Another simple way of explaining the properties of a modulated signal is by using a simplistic photon analogy. Transmitters are usually limited in how many photons of energy they can emit per unit time. Thus, in order to emit a large amount of energy, the transmitter has to be on for a long time. By modulating the signal, we are "labeling" the different photons as a function of emission time (see Fig. 6-19). Let us say that the first photons are colored red and the next ones emitted after time Δt yellow. Then, after $2\Delta t$ the color changes to green, and after $3\Delta t$ to blue. When the echo is received, a "magic box" is used

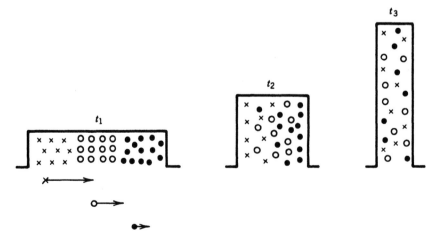

Figure 6-19. The compression of a frequency modulated signal. The differently labeled photons will move at a different speeds (arrows) through the compression box and exit it at the same time.

that first identifies the red photons (which enter it first) and slows them down by a time $3\Delta t$. The yellow ones are slowed down by a time $2\Delta t$ and the green ones by a time Δt. In this way, all the photons will exit the "magic box" at the same time and the instantaneous energy (i.e., power) will be significantly higher.

In the case of binary phase modulation, the transmitted signal phase is changed by 180° at specific times (see Fig. 6-20) according to a predetermined binary code. At reception, a tapped delay line is used. Some of the taps have 180° phase shifts injected in them with a pattern identical to the one in the transmitted pulse. So when the received echo is fully in the delay line, the output is maximum. One cycle earlier or later, there is mismatch between the received signal and the tapped delay line, leading to a much weaker output (see Fig. 6-21). In the case illustrated, the point response has a central peak of 7 with sidelobe of 1. In power, this is a ratio of 49 (~17 dB). The code used is a Barker code. The compression ratio is equal to τ/τ_c. In general, for a Barker code of length N, the level of the sidelobes is $10 \log N^2$.

6-2-4 Range Measurements and Discrimination

If an infinitely short pulse is transmitted toward a point target a distance R away, an infinitely short pulse echo will be received at time $t = 2R/c$ later, where c is the speed of light. The factor 2 represents the fact that the signal travels a distance $2R$. If the pulse has a length τ, the echo will have a length τ. If there are two targets separated by a distance ΔR, the shortest separation ΔR that can be measured is given by (Figs. 6-17 and 6-22):

$$\Delta R = \frac{c\tau}{2} = \frac{c}{2B} \qquad (6\text{-}59)$$

For targets closer than this value, the trailing edge of the pulse returned from the first target will arrive at the radar after the leading edge returned from the second target. This separation, given by Equation (6-59), is called the range resolution. Thus, in order to get good range resolution, we need a short pulse or a wide bandwidth pulse that could be compressed into a short pulse.

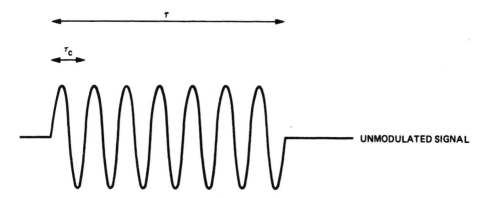

Figure 6-20. Binary phase modulation. When the binary code is indicated by (+), the output signal is in phase with the reference unmodulated signal. When the code is indicated by (–), a 180° phase shift is injected.

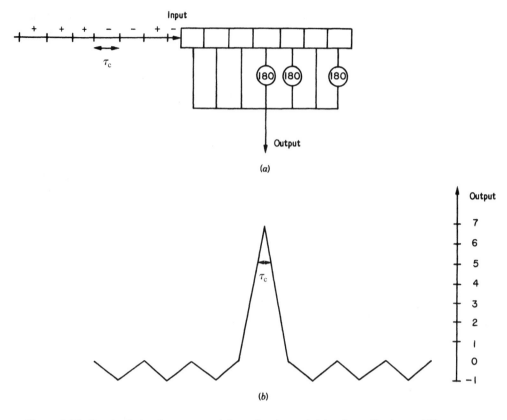

Figure 6-21. Received signals are passed through a tapped delay line with phase shifters corresponding to the binary code on the transmitted pulse. At exact matching, the output peaks. The compressed signal has a width of τ_c even though the input signal has a length of τ.

Range discrimination can also be achieved by using frequency modulation as mentioned before. In this case, a long continuous signal is transmitted with a linearly swept frequency (see Fig. 6-22b). Here, we provide a simple description to illustrate how echoes from different targets can be discriminated in the case of a long continuous FM signal. Consider a system in which the echo is mixed with a signal identical to the signal being transmitted with an appropriate time delay. The resulting low-passed signal has a frequency proportional to the range (see Fig. 6-22b). Mathematically, the transmitted signal $V_t(t)$ and received signal $V_r(t)$ can be written as

$$V_t(t) \sim \cos(\omega_0 t + \pi a t^2)$$

$$V_r(t) \sim V_t\left(t - \frac{2R}{c}\right) \sim \cos\left[\omega_0\left(t - \frac{2R}{c}\right) + \pi a\left(t - \frac{2R}{c}\right)^2\right]$$

After mixing, the output signal $V_0(t)$ is the product of the two signals:

$$V_0(t) \sim \cos \omega_t t \cos \omega_r t = \tfrac{1}{2}\cos(\omega_t + \omega_r)t - \tfrac{1}{2}\cos(\omega_t - \omega_r)t$$

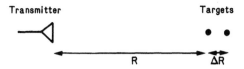

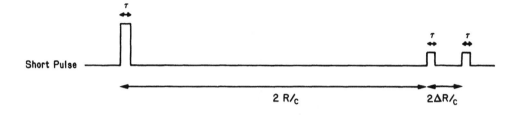

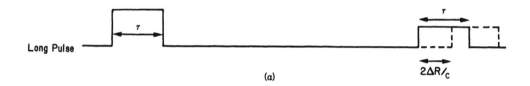

(a)

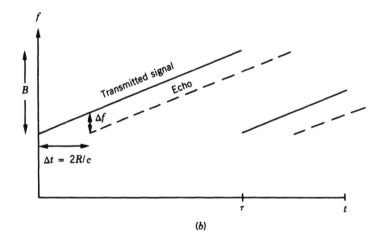

(b)

Figure 6-22. (a) The range measurement technique using pulsed radar. Separation of targets would require a short pulse such that $\tau < 2\ \Delta R/c$. (b) The range measurement technique using FM sweep radar.

where φ_t and φ_r are the instantaneous angular frequencies of the transmitted and received signals, respectively. After the sum frequency is filtered out, the resulting signal is given by

$$V_0(t) \sim \cos(\omega_t - \omega_r)t = \cos\left(\frac{2\omega_0 R}{c} + \frac{4\pi a R t}{c} - \frac{4\pi a R^2}{c^2}\right)$$

which has a frequency $f = 2aR/c$. If there are two targets separated by ΔR, the output signals will be separated in frequency by

$$\Delta f = \frac{2a\Delta R}{c} \tag{6-60}$$

This should be larger than $1/\tau$:

$$\frac{2a\Delta R}{c} > \frac{1}{\tau}$$

$$\Rightarrow \Delta R > \frac{c}{2a\tau} = \frac{c}{2B}$$

This gives the range resolution of such a system, which is identical to Equation (6-53). In the case of multiple targets, the received echo is mixed with the transmitted signal, then passed through a bank of filters. This scheme is used in high-resolution altimeters.

6-2-5   Doppler (Velocity) Measurement and Discrimination

Doppler shifts result from the motion of the sensor and/or motion of the targets, as discussed in Chapter 2. This leads to the received echo having frequencies shifted away from the transmitted frequency by the corresponding Doppler frequencies f_{dn}. The received signal will then be

$$V_r(t) \sim \sum_n a_n \cos(\omega_0 t + \omega_{dn} t) \tag{6-61}$$

After down conversion by the carrier frequency, the signal is then passed in a filter bank that will separate the signals with different Doppler frequencies.

In the case of a pulse signal, the echo will have the same spectrum as the transmitted signal (Fig. 6-16) but shifted by $f_d = 2(v/c)f_0$, where v is the apparent relative velocity of the target. If there are two targets with apparent velocities v and $v + \Delta v$, the echo will contain Doppler shifts at f_d and $f_d + \Delta f_d$. In order to separate the two targets, we need (see Fig. 6-16c)

$$\Delta f_d > \frac{1}{nT} \Rightarrow \frac{2\Delta v}{c} f_0 > \frac{1}{nT} \tag{6-62}$$

or

$$\Delta v \geq \frac{\lambda \text{PRF}}{2n} \tag{6-63}$$

This is called the Doppler resolution. To illustrate, for $\lambda = 3$ cm and $\Delta v = 15$ m/sec, $nT >$ 1 millisecond.

Another limit is that the maximum Doppler shift cannot exceed $1/T = $ PRF, otherwise ambiguity occurs (see Fig. 6-23). Thus, if we require $v(\text{max}) = 600$ m/sec with $\lambda = 3$ cm, then $f_d(\text{max}) = 40$ kHz $\Rightarrow$ PRF > 40 kHz.

If for other reasons a lower PRF is required, a number of techniques, such as PRF hopping, have been used to resolve the ambiguity.

6-2-6 High-Frequency Signal Generation

Radar signals are generated by a series of up conversions in order to use a low-frequency stable oscillator as a source for the output high-frequency signal. At reception, a series of down conversions are used to bring the signal back to a low frequency for more efficient detection.

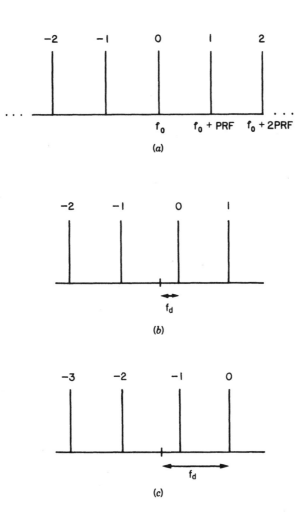

Figure 6-23. Spectrum of pulsed train with increasing Doppler shifts. It is apparent that when $f_d >$ PRF, there is ambiguity. Cases *b* and *c* have identical apparent spectra.

Generation of high-frequency signals can be achieved by a number of upconversion techniques. Let us assume we have two signals V_1 and V_2:

$$V_1 = \cos(\omega_1 t)$$
$$V_2 = \cos(\omega_2 t)$$

$$(6\text{-}64)$$

If we pass these two signals through a mixer that generates an output V_0 proportional to the product of the inputs, then

$$V_0 \sim \cos(\omega_1 t) \cos(\omega_2 t) = \tfrac{1}{2} \cos(\omega_1 + \omega_2)t + \tfrac{1}{2} \cos(\omega_1 - \omega_2)t \qquad (6\text{-}65)$$

If the output is passed through a high-pass filter, then the higher-frequency signal is passed through. This leads to up conversion from ω_1 or ω_2 to $\omega_1 + \omega_2$. This process can be repeated multiple times, leading to very high frequencies.

If V_1 is a modulated signal, let us say

$$V_1 = \cos(\omega_1 t + at^2) \qquad (6\text{-}66)$$

and we mix it with $\cos(\omega_2 t)$, then

$$V_0 \sim \cos[(\omega_1 + \omega_2)t + at^2] \qquad (6\text{-}67)$$

This shows that the carrier frequency is up converted but the bandwidth stays the same (Fig. 6-24).

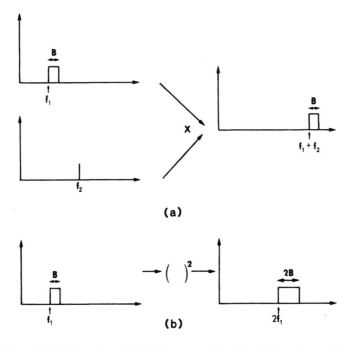

Figure 6-24. Generation of a high-frequency signal by mixing and squaring.

If we pass the V_1 signal into a nonlinear device that provides an output proportional to the square of the input, then

$$V_0 \sim V_1^2 = \cos^2(\omega_1 t) = \tfrac{1}{2} + \tfrac{1}{2} \cos(2\omega_1 t) \tag{6-68}$$

and the output signal has double the frequency of the input signal. If V_1 is modulated, then

$$V_0 \sim V_1^2 = \cos^2(\omega_1 t + at^2) = \tfrac{1}{2} + \tfrac{1}{2} \cos(\omega_1 t + 2at^2) \tag{6-69}$$

In this case, the bandwidth is also doubled (see Fig. 6-24).

At reception, mixers are used to generate the sum and difference frequencies. A low-pass filter will then filter out the high-frequency component and leave only the low-frequency component. This allows us to reduce the frequency of a received signal.

6-3 IMAGING SENSORS: REAL-APERTURE RADARS

The real-aperture imaging technique leads to an azimuth resolution that is linearly proportional to the distance between the sensor and the surface. Therefore, this technique is not used from spaceborne platforms if the objective is to have high-resolution imaging. It is commonly used for scatterometry and altimetry, which will be discussed later in this chapter. In this section, we will particularly emphasize some of the elements common to any imaging radar system, be it real aperture or synthetic aperture.

6-3-1 Imaging Geometry

The configuration of an imaging radar system is shown in Figure 6-25. The radar antenna illuminates a surface strip to one side of the nadir track. This side-looking configuration is necessary to eliminate right–left ambiguities from two symmetric equidistant points. As the platform moves in its orbit, a continuous strip of swath width SW is mapped along the flight line. The swath width is given by

$$SW \simeq \frac{h\beta}{\cos^2 \theta} = \frac{\lambda h}{W \cos^2 \theta} \tag{6-70}$$

where β is the antenna beam width in elevation, W is the width of the antenna, and θ is called the look angle. This expression assumes that $\beta \ll 1$ and does not take into account the Earth's curvature.

To illustrate, for $\lambda = 27$ cm, $h = 800$ km, $\theta = 20°$, and $W = 2.1$ m, the resulting swath width is equal to 100 km.

6-3-2 Range Resolution

The range resolution corresponds to the minimum distance between two points on the surface that can still be separable. If two points are separated by a distance X_r, their respective echoes will be separated by a time difference Δt equal to

$$\Delta t = \frac{2X_r}{c} \sin \theta \tag{6-71}$$

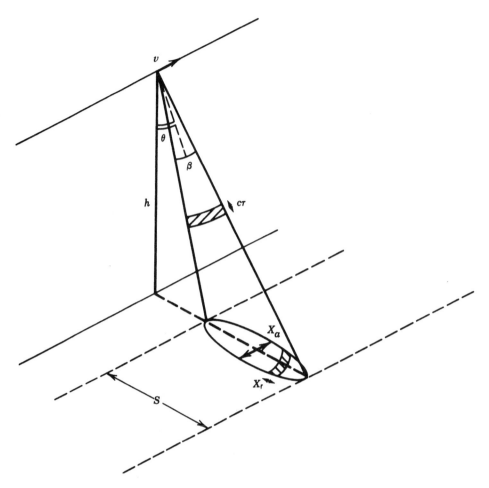

Figure 6-25. Geometry of a real-aperture imaging radar.

As discussed earlier, the smallest discriminable time difference is equal to τ or $1/B$. Thus, the range resolution is given by

$$2\frac{X_r}{c}\sin\theta = \tau$$

$$\Rightarrow X_r = \frac{c\tau}{2\sin\theta} = \frac{c}{2B\sin\theta}$$

(6-72)

Thus, a signal of bandwidth $B = 20$ MHz will provide a range resolution equal to 22 m for $\theta = 20°$.

We note that the resolution as described by Equation (6-72) is known as the *ground range* resolution, as it refers to the closest separation on the surface at which two targets can still be discriminated. Earlier, in Equations (6-53) and (6-59), we derived the resolution of two targets along the direct line of sight of the radar. This is commonly known as the *slant range* resolution of the radar. As can be seen from the expressions in Equation

(6-53) and (6-72), the ground range and slant range resolutions differ only in the sin θ term in the denominator of Equation (6-72), which represents the projection of the pulse on the surface. In practice, the range resolution is also a function of the surface slope. Surfaces that are sloped toward the radar have local angles of incidence less than that of an untilted surface, and the ground range resolution will be poorer for the tilted surfaces.

6-3-3 Azimuth Resolution

The azimuth resolution corresponds to the two nearest separable points along an azimuth line, that is, on a constant delay line. It is obvious that this is equal to the width of the antenna footprint because the echoes from all the points along a line spanning that width are returned at the same time. Thus, the azimuth resolution is equal to

$$X_a = \frac{h\beta'}{\cos\theta} = \frac{h\lambda}{L\cos\theta} \tag{6-73}$$

where β' is the antenna beam width in the azimuth.

To illustrate, if $h = 800$ km, $\lambda = 23$ cm, $L = 12$ m, and $\theta = 20°$, then $X_a = 16.3$ km. Even if λ is as short as 2 cm, X_a will still be equal to about 1.4 km, which is considered a low resolution for imaging applications. This is the reason why the real-aperture technique is not used from orbiting platforms when high resolution is desired.

It should be pointed out that the expression for the azimuth resolution is similar to the expression of the theoretical resolution in optical sensors. However, in the case of optical sensors, λ is extremely small (on the order of a micron), thus allowing resolutions of a few tens of meters from orbit with aperture size of only a few centimeters.

6-3-4 Radar Equation

One of the factors that determines the quality of the imagery acquired with an imaging radar sensor is the signal-to-noise ratio (SNR) for a pixel element in the image. In this section, we will consider only the thermal noise. Speckle noise will be considered in a later section.

Let P_t be the transmitted peak power and $G = 4\pi A/\lambda^2$ the gain of the antenna. The power density per unit area incident on the illuminated surface is

$$P_i = \frac{P_t G}{4\pi r^2}\cos\theta \tag{6-74}$$

The backscattered power is then equal to

$$P_s = P_i S\sigma \tag{6-75}$$

where S is the area illuminated at a certain instant of time (e.g., $S = X_a X_r$) and σ is the surface backscatter cross section (similar to the surface albedo in the visible region). The reflected power density per unit area in the neighborhood of the sensor is

$$P_c = \frac{P_s}{4\pi r^2} \tag{6-76}$$

and the total power collected by the antenna is

$$P_r = AP_c$$

or

$$P_r = \left[\frac{P_t G}{4\pi r^2} \cos\theta \right] [X_a X_r \sigma] \left[\frac{A}{4\pi r^2} \right] = \frac{P_t}{8\pi} \frac{W^2 L}{\lambda h^3} \frac{c\sigma}{B} \frac{\cos^4\theta}{\sin\theta} \tag{6-77}$$

The thermal noise in the receiver is given by

$$P_N = kTB \tag{6-78}$$

where k is the Boltzmann constant ($k = 1.380^{-23}$ W/°K Hz) and T is the total noise temperature (including the receiver thermal temperature and the illuminated surface temperature). The resulting signal-to-noise ratio is then

$$SNR = \frac{P_r}{P_N} = \frac{P_r}{kTB} \tag{6-79}$$

A simple way of characterizing an imaging radar sensor is to determine the surface backscatter cross section that gives a signal-to-noise ratio equal to one. This is called the noise equivalent backscatter cross section σ_n.

There are a number of factors that would improve the *SNR* (i.e., decrease σ_n) beyond the value given in Equation (6-79). One such factor is the use of a dispersed pulse with a compression ratio $l = \tau'/\tau = \tau'B$, where τ' is the dispersed pulse length. This improves the signal-to-noise ratio by l, which can commonly exceed 20 dB. In this case, the *SNR* becomes

$$SNR = \frac{P_t l}{8\pi kTB} \frac{W^2 L}{\lambda h^3} \frac{c\sigma}{B} \frac{\cos^4\theta}{\sin\theta} \tag{6-80}$$

6-3-5 Signal Fading

At every instant of time, the radar pulse illuminates a certain surface area that consists of many scattering points. Thus, the returned echo is the coherent addition of the echoes from a large number of points. The returns from these points add vectorially and result in a single vector that represents the amplitude V and phase ϕ of the total echo (see Fig. 6-26). The phase ϕ_i of each elementary vector is related to the distance between the sensor and the corresponding scattering point. If the sensor moves by a small amount, all the phases ϕ_i will change, leading to a change in the composite amplitude V. Thus, successive observations of the same surface area as the sensor moves by will result in a different value V. This variation is called fading. In order to characterize the backscatter properties of the surface area, many observations will be needed and then averaged. Similarly, if we take two neighboring areas which have the same backscatter cross section σ but have somewhat different fine details, the returned signals from the two areas will be different. Thus an image of a homogeneous surface with a constant backscatter cross section will show brightness variations from one pixel to the next. This is called speckle. In order to

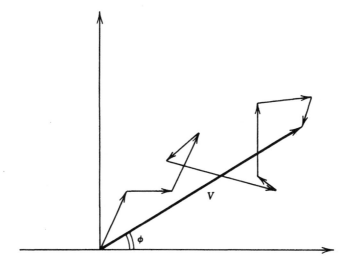

Figure 6-26. Composite return from an area with multiple scatters.

measure the backscatter cross section of the surface, the returns from many neighboring pixels will have to be averaged.

Let us assume that the illuminated area contains only two reflecting identical points A and B (Fig. 6-27) separated by a distance d. The received voltage at the radar is given by

$$V = V_0 e^{-i2kr_1} + V_0 e^{-i2kr_2} \tag{6-81}$$

and assuming that $d \ll r_0$, then

$$V = V_0 e^{-i2kr_0} [e^{-ikd\sin\theta} + e^{+ikd\sin\theta}]$$
$$\Rightarrow |V| = 2V_0 \left| \cos\left(\frac{2\pi d}{\lambda} \sin\theta\right) \right| \tag{6-82}$$

If the radar is moving at a constant velocity v, and assuming that θ is small, then

$$|V| = 2V_0 \left| \cos\left(\frac{2\pi dvt}{\lambda h}\right) \right| \tag{6-83}$$

This function is plotted in Figure 6-27. It shows that the received voltage varies periodically with a frequency

$$f_M = \frac{dv}{\lambda h} \tag{6-84}$$

If we have a very large number of scattering points between A and B, the resulting oscillation will have frequencies up to f_M. Thus, if we have a continuous distribution of points from A to B, the returned echo will oscillate as a function of time with an oscillation spectrum containing frequencies from zero to f_M.

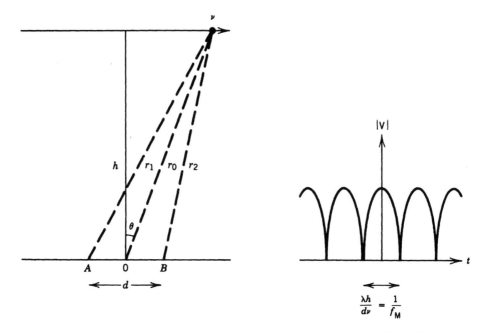

Figure 6-27. Geometry illustrating the return from two point scatters A and B.

Let us assume that the radar antenna has a length L, then the maximum area illuminated by the central beam has a length X_a equal to

$$X_a = \frac{2\lambda h}{L} \tag{6-85}$$

In this case, $d = X_a$; thus, from Equation (6-84),

$$f_M = \frac{2v}{L} \tag{6-86}$$

If $T = X_a/v$ is the total time during which the surface pixel is observed, then the product $N_a = f_M T$ is equal to:

$$N_a = \frac{2v}{L}\frac{X_a}{v} = \frac{X_a}{L} \tag{6-87}$$

This product represents the maximum number of cycles in the fading signal that result from a certain pixel. This corresponds approximately to the number of independent samples that correspond to the imaged pixel.

6-3-6   Fading Statistics

The instantaneous voltage V due to a large number of scatterers N_s is

$$V = V_e e^{i\phi} = \sum_{n=1}^{N_s} V_n e^{i\phi_n} \tag{6-88}$$

If N_s is very large and V_n and ϕ_n are independent random variables with ϕ_n uniformly distributed over the range $(0, 2\pi)$, then V_e has a Rayleigh distribution

$$p(V_e) = \frac{V_e}{s^2} \, e^{-V_e^2/2s^2} \tag{6-89}$$

and ϕ has a uniform distribution

$$p(\phi) = 1/2\pi \tag{6-90}$$

where s^2 is the variance of the signal. The mean of the Rayleigh distribution is given by

$$\overline{V}_e = \sqrt{\frac{\pi}{2}} \, s \tag{6-91}$$

and the second moment is

$$\overline{V_e^2} = 2s^2 \tag{6-92}$$

The resulting variance for the fluctuating component is

$$\overline{V_f^2} = \overline{V_e^2} - (\overline{V}_e)^2 = \left(2 - \frac{\pi}{2}\right)s^2 \tag{6-93}$$

and the ratio of the square of the mean of the envelope (i.e., the dc component) to the variance of the fluctuating component (i.e., the ac component) is given by

$$S = \frac{(\overline{V}_e)^2}{\overline{V_f^2}} = \frac{\pi}{4 - \pi} = 3.66 \text{ or } 5.6 \text{ dB}$$

The ratio S corresponds to an inherent "signal-to-noise ratio" or "signal-to-fluctuation ratio" for a Rayleigh fading signal, even in the absence of additional thermal noise.

If the radar receiver measures the power P of the echo instead of the voltage V, the corresponding probability distribution is given by

$$p(P)dP = p(V)dV \tag{6-94}$$

and

$$dP = 2V \, dV$$

Then

$$p(P) = \frac{1}{2s^2} \, e^{-P/2s^2} \tag{6-95}$$

which is an exponential distribution. The mean value is

$$\overline{P} = 2s^2 \tag{6-96}$$

Figure 6-28*a* shows the comparison between the Rayleigh, exponential, and Gaussian (thermal noise) distributions.

In order to reduce the fading (or speckle) effect, fading signals or speckled pixels are averaged or low-pass filtered. If *N* signals are averaged, the variance of the sum is equal to the mean squared ac component divided by *N*. Thus, for a square-law detector, the standard deviation is given by

$$s_N^2 = \frac{(\overline{P})^2}{N} \qquad \text{and} \qquad s_N = \frac{\overline{P}}{\sqrt{N}} \qquad (6\text{-}97)$$

Figure 6-28*b* shows the density functions of *N* variables with Rayleigh distribution.

Let us assume a fading signal of bandwidth *B* is continuously integrated over a time *T* much longer than the time for the autocovariance function to reduce to zero. If $BT \gg 1$, it can be shown that the integration is equivalent to averaging of *N* signals, where

$$N \simeq BT \qquad (6\text{-}98)$$

N represents the maximum number of cycles in the fading signal.

The number of independent samples in an imaging radar may be increased by transmitting a signal with a larger bandwidth *B'* than the bandwidth *B* required for the range resolution. In this case, the actual range resolution will then be

$$X_r' = \frac{B}{B'} X_r = \frac{X_r}{N_r} \qquad (6\text{-}99)$$

By averaging N_r neighboring pixels, the desired resolution is achieved with a reduction in speckle equivalent to the averaging of N_r signals. Thus, if a radar system uses excess bandwidth and also averages N_a successive signals, the total effective value for *N* is

$$N = N_a N_r = \frac{2X_a}{L} \frac{B}{B'} \qquad (6\text{-}100)$$

This is called the "number of looks." For a detailed analysis of radar signal measurements in the presence of noise, the reader is referred to the text by Ulaby et al., *Microwave Remote Sensing* (1982, Volume 2). Statistical distributions of SAR signals are discussed in great detail by Oliver and Quegan (1998).

The larger the number of looks *N*, the better the quality of the image from the radiometric point of view. However, this degrades the spatial resolution of the image. It should be noted that for *N* larger than about 25, a large increase in *N* leads to only a small decrease in the signal fluctuation. This small improvement in the radiometric resolution should be traded off against the large increase in the spatial resolution. For example, if one were to average 10 resolution cells in a four-look image, the speckle noise would be reduced to about 0.5 dB. At the same time, however, the image resolution would be reduced by an order of magnitude. Whether this loss in resolution is worth the reduction in speckle noise depends on both the aim of the investigation as well as the kind of scene imaged.

Figure 6-29 shows the effect of multilook averaging. The same image, acquired by the NASA/JPL AIRSAR system, is shown displayed at one, four, 16, and 32 looks. This figure clearly illustrates the smoothing effect, as well as the decrease in resolution resulting

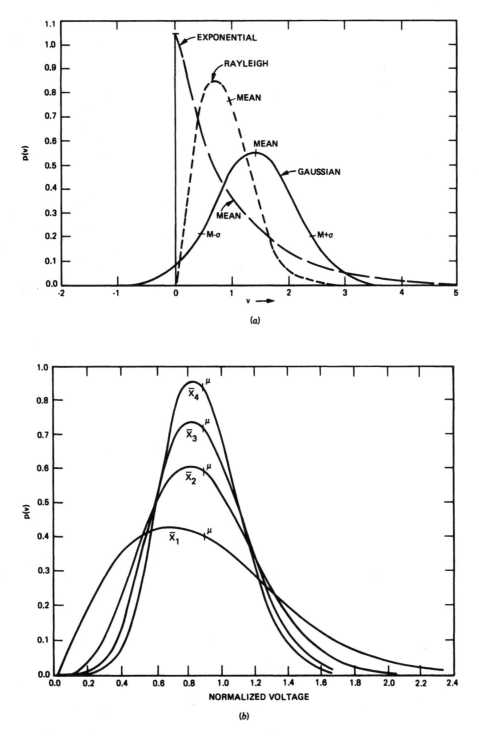

Figure 6-28. (*a*) Exponential, Rayleigh, and Gaussian distributions. (*b*) Density functions for an *N* variable Rayleigh distribution. Reprinted, by permission, from Fawwaz T. Ulaby et al., *Remote Sensing and Surface Scattering and Emission Theory,* Fig. 7-13. © 1982 Artech House, Inc.

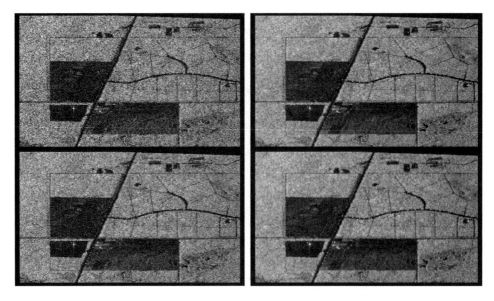

Figure 6-29. The effects of speckle can be reduced by incoherently averaging pixels in a radar image, a process known as multilooking. Shown in this image is the same image, processed at one look (upper left), four looks (lower left), 16 looks (upper right), and 32 looks (lower right). Note the reduction in the granular texture as the number of looks increase, while at the same time the resolution of the image decreases. (Courtesy of NASA/JPL-Caltech.)

from the multilook process. In one early survey of geologists done by Ford (1982), the results showed that even though the optimum number of looks depended on the scene type and resolution, the majority of the responses preferred two-look images. However, this survey dealt with images that had rather poor resolution to begin with, and one may well find that with today's higher-resolution systems, analysts may be asking for a larger number of looks.

6-3-7 Geometric Distortion

Cameras and the human eye image a scene in an angle–angle format, that is, neighboring pixels in the image plane correspond to areas in the scene viewed within neighboring solid angles (see Fig. 6-30). In the case of real-aperture imaging radars, the angle format is still valid in the azimuth dimension, but a time delay format is used in the range dimension. This means that two neighboring pixels in the image plane correspond to two areas in the scene with slightly different range to the sensor (see Fig. 6-30). In this case, the scene is projected in a cylindrical geometry on the imaging plane. This leads to distortions, as illustrated in Figure 6-31. Three segments of equal length but with different slopes are projected in the image plane as segments of different lengths. The side of a hill toward the sensor will be shortened while the other side is stretched. This is called *foreshortening*. If the topography is known, this distortion can be corrected. Figure 6-31*b* shows a radar image acquired with the NASA/JPL AIRSAR system that shows examples of foreshortening and shadowing. Shadowing occurs in regions where the surface slope is such that a portion of the terrain is not illuminated by the radar signal.

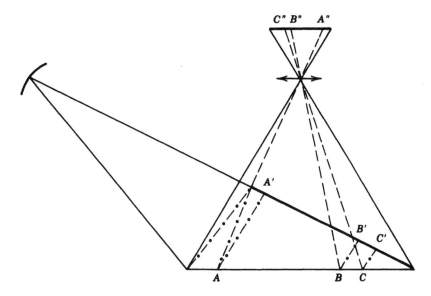

Figure 6-30. Imaging geometry for cameras and radar sensors.

In the extreme case in which the surface slope α is larger than the incidence angle θ, $\alpha > \theta$, a *layover* results—the top of a hill will apear to be laid over the region in front of it. In this extreme case, it is not possible to correct for the distortion.

6-4 IMAGING SENSORS: SYNTHETIC-APERTURE RADARS

As indicated in Equation 6-61, the azimuth resolution of a real-aperture radar is given by

$$X_a = \frac{\lambda h}{L \cos \theta} \tag{6-101}$$

For a spaceborne radar, X_a is typically many hundreds of meters to many kilometers, even if L is large. To illustrate, if $\lambda = 3$ cm, $h = 800$ km, $L = 10$ m, and $\theta = 20°$, then $X_a = 2.5$ km.

In order to improve the azimuth resolution, a synthetic-aperture technique is used. This technique is based on the fact that the target stays in the beam for a significant amount of time, and it is observed by the radar from numerous locations along the satellite path. The concept of the synthetic-aperture technique can be explained in two ways that lead to the same result: the synthetic-array approach or the Doppler synthesis approach.

6-4-1 Synthetic-Array Approach

In Section 6-2-1 and in Figure 6-15c it was pointed out that an array of antennas is equivalent to a single antenna moving along the array line as long as the received signals are coherently recorded and then added in the same way that the signals from the antenna array would have been combined in the waveguide network. In addition, the target is assumed to be static (or its behavior known) during this period.

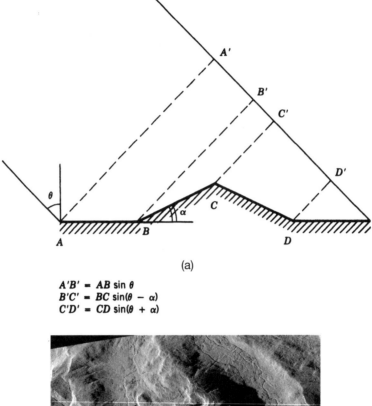

(a)

$A'B' = AB \sin\theta$
$B'C' = BC \sin(\theta - \alpha)$
$C'D' = CD \sin(\theta + \alpha)$

(b)

Figure 6-31. Distortions in radar images resulting from surface topography Data acquired by NASA/JPL AIRSR System.

Let us assume that a radar sensor is moving at a velocity v and has an antenna of length L. The antenna main beam footprint on the surface is equal to

$$\mathcal{L} = \frac{2\lambda h}{L} \tag{6-102}$$

As the sensor moves, successive echoes are recorded at points $x_1, x_2, \ldots, x_i$, along the flight line (see Fig. 6-32). An onboard stable oscillator is used as a reference and the echoes are recorded coherently, that is, amplitude and phase as a function of time. These echoes are then combined in a processor to synthesize a linear array. From Figure 6-32, it can be seen that the maximum array length that could be achieved is equal to $\mathcal{L}$. The synthesized array will have a beam width equal to

$$\theta_s = \frac{\lambda}{\mathscr{L}} = \frac{L}{2h} \tag{6-103}$$

and the resulting array footprint on the ground is

$$X_a = h\theta_s = \frac{L}{2} \tag{6-104}$$

This corresponds to the finest resolution that could be achieved using the synthetic array. At first glance, this result seems most unusual. It shows that the ultimate resolution is independent of the distance between the sensor and the area being imaged. In addition, finer resolution can be achieved with a smaller antenna. This can be explained in the following way. The farther the sensor is, the larger the footprint is on the ground, thus the longer the synthetic array. This leads to a finer synthetic beam, which exactly counterbalances the increase in distance. The smaller the antenna is, the larger the footprint and the synthetic array. This leads to a finer synthetic beam and, therefore, a finer resolution.

6-4-2   Focused Versus Unfocused SAR

In order to achieve the full capability of a synthetic array, each received echo should be phase shifted in order to take into account the fact that the distance between the sensor and the target is variable. The phase shift that needs to be added to the echo received at location x_i in order to focus on point P is (see Fig. 6-33)

$$\phi_i = 2k(h - r_i) = \frac{4\pi}{\lambda}(h - r_i) \tag{6-105}$$

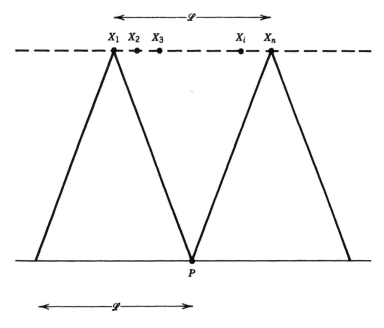

Figure 6-32. Geometry showing the formation of a synthetic array by moving a single antenna along a track.

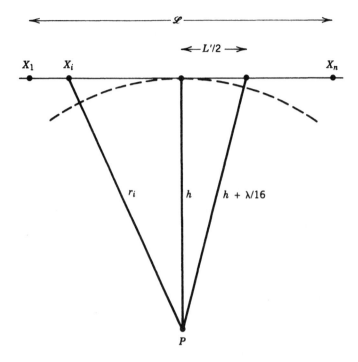

Figure 6-33. Geometry illustrating the range change to a point P during the formation of the synthetic aperture. L' corresponds to the synthetic-aperture length for an unfocused SAR.

Thus, during processing, the phase shift ϕ_i is added to the echo received at x_i (Fig. 6-34a). In order to focus on a point P' at distance h', another set of phase shifts ϕ'_i is added, where

$$\phi'_i = \frac{4\pi}{\lambda}(h' - r'_i)$$

In order to simplify the processing, one can shorten the synthetic array length used to the point at which all the echos can be added with no phase additions. In the case of the unfocused SAR, we assume that phase shift corrections of less than $\lambda/4$ can be neglected. From Figure 6-33 we can see that the corresponding array length L' can be derived from

$$2k\left[\sqrt{h^2 + \frac{L'^2}{4}} - h\right] = \frac{\pi}{4} \tag{6-106}$$

For large h, this can be reduced to

$$kh\left(\frac{L'^2}{4h^2}\right) = \frac{\pi}{4}$$

$$\Rightarrow L' = \sqrt{\frac{\lambda h}{2}} \tag{6-107}$$

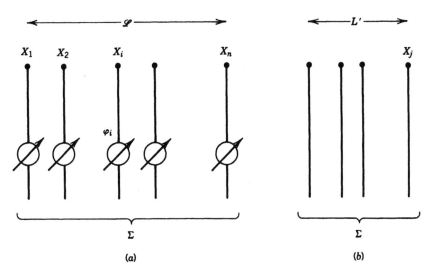

Figure 6-34. Echoes summing in the case of focused (*a*) and unfocused (*b*) SAR.

and the corresponding resolution is

$$X_a = \frac{\lambda h}{L'} = \sqrt{2\lambda h} \qquad (6\text{-}108)$$

In this case, the resolution does depend on the range to the surface. However, the processing involved in synthesizing the aperture is very simple (Fig. 6-34*b*).

6-4-3 Doppler-Synthesis Approach

As the radar sensor flies over a target P, the echo from P will first have a positive Doppler shift (when P enters the beam) that decreases down to zero, then becomes increasingly negative by the time P exits the beam (see Fig. 6-35). The spectrum of the echo from P covers the region $f_0 \pm f_D$, where

$$f_D = \frac{2v}{\lambda} \sin \frac{\theta}{2} \simeq \frac{v\theta}{\lambda} = \frac{v}{L} \qquad (6\text{-}109)$$

and f_0 is the transmitted signal frequency.

If a neighboring target P' is displaced from P by a distance X_a along the azimuth dimensions, the Doppler history from P' will be a replica of the one from P but with a time displacement $t = X_a/v$. The shortest time displacement that can be measured after processing the signal with a spectrum bandwidth $B_D = 2f_D$ is equal to

$$t_m = \frac{1}{B_D} = \frac{1}{2f_D} = \frac{L}{2v} \qquad (6\text{-}110)$$

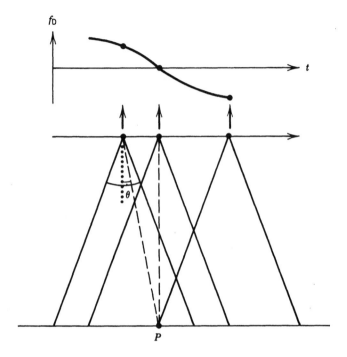

Figure 6-35. Doppler history of a point target as the sensor passes by.

This allows us to derive the finest possible resolution:

$$X_a = vt_m = \frac{L}{2} \tag{6-111}$$

which is the same as that derived using the synthetic-array approach [Equation (6-104)].

Actual imaging SARs use a series of pulses instead of a continuous signal. The pulse repetition frequency (PRF) should be high enough to make sure that the Doppler spectrum is sufficiently well sampled in order to do the processing. From Equation 6-109 and the Nyquist sampling criterium, it can then be stated that

$$PRF > 2f_D = \frac{2v}{L} \tag{6-112}$$

if we assume that the carrier frequency is down converted to zero (called zero offset). If the carrier frequency is converted down to f_D (called range offset), then the minimum PRF is twice the one in Equation (6-112).

Equation (6-112) indicates that at least one sample (i.e., one pulse) should be taken every time the sensor moves by half an antenna length. To illustrate, low-altitude Earth-orbiting satellites have an orbital velocity of about 7 km/sec. For an antenna length of 10 m, the corresponding minimum PRF is equal to 1.4 kHz.

The result in Equation (6-11) is derived assuming that the signal is integrated for the maximum time given in Equation (6-110). This required resolution then drives the value of the PRF as given in (6-112). If the maximum resolution is not required, one may in fact use a smaller fraction of the total Doppler bandwidth during the SAR processing, with a corresponding decrease in resolution. In this case, the requirement on the PRF can also be relaxed somewhat. As long as the ambiguous signal falls outside that part of the Doppler bandwidth used in the processing, the image quality will not be affected.

One can also trade off azimuth resolution and transmitted power using a scheme known as burst mode SAR. Instead of transmitting pulses continuously in time, a burst mode SAR breaks the synthetic-aperture time t_m into shorter blocks known as burst cycles. Inside each burst cycle, the radar transmits a series of pulses [at the appropriate PRF shown in Equation (6-112)] for only a portion of the burst cycle, known as the burst period. In this way, only portions of the phase history of a scatterer are recorded. Special processing algorithms have been developed that take the burst mode imaging geometry into account when compressing the data in the azimuth direction. Since only part of the total integration time is utilized, it follows that the azimuth resolution of a burst mode SAR will be degraded relative to that of a continuously mapping SAR. In fact, the resolution is reduced by a factor equal to the ratio of the synthetic-aperture time to the burst period time. On the other hand, since the radar operates at a duty cycle equal to the ratio of the burst period relative to the burst cycle time, the required average power will also be reduced by this amount. In addition, the total data volume is also reduced because radar echoes are only sampled during the bursts. This reduces the data rate by the same factor as the average power. Severe power and data rate limitations have forced the designs of the planetary SAR sensors on the Magellan and Cassini missions to operate in the burst mode.

The concept of a burst mode SAR can be extended to increase the size of the swath that can be imaged. This implementation, known as ScanSAR, requires fast switching of the antenna beam in the elevation direction. The switching of the antenna beam is done in such a way that a number of parallel swaths are imaged. For example, in the case of the Shuttle Radar Topography Mission (SRTM) four such adjacent swaths were imaged. The data collection for a ScanSAR operation involves pointing the antenna beam to subswath 1 and then transmitting and recording a burst of pulses. The beam is then switched to subswath 2 and another burst of pulses are transmitted and recorded. The process is repeated for all the other subswaths before returning to subswath 1. Therefore, even though the radar is continuously transmitting and receiving, for each individual subswath, the radar operates in a burst mode, with the corresponding reduction in azimuth resolution. The ScanSAR mode was first demonstrated in space by the SIR-C system, and the Canadian RadarSat system is the first operational SAR sensor to utilize the ScanSAR mode routinely to increase its swath width. The Cassini radar also operates in the ScanSAR mode with five subswaths. Because of the changing geometry between subswaths, the PRF must be changed for each subswath to ensure optimum image quality.

6-4-4 SAR Imaging Coordinate System

As mentioned earlier, the human eye and the optical camera have an angle–angle format. Real-aperture radars have an angle–time delay format. Synthetic-aperture radars have a Doppler–time delay format.

Let us consider the configuration shown in Figure 6-36. The radar transmits a pulse that travels as a spherical shell away from the radar. Therefore, points located on a sphere centered at the radar location have their echoes received simultaneously by the radar. To process the radar signals into images, it is typically assumed that all scatterers are located on a reference plane (see Section 6-4-7 for a discussion of SAR processing). The intersection of a sphere centered on the radar position and a horizontal plane is a circle with its center at a position directly underneath the radar (the so-called nadir point—see Figure 6-36b). Thus, a family of concentric spheres intersecting with the reference plane gives a series of concentric circles, centered at the nadir point, which define lines of equidistance (or equal time delay) to the radar.

Points located on coaxial cones, with the flight line as the axis and the radar location as the apex, have returned echoes with identical Doppler shifts. The intersection of these cones with the reference surface plane gives a family of hyperbolas. Objects on a certain hyperbola will provide equi-Doppler returns. Thus, the surface can be referenced to a coordinate system of concentric circles and coaxial hyperbolas, and each point in the imaged plane can be uniquely identified by its time delay and Doppler shift. The only potential ambiguity is due to the symmetry between the right and the left parts of the flight track. However, this is easily avoided by illuminating only one side of the track, that is, by using side illumination.

The brightness that is assigned to a specific pixel in the radar image is then proportional to the echo energy contained in the time delay bin and the Doppler bin, which correspond to the equivalent point on the surface being imaged. The achievable resolution in the azimuth is (from Equation 6-104)

$$X_a = \frac{L}{2}$$

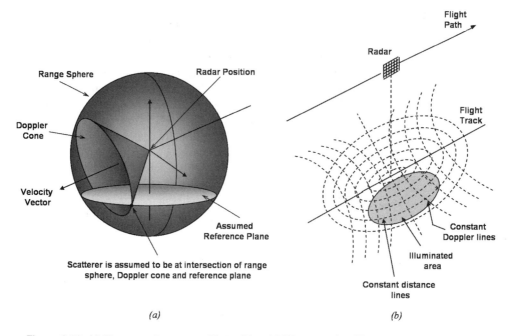

(a)

(b)

Figure 6-36. (a) The geometry assumed in traditional SAR processing. The scatterer is assumed to be at the intersection of the range sphere, the Doppler cone, and an assumed flat reference plane. The case shown is for a left-looking SAR. (b) Coordinate system (equi-Doppler and equirange lines) for synthetic-aperture radar imaging.

and the achievable range resolution is the same as in the case of real-aperture radar (see Equation 6-60):

$$X_r = \frac{c}{2B \sin \theta}$$

It should be pointed out that the resolution of the SAR is not dependent on the altitude of the sensor. This can be explained by the fact that the imaging mechanism uses the Doppler shifts in the echo and the differential time delays between surface points, neither of which is a function of the distance between the sensor and the surface. Of course, the altitude still plays a major role in determining the power required to acquire a detectable echo and in determining the size of the antenna, as will be discussed later.

6-4-5 Ambiguities and Artifacts

A radar images a surface by recording the echoes line by line with successive pulses. For each transmitted pulse, the echos from the surface are received as a signal that has a finite duration in time. The echo starts at a time determined by the round-trip time to that part of the surface that is closest to the radar, and, theoretically, would continue until we encounter the end of the surface. In practice, the strength of the echo is modulated by the antenna pattern in the elevation plane, resulting in an echo that has a finite time duration. Therefore, the leading edge of each echo corresponds to the near edge of the largest possible image scene, and the tail end of the echo corresponds to the far edge of the largest possible scene. The length of the echo (i.e., swath width of the scene covered) is determined by the antenna beam width and the size of the time window during which the data are recorded. The exact timing of the echo reception depends on the range between the sensor and the surface being imaged. If the timing of the pulses or the extent of the echo are such that the leading edge of one echo overlaps with the tail end of the previous one, then the far edge of the scene is folded over the near edge of the scene. This is called *range ambiguity*. If the full echo determined by the antenna pattern is recorded, the temporal extent of the echo is equal to (see Fig. 6-37)

$$T_e \approx 2\frac{R_s}{c}\beta \tan \theta = 2\frac{h\lambda}{cW}\frac{\sin \theta}{\cos^2 \theta} \tag{6-113}$$

where we assumed that β is small and neglected any Earth curvature. To avoid overlapping echoes, this time extent should be shorter than the time separating two pulses (i.e., $1/PRF$). Thus, we must have

$$PRF < \frac{cW}{2h\lambda}\frac{\cos^2 \theta}{\sin \theta} \tag{6-114}$$

In addition, the timing of the pulses should be selected such that the echo is completely within an interpulse period, that is, no echoes should be received during the time that a pulse is being transmitted.

The above equation gives an upper limit for the PRF. Another kind of ambiguity present in SAR imagery also results from the fact that the target's return in the azimuth direction is sampled at the PRF. This means that the azimuth spectrum of the target return

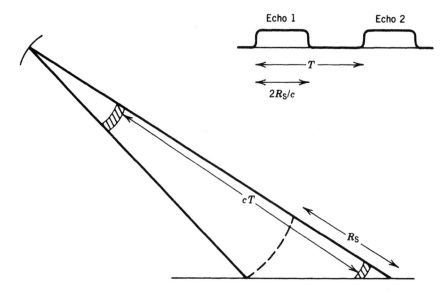

Figure 6-37. Geometry illustrating the range configuration.

repeats itself in the frequency domain at multiples of the PRF. In general, the azimuth spectrum is not a band-limited signal; instead, the spectrum is weighted by the antenna pattern in the azimuth direction. This means that parts of the azimuth spectrum may be aliased, and high-frequency data of one version of the spectrum will actually appear in the low-frequency part of the neighboring spectrum. In actual images, these *azimuth ambiguities* appear as ghost images of a target repeated at some distance in the azimuth direction as shown in Figure 6-38. To reduce the azimuth ambiguities, the PRF of a SAR has to exceed the lower limit given by Equation (6-112).

In order to reduce both range and azimuth ambiguities, the PRF must satisfy both the conditions expressed by Equations (6-112) and (6-114). Therefore, we must insist that

$$\frac{cW}{2h\lambda}\frac{\cos^2\theta}{\sin\theta} > \frac{2v}{L} \qquad (6\text{-}115)$$

from which we derive a lower limit for the antenna size as

$$LW > \frac{4vh\lambda}{c}\frac{\sin\theta}{\cos^2\theta} \qquad (6\text{-}116)$$

We note that Equation (6-116) was derived assuming that the entire echo in time as determined by the elevation antenna pattern is used, *and* the entire Doppler spectrum as determined by the azimuth antenna pattern is used. This, in effect, means that the widest swath possible is imaged at the highest possible azimuth resolution. If either of these requirements is relaxed, the antenna can actually be smaller than that given in Equation (6-116). For example, if we use a smaller fraction of the Doppler bandwidth in the processor, we can actually afford to have neighboring Doppler spectra overlap partially, as long as the overlapping portions fall outside that portion of the Doppler bandwidth that is used in

Figure 6-38. Azimuth ambiguities result when the radar pulse repetition frequency is too low to sample the azimuth spectrum of the data adequately. In this case, the edges of the azimuth spectrum fold over themselves, creating ghost images as shown in this figure.

the processor. For a fixed PRF, this means that the azimuth beamwidth may actually be larger than the value $2v/PRF$, that is, the antenna length may be smaller than what is shown in Equation (6-116). The penalty, of course, is that the azimuth resolution will be poorer than if we utilized the entire Doppler bandwidth. Similarly, if we only utilize a portion of each echo in the interpulse period, the echoes from successive pulses can actually overlap, as long as the overlapping portions fall outside of that portion of the echo that we actually use in the processing, meaning that the elevation beamwidth may be larger (i.e., the antenna width may be smaller) than that assumed in Equation (6-104). In this case, the penalty is that the swath will be narrower, that is, we will image a smaller portion of the surface. See Freeman et al. (2000) for a more detailed discussion of this issue.

Another type of artifact that is encountered in radar images results when a very bright surface target is surrounded by a dark area. As the image is being formed, some spillover from the bright target, called sidelobes, although weak, could exceed the background and become visible, as shown in Figure 6-39. It should be pointed out that this type of artifact is not unique to radar systems. They are common in optical systems, where they are known as the sidelobes of the point spread function. The difference is that in optical systems, the sidelobe characteristics are determined by the characteristics of the imaging optics, that is, the hardware, whereas in the case of a SAR, the sidelobe characteristics are

Figure 6-39. Sidelobes from bright targets (indicated by the arrows in the image) can exceed the signals from neighboring pixels.

determined by the characteristics of the processing filters. In the radar case, the sidelobes may, therefore, be reduced by suitable weighting of the signal spectra during matched filter compression. The equivalent procedure in optical systems is apodization of the telescope aperture.

The vast majority of these artifacts and ambiguities can be avoided with proper selection of the sensor's and processor's parameters. However, the interpreter should be aware of their occurrence because in some situations they might be difficult, if not impossible, to suppress.

6-4-6   Point Target Response

One simple way to visualize the basic principle behind SAR processing and the effects of complicating factors (planet rotation, attitude drift, etc.) is to study the response of a point target. Many different SAR processing algorithms are in use, and their detailed discussion is beyond the scope of this text. The interested reader is referred to the texts and articles cited in the references. Here, we shall illustrate SAR processing through the use of the so-called rectangular imaging algorithm.

Let us consider the case in which a chirp is used and the transmitted wave form is given by

$$W(t) = A(t) \exp\left[i2\pi\left(f_0 t + \frac{Bt^2}{2\tau'}\right)\right] \qquad (6\text{-}117)$$

where

$$A(t) = 1 \text{ for } nT - \frac{\tau'}{2} < t < nT + \frac{\tau'}{2}$$

$$A(t) = 0 \text{ otherwise}$$

and n is an integer. We shall now consider the return from a single-point scatterer that is assumed to be at a range $r(s)$ from the radar. We use the symbol s to denote the time at which the pulse is transmitted to be consistent with most radar texts. This is sometimes referred to as the "slow time" to distinguish it from the time measured inside the interpulse period, which is referred to as the "fast time" and is typically represented by the symbol t. As the radar moves along its path, a two-dimensional array of values is recorded. One dimension corresponds to the time within an individual interpulse period, and represents the "fast time" mentioned earlier. This process is repeated for each transmitted pulse, forming the second dimension of the array, which corresponds to the "slow time."

During data collection, the point target will be in view of the antenna for a time period T_i. During this period, the range between the sensor and the target is equal to (see Fig. 6-40)

$$r(s) = \sqrt{r_0^2 + v^2 s^2} = \sqrt{h^2 + D^2 + v^2 s^2} \tag{6-118}$$

where we neglected the effect of the Earth's curvature. Usually, $vs \ll r_0$, then

$$r(s) = r_0 + \frac{v^2 s^2}{2 r_0} \tag{6-119}$$

and the returned echo has a phase shift $\phi(t)$ equal to

$$\phi(s) = 2\pi \frac{2r(s)}{\lambda} = \frac{4\pi r_0}{\lambda} + \frac{2\pi v^2 s^2}{\lambda r_0} \tag{6-120}$$

Comparing this phase to that of the transmitted signal, we note that Equation (6-120) represents a linear FM signal with a carrier frequency that is zero, with a constant phase

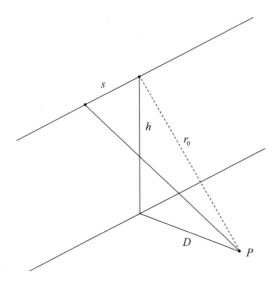

Figure 6-40. Geometry showing the distance to a point target.

added to it. In addition, the leading edge of the echo will arrive back with a time delay t_d relative to the transmitted pulse:

$$t_d = \frac{2r(s)}{c} = \frac{2r_0}{c} + \frac{v^2s^2}{r_0c} = t_0 + \Delta t_d \qquad (6\text{-}121)$$

Neglecting the effects of the antenna pattern and $1/r^3$ change in the amplitude, the resulting signal amplitude pattern (i.e., point response) is shown in Figure 6-41. In the figure, Z represents vs and Y represents time referred to the start of each pulse. In this representation, the echo amplitude is then given by

$$E(Y, Z) \sim \exp\frac{i2\pi B(Y - Z^2/r_0c)^2}{2\tau'} + i\frac{2\pi Z^2}{\lambda r_0} \qquad (6\text{-}122)$$

for

$$\frac{Z^2}{r_0c} - \frac{\tau'}{2} < Y < \frac{Z^2}{r_0c} + \frac{\tau'}{2}$$

and

$$\frac{-vT_i}{2} < Z < \frac{vT_i}{2}$$

If we take a cut along $Z = 0$, then the received signal amplitude is an exact duplicate (within a constant factor) of the transmitted signal amplitude shifted down to zero offset. If we take a cut at $Z = vt$, then the received signal is similar to the one at $Z = 0$, with an upshift of frequency equal to $2v^2t/\lambda r_0$ and a start-time delay of v^2t^2/r_0c.

The role of the processor is to match the returned echo from a large number of targets to a family of templates similar to the one in Figure 6-41 with the parameters corresponding to each and every point in the swath. It should be noted that points along a constant distance D have identical templates but are displaced in time (i.e., along the Z axis). However, points at different distances D (e.g., different r_0) will have slightly different templates.

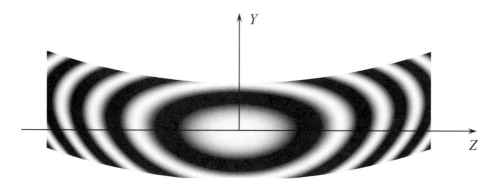

Figure 6-41. Point target response.

6-4-7 Correlation with Point Target Response

The received signal is a combination of the echoes from a large number of point targets in the scene. When this signal is correlated with the reference response corresponding to one pixel in the imaged scene, the output will be maximum when the echo from the targets in that pixel is matched exactly with the reference. All other echoes from other pixels will not match exactly with the reference, thus resulting in only very weak output.

Let $V_s(t)$ be the received signal from a target A at distance r_0. In Appendix D, we show that if we compress a single pulse of the form given in Equation (6-117) using a matched filter, the resulting signal is of the form

$$V_s(s, t) = A_0 \exp[-i4\pi r(s)/\lambda] \frac{\sin[\pi B(t - 2r(s)/c)]}{\pi B(t - 2r(s)/c)} \tag{6-123}$$

This expression shows that each pulse will compress to have its maximum at a time equal to the round-trip time of the pulse to the point target. In Appendix D, we derive Equation (6-123) using a convolution between the received signal and a replica of the transmitted signal. In practice, this convolution is performed in the frequency domain; a convolution of two signals in the time domain is equivalent to the product of the Fourier transforms of the two signals in the frequency domain. Using the speed of digital fast Fourier transforms (FFTs), this convolution can be accomplished efficiently in this manner. This process is known as range compression. What now remains is to compress the signal in the azimuth, or slow time, direction.

The expression of $V_s(s, t)$ is [from Equation (6-123)]

$$V_s(s, t) = A_0 \exp\left[\frac{4\pi r_0}{\lambda}\right] \frac{\sin[\pi B(t - 2r(s)/c)]}{\pi B(t - 2r(s)/c)} \exp\left[-\frac{i2\pi v^2 s^2}{\lambda r_0}\right] \tag{6-124}$$

The appropriate reference function $V_r(t)$ for a point target at the same location as A is, of course, the same as the last term in V_s. So, the output of the correlator is:

$$V_0(\xi) = \int_{-T_i/2}^{T_i/2} V_s(s) V_r^*(s + \xi) ds \tag{6-125}$$

where V_r^* is the complex conjugate of V_r. Conceptually, the compression of the signals in the along-track direction therefore is also a matched filter operation, which again can be implemented using FFTs. In practice, there is a bit more work resulting from the fact that the change in the range to the point target may be larger than the "fast time" bins used in the sampling. The result of this is that the range-compressed signals fall on an arc in the data array as shown in Figure 6-41, and must first be resampled to ensure that the appropriate signal is used as the input to the matched filter. This is known as "range walk" and is more pronounced for spaceborne systems. Replacing V_s and V_r by their expressions, similar to what we did in Appendix D, we get

$$V_0(\xi, t) = A_0 \exp\left[\frac{4\pi r_0}{\lambda}\right] \frac{\sin[\pi B(t - 2r_0/c)]}{\pi B(t - 2r_0/c)} \int_{-T_i/2}^{T_i/2} \exp\left[\frac{-i2\pi v^2}{\lambda r_0} \{s^2 - (s + \xi)^2\}\right] ds \tag{6-126}$$

The integration limits are the time when the target is in the antenna beam. The reference function usually extends over a longer time, so it does not affect the integration limits. Following the same procedure as in Appendix D, we find that the resulting output is

$$V_0(\xi, t) = A_0 T_i \exp\left[\frac{4\pi r_0}{\lambda}\right] \frac{\sin[\pi B(t - 2r_0/c)]}{\pi B(t - 2r_0/c)} \frac{\sin(aT_i\xi)}{(aT_i\xi)} \tag{6-127}$$

where $a = 2\pi v^2/\lambda r_0$. This output is maximum at $\xi = 0$, that is, when point A is aligned with the point target of the reference.

If we had two targets (A and B) at the same range but displaced by a distance X (i.e., time displacement $\Delta t = X/v$), the correlator output would be

$$V_0(\xi, t) = \exp\left[\frac{4\pi r_0}{\lambda}\right] \frac{\sin[\pi B(t - 2r_0/c)]}{\pi B(t - 2r_0/c)} \left\{ AT_i \frac{\sin(aT_i\xi)}{(aT_i\xi)} + BT_i \frac{\sin[aT_i(\xi - \Delta t)]}{[aT_i(\xi - \Delta t)]} \right\} \tag{6-128}$$

which shows two peaks displaced by a time distance Δt. The width of each peak response is equal to

$$\Delta\xi = \frac{2\pi}{aT_i} = \frac{\lambda r_0}{v^2 T_i} = \frac{\lambda r_0}{\mathcal{L}v} \tag{6-129}$$

This corresponds to a spatial resolution X_a equal to

$$X_a = v\Delta\xi = \frac{\lambda r_0}{\mathcal{L}} \tag{6-130}$$

which is the same as Equation (6-104) with $\mathcal{L}$ replaced by $2\lambda r_0/L$ [from Equation (6-102)].

The rectangular algorithm is shown in Figure 6-42, and consists of three main steps. The first is the range compression, followed by a resampling to compensate for range walk, and then followed by azimuth compressions. As mentioned before, significant reduction in the number of computations is achieved by using the fact that when transformed to the frequency domain a correlation becomes a multiplication. Thus, the same function can be achieved by taking the Fourier transform of V_s, multiplying it by the reference function in

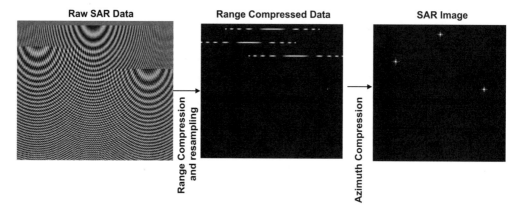

Figure 6-42. This figure illustrates how the rectangular SAR processing algorithm is implemented. Three point scatterers are at different range and azimuth positions. The simulation is for an airborne SAR with negligible range walk.

the frequency domain (i.e., its Fourier transform) and then doing an inverse transform. If the reference function has U elements, the correlation in the time domain requires $U \times U$ multiplications. In the case of the FFT approach, only $2U \log U$ operations are needed. Considering that U is usually many thousands, the frequency domain approach reduces the number of operations required by more than two orders of magnitude.

The description of the rectangular algorithm assumed that the radar moved along a straight line. This is seldom the case for airborne systems. In these systems, the processor includes another resampling step to take into account the actual flight path of the radar. The signals are shifted in time to reference their time origin to an assumed reference flight path (typically a straight line). However, the signal phase should also be changed to compensate for the differential change in range between the actual and the reference flight paths. This process is known as motion compensation.

Note from Equation (6-127) that the point target response is the product of two $\sin x/x$ functions, one aligned with the range direction, and one aligned with the azimuth or along-track direction. This response is similar to what would be measured with a rectangular aperture in an optical system. Compare for example the responses shown in Figure 6-42 to the point spread function of a rectangular aperture shown in Chapter 3. The point target response of the radar system as measured in the image is, therefore, strongly dependent on the filters used in the processor.

The rectangular algorithm is the simplest to implement but is really only applicable if the antenna beamwidth is relatively narrow in the along-track direction. If a wide beam is used, such as in very high resolution radars, the full two-dimensional reference function must be used in the SAR processing. Several algorithms are in use that are appropriate for this case. The reader is referred to the references for more details.

6-4-8 Advanced SAR Techniques

The field of synthetic-aperture radar changed dramatically over the past two decades with the operational introduction of advance radar techniques such as polarimetry and interferometry. Although both of these techniques had been demonstrated much earlier, radar polarimetry only became an operational research tool with the introduction of the NASA/JPL AIRSAR system in the early 1980s, and reached a climax with the two SIR-C/X-SAR flights on board the space shuttle Endeavour in April and October 1994. Radar interferometry received a tremendous boost when the airborne TOPSAR system was introduced in 1991 by NASA/JPL, and when data from the European Space Agency ERS-1 radar satellite became routinely available in 1991. The Shuttle Radar Topography Mission, flown on the space shuttle Endeavour in February 2000, was the first spaceborne application of a fixed-baseline interferometer and mapped 80% of the land mass of the Earth in a single space shuttle mission. In this section, we shall describe these two techniques in more detail.

SAR Polarimetry. As mentioned before, electromagnetic wave propagation is a vector phenomenon, that is, all electromagnetic waves can be expressed as complex vectors. Plane electromagnetic waves can be represented by two-dimensional complex vectors. This is also the case for spherical waves when the observation point is sufficiently far removed from the source of the spherical wave. Therefore, if one observes a wave transmitted by a radar antenna when the wave is a large distance from the antenna (in the far field of the antenna), the radiated electromagnetic wave can be adequately described by a two-dimensional complex vector. If this radiated wave is now scattered by an object, and one observes this wave in the far field of the scatterer, the scattered wave can again be adequately described by a two-dimensional vector. In this abstract way, one can consider the scatterer as a mathematical operator that takes one two-dimensional complex vector (the wave impinging upon the object) and changes it into another two-dimensional vector (the scattered wave). Mathematically, therefore, a scatterer can be characterized by a complex 2 × 2 scattering matrix:

$$\mathbf{E}^{sc} = \begin{pmatrix} S_{hh} & S_{hv} \\ S_{vh} & S_{vv} \end{pmatrix} \mathbf{E}^{tr} = [\mathbf{S}]\mathbf{E}^{tr} \tag{6-131}$$

where $\mathbf{E}^{tr}$ is the electric field vector that was transmitted by the radar antenna, $[\mathbf{S}]$ is the 2 × 2 complex scattering matrix that describes how the scatterer modified the incident electric field vector, and $\mathbf{E}^{SC}$ is the electric field vector that is incident on the radar receiving antenna. This scattering matrix is a function of the radar frequency and the viewing geometry. The scatterer can, therefore, be thought of as a polarization transformer, with the transformation given by the scattering matrix. Once the complete scattering matrix is known and calibrated, one can synthesize the radar cross section for any arbitrary combination of transmitting and receiving polarizations. Figure 6-3 shows a number of such synthesized images for the San Francisco Bay area in California. The data were acquired with the NASA/JPL AIRSAR system.

The voltage measured by the radar system is the scalar product of the radar antenna polarization and the incident wave electric field:

$$V = \mathbf{p}^{rec} \cdot [\mathbf{S}]\mathbf{p}^{tr} \tag{6-132}$$

Here, $\mathbf{p}^{tr}$ and $\mathbf{p}^{rec}$ are the normalized polarization vectors describing the transmitting and receiving radar antennas. The power received by the radar is the magnitude of the voltage squared:

$$P = VV^* = |\mathbf{p}^{rec} \cdot [\mathbf{S}]\mathbf{p}^{rad}|^2 \tag{6-133}$$

Expanding the expression inside the magnitude sign, it can be shown that the received power can also be written in terms of the scatterer covariance matrix as follows:

$$P = VV^* = (\tilde{\mathbf{A}}\mathbf{T})(\tilde{\mathbf{T}}\mathbf{A})^* = \tilde{\mathbf{A}}\mathbf{T}\mathbf{T}^*\mathbf{A}^* = \mathbf{A} \cdot [\mathbf{C}]\mathbf{A}^*; \; [\mathbf{C}] = \mathbf{T}\tilde{\mathbf{T}}^* \tag{6-134}$$

where $\tilde{\mathbf{A}} = (p_h^{rec}p_h^{rad} \; p_h^{rec}p_v^{rad} \; p_v^{rec}p_h^{rad} \; p_v^{rec}p_v^{rad})$ represents the transpose of the antenna polarization vector elements, and $\tilde{\mathbf{T}} = (S_{hh} \; S_{hv} \; S_{vh} \; S_{vv})$ represents only the scatterer. The superscript * denotes complex conjugation. The covariance matrix characterization is particularly useful when analyzing multilook radar images, since the covariance matrix of a multilook pixel is simply the average covariance matrix of all the individual measurements contained in the multilook pixel. Equation (6-134) shows the covariance matrix to be a 4 × 4 complex Hermetian matrix. In the case of radar backscatter, reciprocity dictates that $S_{hv} = S_{vh}$ and the covariance matrix can in general be written as a 3 × 3 complex Hermetian matrix. Also note that it is always possible to calculate the covariance matrix from the scattering matrix. However, the inverse is not true; it is not always possible to calculate an equivalent scattering matrix from a knowledge of the covariance matrix. This follows from the fact that the off-diagonal terms in the covariance matrix involve cross products of the scattering matrix elements, for example $S_{hh}S_{hv}^*$. For a single scattering matrix there is a definite relationship between this term and the two diagonal terms $S_{hh}S_{hh}^*$ and $S_{hv}S_{hv}^*$. However, once the covariance matrix elements are averaged spatially, such as during multilooking of an image, this definite relationship no longer holds, and we cannot uniquely find an equivalent S_{hh} and S_{hv} that would satisfy all three cross-products $\langle S_{hh}S_{hv}^*\rangle$, $\langle S_{hh}S_{hh}^*\rangle$ and $\langle S_{hv}S_{hv}^*\rangle$. (The angular brackets $\langle\ \rangle$ denote spatial averaging.)

The typical implementation of a radar polarimeter involves transmitting a wave of one polarization and receiving echoes in two orthogonal polarizations simultaneously. This is followed by transmitting a wave with a second polarization, and again receiving echoes with both polarizations simultaneously, as shown in Figure 6-43. In this way, all four elements of the scattering matrix are measured. This implementation means that the transmitter is in slightly different positions when measuring the two columns of the scattering matrix, but this distance is typically small compared to a synthetic aperture and, therefore, does not lead to a significant decorrelation of the signals. The more important aspect of this implementation is to remember that the PRF must be high enough to ensure that each polarimetric channel is sampled adequately. Therefore, each channel independently has to satisfy the minimum PRF as given by Equation (6-100). Since we are interleaving two measurements, this means that the master PRF for a polarimetric system runs at twice the rate of a single-channel SAR. The NASA/JPL AIRSAR system pioneered this implementation for SAR systems (Zebker et al., 1987) and the same implementation was used in the SIR-C part of the SIR-C/X-SAR radars (Jordan et al., 1995).

Equations (6-133) and (6-134) show that once the scattering matrix or covariance matrix is known, the response of the scene can be calculated for any arbitrary polarization combinations. This is known as *polarization synthesis* and is discussed in more detail in Chapter 2 of Ulaby and Elachi (1990). Figure 6-3 shows an example of images synthesized at various polarization combinations. Note that if we allow the polarization of the

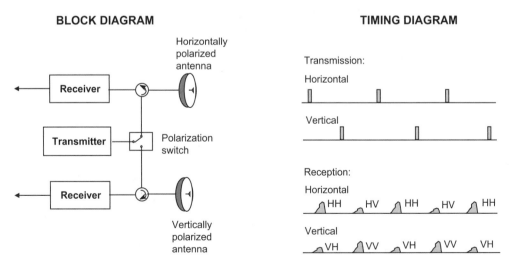

Figure 6-43. A polarimetric radar is implemented by alternatively transmitting signals out of horizontally and vertically polarized antennas, and receiving at both polarizations simultaneously. Two pulses are needed to measure all the elements in the scattering matrix.

transmitting and receiving antennas to be varied independently, the polarization response of the scene would by a four-dimensional space. This is most easily understood by representing each of the two polarizations by the orientation and ellipticity angles of the respective polarization ellipses, as shown in Chapter 2. The polarization response is, therefore, a function of these four angles. Visualizing such a four-dimensional response is not easy. To simplify the visualization, the so-called *polarization response* (van Zyl, 1985; Ulaby and Elachi, 1990) was introduced. The polarization response is displayed as a three-dimensional figure, and the transmitting and receiving polarizations are either the same (the *co-polarized* response), or they are orthogonal (the *cross-polarized* response). Figure 6-44 shows the polarization responses of a trihedral corner reflector, a device commonly used to calibrate radar images. Many examples of polarization responses can be found in Ulaby and Elachi (1990).

SAR Interferometry. SAR interferometry refers to a class of techniques by which additional information is extracted from SAR images that are acquired from different vantage points or at different times. Various implementations allow different types of infor-

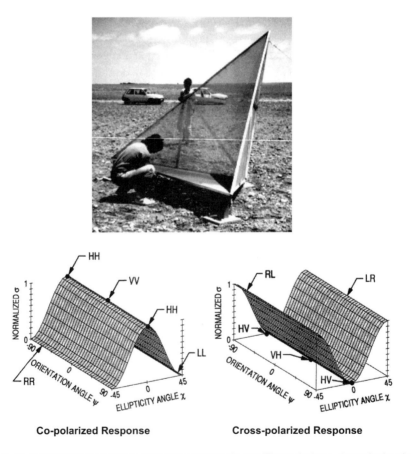

Figure 6-44. Polarization responses of trihedral corner reflector. These devices, shown in the photograph at the top, are commonly used to calibrate radar systems. The positions of some of the commonly used transmitting and receiving polarization combinations are shown on the two responses.

mation to be extracted. For example, if two SAR images are acquired from slightly different viewing geometries, information about the topography of the surface can be inferred. On the other hand, if images are taken at slightly different times, a map of surface velocities can be produced. Finally, if sets of interferometric images are combined, subtle changes in the scene can be measured with extremely high accuracy.

Radar Interferometry for Measuring Topography. SAR interferometry was first demonstrated by Graham (1974), who demonstrated a pattern of nulls or interference fringes by vectorally adding the signals received from two SAR antennas, one physically situated above the other. Later, Zebker and Goldstein (1986) demonstrated that these interference fringes can be formed after SAR processing of the individual images if both the amplitude and the phase of the radar images are preserved during the processing.

The basic principles of interferometry can be explained using the geometry shown in Figure 6-45. Using the law of cosines on the triangle formed by the two antennas and the point being imaged, it follows that

$$(R + \delta R)^2 = R^2 + B^2 - 2BR \cos\left(\frac{\pi}{2} - \theta + \alpha\right) \qquad (6\text{-}135)$$

where R is the slant range to the point being imaged from the reference antenna, δR is the path length difference between the two antennas, B is the physical interferometric baseline length, θ is the look angle to the point being imaged, and α is the baseline tilt angle with respect to the horizontal.

From Equation (6-135) it follows that we can solve for the path length difference δR. If we assume that $R \gg B$ (a very good assumption for most interferometers), we find that

$$\delta R \approx -B \sin(\theta - \alpha) \qquad (6\text{-}136)$$

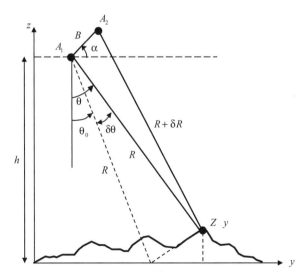

Figure 6-45. Basic interferometric radar geometry. The path length difference between the signals measured at each of the two antennas is a function of the elevation of the scatterer.

The radar system does not measure the path length difference explicitly, however. Instead, what is measured is an interferometric phase difference that is related to the path length difference through

$$\phi = \frac{a2\pi}{\lambda}\delta R = -\frac{a2\pi}{\lambda}B\sin(\theta_0 - \alpha) \qquad . \qquad (6\text{-}137)$$

where $a = 1$ for the case in which signals are transmitted out of one antenna and received through both at the same time, and $a = 2$ for the case in which the signal is alternately transmitted *and received* through one of the two antennas only. The radar wavelength is denoted by λ.

From Figure 6-45 it also follows that the elevation of the point being imaged is given by

$$z(y) = h - R\cos\theta \qquad (6\text{-}138)$$

with h denoting the height of the reference antenna above the reference plane with respect to which elevations are quoted. From Equation (6-138) one can infer the actual radar look angle from the measured interferometric phase as

$$\theta = \alpha - \sin^{-1}\left(\frac{\lambda\phi}{a2\pi B}\right) \qquad (6\text{-}139)$$

Using Equations (6-139) and (6-138) one can now express the inferred elevation in terms of system parameters and measurables as

$$z(y) = h - R\cos\left[\alpha - \sin^{-1}\left(\frac{\lambda\phi}{a2\pi B}\right)\right] \qquad (6\text{-}140)$$

This expression is the fundamental interferometric SAR equation for broadside imaging geometry.

To understand better how the information measured by an interferometer is used, consider again the expression for the measured phase difference given in Equation (6-137). This expression shows that even if there is no relief on the surface being imaged, the measured phase will still vary across the radar swath, as shown in Figure 6-46a for a radar system with parameters similar to those of the Shuttle Radar Topography Mission, which is discussed in more detail below. To show how this measurement is sensitive to topographical relief, consider now the case in which there is indeed topographical relief in the scene. For illustration purposes, we used a digital elevation model of Mount Shasta in California, shown in Figure 6-46b, and calculated the expected interferometric phase of the SRTM system for this area. The result, shown in Figure 6-46c, shows that the parallel lines of phase difference for a scene without relief are distorted by the presence of the relief. If we now subtract the expected "smooth" Earth interference pattern shown in Figure 6-36a from the distorted pattern, the resulting interference pattern is known as the *flattened interferogram*. For the Mount Shasta example, the result is shown in Figure 6-46d. This figure shows that the flattened interferogram resembles a contour map of the topography.

The distortion of the "smooth" Earth interferogram by the topographical relief is a consequence of the fact that the presence of topography slightly modifies the radar look angle from the value in the absence of topography (see Fig. 6-45). If we denote the look angle in

Figure 6-46. This figure shows how the topography of a scene is expressed in the interferometric phase. If there is no topography, all interferometric fringes will be parallel to the radar flight path as shown in (a). Using the topography of Mount Shasta, California, shown in perspective view in (b), the expected fringes for the SRTM system are shown in (c). Once the contribution from a smooth Earth as shown in (a) is subtracted from (c), the resulting flattened interferogram (d) resembles a contour map of the topography.

the absence of relief for a given range by θ_0, and z is the elevation of the pixel including the topography at the same range, it follows from the geometry in Figure 6-45 that the change in the look angle introduced by the relief is

$$\delta\theta \approx \frac{z}{r_0 \sin \theta_0} \qquad (6\text{-}141)$$

From Equation (6-137) we can write the phase of the pixel as

$$\phi = -\frac{a2\pi}{\lambda} B \sin(\theta_0 + \delta\theta - \alpha) \approx -\frac{a2\pi}{\lambda} B \sin(\theta_0 - \alpha) - \frac{a2\pi}{\lambda} B \cos(\theta_0 - \alpha)\delta\theta \quad (6\text{-}142)$$

The first term on the right in Equation (6-142) is simply the phase one would measure in the absence of relief, that is, the phase shown in Figure 6-44a in Mount Shasta scene. If this phase field due to a smooth Earth is subtracted from the actual interferometric phase, the resulting phase difference is the flattened interferogram

$$\phi_{\text{flat}} \approx -\frac{a2\pi}{\lambda} B \cos(\theta_0 - \alpha) \frac{z}{r_0 \sin \theta_0} \tag{6-143}$$

where we have combined Equation (6-141) and the second term on the right in Equation (6-142). This expression now clearly shows the sensitivity of the measured phase in the flattened interferogram to the topographical relief. The interferometric *ambiguity height* is defined as that elevation for which the interferometric phase in the flattened interferogram changes by one cycle. This is easily shown from Equation (6-143) to be

$$e = \frac{\lambda r_0 \sin \theta_0}{aB \cos(\theta_0 - \alpha)} \tag{6-144}$$

The ambiguity height can be interpreted as the sensitivity of the radar interferometer to relief. The smaller the ambiguity height, the more the measured interferometric phase will change for a given change in surface elevation. On the other hand, the radar system only measures the phase modulo 2π. Therefore, if the total relief in the scene exceeds the ambiguity height, the phase will be "wrapped," and the interferogram will appear as a contour map with the interferometric phase changing through multiple cycles of 2π, as shown in Figure 6-46. To reconstruct the topography, one has to "unwrap" the phase. We shall discuss this operation in more detail below.

From Equation (6-144) it is clear that the best sensitivity to elevation is achieved by maximizing the electrical length of the interferometric baseline. In practice, however, it is not possible to arbitrarily increase this length. In order to unwrap the phase, the interferometric phase must be sampled often enough. In the radar image, this sampling happens at each pixel, so individual samples are separated in slant range by the slant range pixel spacing ρ_s. If the baseline becomes so large that the interferometric phase changes by more than 2π across one pixel, the resulting interferogram will appear to have a random phase, and we will not be able to unwrap the phase to reconstruct the elevation profile. This situation will be even worse if the terrain slope is such that the surface is tilted toward the radar.

To explain this in more detail, consider the geometry of a tilted surface as shown in Figure 6-47 for the case of an airborne system for which we can ignore the curvature of the Earth. From Equation (6-137) we see that the interferometric phases for the two pixels can be written as

$$\phi_1 = -\frac{a2\pi}{\lambda} B \sin(\theta - \alpha); \qquad \phi_2 = -\frac{a2\pi}{\lambda} B \sin(\theta + \delta\theta - \alpha) \tag{6-145}$$

If the slant range pixel spacing ρ_s is small, the change in the look angle will be small, and we can approximate the second phase as

$$\phi_2 \approx -\frac{a2\pi}{\lambda} B \sin(\theta - \alpha) - \frac{a2\pi}{\lambda} B \cos(\theta - \alpha)\delta\theta = \phi_1 - \frac{a2\pi}{\lambda} B \cos(\theta - \alpha)\delta\theta \tag{6-146}$$

From Figure 6-47 we find that

$$\delta\theta = \frac{\rho_s}{r_0 \tan(\theta - \beta)} \tag{6-147}$$

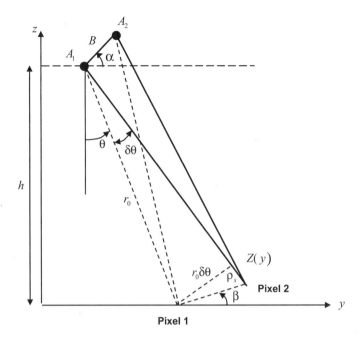

Figure 6-47. The rate at which the interferometric phase changes from pixel to pixel in the range direction is a function of the local slope of the surface. This figure shows a single pixel with a local slope β; positive values of β mean the surface element is tilted toward the radar.

with β being the terrain slope for the pixel; positive values indicate that the terrain is tilted toward the radar. Combining Equations (6-145)–(6-147) we find that the change in phase across the pixel is

$$\Delta\phi = \phi_2 - \phi_1 = -\frac{a2\pi}{\lambda} B \cos(\theta - \alpha)\frac{\rho_s}{r_0 \tan(\theta - \beta)} \qquad (6\text{-}148)$$

Note that this expression indicates that the interferometric phase will change more rapidly for surfaces tilted toward that radar, and less so for surfaces tilted away from the radar. This effect is clearly noticeable in Figure 6-46 on Mount Shasta.

If the change in phase from pixel to pixel in the range direction has to be less than a given value, Equation (6-145) clearly limits the length of the baseline to some maximum value. The rapid change in phase observed when the baseline length increases causes the interferometric phase to appear noisy in the image. This random phase is generally known as *baseline decorrelation*. To understand what is meant by baseline decorrelation, consider a sinusoidal surface with a spatial wavelength Λ that is illuminated by an electromagnetic wave with a wavelength λ_0. We further assume that the surface is tilted toward the source of illumination by an angle β, and that the angle of incidence of the electromagnetic wave is θ, as shown in Figure 6-48. The path length difference for the two "rays" shown, d, is

$$d = 2\Lambda \sin(\theta - \beta) \qquad (6\text{-}149)$$

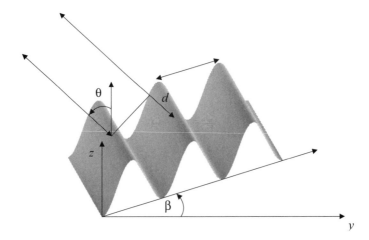

Figure 6-48. A sinusoidal surface with spatial wavelength Λ is tilted towards the radar by an angle β.

where the factor 2 denotes the two-way travel of the electromagnetic wave. The maximum return will be measured if there is a multiple of 2π phase difference between the contributions represented by the two "rays." Considering only the first multiple of 2π, this will happen if

$$\frac{2\pi d}{\lambda_0} = 2\pi, \Rightarrow \lambda_0 = 2\Lambda \sin(\theta - \beta) \qquad (6\text{-}150)$$

This is the well-known Bragg relationship between the spatial wavelength of the surface and the wavelength of the illuminating wave [see Eq. (6-2)]. Now suppose both the electromagnetic source and observation are moved slightly so that the new angle of incidence is $\theta + \delta\theta$. The new electromagnetic wavelength for maximum return would be

$$\lambda_1 = 2\Lambda \sin(\theta + \delta\theta - \beta) \approx 2\Lambda \sin(\theta - \beta) + 2\Lambda\delta\theta \cos(\theta - \beta) \qquad (6\text{-}151)$$

which, using Equation (6-150), can be written as

$$\lambda_1 = \lambda_0 + \frac{\lambda_0\delta\theta}{\tan(\theta - \beta)} = \lambda_0 + \delta\lambda \qquad (6\text{-}152)$$

The change in wavelength required to measure the maximum return is therefore

$$\delta\lambda = \frac{\lambda_0\delta\theta}{\tan(\theta - \beta)} \qquad (6\text{-}153)$$

The equivalent change in frequency required is

$$\delta f = \frac{f_0\delta\theta}{\tan(\theta - \beta)} \qquad (6\text{-}154)$$

As long as the illuminating signal contains both frequencies f_0 and $f_0 + \delta f$, and each receiver is capable of receiving both these frequencies, both observations will contain the reflection from the surface, and the resulting interferogram formed by combining the two individual measurements from the different angles will show good correlation between the signals. If the total bandwidth of the electromagnetic wave and associated receiving system is B_f, it follows that the largest change in angle that would still result in correlation between the signals would be

$$|\delta\theta_{max}| = B_f \tan(\theta - \beta)/f_0 \qquad (6\text{-}155)$$

If we assume that the distance between the electromagnetic source and the surface being illuminated is R, we can translate the maximum change in angle derived above to the maximum baseline projected orthogonally to the look direction as follows:

$$B_{\perp c} = B_c \cos(\theta - \alpha) = R|\delta\theta_{max}| = \left| \frac{\lambda_0 R B_f \tan(\theta - \alpha)}{c} \right| \qquad (6\text{-}156)$$

where c is the speed of light.

This is known as the *critical baseline*. Baselines longer than this would lead to a complete loss of coherence between the two interferometric signals and, consequently, result in an interferogram with random phases that cannot be unwrapped to produce a reliable estimate of the terrain elevation. As an example of the critical baseline length, consider the ERS radar system operating at an altitude of 785 km at an incidence angle of 23°. The range bandwidth is 13.5 MHz, with a center frequency of 5.3 GHz. Using these parameters, one finds that $B_{\perp c} = 1.1$ km.

The preceding analysis was done assuming that the surface shape is described by a single-frequency sinusoid. Natural surfaces usually have more complex shapes than this, and are generally characterized by a roughness spectrum comprised of the sum of different spatial frequencies with different amplitudes. However, the same argument as before can still be applied to each roughness spectrum component. The electromagnetic wave will preferentially interact with that particular component of the roughness spectrum that satisfies the Bragg relationship derived earlier [see also Eq. (6-2)]. A necessary condition for coherence between the interferometric channels to exist is that the received signal from *the same* roughness spectrum component be present in both interferometric channels. A careful consideration of this fact leads to the same result as the one we derived earlier, even if more than one spatial frequency component is present in the roughness spectrum of the surface.

The preceding analysis also assumes that the source and observation point are at the same position. This is the case for repeat-track interferometry, and also in the case in which a simultaneous baseline system is operated in the so-called ping-pong mode, where the signal is transmitted alternately out of each of the two antennas. In the case in which there is only one illumination source but the signals are received at two different locations, a similar argument shows that the critical baseline is exactly twice the value we derived before in Equation (6-156).

This analysis suggests that any nonzero baseline would result in only a portion of the two individual bandwidths overlapping. Those portions of the bandwidth that do not overlap would still contain signals, but those signals interact with different spatial components of the surface, and hence will not correlate with signals in the other interferometric channel, essentially reducing the signal-to-noise ratio of the interferogram. One way, therefore, to reduce fluctuations in the interferogram would be to utilize only those portions of the individual bandwidths that actually overlap when forming the interferogram by pre-filtering the individual signals to only retain the overlapping parts of the spectra before forming the interferogram. This technique of increasing the signal-to-noise ratio of inter-ferograms was first introduced by Gatelli et al. (1994). This spectral filtering, however, is a function of the angle of incidence and the range to the scene. In addition, the amount of frequency shift is strictly speaking a function of the terrain slope angle β. Since the local slope changes throughout the scene, there will be some uncompensated spectral shifts that are the result of the frequency shift introduced by the terrain slope itself. If there are large changes in the local slope over relatively small areas, the result would be local variations in phase noise of the interferogram.

SAR interferometers for the measurement of topography can be implemented in one of two ways. In the case of single-pass or fixed-baseline interferometry, the system is con-figured to measure the two images at the same time through two different antennas usually arranged one above the other. The physical separation of the antennas is referred to as the baseline of the interferometer. In the case of repeat-track interferometry, the two im-ages are acquired by physically imaging the scene at two different times using two differ-ent viewing geometries.

So far, most single-pass interferometers have been implemented using airborne SARs. The Shuttle Radar Topography Mission (SRTM), a joint project of the United States Na-tional Imagery and Mapping Agency (NIMA) and the National Aeronautics and Space Administration (NASA), was the first spaceborne implementation of a single-pass inter-ferometer (Farr and Kobrick, 2000). Launched in February 2000 from Cape Canaveral in Florida on the Space Shuttle Endeavour, SRTM used modified hardware from the C-band radar of the SIR-C system, with a 62 m long boom and a second antenna to form a single-pass interferometer. The SRTM mission acquired digital topographic data of the globe be-tween 60° north and south latitudes during one eleven-day shuttle mission. The SRTM mission also acquired interferometric data using modified hardware from the X-band part of the SIR-C/X-SAR system in a collaboration between NASA and the Deutches Zentrum für Luft und Raumfahrt (DLR) in Germany. The swaths of the X-band system, however, were not wide enough to provide global coverage during the mission.

Most of the SAR interferometry research has focused on into understanding the vari-ous error sources and how to correct their effects during and after processing. As a first step, careful motion compensation must be performed during processing to correct for the actual deviation of the aircraft platform from a straight trajectory (Madsen et al., 1993). The single-look SAR processor must preserve both the amplitude and the phase of the im-ages. After single-look processing, the images are carefully coregistered to maximize the correlation between the images. The so-called *interferogram* is formed by subtracting the phase in one image from that in the other on a pixel-by-pixel basis. In practice, this is done by multiplying one image by the complex conjugate of the other image and extract-ing the resulting phase.

The interferometric SAR technique is better understood by briefly reviewing the dif-ference between traditional and interferometric SAR processing. In traditional (noninter-ferometric) SAR processing, it is assumed that the imaged pixel is located at the intersec-tion of the Doppler cone (centered on the velocity vector), the range sphere (centered at the antenna), and an assumed reference plane, as shown in Figure 6-36a. Since the Doppler cone has its apex at the center of the range sphere and its axis of symmetry is aligned with the velocity vector, it follows that all points on the intersection of the Doppler cone and the range sphere lie in a plane orthogonal to the velocity vector. However, we do not know exactly where on this circle that forms the intersection of the cone and the plane orthogonal to the velocity vector the actual scatterer is. Therefore, the traditional SAR processor assumes that all scatterers lie in some reference plane, and places their images at the intersection of this reference plane and the circle forming the intersection of the Doppler cone and the plane orthogonal to the velocity vector.

The additional information provided by cross-track interferometry is that the imaged point also has to lie on the cone described by a constant phase, which means that one no longer has to assume an arbitrary reference plane. This cone of equal phase has its axis of symmetry aligned with the interferometer baseline and also has its apex at the center of the range sphere. It then follows that the imaged point lies at the intersection of the Doppler cone, the range sphere and the equal phase cone, as shown in Figure 6-49. It should be pointed out that in actual interferometric SAR processors, the two images acquired by the two interferometric antennas are actually processed individually using the traditional SAR processing assumptions. The resulting interferometric phase then represents the elevation with respect to the reference plane assumed during the SAR processing. This phase is then used to find the actual intersection of the range sphere, the Doppler cone, and the phase cone in three dimensions.

Once the images are processed and combined, the measured phase in the interferogram must be *unwrapped* before the topography can be reconstructed. During this procedure, the measured phase, which only varies between 0° and 360°, must be unwrapped to retrieve the original phase by adding or subtracting multiples of 360°. The earliest phase unwrapping routine was published by Goldstein et al. (1988). In this algorithm, areas in which the phase will be discontinuous due to layover or poor signal-to-noise ratios are identified by branch cuts, and the phase unwrapping routine is implemented such that

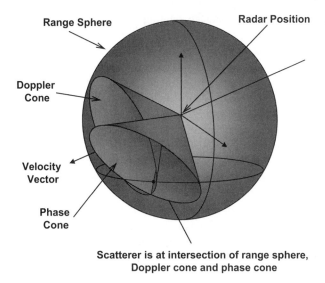

Scatterer is at intersection of range sphere, Doppler cone and phase cone

Figure 6-49. Interferometric SAR processing geometry. The scatterer must be at the intersection of the range sphere, Doppler cone, and phase cone, so we no longer have to assume that all scatterers lie on a reference plane.

branch cuts are not crossed when unwrapping the phases. A second class of phase un-wrapping algorithms is based on a least-squares fitting of the unwrapped solution to the gradients of the wrapped phase. This solution was first introduced by Ghiglia and Romero (1989). The major difference between these two classes of phase unwrapping algorithms lies in how errors are distributed in the image after phase unwrapping. Branch cut algo-rithms localize errors in the sense that areas with low correlation (resulting in high so-called residue counts) are fenced off and the phase is not unwrapped in these areas, leav-ing holes in the resulting topographic map. In the case of least-squares algorithms, the unwrapping is the result of a global fit, resulting in unwrapping even in areas with low correlation. The errors, however, are no longer localized but instead are distributed through the image. Phase unwrapping remains one of the most active areas of research, and many algorithms remain under development. Detailed discussions and additional ref-erences on phase unwrapping can be found in van Zyl and Hjelmstadt (1999) and Rosen et al. (2000).

Even after the phases have been unwrapped, the absolute phase is still not known. This absolute phase is required to produce a height map that is calibrated in the absolute sense. One way to estimate this absolute phase is to use ground control points with known eleva-tions in the scene. However, this human intervention severely limits the ease with which interferometry can be used operationally. Madsen et al. (1993) reported a method by which the radar data itself is used to estimate this absolute phase. The method breaks the radar bandwidth up into upper and lower halves, and then uses the differential interfero-gram formed by subtracting the upper-half spectrum interferogram from the lower-half spectrum interferogram to form an equivalent low-frequency interferometer to estimate the absolute phase. Unfortunately, this algorithm is not robust enough in practice to fully automate interferometric processing. This is one area in which significant research is needed if the full potential of automated SAR interferometry is to be realized.

Absolute phase determination is followed by height reconstruction. Once the elevations in the scene are known, the entire digital elevation map can be geometrically rectified. Madsen et al. (1993) reported accuracies ranging between 2.2 m r.m.s. for flat terrain and 5.5 m r.m.s. for terrain with significant relief for the NASA/JPL TOPSAR interferometer.

An alternative way to form the interferometric baseline is to use a single-channel radar to image the same scene from slightly different viewing geometries. This technique, known as *repeat-track* interferometry, has been mostly applied to spaceborne data start-ing with data collected with the L-band SEASAT SAR. Other investigators used data from the L-band SIR-B, SIR-C, and JERS, and the C-band ERS-1/2, Radarsat, and En-visat ASAR radars. Repeat-track interferometry has also been demonstrated using air-borne SAR systems (Gray and Farris-Manning, 1993).

Two main problems limit the usefulness of repeat-track interferometry. The first is due to the fact that, unlike the case of single-pass interferometry, the baseline of the repeat-track interferometer is not known accurately enough to infer accurate elevation informa-tion form the interferogram. Zebker et al. (1994) show how the baseline can be estimated using ground control points in the image. The second problem is due to differences in scattering and propagation that result from the fact that the two images forming the inter-ferogram are acquired at different times. The radar signal from each pixel is the coherent sum (i.e., amplitude and phase) of all the voltages from the individual scatterers contained in the radar pixel. If there are many such individual scatterers in each pixel, and they move relative to each other between observations, the observed radar signal amplitude and phase will also change between observations. If we consider an extended area with uniform average scattering properties, consisting of several radar pixels, and the relative movement of scatterers inside each pixel exceeds the radar wavelength, then the observed radar interferometric phase will appear quite noisy from pixel to pixel. The result is that the correlation between interferometric phases of neighboring pixels is lost, and the phase cannot be unwrapped reliably. This is known as *temporal decorrelation,* which is worst at the higher frequencies (Zebker and Villasenor, 1992). For example, C-band images of most vegetated areas decorrelate significantly over as short a time as one day. This is not surprising, since most components of the vegetation can actually move more than the approximately 6 cm of the C-band radar wavelength, even in the presence of gentle breezes. This problem more than any other limits the use of the current operational spaceborne single-channel SARs for topographic mapping, and is the main reason why the SRTM mission was implemented as a fixed-baseline interferometer.

Radar Interferometry for Measuring Surface Velocity. The previous discussion assumed that the surface imaged by the radar is stationary. Now consider the case in which the radar images the surface from the same observation point but at two different times. This implementation is known as *along-track* interferometry and was first described by Goldstein and Zebker (1987). In their experiment, they used two SAR antennas mounted on the body of the NASA DC-8 aircraft such that one antenna was some distance forward of the other. The radar transmitted signals out of one antenna, and the returns were measured through both antennas simultaneously. In this configuration, the one antenna would image the scene as if the phase center were at the transmitting antenna, whereas the phase center for the receive-only antenna lay at the center of the baseline connecting the two antennas. Since the aircraft is moving at a velocity v, this means that two images are acquired, separated in time by

$$T_{\text{obs}} = \frac{B}{2v} \tag{6-157}$$

Note that if the system were implemented by alternately transmitting *and receiving* out of each antenna, the observation time difference would be doubled. Now assume that a scatterer moves with a velocity vector $\boldsymbol{v}_s$ on a horizontal surface, as shown in Figure 6-50. This would be the case, for example, if we were imaging an ocean current or a flowing river. During the time T_{obs}, the position of the scatterer is displaced by

$$\mathbf{D} = \boldsymbol{v}_s T_{\text{obs}} \tag{6-158}$$

The change in range of the scatterer relative to the radar between these observations is then

$$\delta R = \hat{\mathbf{n}} \cdot \mathbf{D} = \hat{\mathbf{n}} \cdot \boldsymbol{v}_s T_{\text{obs}} \tag{6-159}$$

where $\hat{\mathbf{n}}$ is a unit vector pointing from the radar to the original position of the scatterer. This change in range will be recorded as a change in phase of the radar signal from the first image to the second. This change in phase is

$$\phi = \frac{2\pi}{\lambda} \delta R = \frac{2\pi B}{2\lambda v} \hat{\mathbf{n}} \cdot \boldsymbol{v}_s \tag{6-160}$$

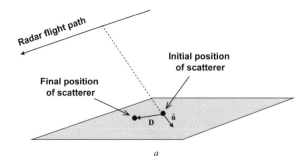

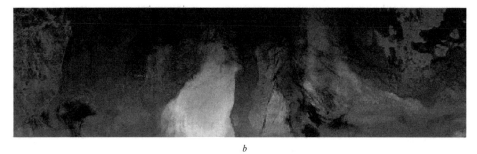

Figure 6-50. (a) Along-track interferometry imaging geometry. (b) Interferogram acquired over the Straits of Juan de Fuca with the NASA/JPL AIRSAR system in 1999. The interferogram shown was acquired with the L-band system, with a baseline of 20 m in the along-track direction. Given the normal flight parameters of the NASA DC-8 aircraft, this image has a velocity ambiguity of 2.4 m/s. (Courtesy of NASA/JPL-Caltech.) See color section.

As long as the projected velocity of the scatterer is low enough such that the phase change between observations is less than 2π, the measured phase field can be unambiguously inverted for the velocity of the scatterer. Note, however, that with a single baseline we would only measure the velocity of the scatterer projected onto the radar line of sight. If the scatterer moves in a direction orthogonal to the radar line of sight, the phase will not change. Figure 6-51*b* shows an example of along-track interferometric phases measured over the Juan de Fuca Straits in the northwestern part of the Pacific Ocean with the along-track interferometry mode of the NASA/JPL AIRSAR system. This image clearly shows the interferometric phase change associated with the tidal outflows from the bays into the straits.

So far, we have discussed the change in phase as if we were observing a single scatterer. In reality, each radar pixel contains many scatterers, and they may be moving relative to each other during the time between observations, as discussed in the previous section. This temporal decorrelation limits the largest separation, or baseline, we can use between the antennas. This concept is analogous to the critical baseline discussed before. The difference here is that the maximum baseline for along-track interferometry is a function of the relative movement of scatterers inside each pixel. This is an expression of the way in which the surface is changing over time, and can be characterized by a surface coherence time.

Differential Interferometry for Surface Deformation Studies. One of the most exciting applications of radar interferometry is implemented by subtracting two interferomet-

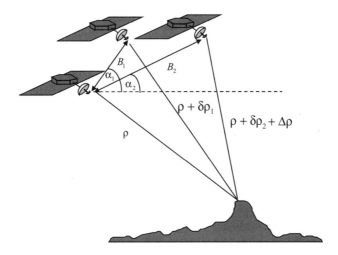

Figure 6-51. Three-pass differential interferometry imaging geometry. The surface deformation occurs between the second and third data acquisitions.

ric pairs separated in time from each other to form a so-called *differential interferogram*. In this way, surface deformation can be measured with unprecedented accuracy. This technique was first demonstrated by Gabriel et al. (1989) using data from SEASAT data to measure centimeter-scale ground motion in agricultural fields. Since, then this technique has been applied to measure centimeter-scale coseismic displacements and to measure centimeter-scale volcanic deflation. The added information provided by high-spatial-resolution coseismic deformation maps was shown to provide insight into the slip mechanism that would not be obtainable from the seismic record alone. See the references at the end of this chapter for many examples. A summary of the results can be found in Rosen et al. (2000).

Differential interferometry is implemented using repeat-pass interferometric measurements made at different times. If the ground surface moved between observations by an amount Δr toward the radar, the phase difference in the repeat pass interferogram would include this ground displacement in addition to the topography. In that case, the phase difference in the flattened interferogram becomes

$$\Delta\phi_{\text{flat}} \approx -\frac{4\pi}{\lambda}B_{\perp}\frac{z}{r_1 \sin\theta_0} + \frac{4\pi}{\lambda}\Delta r \qquad (6\text{-}161)$$

This expression shows that the phase difference is far more sensitive to changes in topography, that is, surface displacement or deformation, than to the topography itself. The elevation has to change by one ambiguity height to cause one cycle in phase difference, whereas a surface displacement of $\lambda/2$ would cause the same amount of phase change. This is why surface deformations of a few centimeters can be measured from orbital altitudes using SAR systems.

To extract the surface deformation signal from the interferometric phase, one has to separate the effects of the topography and the surface deformation. Two methods are commonly used to do this. If a good DEM is available, one can use that to form synthetic fringes and subtract the topography signal from the measured phase difference. The remaining signal then is due only to surface deformation. In this case, the relative lack of sensitivity to topography works to our benefit. Therefore, DEM errors on the order of several meters usually translate to deformation errors of only centimeters in the worst case.

The second way in which the deformation signal is isolated is to use images acquired during three overpasses, as shown in Figure 6-51. This method generates a DEM from one pair of images, and then uses that as a reference to subtract the effects of topography from the second pair. For this case, the differential phase can be written as

$$\Delta\phi_{3\ \text{pass}} = \Delta\phi_{\text{flat }1} - \frac{B_1\cos(\theta - \alpha_1)}{B_2\cos(\theta - \alpha_2)}\Delta\phi_{\text{flat }2} \tag{6-162}$$

where the subscripts refer to the first pair and second pair. If the motion only occurred between the first and second passes, one interferogram will include the deformation signal, whereas the other will not. The resulting phase difference in Equation (6-162) will then contain only the deformation signal.

This process is illustrated in Figure 6-52. Here we artificially added a deformation signal to the digital elevation model of Mount Shasta previously shown in Figure 6-46. The deformation signal has an amplitude of 30 cm, and the center of the deformation is situated on the side of Mount Shasta. Figure 6-52 shows the two individual flattened interferograms. The one on the left was calculated without the deformation signal, whereas the one in the middle includes the deformation signal. These two interferograms would be acquired if the three-pass scheme shown in Figure 6-51 were used. If the two flattened interferograms are subtracted from each other, the deformation signal is clearly visible in the image on the right in Figure 6-52.

Polarimetric Interferometry. In the preceding discussion on radar interferometry, it was assumed that only one polarization was transmitted and only one polarization mea-

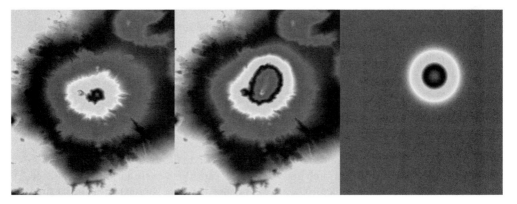

Before deformation   Including deformation   Deformation signal

Figure 6-52. An illustration of differential interferometry. A deformation signal of 30 cm amplitude was added to the digital elevation model of Mount Shasta shown in Figure 6-46, and the resulting interferograms were calculated. The surface deformation occurs between the second and third data acquisitions.

sured upon reception of the radar waves. However, as mentioned before, electromagnetic wave propagation is by nature a vector phenomenon. Therefore, in order to capture the complete information about the scattering process, interferometric measurements should really be made in the full polarimetric implementation of a radar system. In this case, there are really three different measurements being made at the same time. First, there are the two polarimetric radar measurements at each end of the baseline, represented below by the two covariance matrices $[\mathbf{C}_{11}]$ and $[\mathbf{C}_{22}]$. Since the baseline is generally short compared to the distance to the scene, these two measurements can be expected to be nearly identical, except for the very small change in the angle of incidence from one end of the baseline to the other. (The exception, of course, is if the two measurements are made in the repeat-track implementation. In that case, temporal changes could cause the two covariance matrices to be quite different.) The third measurement, of course, is the full vector interferogram as opposed to the scalar implementation described earlier.

The vector interferogram, which is the complex cross-correlation of the signal from one end of the baseline with that from the other end of the baseline, can be described as

$$V_1 V_2^* = \widetilde{\mathbf{A}}_1 \mathbf{T}_1 \widetilde{\mathbf{T}}_2^* \mathbf{A}_2^* = \mathbf{A}_1 \cdot [\mathbf{C}_{12}] \mathbf{A}_2^* \tag{6-163}$$

The correlation of the two signals after averaging is

$$\mu = \frac{\langle V_1 V_2^* \rangle}{\sqrt{\langle V_1 V_1^* \rangle \langle V_2 V_2^* \rangle}} = \frac{\mathbf{A}_1 \cdot \langle [\mathbf{C}_{12}] \rangle \mathbf{A}_2^*}{\sqrt{(\mathbf{A}_1 \cdot \langle [\mathbf{C}_{11}] \rangle \mathbf{A}_1^*)(\mathbf{A}_2 \cdot \langle [\mathbf{C}_{22}] \rangle \mathbf{A}_2^*)}} \tag{6-164}$$

where the angular brackets $\langle \ \rangle$ denote averaging. The interferometric phase is the phase angle of the numerator of Equation (6-164):

$$\phi_{\text{int}} = \arg(\mathbf{A}_1 \cdot \langle [\mathbf{C}_{12}] \rangle \mathbf{A}_2^*) \tag{6-165}$$

Using this formulation, Cloude and Papathanassiou (1999) showed, using repeat-track SIR-C interferometric data, that polarization diversity can be used successfully to optimize the correlation between images. They also showed significant differences in the measured elevation in forested areas when using polarization optimization. At present, polarimetric interferometry is a very active research area. Unfortunately, progress is hampered severely by lack of availability of well-calibrated data, as only a handful of radar systems have been upgraded to full polarimetric interferometry capability.

6-4-9 Description of SAR Sensors

Several civilian SAR sensor have been flown in space over the past two decades. Here we shall summarize the characteristics of a few of these sensors. We shall first discuss three SAR instruments that flew on the U.S. space shuttles, followed by a brief summary of some of the SAR missions that were launched during the 1990s. We conclude this section with brief summaries of two planetary SAR missions.

Shuttle Imaging Radar Missions: SIR-A, SIR-C/X-SAR, and SRTM. A typical synthetic-aperture imaging radar can be described by the Shuttle Imaging Radar (SIR-A), which was flown on the shuttle Columbia in November 1981. The SIR-A antenna consisted of a 2.16 × 9.4 m phased array s coupled to the SIR-A sensor. The sensor provides the antenna with a series of high-power coherent pulses of energy at the L band and amplifies the weak return echoes received by the antenna. The radar sensor consists of four subassemblies: transmitter, receiver, logic and control, and power converter. A diagram of the sensor is shown in Figure 6-53.

To obtain an adequate signal-to-noise ratio from a system whose range resolution is 40 m on the surface and which utilizes a solid-state transmitting device, it is necessary to use a long transmitted pulse and pulse compression techniques to reduce the peak power requirement. The output of the transmitter assembly is, as a result, a linearly swept, frequency-modulated pulse (or chirp) having a 211-to-1 compression ratio. It is generated in a surface acoustic wave (SAW) device located in the chirp generator subassembly of the transmitter assembly. The output of the transmitter is coupled to the antenna subsystem through an output combiner.

Echo returns are coupled into the receiver assembly through the output network in the transmitter. Because the echo's intensity is expected to vary in proportion to the variation of antenna gain with angle, a sensitivity time control (STC) has been incorporated in the receiver. The STC action linearly decreases the receiver gain by 9 dB during the first half of the return echo period, and then increases the gain until the end of the echo has been received. The application of the STC results in a nearly uniform signal (echo) return for a uniform scattering field, and, as a result, the dynamic range required to record the data is reduced by 9 dB.

The received signal is recorded on an optical film that is retrieved after landing. The film (called signal film) is then processed in an optical correlator to generate the final image. Alternatively, the received signal can be digitized and then recorded on board or transmitted to the ground via a digital data link. This was the case with the SIR-B sensor flown in 1984.

The SIR-C/X-SAR SAR systems were flown on the space shuttle Endeavour in April and October 1994, and were the first multiparameter civilian SAR systems flown in space. The SIR-C system incorporated fully polarimetric radars operating simultaneously at the L band (24 cm wavelength) and C band (6 cm wavelength), whereas the X-SAR system acquired VV polarized images at the X band (3 cm wavelength). This system, therefore, acquired both multifrequency and multipolarization images simultaneously. The SIR-C system was provided by NASA, whereas the X-SAR system was provided jointly by the German and Italian Space Agencies.

The antennas for these three radars were mounted side by side inside the shuttle cargo bay, and had a combined size of 12 m × 4.4 m. The individual antennas measured 12 m × 2.9 m for the L band, 12m × 0.7 m for the C band, and 12 m × 0.4 m for the X band. The phase array C- and L-band antennas were fixed to the structure and utilized electronic beam steering to image at different look angles, utilizing a uniform grid of transmit and receive modules with phase shifters for beam steering. The X-band antenna was a passive, slotted waveguide array and used a traveling wave tube amplifier as the transmitter. The X-SAR antenna was mechanically steered using a tilt mechanism, with a look angle range of 15–60°, in alignment with SIR-C. The SIR-C system also utilized electronic beam steering to demonstrate ScanSAR imaging from space for the first time.

The received signals in the SIR-C system were first amplified and down-converted using solid-state receivers. Four channels of data were received simultaneously; two polarizations each for L and C band. The output of the receivers was digitized, formatted, and then recorded on high-speed digital recorders on the shuttle. Each of the four channels produced data at a 45 Mbps rate, and the recorders accepted data at a rate of 180 Mbps.

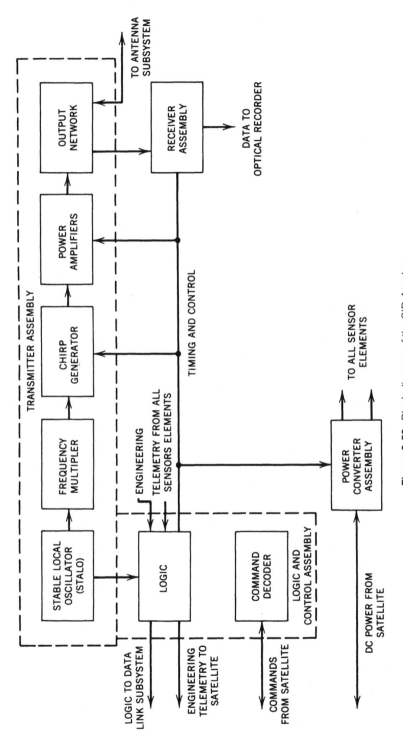

Figure 6-53. Block diagram of the SIR-A radar.

285

Because much of the fully polarimetric imaging swath was limited by the available data rate, a special mode was included that recorded only two data channels per recorder, effectively doubling the available swath. This, however, required the use of two of the three available data recorders simultaneously. The X-band data were similarly amplified, down-converted, digitized, and recorded on the shuttle. A small amount of data was sent to the ground via the Tracking and Data Relay Satellite System during the mission, and processed on the ground to verify proper operation of the radar systems.

During the last few days of the second flight in October 1994, the SIR-C/X-SAR system acquired repeat-track interferometric data at all three frequencies, with across-track interferometric baseline separations of 10 to 4700 m. Day-to-day repeats were accomplished during the second flight as well as 6 month repeats between flights. SIR-C also had an along-track interferometric mode for detection of motion in the azimuth direction. This mode was achieved by simultaneously operating the outermost six C-band panels at each end as separate antennas.

The Shuttle Radar Topography Mission (SRTM) was designed to produce a three-dimensional image of 80% of the Earth's land surface in a single 10 day space shuttle flight. It flew in February 2000 on the space shuttle Endeavour, and was the first spaceborne single-pass interferometric SAR. The SRTM payload used the maximum resources that the space shuttle could offer with respect to energy use, flight duration, and payload mass, with the result that the mapping phase of the flight operations was limited to 159 orbits. With a 60-meter-long mast and more than 13 tons of payload, SRTM was the largest structure ever flown in space.

The SRTM actually flew two interferometers: a C-band system provided by the United States, and an X-band system provided by Germany. In both cases, the transmitter was located in the shuttle cargo bay. The baseline of the interferometer was formed by a specially designed 60-meter-long mast, which extended from the shuttle cargo bay and carried receive-only antennas on the other end. The mast was a carbon fiber truss structure that consisted of 87 cube-shaped sections, called bays. The entire mast was stored in a 3-m-long canister, and small electrical motors were used to deploy and retract the mast prior to and after mapping operations. The mast was deployed and retracted by a specially designed rotating mechanism housed inside the canister. Unique latches on the diagonal members of the truss structure of the mast snapped into place and provided rigidity as the mast was deployed bay by bay out of the canister. Figure 6-54 shows a number of photographs of the SRTM hardware.

A swath width of greater than approximately 218 km is required to cover the Earth completely in 159 orbits. The existing SIR-C ScanSAR mode, combined with the dual-polarization capability, allowed a 225 km swath at the C band. This provided complete land coverage within 57° north and south latitude. The X-band interferometer swath was limited to 45–50 km, as it was not capable of operating in the ScanSAR mode.

In order to reconstruct the topography accurately from the radar data, one needs to know the absolute length and orientation of the baseline, as well as the absolute position of the shuttle. The Attitude and Orbit Determination Avionics (AODA) system combined the functions of metrology and attitude and orbit determination to provide this information. The attitude of the outboard antenna structure was measured using an array of three red-light-emitting diodes, whose relative positions were measured using the ASTROS target tracker. An inertial reference unit (IRU) was used to measure changes in the attitude very precisely. Data from the IRU and a star tracker were combined to provide an absolute reference of the baseline attitude relative to known stars. The length of the baseline was measured using an electronic distance-measurement unit mounted in the shuttle cargo

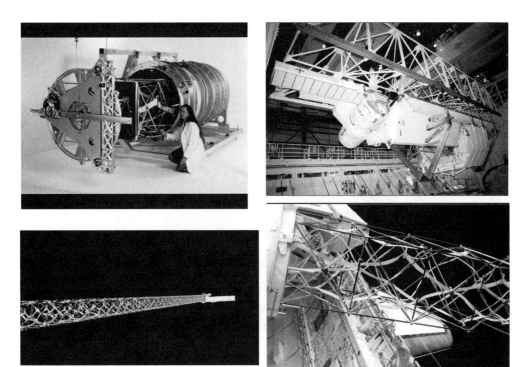

Figure 6-54. The photograph on the top left shows the SRTM mast and canister during development. The outboard antenna mounts to the large plate visible in the front. The top-right photo shows the SRTM hardware being lowered into the cargo bay of the space shuttle Endeavour. The canister and outboard antenna are clearly visible. Note the size of the people guiding the hardware installation in the bottom of the photo. The photograph on the bottom left shows an on-orbit view of the mast and outboard antenna from the shuttle. The photo on the bottom right is a close-up of part of the mast showing three bays and cables. The aft wing of the Endeavour is visible in the background. (Cortesy of NASA/JPL-Caltech.)

bay and a corner-cube reflector mounted on the outboard antenna structure. This system measured the length of the mast to an accuracy of 1 mm. Finally, the absolute position of the shuttle was measured using global positioning satellite receivers with the antennas mounted on the outboard antenna structure to provide the best view of the GPS satellites.

All the SRTM data have been processed into digital elevation maps. Figure 6-55 shows a shaded relief map of the state of California derived from the SRTM data. Processed data and general information about the SRTM mission can be found on the Web at http://www.jpl.nasa.gov/srtm/ and at http://srtm.usgs.gov/. The vertical accuracy (90%) of the SRTM data is estimated to be about 7 m.

International SAR Missions. The international use of spaceborne imaging SARs for long-term Earth observation dramatically increased in the 1990s. Whereas the U.S. NASA missions were shuttle-based, there are four SAR imaging satellites operated by the Europeans, Japanese, and the Canadians. These satellites transmit the data to Earth at tracking stations distributed throughout the world. The data are then processed at central processing facilities in Europe, Japan, Canada, and the United States.

Figure 6-55. Shaded relief map of California derived from SRTM C-band radar interferometry measurements.

The European Space Agency (ESA) launched a SAR aboard the European Remote-Sensing Satellite (ERS) in August 1991 and another in April 1995. Each ERS employs a C-band SAR operating at a fixed look angle of 23° and with VV polarization. The antenna is a 10 m × 1 m planar waveguide array and the transmitter utilizes a travelling wave tube amplifier. The swath is 100 km and the resolution is 24 m. Global land and ocean coverage have been provided by ERS-1 and 2. Of particular note to scientists are the Arctic ice coverage products. Multiyear, time-sequenced data depict the polar ice pack motion, ice thickness, and glacier motion. Pairs of ERS images, including tandem ERS-1/2, have been used to generate interferometric data products of surface topography and surface deformation. These data are especially useful for analyzing postseismic events and detecting surface deformations associated with volcano preeruptions.

The National Space Development Agency (NASDA) launched a SAR aboard the Japan Earth Resources Satellite (JERS-1) in February 1992. JERS employed a L-band SAR operating at a fixed look angle of 35° and with HH polarization. The swath is 75 km and the resolution is 18 m. JERS has provided gobal land-cover mapping. Significant JERS products include a digital map database of the South American rainforests, which are often cloud covered. Repeat-pass interferometric data have been used to measure surface deformations on the order of millimeters over extended periods of time.

The Canadian Space Agency launched a SAR aboard RADARSAT in September 1995. RADARSAT employs a C-band SAR capable of multiple modes. The antenna is a 1.5 m × 15 m planar slotted waveguide array. Ferrite variable phase shifters are employed for electronic beam steering, which is a key feature that allows significant control over the positioning of the imaging swaths, and also enables ScanSAR mapping. The transmitter utilizes two traveling wave tube amplifiers for redundancy. The amplifiers are derived from those developed for ERS. The RADARSAT modes include a variety of swath widths and resolutions. A fine-resolution mode is capable of 9 m resolution over a 49 km swath. The use of SCANSAR allows swath coverage as wide as 500 km, with resolution as good as 40 m, to achieve very large area mapping.

RADARSAT has provided global SAR products for commercial and scientific customers. RADARSAT International distributes data to the commercial users. Arctic ice coverage products have been utilized for ship operations and for research of sea ice motion. The RADARSAT Antarctic Mapping Project (RAMP) has mapped Antarctica, providing critical benchmarks for gauging future changes in the extent, shape, and dynamics of the great Antarctic Ice Sheet. It will also contribute to the understanding of iceberg formation and the geologic history of the Antarctic continent.

Planetary SAR Sensors: Magellan Radar and Cassini Radar. The concept of an imaging radar mission to Venus, using SAR to peer through the dense cloud cover, was first put forward in the late 1960s. The scientific objectives for a NASA SAR mission were established in a 1972 study, which defined a radar imaging mission patterned after the optical imaging of the Mariner 9 Mars mapping mission. This proposed mission, known as the Venus Orbiting Imaging Radar (VOIR), was deemed too costly. A scaled-down mission, with only radar (imaging, altimetry, and radiometry) and Doppler gravity experiments, was designed in 1982. The redesigned mission, known as Magellan, was launched from the space shuttle Atlantis in 1989.

The design of the Magellan radar system was constrained in many ways by the choice of spacecraft and its orbit. Because of limited resources, the SAR shared the high-gain antenna with the telecommunications subsystem. This 3.7 m parabolic dish antenna had neither sufficient size nor the proper beam pattern for easy adaptation to SAR operations. In addition, the data rate and volume were severely limited by the communications link. During each orbit, data could be played back at a rate of 269 kb/s for a maximum duration of 112 out of the 196 minutes per orbit. Finally, the elliptical orbit (radar altitudes varied between 290 km and 3000 km) meant that the imaging geometry changed rapidly and, consequently, radar parameters had to updated frequently. In fact, it took nearly 4000 commands per orbit to operate the SAR. These operating sequences were uploaded every 3 to 4 days.

The Magellan radar operated at S band, with a radar frequency of 2385 MHz. To cope with the severe constraints in data rate and volume, the radar was operated in the burst SAR mode, in which a synthetic aperture is broken into sequences of pulses (known as "bursts") with period of no transmission interleaved with the bursts. In this way, the along-track integration time is decreased, resulting in a poorer resolution. However, both the data rate and average power requirement are reduced by the duty cycle of the bursts. The burst mode alone, however, was not enough to fit the Magellan SAR data in the limited data rate and volume offered by the spacecraft. To reduce data rate and volume further, echoes were digitized to 2 bits using a block adaptive quantizer (BAQ). This type of quantizing makes use of the slowly varying nature of the echo strength between echoes and between bursts. By averaging the returns over blocks of data, the quantizer chooses a threshold level to compare successive echoes to. If the echo strength exceeds the threshold, a value of "1" is assigned; otherwise a value of "0" is assigned. The sign of the voltage is preserved in a second bit. In this way, the original 8-bit data are reduced to 2 bits, while preserving the dynamic range of the original 8-bit data. Even with both the burst mode and the BAQ, the SAR swath width was 20–35 km. Range resolutions varied between 120–360 m, and the along-track resolution was 120 m.

The primary objective of the Magellan mission was to map 70% of Venus during one rotation of the planet (243 Earth days). The Magellan mission met all of its requirements and most ambitious goals. The imaging coverage requirement of 70% was surpassed with 98% coverage. Similar successes were achieved for the altimetry and radiometry coverage.

The Cassini radar, a Ku-band (13.78 GHz frequency, 2.17 cm wavelength) radar instrument designed to map much of the surface of Saturn's moon Titan, was launched on October 15, 1997 and went into orbit around Saturn in July 2004. The radar design accommodates a wide range of approach geometries, terrain types, and operating conditions and in can be operated in four main modes. The SAR mode can operate at altitudes less than 4000 km at highest resolution (about 400 m, varying with altitude) or lower resolution (about 1 km). Images are produced at incidence angles up to 30°, either left or right of the nadir, and at 2 to 7 looks. A swath of 120–450 km width is created from five antenna beams, with incidence angle typically maximized consistent with the desired signal-to-noise ratio. In the altimetry mode, the central 0.35° beam is used for time-of-flight measurements, which are converted into Titan radii with relative accuracy near 150 m, although actual accuracy may be further limited by spacecraft position/attitude uncertainties. The scatterometry mode measures surface backscatter coefficient σ^0 variation with incidence angle at distances up to 22000 km, and the radiometry mode measures brightness temperature. Although coverage using SAR and altimetry modes is limited by both orbital geometry and range, scatterometry and radiometry modes can achieve near full-disk coverage on a single pass and global coverage through the mission. The radiometry mode can generate rastered "images" by turning the entire spacecraft in a raster pattern.

The Cassini radar uses the 3.66 m telecommunications on the spacecraft with a custom feed that allows the antenna beam to be scanned in the cross-track direction. Five sub-swaths are used in the ScanSAR mode to provide an adequate swath width. Since SAR imaging is limited by SNR considerations to less than 4000 km altitudes, this allows for about 32 minutes of imaging (16 minutes on either side of the closest approach) during a typical Titan flyby. Since the spacecraft ephemeris will not be known accurately in real time, the radar design must accommodate uncertainties in the imaging geometry. The Cassini radar design solves this problem by utilizing a burst mode of operation. A burst of pulses is transmitted for a period slightly shorter than the expected round trip travel time for the radar pulses to the Titan surface. The radar then switches to a receive mode and records the echoes from the burst of pulses. The uncertainty in the spacecraft position and pointing can, therefore, be accommodated by adjusting the burst period and the record window timing. To reduce the data volume, a 2-bit block-adaptive quantizer, similar to that used by the Magellan radar, is used. The Cassini radar transmits about 65 W of power and uses a signal bandwidth that varies between 0.25 and 1.0 MHz.

The Cassini spacecraft will fly past Titan a total of 44 times, with additional passes planned if the mission is extended past 2008. Because Titan is used for gravitational assists that shape the entire tour through the Saturn system, both ground tracks and closest-approach distances vary, but some will come as close as 950 km. On approximately 20 of these passes, the radar will acquire data in at least one of its active modes. In total, some 20% of the surface will be observed in SAR mode, and since its observations are not affected by the visibly opaque atmosphere, these will be the highest resolution views of Titan's surface. Observations of the other moons, rings, and atmosphere of Saturn are also planned.

6-4-10 Applications of Imaging Radars

Imaging radar data are being used in a variety of applications, including geologic mapping, ocean surface observation, polar ice tracking, and vegetation monitoring. Qualitatively, radar images can be interpreted using the same photo interpretation techniques used with visible and near infrared images.

One of the most attractive aspects of radar imaging is the ability of the radar waves to penetrate dry soils to image subsurface features. This capability was first demonstrated with SIR-A imagery of southern Egypt and northern Sudan in the eastern Sahara desert. Figure 6-56 shows a comparison of a LandSat image acquired in the visible and infrared part of the spectrum to a mutlifrequency SIR-C/X-SAR image acquired over south-central Egypt. This image was taken in the same area where SIR-A data previously showed buried ancient drainage channels. The Safsaf Oasis is located near the bright yellow feature in the lower left center of the Landsat image. While some features at the surface are visible in both images, in much of the rest of the image, however, the radar waves show a wealth of information about the subsurface geologic structure that is not otherwise visible in the LandSat image. For example, the dark drainage channels visible in the bottom of the radar image are filled with sand as much as 2 m thick. These features are dark in the radar image, because the sand is so thick that even the radar waves cannot penetrate all the way to the bottom of these channels. Only the most recently active drainage channels are visible in the LandSat image. Another example is the many rock fractures visible as dark lines in the radar image. Also visible in the radar image are several blue circular granite bodies. These show no expression in the LandSat image. Field studies conducted

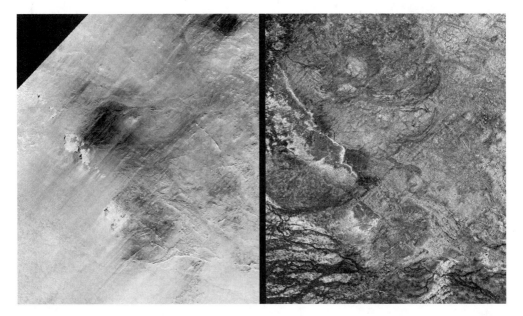

Figure 6-56. A comparison of images of the Safsaf Oasis area in south-central Egypt. On the left is a LandSat Thematic Mapper image showing bands 7, 4, and 1 displayed as red, green, and blue. The image on the right is from the SIR-C/X-SAR system, displaying L-band HH, C-band HH, and X-band VV as red, green, and blue, respectively. Each image represents an area of approximately 30 km by 25 km. (Courtesy of NASA/JPL-Caltech.) See color section.

previously indicate that at the L band the radar waves can penetrate as much as 2 m of very dry sand to image subsurface structures.

Figure 6-57 shows the effects of seasonal changes as they are manifested in the radar responses. These images of the Prince Albert region in Saskatchewan, Canada were acquired with the SIR-C/X-SAR system in 1994. The top image was taken in April, with much of the area still frozen. In particular, the small frozen lakes show as bright blue areas, because of the enhanced X-band scattering from the small-scale roughness on the surface of the ice. In October, these lakes shows little return from the relatively smooth water surface. The April image show little difference between different forest types, or even between mature trees and cleared areas. These differences are quite pronounced in the October image, however. The cleared areas are now clearly visible as green, irregular polygons. The red areas in the October image contain old jack pine trees. The change in scattering is due to the increased moisture in these trees in late summer, as opposed to the frozen conditions in April. The area in the middle of the image that shows as blue/green in October and red in April contains a mixture of black spruce, jack pines, and aspen trees. The relative increase in scattering from the shorter wavelengths in October indicates that the canopy was mostly frozen in April and moist in October.

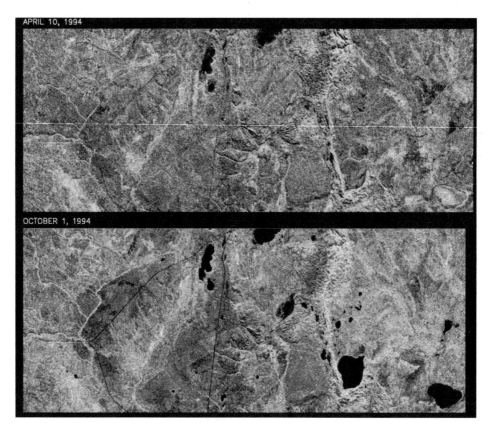

Figure 6-57. Seasonal images of the Price Albert area in Canada. Both images were acquired with the SIR-C/X-SAR system and display the L-band return in red, the C-band return in green, and the X-band return in blue. The image on the top was acquired on April 10, 2004, and the one on the bottom on October 1, 1994. (Courtesy of NASA/JPL-Caltech.) See color section.

Long-wavelength radar signals are capable of penetrating forest canopies. This is illustrated with the two false-color images of the Manaus region of Brazil in South America shown in Figure 6-58. These images were acquired by the SIR-C/X-SAR system on April 12, and October 3, 1994. The two large rivers visible in this image are the Rio Negro at the top and the Rio Solimoes at the bottom. These rivers combine at Manaus (west of the image) to form the Amazon River. The false colors were created by displaying three L-band polarization channels from the SIR-C system: red areas correspond to high HH backscatter, whereas green areas correspond to high HV backscatter. Blue areas show low VV returns; hence, the bright blue colors of the smooth river surfaces can be seen. The relatively large HV returns indicate that green areas in the image are heavily forested, whereas blue areas are either cleared forest or open water. Double-bounce scattering hap-

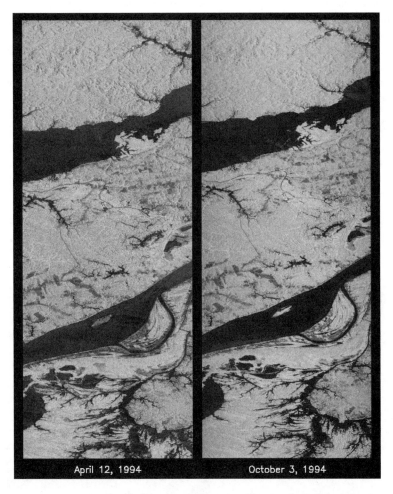

Figure 6-58. Seasonal images of the Amazon rain forest near Manaus, Brazil. Both images were acquired with the SIR-C/X-SAR system, and display the L-band HH return in red, the L-band HV return in green, and the inverse of the L-band VV return in blue. The image on the left was acquired on April 12, 2004, and the one on the right on October 3, 1994. The images are about 8 km wide, and 25 km long. The yellow and red areas represent flooding under the forest canopy. (Courtesy of NASA/JPL-Caltech.) See color section.

pens when radar signals penetrate the forest canopy, reflect off the ground, and then reflect again off the tree trunks (or vice versa) before returning to the radar. This effect is strongly enhanced if the forest floor is flooded, mainly because the smooth water surface causes very strong specular reflection. This effect is most pronounced for the HH polarization because of the two forward specular (Fresnel) reflections involved (see Figure 3-5). Therefore, the yellow and red areas in these images are interpreted as flooded forest or floating meadows. Note that the extent of the flooding is much greater in the April image than in the October image and appears to follow the 10 meter (33 foot) annual rise and fall of the Amazon River. Field studies by boat, on foot, and in low-flying aircraft by the University of California at Santa Barbara, in collaboration with Brazil's Instituto Nacional de Pesguisas Estaciais, during the first and second flights of the SIR-C/X-SAR system have validated the interpretation of the radar images.

The increased availability of calibrated data starting in the early 1990s allowed much more quantitative interpretation of SAR data, and the modeling of radar returns from many different types of terrain became much more sophisticated as a result. At the same time, it was shown that polarimetric SAR data provide a very powerful tool for interpreting radar scattering from different terrain types. As an example, Figure 6-59 shows the same image of San Francisco previously shown in Figure 6-3, with polarization responses extracted for the ocean, a portion of the urban area, and a portion of Golden Gate Park as indicated. It is immediately clear that these polarization responses are quite different. In particular, the response of the park appears to sit on a large pedestal, whereas the ocean response shows virtually no such pedestal. The urban area, on the other hand, shows an intermediate value for the pedestal. Also note that the urban response has its maximum return at HH (see Figure 6-44 for the definitions), with a local maximum at VV polarization. This is characteristic of a double-bounce scattering geometry in which the two bounces both occur from dielectric interfaces. In this scene, the double-bounce scattering is a result of a reflection from the street, followed by a reflection from the face of a build-

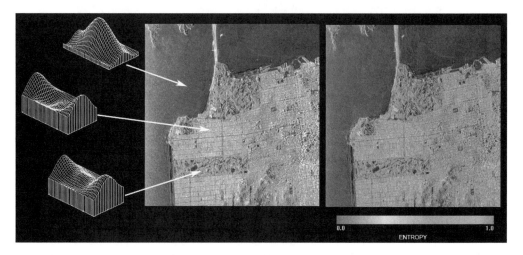

Figure 6-59. The polarization responses for three different areas in the San Francisco L-band image (acquired with the NASA/JPL AIRSAR system) as indicated by the arrows are shown here for comparison. The figure on the right shows the entropy of the covariance matrix for each multilook pixel. The entropy is a measure of the randomness contained in the covariance matrix. See color section.

ing. Comparisons to theoretical models have shown the L-band ocean response to be well modeled by a single reflection as predicted by the small-perturbation or other smooth-surface models.

The relative height of the "pedestal" in a polarization response is a measure of how much the scattering properties vary from pixel to pixel. If a scatterer is characterized by a scattering matrix as shown in Equation (6-132), it can be shown that the copolarized response will typically exhibit two polarization combinations for which the received power is zero. This does not mean that there is no radar echo; instead, for these two cases the radar echo has a polarization that is orthogonal to the receiving antenna polarization. In a typical polarimetric implementation as described earlier, each pixel will be represented by a scattering matrix in the single-look data. The polarization responses shown in Figure 6-59 were calculated by averaging the power of a block of neighboring pixels. The resulting polarization response can then be thought of as the average of the polarization responses of the individual pixels. If all the individual pixels are identical, the resulting polarization response will also exhibit two nulls in the copolarization response. If the individual polarization responses are different, however, it means that the exact positions of the individual copolarization nulls will also be different. The average copolarization response in this case will exhibit a minimum rather than a null, and the copolarization response appears to sit on a "pedestal." The observed polarization response of the ocean shown in Fig. 6-59 exhibits a very small pedestal, indicating very little variation on the individual responses that were averaged to provide the response shown. The polarization response of Golden Gate Park shows a large pedestal, a consequence of the randomly oriented branches of the vegetation, which leads to polarization responses that vary significantly from pixel to pixel. This is commonly referred to as "diffuse scattering" in the literature.

The average polarization response of an extended area can be calculated by first calculating the average covariance matrix of all the pixels included in the area of interest. In general, this average covariance matrix has no scattering matrix equivalent unless all the individual covariance matrices are identical. Instead, this average covariance matrix can generally be broken into the sum of three individual covariance matrices using an eigenvector decomposition first introduced in the context of radar imaging by Cloude (1988). This decomposition, for the backscatter case, is

$$\langle [\mathbf{C}] \rangle = \lambda_1 [\mathbf{C}_1] + \lambda_2 [\mathbf{C}_2] + \lambda_3 [\mathbf{C}_3] \tag{6-166}$$

where λ_i ($i = 1, 2,$ or 3) are the eigenvalues of the average covariance matrix and

$$[\mathbf{C}_i] = \mathbf{e}_i \tilde{\mathbf{e}}_i^* \qquad i = 1, 2, 3 \tag{6-167}$$

where $\mathbf{e}_i$ ($i = 1, 2,$ or 3) are the eigenvectors of the average covariance matrix. For most applications, van Zyl and Burnette (1992) showed that this decomposition can be interpreted in terms of single reflections, double reflections, and diffuse scattering. Note that this decomposition provides the equivalent of a "polarization unmixing" in the sense that the average response is broken into fractions of the power contributed by each of the individual scattering mechanisms. In a mathematical sense, this process is equivalent to a principal component analysis. The eigenvalues provide a simple way to express the height of the "pedestal" discussed earlier. Durden et al. (1990) showed that the pedestal height is related to the ratio of the minimum eigenvalue to the maximum eigenvalue:

$$\text{pedestal height} = \lambda_{\min} / \lambda_{\max} \tag{6-168}$$

Cloude (1998) pointed out that that the entropy of the average covariance matrix, defined as

$$H_T = \sum_{i=1}^{3} - P_i \log_3(P_i) \qquad (6\text{-}169)$$

with $P_i = \lambda_i/(\lambda_1 + \lambda_2 + \lambda_3)$ representing the fraction of the total power contained in each eigenvalue, is a measure of the randomness of the average covariance matrix. Entropy values of zero mean no randomness, whereas entropy values of 1 mean complete randomness. (Note that in the most general case, there are four eigenvalues; these reduce to three in the backscatter case.) The image on the right in Figure 6-59 shows the entropy of the San Francisco image. The vegetated areas have the largest entropy, indicating the most randomness. The urban areas have intermediate entropy values, and the ocean shows the lowest entropy values, consistent with slightly rough surface scattering.

The elements of the scattering matrix are complex numbers. Therefore, in addition to the magnitudes of the individual matrix elements, a polarimetric radar also measures the phase for each element. This allows one to study phase differences between the elements of the scattering matrix. As shown in Chapters 2 and 7 of Ulaby and Elachi (1990), this phase difference can be used to identify scattering mechanisms in polarimetric images. For example, areas dominated by slightly rough surface scattering typically have an HH–VV phase difference near 0°, whereas areas dominated by double reflections typically show HH–VV phase differences near 180°. Areas with large pedestals in their polarization responses typically show random HH–VV phase differences. Figure 6-60 shows the HH–VV phase difference for the San Francisco image. The ocean shows phase differ-

Figure 6-60. The image on the left shows the HH-VV phase difference for the San Francisco L-band image. The display on the right highlights areas with phase differences near 0° in blue, those with phase differences near 180° in red, and areas with random phases in green. See color section.

ences near 0°, the urban areas near 180°, and the vegetated areas show random phases. A simple way to show this information using amplitude data was introduced by Lee et al. (2004), in which the magnitude of $S_{hh} + S_{vv}$ is typically displayed in blue, the magnitude of $S_{hh} - S_{vv}$ is typically displayed in red, and the magnitude of S_{hv} is displayed in green. Therefore, areas with phase differences near zero would show strong returns in the blue channel, whereas areas in which double reflections dominate will show strong reflections in red. Areas with significant random scatter will show strong reflections in the green image. This display for the San Francisco image is shown in Figure 6-60. Note the strong red returns in the urban areas and mostly blue returns in the ocean. All the vegetated areas show strong reflections in the green channel.

We note that a significant portion of the urban area in Figure 6-60 seems to have phase differences that are quite random, even if the average is near 180°. This is related to the orientation of the streets and buildings relative to the radar look direction. In these images, the radar illumination is from the left, so streets that run vertically in the image, such as those on the left portion of the city, have buildings that face back toward the radar, providing efficient double reflections. When the streets are not orthogonal to the radar look direction, the double-reflection term decreases, as the specularly reflected energy is directed away from the radar. This leads to an increase in the randomness of the scattering mechanisms, with a corresponding increase in the phase variation from pixel to pixel.

As an application of this interpretation to a forested area, consider the three images of a portion of the Black Forest in Germany shown in Figure 6-61. These are displayed to show the single reflections in blue, double reflections in red, and random scattering in green. The town of Villingen is visible as the red areas on the left in all three images. The left part of the image, with generally lower backscatter, is covered with agricultural fields, and the right half of the image is covered with coniferous forest. The biomass range of this forest is on the order of 200 tons/hectare. The L-band image in the middle shows the agricultural area to be mostly surface scattering, and the forest to be dominated by ran-

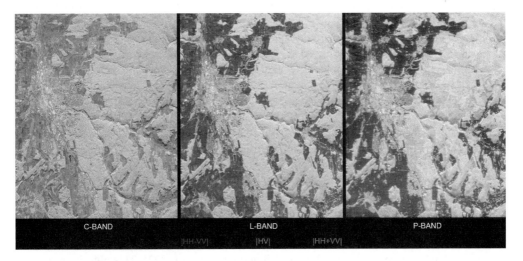

Figure 6-61. Three frequency images of a portion of the Black Forest in Germany (acquired with the NASA/JPL AIRSAR system). The town of Villingen is shown as the red areas in the left half of the images. The dark areas surrounding the town are agricultural fields, and the brighter areas in the right half of the images are coniferous forests. Radar illumination is from the top. (Data Courtesy of NASA/JPL-Caltech.) See color section.

dom scattering. Modeling studies have shown the L-band return to be dominated by scattering from the randomly oriented branches in the forest canopy. The agricultural areas are generally blue in the L-band image, indicating that the 24 cm L-band wavelength interacts little with the low-biomass vegetation. The C-band image shows less green and an increase in blue in the forest. This is a consequence of the fact that even the secondary branches are not thin compared to the C-band wavelength of 5.6 cm. Scattering models show that when the branches become thick compared to the radar wavelength, the scattering appears less random (van Zyl and Burnette, 1992). Note also the increase in the amount of green visible in the agricultural area in the C-band image. This shows that even though the biomass in the agricultural fields is generally low, the shorter C-band wavelength interacts strongly with the vegetation and shows random scattering as a result. The P-band (68 cm wavelength) image shows large areas in the forest that are dominated by double-reflection scattering with a reflection off the ground followed by a reflection from the tree trunks, and vice versa. This clearly shows the increased penetration of the longer-wavelength radar signals. When the ground surface under the forest is tilted either toward the radar or away from the radar, this double-reflection component of the total scattering decreases rapidly (van Zyl, 1990), and scattering by the randomly oriented branches dominate. This is clearly visible in the P-band image near the streams.

One of the very active research areas in radar image analysis is aimed at estimating forest biomass. Earlier works correlated polarimetric SAR backscatter with total above-ground biomass, and suggested that the backscatter saturates at a biomass level that scales with frequency, a result also predicted by theoretic models. This led some investigators to conclude that these saturation levels define the upper limits for accurate estimation of biomass (Imhoff, 1995), arguing for the use of low-frequency radars to be used for monitoring forest biomass (Rignot et al., 1995a).

More recent work suggests that some spectral gradients and polarization ratios do not saturate as quickly and may, therefore, be used to extend the range of biomass levels for which accurate inversions could be obtained (Ranson and Sun, 1994). Rignot et al. (1995a) showed that inversion results are most accurate for monospecies forests, and that accuracies decrease for less homogeneous forests. They conclude that the accuracies of the radar estimates of biomass are likely to increase if structural differences between forest types are accounted for during the inversion of the radar data.

Such an integrated approach to retrieval of forest biophysical characteristics is reported in Ranson et al. (1995) and Dobson et al. (1995). These studies first segment images into different forest structural types, and then use algorithms appropriate for each structural type in the inversion. Furthermore, Dobson and coworkers (1995) estimate the total biomass by first using the radar data to estimate tree basal area and height and crown biomass. The tree basal area and height are then used in allometric equations to estimate the trunk biomass. The total biomass, which is the sum of the trunk and crown biomass values, is shown to be accurately related to allometric total biomass levels up to 25 kg/m^2, whereas Kasischke and coworkers (1995) estimate that biomass levels as high as 34 to 40 kg/m^2 could be estimated with an accuracy of 15–25% using multipolarization C-, L-, and P-band SAR data.

Research in retrieving geophysical parameters from nonvegetated areas is also an active research area. One of the earliest algorithms to infer soil moisture and surface roughness for bare surfaces was published by Oh and coworkers (1992). This algorithm uses polarization ratios to separate the effects of surface roughness and soil moisture on the radar backscatter, and an accuracy of 4% for soil moisture is reported. Dubois et al. (1995) reported a slightly different algorithm, based only on the copolarized backscatters measured at the L band. Their results, using data from scatterometers, airborne SARs, and spaceborne SARs (SIR-C) show an accuracy of 4.2% when inferring soil moisture over bare surfaces. Shi and Dozier (1995) reported an algorithm to measure snow wetness, and demonstrated accuracies of 2.5%.

The past few years have seen the application of differential interferometry to a number of different geophysical processes. Earlier results focused on measuring the coseismic deformation signal of earthquakes, introducing the technique. Figure 6-62 shows the deformation signal resulting from a magnitude 6.1 earthquake that was centered in the Eureka Valley area of California (Peltzer and Rosen, 1995). The earthquake occurred on May 17, 1993 and caused a few centimeters of subsidence over an area of about 35 km by 20 km. Profiles through the deformation signal show that the noise on the measurement is approximately 3 mm. Using the measured deformation signal, Peltzer and Rosen (1995) argued that the rupture started at depth, and then propagated diagonally upward and southward on a north-east fault plane that dips to the west. They came to this conclusion based on the fact that the observed subsidence signature is elongated in the north-northwest direction.

Deformation signals measured with differential interferometry are now coupled with models of buried fault displacements to estimate the parameters of these faults. For example, Tobita and coworkers (1998) used Okada's (1985) model and deformation fields measured using JERS-1 differential interferometry to estimate the parameters of 12 subsections of the earthquake fault that caused the North Sakhalin earthquake in Russia. They estimated the parameters required in Okada's model by comparing the predicted deformation field with the measured one and performing a least squares optimization to minimize the differences. Similarly, Peltzer and coworkers (1996) used the measured deformation fields from ERS-1 interferometry in southern California to argue that the process responsible for the observed vertical rebound in the fault step-overs is not governed by viscous flow of the lower crust as previously proposed. Instead, they argue that

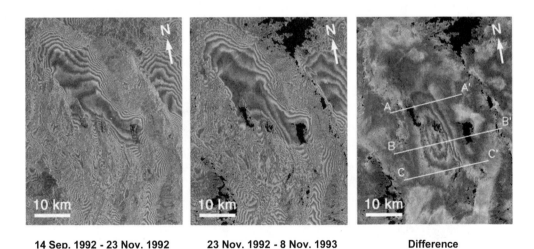

14 Sep. 1992 - 23 Nov. 1992   **23 Nov. 1992 - 8 Nov. 1993**   **Difference**

Figure 6-62. Deformation signals measured at C-band following the M 6.1 Eureka Valley earthquake in California. The left two images are the individual interferograms constructed from three acquisitions. The earthquake occurred between the second and third acquisitions. (Reprinted from *Science.* Peltzer and Rosen, 1995.) See color section.

the relaxation times observed in the interferometric data are more characteristic of post-seismic phenomena that could be explained by pore fluid flow in the upper crust.

An even more challenging application of interferometric SAR is the measurement of slow deformation processes such as postseismic or interseismic strain occurring in the absence of earthquakes. Since these processes do not radiate seismic waves, they are not as well documented and understood as earthquakes. Using ERS-1 data, Peltzer and coworkers (1997) measured the aseismic creep along the Eureka Peak fault and the Burnt Mountain fault following the 1992 Landers earthquake. They showed that the geodetic moment released by the creep was two orders of magnitude larger than that released by aftershocks during the period from January 10, 1993 to May 23, 1995. Aseismic creep along the San Andreas Fault near Parkfield California was also reported by Rosen and coworkers (1998). Their result not only shows creep, but it also shows local variability in the slip along the fault that had not been measured previously. Fujiwara and coworkers (1998) applied differential interferometry using JERS-1 data to measure crustal deformation near the Izu Peninsula in Japan. The interferometric results suggest a balloon-like crustal inflation from a localized magma source at a depth of 5–10 km. This interpretation, different from previous theories, is driven by the interferometric results with their much better spatial coverage compared to sparse leveling data. Crustal deformation in the vicinity of the Pu'u O'o lava vent was also reported by Rosen et al. (1996) using SIR-C L-band data. Figure 6-63 shows an example of a deformation signal measured on Darwin volcano in the Galapagos Islands (Amelung et al., 2000a). The image on the left shows the deformation signal, and the image on the right shows the predicted deformation if a point source at 3 km depth is responsible for the deformation. Amelung and coworkers (2000a) show several examples in this paper in which the deformation cannot be explained by the traditional Mogi point sources. Instead, sills of lava extending over a relatively large area are required to explain the observed deformation.

Differential interferometry also allows one to monitor effects related to human activity such as water and oil withdrawal. In a study reported by Peltzer and coworkers (1997), an area of ground subsidence was detected using ERS-1 interferometry in Pomona, California. They reported that in all the interferograms that they have analyzed, this area showed

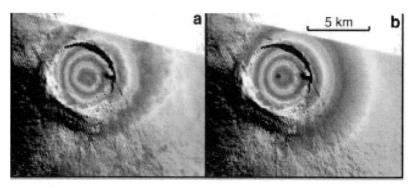

Interferogram 1992-1998   **Predicted deformation**

Figure 6-63. Observed and predicted deformation signals on Darwin volcano in the Galapagos Islands. The prediction assumes a Mogi point source at 3 km depth. (Reprinted fron *Nature*. Amelung et al., 1995.) See color section.

a deficit in volume, suggesting that over the years, seasonal water recharges do not keep up with the intensive use of water. They also report local deformations related to oil pumping in the Beverly Hills oil field, and in the San Pedro and Long Beach areas. The power of the interferometric technique is that it provides a synoptic and spatially continuous image of such surface deformation. Ground subsidence such as that reported by Peltzer and coworkers (1997) may become a hazard to structures and people, and can produce damage to buildings, freeway bridges, and pipelines. Differential interferometry is a relatively simple and effective way to monitor these types of hazards.

Differential SAR interferometry has also led to spectacular applications in polar ice sheet research by providing information on ice deformation and surface topography at an unprecedented level of spatial detail. Goldstein and coworkers (1993) observed ice stream motion and tidal flexure of the Rutford Glacier in Antarctica with a precision of 1 mm per day and summarized the key advantages of using SAR interferometry for glacier studies. Joughin and coworkers (1996a) studied the separability of ice motion and surface topography in Greenland and compared the results with both radar and laser altimetry. Rignot and coworkers (1995b) estimated the precision of the SAR-derived velocities using a network of in-situ velocities, and demonstrated, along with Joughin (1995), the practicality of using SAR interferometry across all the different melting regimes of the Greenland Ice Sheet. An example of monitoring the velocity of the Ryder Glacier in Greenland was reported by Joughin and coworkers (1996b). In this study, they used interferometric data acquired with the ERS-1 system and inferred that the speed of the Ryder Glacier increased roughly three times over a 7 week period during the melting season in 1995. The resulting interferograms are shown in Figure 6-64.

Satellite interferometry only measures ice sheet movement in the direction of the radar line of sight. Using interferometric pairs acquired during ascending and descending passes, Joughin and coworkers (1996c) reconstructed the three dimensional flow pattern of the Greenland ice sheet in the vicinity of the Humboldt Glacier. To reconstruct the three-dimensional flow field, they made the assumption that the ice velocity is parallel to the surface of the ice. They constructed a DEM of the area using the differential interferograms, and using the geometry inferred from the DEM plus their two line of sight measurements of the velocity from the ascending and descending passes, reconstructed the total flow of the surface of the ice sheet.

Another exciting application of differential interferometry is the measurement of the position of glacier grounding lines. The grounding line of a glacier is defined as that position at which the glacier detaches from its bed and becomes afloat in the ocean. The position of this grounding line is extremely sensitive to small changes in sea level or glacier thickness. Rignot (1996) used a technique known as quadruple differencing to identify the position of glacier grounding lines to an accuracy of less than 100 m. In the case of a floating glacier, each interferogram contains contributions from the topography, the velocity of the glacier, and the motion of the glacier due to tidal flexure. By properly scaling the interferograms for the baselines, two pairs of interferograms can be used to eliminate the contribution due to the topography, as described earlier in the surface deformation section. The resulting differential interferogram only contains contributions from the glacier motion and the tidal flexure. If one now makes the additional assumption that the glacier velocity remains steady and continuous in between observations, two such differential interferograms can be subtracted to eliminate the contribution of the glacier movement, leaving only the contribution from the tidal flexure. Using this technique, Rignot and coworkers (1997) measured the grounding line of 14 outlet glaciers in Greenland. By

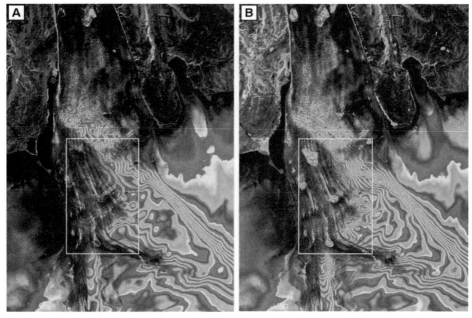

21-22 September 1995   26-27 October 1995

Figure 6-64. Two interferograms over the Ryder Glacier in Greenland acquired with the ERS-1 radar system. The interferogram on the right shows a dramatic increase in the speed of the glacier, as evidenced by the closer spacing in the interference fringes in this image as compared to the one on the left. (Reprinted with permission from Joughin et al., 1996b. Copyright 1996 AAAS). See color section.

comparing the ice discharge at the grounding line from knowledge of the glacier speed and thickness to that at the calving front of iceberg production, they concluded that the ice discharge at the grounding line is 3.5 times that at the calving front. They concluded that the difference is due to basal melting at the underside of the floating glaciers. Their results suggest that the north and northeast parts of the Greenland ice sheet may be thinning and contributing positively to sea level rise. Rignot (1998) also reported that the grounding line of the Pine Island Glacier in West Antarctica has retreated by 1.2 km per year between 1992 and 1996. He contributes this retreat to a thinning of 3.5 m per year at the glacier grounding line, caused by a bottom melting rate that exceeds the rate at which the floating glacier tongue will remain in mass balance.

Ocean waves are visible in radar images as a periodic pattern of the image brightness. Wave packets corresponding to internal waves (or solitons) are seen near the coast in Figure 6-65 (left), whereas 300 m surface swells are seen in 6-65 (right). Ice floes are imaged as bright/gray tone features separated by bright ridges and dark open-water channels (Fig. 6-66). Long-term repetitive imaging allows accurate tracking of the ice motion.

The application of along-track interferometry so far has been limited by the availability of data. Nevertheless, several investigations have shown the great promise of this technique. For example, Goldstein and coworkers (1989) used this technique to measure ocean currents with a velocity resolution of 5 to 10 m/s. Along-track interferometry was also used by Marom and coworkers (1990; 1991) to estimate ocean surface current velocity and wave-number spectra. This technique was also applied to the measurement of

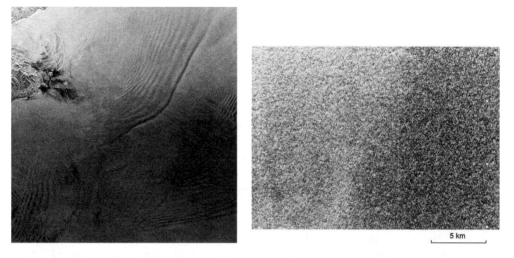

Figure 6-65. Left, Seasat image of ocean internal waves in the Gulf of California acquired on September 29, 1978. Image size is 70 km × 70 km. Right, Typical example of ocean surface waves imaged by Seasat SAR. (Courtesy of Jet Propulsion Laboratory.)

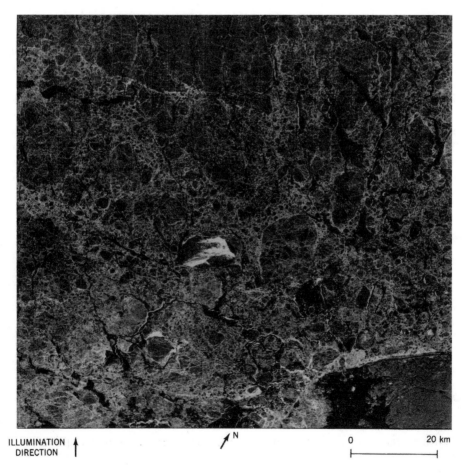

ILLUMINATION DIRECTION

N

0   20 km

Figure 6-66. Seasat image of polar ice floes near Banks Island (lower-right corner) acquired on October 3, 1978. (Courtesy of Jet Propulsion Laboratory.)

ship-generated internal wave velocities by Thompson and Jensen (1993). Figure 6-67 shows an example of along-track interferometric results. The figure shows an image of the coast of Hawaii acquired with the NASA/JPL AIRSAR system operating in the along-track interferometric mode. The radar backscatter image shows little information about the ocean surface, whereas the complicated wave patterns of the swell interacting with the coastline are readily visible in the along-track interferometric phase.

In addition to measuring ocean surface velocities, Carande (1994) reports a dual-baseline implementation, implemented by alternately transmitting out of the front and aft antennas, and receiving through both at the same time, to measure ocean coherence time. By using this implementation, two baselines are realized simultaneously, one with physical length equal to the antenna separation, and one with length equal to half this separation. By measuring the change in the coherence between the short and long baselines, one can estimate the time at which the coherence will drop to zero. Carande estimated typical ocean coherence times for the L band to be about 0.1 second. Shemer and Marom (1993) proposed a method to measure ocean coherence time using only a model for the coherence time and one interferometric SAR observation.

6-5 NONIMAGING RADAR SENSORS: SCATTEROMETERS

Scatterometers are radar sensors that provide the backscattering cross section of the surface area illuminated by the sensor antenna. They usually have a large footprint and are for the most part used in studying average scattering properties over large spatial scales. They have been particularly useful in measuring the ocean backscatter cross section as a means of deriving the near-surface wind vector. The physical basis for this technique is that the strength of the radar backscatter is proportional to the amplitude of the surface capillary and small gravity waves (Bragg scattering), which in turn is related to the wind speed near the surface. Moreover, the radar backscatter can be measured at different azimuth angles, allowing the determination of the wind direction. Figure 6-68 shows the behavior of $\sigma(\theta)$ for different wind velocities.

 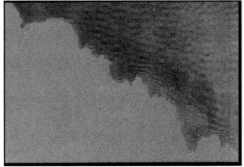

Radar Backscatter   **Along-Track Interferometric Phase**

Figure 6-67. Radar backscatter (left) and along-track interferometric phase (right) of a section of the Hawaiian coastline. Red colors mean the surface is moving away from the radar; illumination is from the top. The complicated wave patterns are only visible in the phase image. (Courtesy NASA/JPL-Caltech.) See color section.

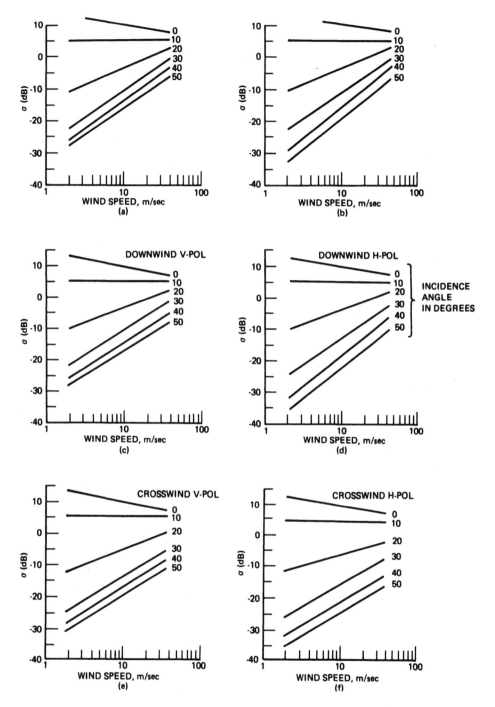

Figure 6-68. σ versus wind speed for constant incidence angles and wind direction. (*a*) upwind. *V* polarization; (*b*) upwind. *H* polarization; (*c*) downwind. *V* polarization; (*d*) downwind. *H* polarization; (*e*) crosswind. *V* polarization; (*f*) crosswind. *H* polarization. (From Jones et al., 1977.)

   Scatterometers use a number of different configurations (see Fig. 6-69). The side-look-ing fan beam scatterometer (Fig. 6-69a) allows sensing of a wide swath. The forward-looking fan beam scatterometer (Fig. 6-69b) allows measurement only along the flight track; however, for each point on the surface, the backscattered return can be measured for a variety of incidence angles. The tilted fan beam (Fig. 6-67c) allows wide swath sens-ing and, with multiple beams, multiple look direction measurement. The scanning pencil beam scatterometer (Fig. 6-67d) provides backscatter measurement at a constant angle and dual look direction over a wide swath.

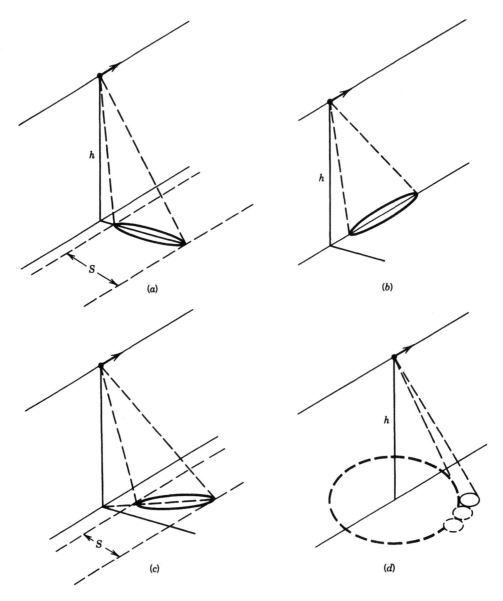

Figure 6-69. Different scatterometer configurations. (*a*) Side-looking fan beam; (*b*) forward-looking fan beam; (*c*) tilted fan beam, and (*d*) scanning pencil beam.

6-5-1 Examples of Scatterometer Instruments

Several scatterometers have been flown in space, dating back to the scatterometer flown on the Seasat satellite in 1978, which used a dual-fan-beam tilted configuration that allowed the acquisition of global wind fields. Table 6-1 shows the parameters of several of these systems. Here we shall briefly describe the Seasat scatterometers and the SeaWinds instrument hat was launched on the QuikScat satellite in 1999.

Seasat Scatterometer. The separation of surface pixels in the Seasat scatterometer system is achieved by using the antenna physical footprint on the surface and the Doppler shift in the returned echo. As was discussed in the synthetic-aperture radar case, an analysis of the echoes' Doppler shift allows us to subdivide the surface into equi-Doppler lines, which correspond to the intersection of a family of cones with the surface. If we neglect the surface curvature, the intersection is a family of hyperbolas. When the antenna footprint is superimposed on this pattern, the narrow dimension defines a series of identifiable cells that are the surface measurement pixels (see Fig. 6-70).

The Seasat sensor, commonly called SASS (Seasat-A Satellite Scatterometer), had two tilted fan beams on each side. In this way, as the satellite moved, each surface point was observed twice, once with the forward antenna and once with the backward antenna. This provided measurements at two different look directions.

The SASS sensor operated at a frequency of 14.6 GHz (i.e., wavelength of 2 cm). It incorporated four dual-polarized fan beam antennas that produced an X-shaped pattern of illumination on the surface (Fig. 6-71). Twelve Doppler filters were used to subdivide the antenna footprint electronically into resolution cells approximately 50 km on the side. The total swath covered was 750 km, with the incidence angle ranging from 25° to 65° from vertical. Three additional Doppler cells provided measurement near the satellite track at incidence angles of 0°, 4°, and 8°.

A simplified diagram of the sensor is shown in Figure 6-72. A frequency synthesizer provided an excitation signal at 14.6 GHz to the transmitter and reference signals to the receiver. The travelling wave tube amplified the signal to a 100 W peak level before transmission to the antenna switching matrix (ASM). The ASM selects the antenna for each set of backscatter measurement by switching in a periodic fashion according to the selected instrument operating mode.

In all operating modes, 15 backscatter measurements were made every 1.89 seconds, with an antenna switching cycle completed every 7.56 seconds. This timing was designed to provide measurements that were located on a 50 km spacing in the along-track direction. The noise source provided a periodic receiver gain calibration every 250 seconds.

Table 6-1. Spaceborne Scatterometer Parameters

Parameter	SASS on Seasat	ERS-1	NSCAT	SeaWinds on QuikScat
Frequency	14.6	5.3	14	13.4
Spatial resolution	50 km	50 km	25, 50 km	25 km
Swath width	500 km	500 km	600 km	1800 km
Antenna type	Fan beam	Fan beam	Fan beam	Conical scan
Number of antennas	4	3	6	1
Polarization	VV, HH	VV	VV, HH	VV, HH
Orbit altitude	800 km	785 km	820 km	802 km

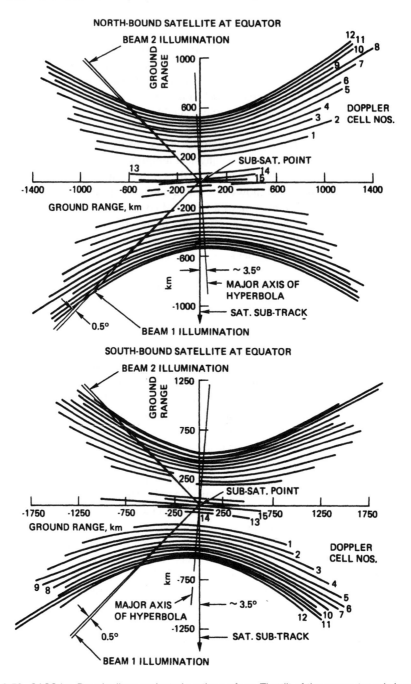

Figure 6-70. SASS iso-Doppler lines projected on the surface. The tilt of the symmetry axis is a result of the Earth's rotation (from Bracalante et al., 1980).

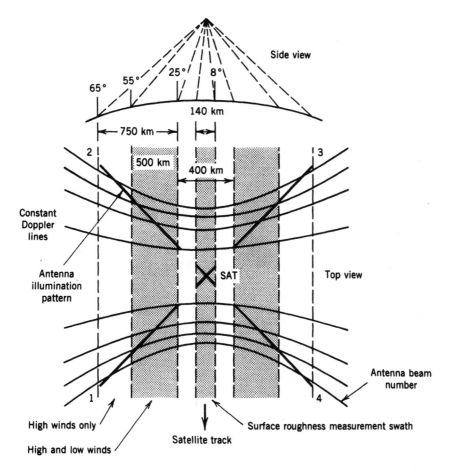

Figure 6-71. SASS measurement geometry.

Using range gating and Doppler filtering techniques in the signal processor, the backscatter return was measured at different incidence angles. Figure 6-70 shows the 15 iso-Doppler lines corresponding to the 15 scatterometer Doppler cell center frequencies. The Doppler lines have a symmetry axis that is rotated relative to the spacecraft subtrack. This is due to the Earth's rotation.

In the receiver, a square-law detector and gated integrator are used to sample the reflected power from an antenna 64 times during a 1.84 second measurement period. The first three samples are used to place the receiver–processor in the most appropriate of four possible gain states, which put the received signal level in the linear portion of the square-law detector. The remaining 61 samples are then processed using the selected gain state. The mean values of the 61 integrated voltage levels are then entered into the data stream for each of the 15 Doppler cells. The 1.84-second measurement interval is repeated continuously, but a different antenna is activated for each consecutive sampling period.

SeaWinds on QuikScat. The SeaWinds scatterometer on the QuikScat satellite is an example of a conically scanning pencil beam system. Fan beam scatterometers, such as

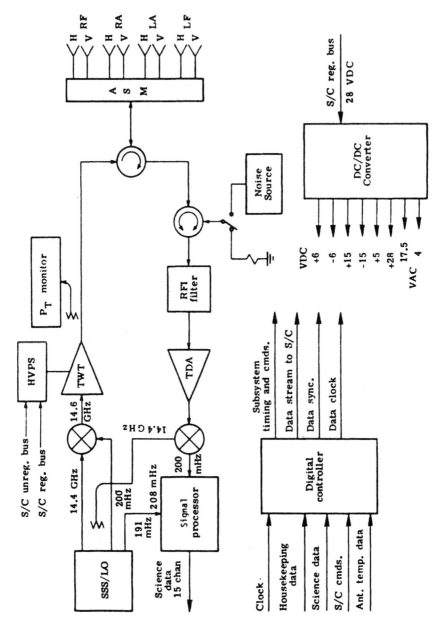

Figure 6-72. SASS block diagram. (From Johnson et al., © 1980 IEEE.)

the Seasat scatterometer, require several fixed antennas pointing at multiple azimuth angles to retrieve ocean wind patterns. These antennas are typically several meters long, and require an unobstructed field of view, making their accommodation on a spacecraft quite challenging. Conically scanning systems like the SeaWinds instrument use a single dish antenna, typically on the order of 1 m in diameter, that is conically scanned about the nadir direction to provide measurements at multiple azimuth angles. A key advantage of such pencil beam systems is that, because of their more compact design, they are much easier to accommodate on spacecraft without the necessity of complex deployment schemes or severe field-of-view constraints.

Another advantage of such a scanning system over a fan beam system such as SASS or NSCAT is that there is no gap between the swaths on either side of the satellite track. In addition, all measurements are made at the same fixed angle of incidence. This significantly simplifies the analysis of the data, since geophysical inversion algorithms only have to be developed for one angle of incidence.

The SeaWinds scan geometry is shown in Figure 6-73. The instrument uses a single, parabolic, 1 m dish antenna with two feeds positioned such that two beams are generated. These beams point at angles of 40° and 46° with respect to the nadir, respectively. The two feeds use different polarizations; the inner angle beam is polarized horizontally, and the outer beam is polarized vertically. Taking the spherical nature of the Earth and the 802 km QuikScat orbit altitude into account, these beams intersect the ocean surface at local angles of incidence of 47° and 55°, respectively. The inner beam images a swath of 1400 km, whereas the outer beam images a swath of 1800 km. This wide swath capability allows QuikScat to image 90% of the ocean surface each day. Figure 6-74 shows a number of ascending and descending data swaths acquired by the SeaWinds instrument in one day.

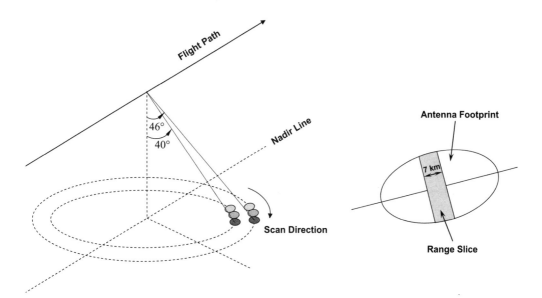

Figure 6-73. SeaWinds instrument imaging geometry. The left image shows rotating pencil beams; the right image shows the beam spot on the ground for one of the beams. Also shown is the resolution slice inside the beam spot.

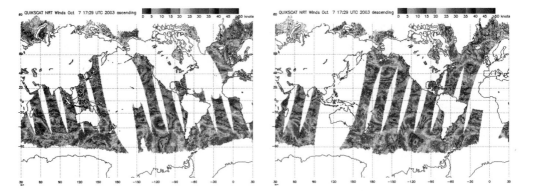

Figure 6-74. Image swaths acquired by the SeaWinds instrument over a 24 hour period ending on October 7, 2003. The left image shows the ascending swaths; the right image shows the descending swaths. (Courtesy of NASA/JPL-Caltech.) See color section.

With this two-beam geometry, each point in the inner swath is seen four times; once in the forward and once in the aft direction with the inner beam, and once each in the forward and aft directions with the outer beam. These four measurements provide improved estimations of the wind direction over the ocean.

The size of the SeaWinds beam spots on the ground are about 24×31 km for the inner beam, and 26×36 km for the outer beam, respectively. The system uses a linear FM chirp signal that employs deramp processing to further process each beam spot into 7 km slices in the range direction, as shown in Figure 6-73. To avoid gaps in the coverage, a scanning pencil beam system has to complete one scan before the satellite has moved a distance equal to the along-track size of a beam spot. The SeaWinds instrument uses a scanning rate of 18 rpm, which means that the satellite nadir point moves 22 km between scans.

6-5-2 Examples of Scatterometer Data

The SASS data acquired during the summer of 1978 were used to prove the capability of spaceborne scatterometers to map the near-surface wind on a regional as well as global scale. Data from the SeaWinds instrument are now routinely processed into global ocean wind fields and are used by a number of different government agencies.

Figure 6-75 shows the derived mean wind pattern over the Pacific Ocean during the month of July 1978. Figure 6-76 shows the near-vertical backscatter cross section across typhoon Carmen. Because of the strong wind and resulting surface roughness under the hurricane, the nadir backscatter decreases drastically by more than 10 dB. Figure 6-77 shows a global view of the ocean winds for September 20, 1999, as measured by the Sea-Winds instrument on the QuikScat satellite. The large storm in the Atlantic off the coast of Florida is Hurricane Gert. Tropical storm Harvey is evident as a high-wind region in the Gulf of Mexico, and farther west in the Pacific is tropical storm Hilary. An extensive storm is also present in the South Atlantic Ocean near Antarctica.

Scatterometer data are being used more frequently to study land processes. Unfortunately, typical resolutions of spaceborne scatterometers are too coarse for most land process studies. However, a technique known as scatterometer image reconstruction with filtering (SIRF) (Long et al., 1993) can be used to improve the spatial resolution of some scatterometers images. This method works best when the scatterometer resolution el-

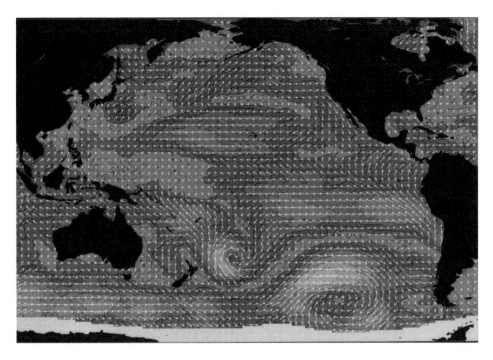

Figure 6-75. Wind patterns over the Pacific for July 1978 derived from the SASS measurement. (Courtesy of P. Woicesyhn, Jet Propulsion Laboratory.) See color section.

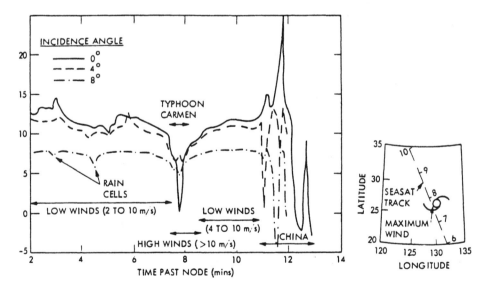

Figure 6-76. Backscatter data near typhoon Carmen for near vertical incidence angles. (From Bracalante et al., © 1980 IEEE.)

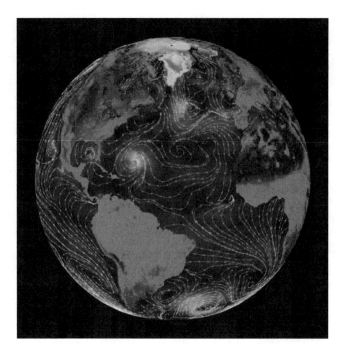

Figure 6-77. Global ocean winds as measured by QuikScat on September 20, 1999. Colors over the ocean indicate wind speed, with orange as the fastest wind speeds and blue as the slowest. White streamlines indicate the wind direction. (Courtesy of NASA/JPL-Caltech.) See color section.

ement is long and narrow (see Fig. 6-74b). Because of their shape and orientation, these long, narrow elements overlap only partially for the measurements taken with a forward-looking beam and an aft-looking beam. If one now makes the assumption that the radar cross section does not change in the time between the forward-looking and aft-looking acquisitions, and, further, that the radar cross section does not vary as a function of the azimuth look direction, we can place all these measurements on a finer grid, and then deconvolve the observations to provide the measurements on this finer grid. The practical best resolution achievable is roughly equivalent to the narrowest dimension of the measurement cell. Figure 6-78 shows an image of Antarctica constructed using this algorithm using data acquired on October 2, 2000 with the SeaWinds instrument on the QuikScat satellite. The brighter central area in the image is Antarctica. It is bright due to high Ku-band radar return echoes from glacial snow and ice covering the continent. The sea-ice pack surrounding Antarctica is visible as the darker outer area in the center of the image. The black circle in Antarctica is an area where no data is collected due to the orbit of the satellite.

6-6 NONIMAGING RADAR SENSORS: ALTIMETERS

Altimeters use the ranging capability of radar sensors to measure the surface topographic profile. The altitude measurement is given by (Fig. 6-79)

$$h = \frac{ct}{2} \qquad (6\text{-}170)$$

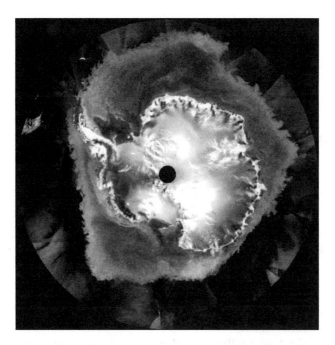

Figure 6-78. Resolution-enhanced image of Antarctica using data acquired on October 2, 2000 with the SeaWinds instrument on the QuikScat satellite. See text for details. (Courtesy NASA/JPL-Caltech.)

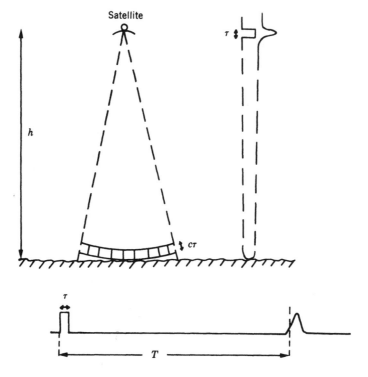

Figure 6-79. Altimeter measurement geometry.

with an accuracy of

$$\Delta h = \frac{c\tau}{2} = \frac{c}{2B} \qquad (6\text{-}171)$$

where t is the time delay between the transmitted signal and the received echo. This altitude measurement is the distance between the sensor and the portion of the surface illuminated by the antenna beam or by the pulse footprint (see Fig. 6-80). In the first case, the footprint is given by

$$X = \frac{\lambda h}{L} \qquad (6\text{-}172)$$

This corresponds to a beam-limited altimeter, commonly used for land topography mapping. In the second case, the footprint is given by

$$X = 2\sqrt{c\tau h} \qquad (6\text{-}173)$$

which corresponds to a pulse-limited altimeter, commonly used over relatively smooth areas such as the ocean surface.

From Equation (6-170) it follows that there are two sources of error for the elevation measurement. The first is the knowledge of the speed of light, or more properly, the group velocity of the pulses transmitted by the radar altimeter. As these signals propagate through the atmosphere, local variations in water vapor will change the index of refraction of the air, which will change the group velocity as a result. The variations in electron density in the ionosphere will also introduce variations in the group velocity. These effects have to be compensated for in the signal processing to infer the correct elevation. The second error source is the measured round-trip time. Ultrastable local oscillators are used in altimeters to reduce this error. The measurement in Equation (6-170) is the elevation relative to the altimeter. To translate this measurement into an accurate representation of the surface elevation, one has to know the position of the spacecraft very accurately. In the case of the Topex/Poseidon mission, the satellite position is reconstructed to an

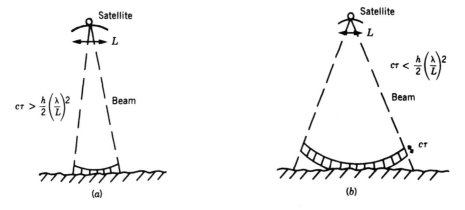

Figure 6-80. Types of altimeters: (*a*) beam limited, and (*b*) pulse limited.

accuracy of a few centimeters using a combination of Global Positioning Satellite (GPS) measurements and laser and microwave tracking of the satellite.

6-6-1 Examples of Altimeter Instruments

Altimeters have been flown on a number of spacecraft, including Seasat, Pioneer Venus Orbiter, Apollo 17, and GeoSAT. The Seasat altimeter, which was flown in 1978, provided profiles of the ocean surface with an accuracy of a fraction of a meter. The Topex/Poseidon mission, launched in 1992, carried the first altimeter instrument suite that was optimized for studying dynamic variations in the ocean topography.

The Seasat Altimeter. The Seasat altimeter operated at a frequency of 13.5 GHz using a 1 m diameter, horn-fed, parabolic dish antenna. A simplified block diagram of the sensor is shown in Figure 6-81.

The local oscillator generates 12.5 nanosecond impulses at a 250 MHz center frequency, which are applied to the chirp generator. The chirp generator is a surface acoustic wave (SAW) device fabricated on a lithium tantalate substrate. The resulting chirped pulse has a linearly decreasing frequency with an 80 MHz bandwidth and a pulse length of 3.2 μsec. The pulse repetition frequency (PRF) is 1020 Hz.

During the transmit mode, the chirp pulse at 250 MHz is upconverted to 3375 MHz, amplified to a 1 W level, and multiplied by 4 to 13.5 GHz. This also multiplies the bandwidth by 4 in order to achieve the desired 320 MHz bandwidth and height measurement accuracy of 0.47 m. In the receive mode, the chirp pulse is upconverted to 3250 MHz, amplified to 0.1 W, multiplied by 4 to 13.0 GHz, and used for mixing with the received echo.

The TWT amplifier amplifes the transmit pulse to 2 kW before it is sent to the five-port circulator that provides for transmit/receive mode switching as well as calibration mode switching. In the receive mode, the first mixer, which is located immediately after the circulator, achieves full deramping by mixing the 13.5 GHz incoming chirp signal with the 13.0 GHz local chirp signal, resulting in a 500 MHz CW signal to form the in-phase and quadrature (I and Q) video signals, which are digitized and stored for use in the digital filtering scheme. The digital filter bank and the adaptive tracking unit are then used for height tracking and receiver automatic gain control (AGC) and wave height estimation.

The TOPEX/Poseidon Altimeter Mission. The TOPEX/Poseidon mission, a joint project involving NASA and the French Space Agency (CNES), was launched in 1992. This mission was designed specifically to make highly accurate measurements of the dynamic components of ocean topography. The measurement error for sea surface height as reported by the TOPEX/Poseidon system has been shown to be on the order of 4.1 cm. This mission is discussed in great detail in the book edited by Fu and Cazenave (2001).

The primary instrument of the TOPEX/Poseidon payload is a dual-frequency (Ku- and C-band) altimeter system that operates at both frequencies simultaneously. This dual-frequency measurement allows one to estimate and correct for the time delay introduced by the ionosphere. A three-frequency microwave radiometer operating at frequencies of 18, 21, and 37 GHz is used to estimate the atmospheric water content. This measurement allows one to estimate the index of refraction of the atmosphere, which is then used to calculate the proper propagation speed to use in the calculation of the distance between the altimeter and the ocean surface.

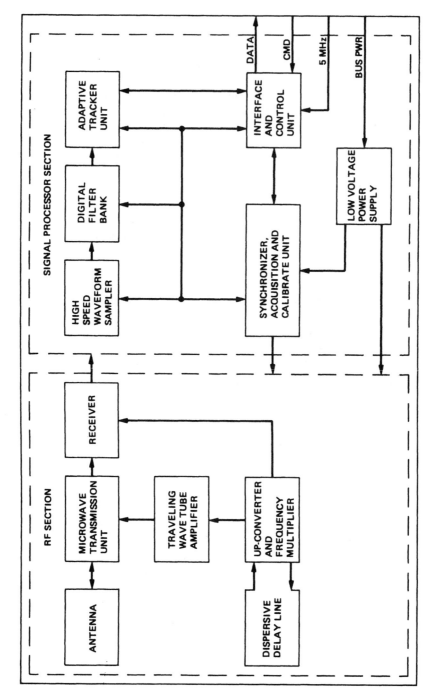

Figure 6-81. Seasat altimeter major functional elements. (From Townsend, © 1980 IEEE.)

The TOPEX/Poseidon satellite orbit is reconstructed using a combination of GPS measurements and laser and microwave tracking. At the time of the TOPEX/Poseidon launch in 1992, GPS tracking of satellite orbits was not yet available operationally, and the six-channel GPS receiver was included in the payload as an experiment to evaluate the utility of using GPS signal to track the satellite. These measurements have since proved to be highly accurate, and have provided an independent verification of the orbit determination. The primary means of tracking the TOPEX/Poseidon satellite is satellite laser ranging and and a system known as Doppler Orbitography and Radiopositioning Integrated by Satellite (DORIS). Although the laser tracking is highly accurate, it is limited to clear atmospheric conditions and the availability of the laser stations. The DORIS system, in contrast, provides all-weather tracking and the temporal and geographical coverage to complement the laser tracking. The TOPEX/Poseidon orbit-reconstruction error has been shown to be less than 2.5 cm.

The TOPEX/Poseidon mission was designed specifically for highly accurate measurements of ocean topography. This even extended to the design of the satellite orbit. As mentioned above, the primary orbit tracking for the TOPEX/Poseidon satellite was planned to provided by ground tracking stations. These are of varied quality and are not available continuously along the satellite orbit. Therefore, orbit reconstruction required dynamical modeling of the satellite orbit. The two dominant errors of satellite orbit modeling are uncertainties in short scales of the geopotential field and variations in the atmospheric drag on the satellite. Both of these are strongly dependent on the satellite orbit altitude, and drop quickly with increasing satellite altitude. On the other hand, two factors argue for a lower orbit. The first is the fact that the radar return decreases as the fourth power of the altitude (see Section 6-3-4 on the radar equation), and the second is the fact that the space radiation environment is worse for higher altitudes. The TOPEX/Poseidon orbit altitude of 1336 km is a compromise between these factors.

In addition to the orbit altitude, the orbit inclination must also be selected carefully. The inclination of the orbit determines the maximum latitude at which the ocean topography can be measured. The TOPEX/Poseidon orbit inclination of 66° was chosen to minimize the effect of ocean tides on the measurements of sea surface topography. The most energetic tides are the diurnal and semidiurnal components, with frequencies near one and two cycles per day. This short timescale variability cannot be resolved in typical altimeter measurements, and, therefore, are aliased to the lower-frequency components of the signals that are of interest to oceanographers. Exactly how the tidal signal aliases into the ocean height signal turns out to be a strong function of the orbit period, the ground track spacing (which is a function of the number of orbits in the repeat period), and the orbit inclination. The TOPEX/Poseidon orbit altitude, and, hence, the repeat period and number of orbits in the period, was chosen mostly as a trade-off between minimizing orbit perturbations due to gravity and drag, the radiation environment, and the radar measurement power considerations. The 10 day repeat orbit of TOPEX/Poseidon is one of the few orbits for which an inclination exists that would place the tidal aliasing frequencies such that they can be separated from the ocean height signals. These considerations led to the selection of the 66° orbit inclination of the TOPEX/Poseidon orbit.

6-6-2 Altimeter Applications

The permanent topography of the ocean surface is a reflection of the geoid determined by the geographical variations in earth's gravity field. These variations in the gravity field are a consequence of the distribution of the mass inside or at the surface of the earth. Measurements of the static ocean totpography are therefore useful in the study of marine geophysics. An example of the Seasat altimeter data is illustrated in Figure 6-82, which shows the sea surface height over the Puerto Rican and Venezuelan Coast trenches. Sea surface depression of about 15 m is observed due to the Puerto Rican trench. The reverse effect (i.e., surface rise) was observed over sea mount-type features (Fig. 6-83).

The permanent topography of the ocean provides much information about the mass distribution of the earth, but it is the dynamic part of the ocean topography that conveys information about many of the processes that affect the climate of the earth. These processes include the circulation in the ocean, sea level rise or fall, and shorter-term meteorological processes such as the El Niño phenomenon. These will be discussed in detail in the next chapter.

Radar altimeters are also able to make two additional useful measurements: ocean wave height and normal surface backscatter. As the pulse is reflected from the surface, energy is contributed from different parts of the surface that are at different heights. Thus, the returned echo has a rise time that is related to the r.m.s height of the surface. In a simplified way, the surface peak-to-peak height variation or wave height is equal to the rise time $\Delta \tau$ multiplied by $c/2$. Figure 6-84 gives an example of the wave height measurement as the satellite track crossed hurricane Fico off the west coast of Mexico.

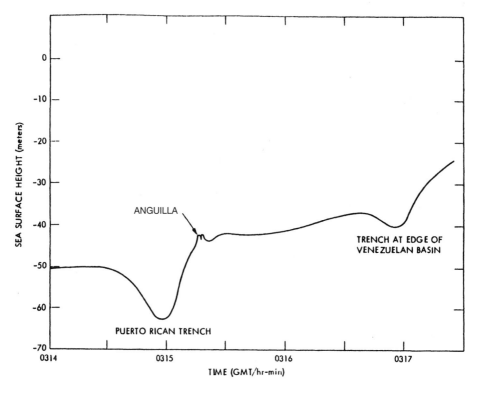

Figure 6-82. Sea surface height over two trenches measured with the Seasat altimeter. (From Townsend, © 1980 IEEE.)

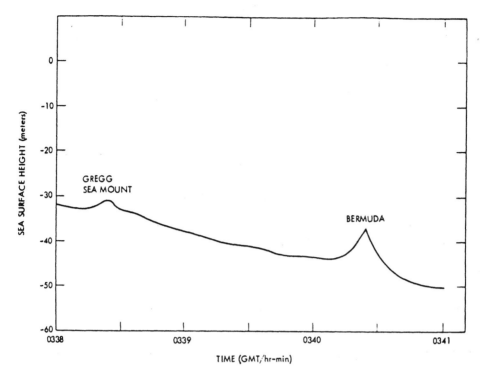

Figure 6-83. Sea surface height over a sea mount as measured with the Seasat altimeter. (From Townsend, © 1980 IEEE.)

By measuring the maximum intensity of the returned echo, the surface normal backscatter σ_N can be derived. It provides information about the slopes distribution. Figure 6-84 also shows the variation of σ_N near the center of hurricane Fico.

6-6-3 Imaging Altimetry

Conventional altimeters provide a profile of the topography along the nadir track. However, if the illuminating beam is very narrow and is scanned across the track, the three-dimensional topography can be acquired by having a series of neighboring profiles. The beam scanning can be achieved electronically by using a phased-array antenna or mechanically by scanning the antenna or spinning the spacecraft (see Fig. 6-85).

An example of the spinning imaging altimeter is the Pioneer Venus Orbiter (PVO) altimeter. The prime objective of the PVO radar was to obtain altimetry measurement over a large percentage of the Venusian surface. The Pioneer Venus Orbiter was put in orbit around Venus in December, 1978. Altimetry data was acquired between latitudes 76°N and 63°S. The altimeter footprint varied from about 7 km to about 100 km, depending on the altitude of the spacecraft, which was in a highly elliptical orbit. The height measurement accuracy was about 200 m. The PVO radar has an operating frequency of 1.757 GHz (i.e., wavelength of 17 cm) and uses a 38 cm dish that is mounted on the spinning spacecraft (spin period is about 12 seconds). The transmitted peak power is 20 W. The transmitted signal is modulated with a 55 element, bilevel phase PN code. The instrument

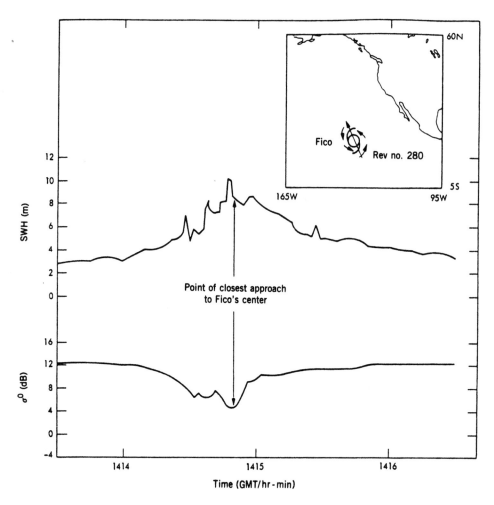

Figure 6-84. Significant wave height and backscatter measurement over hurricane Fico measured by the Seasat altimeter. (From Townsend, © 1980 IEEE.)

operates in two distinct modes: altimetry and imaging. The altimetry mode operates at altitudes below 4700 km. Estimates of the echo's absolute time delay, its time dispersion, and its intensity are obtained simultaneously; the latter two quantities are used to yield information on local surface properties. The imaging mode operates only when the spacecraft is below 550 km and when the radar antenna is on one side or the other of the nadir as the spacecraft rotates around its axis. In this "side-looking" mode, the echo time delay and Doppler characteristics are used to generate a map of the surface radar backscatter at various angles between 30° and 58°.

Figure 6-86 is a false color rendition of the topography of Venus as derived for the PVO altimetry data. This map is cast in a standard Mercator projection and covers the majority of the Venusian surface except for the polar regions. The two main "high" regions are the Ishtar Terra and the Aphrodite Terra, the latter one being spatially larger than South America. Except for the prominent elevated features, the planet is quite flat. The

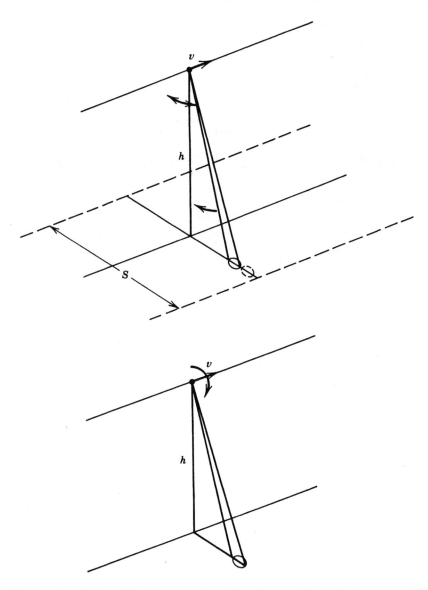

Figure 6-85. Imaging altimeters: (a) using a scanning antenna beam or (b) spinning the spacecraft.

highest point observed has an altitude of 11.1 km relative to a median radius of 6051.2 km. The lowest point observed is at −1.9 km.

6-6-4 Wide Swath Ocean Altimeter

Conventional altimeters measure a profile of the surface height underneath the satellite. The spatial resolution of the resulting surface height map is, therefore, limited by how closely spaced these profiles are. For example, the TOPEX/Poseidon tracks are separated by about 315 km at the equator, limiting the spatial scales of ocean processes that can be

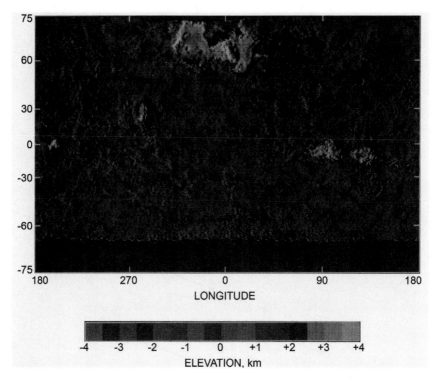

Figure 6-86. Map of the surface of Venus generated from the PVO radar data. See color section.

studied with this system to about this value. This spatial resolution can be improved by adding more satellites with orbits offset such that more ground tracks are covered. This, however, comes at the expense of having to coordinate and launch multiple satellites, as well as cross-calibrating different systematic errors due to orbit and instrument biases. If mesoscale ocean phenomena must be imaged, spatial resolutions on the order of the Rossby radius of deformation (scales on the order of 30 km) must be achieved, requiring a relatively large number of such satellites.

The Wide Swath Ocean Altimeter (Rodriguez and Pollard, 2001) is a proposed coarse resolution imaging altimeter that will image a swath on either side of the satellite track in addition to the traditional nadir altimeter profile. The proposed configuration of the WSOA is shown in Figure 6-87. This concept draws from the experience gained at the Jet Propulsion Laboratory with the Shuttle Radar Topography mission, which used an interferometric SAR to map the topography of the land surfaces of the Earth. A swath of approximately 100 km is imaged on either side of the conventional altimeter profile. The intrinsic cross-track resolution varies from approximately 670 m in the near range to about 100 m in the far range. The along-track resolution is given by the azimuth beamwidth, and is approximately 13.5 km. In order to have spatially uniform resolution cells, and to reduce random measurement error, the final measurements are averaged to 15 km resolution cells.

The 200 km swath width of the WSOA enables near-global coverage with a 10 day repeat orbit such as that of TOPEX/Poseidon. In fact, this arrangement allows most areas to be imaged twice in the 10 day period, and areas at higher latitudes are covered as many as

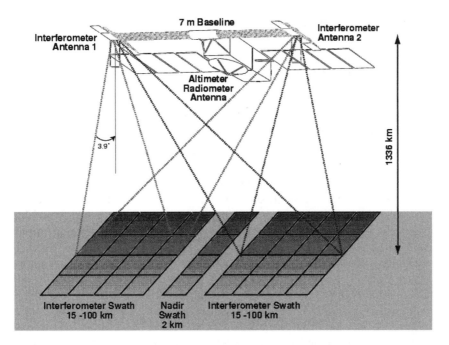

Figure 6-87. The proposed WSOA instrument concept integrated with the Jason altimeter on the Proteus satellite bus. (Courtesy of NASA/JPL-Caltech.) See color section.

four times in each cycle. These multiple "looks" at each surface can be used to average more measurements to reduce the height errors of the measurement. Figure 6-88 shows a comparison of the coverage of the TOPEX/Poseidon altimeter and the WSOA in the same orbit for an area near the equator where the ground track separation is at a maximum. Note the details of the ocean processes visible in the WSOA image.

Conventional Altimeter Data

Wide-Swath Altimeter Data

Figure 6-88. A comparison of the coverage between the TOPEX/Poseidon conventional altimeter (left) and the proposed WSOA instrument (right) in the same orbit. (Courtesy NASA/JPL-Caltech.) See color section.

Using design parameters consistent with the capabilities of the Proteus satellite planned to be used in the joint NASA/CNES Ocean Surface Topography Mission (OSTM), Rodriguez and Pollard showed that a sea surface height accuracy of 5 cm should be achievable with this WSOA instrument.

6-7 NONCONVENTIONAL RADAR SENSORS

Radar sensors have been used in a number of ways to acquire additional information about the surface of the Earth. As discussed earlier, in the case of the SAR, equi-Doppler and equirange surfaces correspond to a family of cones and a family of spheres. The intersection of these surfaces is a family of circles in the vertical plane. The intersection of a circle with the planetary surface gives a unique point (except for the left–right ambiguity). This assumes that the planet surface is well known. Otherwise, a third measurement is required to uniquely determine the location of a target. This can be achieved by using two antennas as an interferometer. In this case, a family of hyberboloid surfaces corresponds to points at which the differential phase between the two antennas is constant. The intersection of the above three sets of surfaces will uniquely identify each and every point in a three-dimensional space. Thus, each point is characterized by a specific set of (1) range to one antenna, (2) Doppler history, and (3) differential range between the two antennas. In actuality, because the phase has a 2π ambiguity, there is some ambiguity in one of the measurements, but this can be resolved by using few reference points.

Another type of radar sensor uses a bistatic configuration to study the surface properties. The bistatic scattering cross section $\sigma(\theta_1, \theta_2)$ provides additional information about the surface. This configuration was applied in some planetary studies by using a transmitter on an orbiting spacecraft and a receiver on the Earth. It was also used to image surface areas illuminated by a transmitter on an orbiting spacecraft with the receiver on an airborne platform.

6-8 SUBSURFACE SOUNDING

In some types of materials, the absorption loss of low-frequency electromagnetic waves is low, thus allowing deep penetration of the signal below the surface. A lossy medium is characterized by the penetration depth L_p, which gives the depth at which the signal power is attenuated by e^{-1}. Thus, the signal power at a depth d is

$$P(d) = P_0 e^{-d/L_p}$$

For a loss tangent of about 0.01, $\varepsilon' = 4$, and $\lambda = 25$ cm, then $L_p = 2$ m.

In very dry regions, the lack of moisture leads to very low values of $\tan \delta$ ($<10^{-2}$), thus allowing penetration of a few meters. Subsurface imaging was achieved from a spaceborne imaging radar (SIR-A) in 1981 of a large region in the Egyptian desert, where the soil moisture is less than 0.5% (see Fig. 6-89).

In the case of polar ice sheets, the loss tangent is less than 10^{-3}, leading to an absorption length at 60 MHz ($\lambda = 5$ m) of more than 400 m. This allowed the use of low-flying airborne radars to sound the Antarctic and Greenland ice sheets down to the bedrock, at depths of more than 4 km (Fig. 6-90).

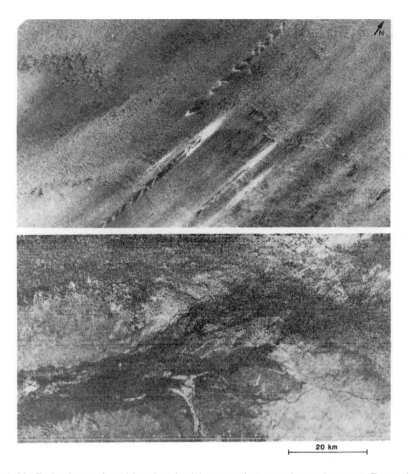

20 km

Figure 6-89. Radar (bottom) and Landsat (top) images of an area in southwestern Egypt clearly showing the penetration achieved with the radar sensor.

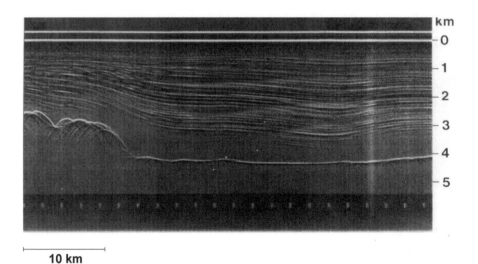

10 km

Figure 6-90. Example of electromagnetic sounding of the Antarctic ice sheet from an airborne sensor. (Courtesy of Technical University of Denmark.)

EXERCISES

6.1. Plot the scattering cross sections σ_{hh} and σ_{vv} as a function of the angle of incidence for a wavelength of $\lambda = 25$ cm for the cases of a Gaussian correlation function and an exponential correlation function. Assume in each case that the r.m.s. height of the surface is 1 cm, and that the correlation length is 10 cm. Assume that the surface dielectric constant is $\varepsilon = 3$. Use the small-perturbation model expressions.

6.2. Use the expressions for the small-perturbation model and plot the ratio σ_{hh}/σ_{vv} as a function of the dielectric constant for angles of incidence $\theta = 20°$, $40°$, and $60°$. Use the range 2–30 for the dielectric constant. Do you think this ratio can be used to measure the dielectric constant? For which dielectric constants will the measurement be most accurate?

6.3. Calculate and plot the spectrum for the following pulse shapes:

(a) $f(t) = \begin{cases} \cos(\omega_0 t) & \text{for } -\tau/2 \le t \le \tau/2 \\ 0 & \text{otherwise} \end{cases}$

(b) $f(t) = \begin{cases} \cos(\omega_0 t) & \text{for } nT - \tau/2 \le t\, nT + \le \tau/2; n = \ldots -2, -1, 0, 1, 2, \ldots \\ 0 & \text{otherwise} \end{cases}$

(c) $f(t) = \begin{cases} \cos(at^2 + \omega_0 t) & \text{for } -\tau/2 \le t\, nT + \le \tau/2 \\ 0 & \text{otherwise} \end{cases}$

6.4. Assume a scatterometer is in orbit at an altitude of 800 km moving at a speed of 7 km/sec. The operating frequency is 14.6 GHz, and the antenna has a length of 5 m and a width of 0.5 m.

(a) Plot the iso Doppler curves on the Earth that correspond to the Doppler frequencies of 0, 100 kHz, 200 kHz, 300 kHz, 400 kHz, and 500 kHz. Neglect the curvature of the Earth, and the fact that the Earth is rotating.

(b) What is the effect of the rotation of the Earth at the equator, at $45°$ latitude, and at the poles?

(c) Assuming that the fan beam has a $45°$ azimuth angle, what are the angles of incidence and the distance to the nadir line that correspond to the Doppler frequencies in part (a)? Neglect the rotation of the Earth.

(d) Plot the width of the antenna footprint as a function of the distance to the nadir line.

6.5. An orbiting SAR has the following characteristics:

- Frequency of 1.25 GHz
- Look angle is $45°$
- Swath width is 50 km
- Bandwidth is 40 MHz
- Receiver noise temperature is 500 K
- Orbit altitude is 200 km
- Peak power is 1 kW
- Pulse length is 30 microseconds

Calculate the other key characteristics required to define the sensor:

(a) Antenna width

(b) Maximum PRF

(c) Minimum antenna length, minimum PRF, and best along-track resolution

(d) Noise equivalent backscatter cross section

(e) Data rate assuming dual polarization (HH and VH) operation.

6.6. Plot the parameters in the previous exercise as a function of the look angle for angles between 10° and 60°

6.7. This exercise involves a simple top-level mission design. We want to fly a radar mission with the main objective of monitoring soil moisture on a global scale. To do this, we require an L-band radar system (24 cm wavelength) to map the entire globe twice every eight days at a resolution of 1 km × 1 km or better. We assume this means once each with ascending and descending passes, so we require an 8 day exact repeat orbit. The soil moisture algorithm requires images at HH, VV, and HV polarizations, and about 50 looks in an incidence angle range of 30° to 50°. Here "looks" will be created by adding pixels in the along-track direction. Calculate:

(a) The orbit altitude and swath width required, assuming that we map on both sides of the spacecraft at the same time in a polar orbit (i.e., inclination 98°)

(b) The antenna width, assuming that we will use the "scansar" approach to acquire data in five subswaths on each side of the spacecraft. This means we only need to illuminate one fifth of the swath at a time.

(c) The antenna length, PRF, and bandwidth required.

(d) The peak power required if we want the noise equivalent sigma zero to be less than −35 dB.

6.8. A thin cylinder of length l and radius a oriented at an angle α relative to the horizontal axis is characterized by the following scattering matrix:

$$[S] = \frac{k_0^2 l^3}{3[\ln(4l/a) - 1]} \begin{pmatrix} \cos^2 \alpha & \sin \alpha \cos \alpha \\ \sin \alpha \cos \alpha & \sin^2 \alpha \end{pmatrix}$$

Derive the expressions for the copolarized and cross-polarized power for linear polarizations. Next, derive expressions for the polarization orientation angles at which the co- and cross-polarized powers are a maximum and minimum, respectively. Plot the copolarized and cross-polarized returns as a function of the polarization orientation angle for linear polarizations if $\alpha = 45°$. Is it possible to measure the orientation of the cylinder using a polarimetric radar?

6.9. The average backscatter covariance matrix of a vegetated area with reflections symmetry can be written as

$$[\mathbf{T}] = \xi \begin{pmatrix} 1 & 0 & \rho \\ 0 & \eta & 0 \\ \rho^* & 0 & \zeta \end{pmatrix}$$

where $\xi = \langle S_{hh}S_{hh}^* \rangle$, $\rho = \langle S_{hh}S_{vv}^* \rangle / \langle S_{hh}S_{hh}^* \rangle$, $\eta = 2\langle S_{hv}S_{hv}^* \rangle / \langle S_{hh}S_{hh}^* \rangle$, and $\zeta = \langle S_{vv}S_{vv}^* \rangle / \langle S_{hh}S_{hh}^* \rangle$. For randomly oriented thin cylinders, these parameters are found to be ρ

= 1/3, $\eta = 2/3$, and $\zeta = 1$. Calculate the eigenvalues and entropy for the case of the randomly oriented thin cylinders.

The parameters for a forested area in Shasta Trinity National Forest in northern California measured with the NASA/JPL AIRSAR system are shown in Table 6-2.

(a) Calculate the three eigenvalues and the entropy for each frequency.

(b) Is it reasonable to assume that the scattering is mostly from the branches in the canopy?

(c) Are the branches thin compared to the radar wavelength in each case? Explain your answers.

6.10. You have been retained by a mapping company to do a quick assessment of the suitability of radar interferometry for floodplain mapping. The required height accuracy is 30 cm at a horizontal spacing of 1 m. They own their own corporate jet, which has a wingspan of 20 m that could be used for the baseline of the interferometer. They also believe that they can buy an X-band radar (3.2 cm wavelength) at a reasonable price, with the option of buying a Ka-band radar (0.8 cm wavelength) at a considerably higher price. Finally, they plan to set the data recording window such that they will acquire data between 25° and 60° angle of incidence.

(a) They would like to put GPS antennas at each wingtip to determine the position of each of the two antennas. According to their GPS expert, the position of each GPS antenna can be reconstructed to an accuracy of 1 cm. What would be the contribution to the height error at either frequency if they fly the jet at 5 km and 10 km altitudes above the terrain?

(b) What should the signal-to-noise ratio be for each radar if the phase noise contribution to the height error should be less than 10 cm?

(c) Would you recommend radar interferometry to them as a viable solution to their problem? If so, which radar should they buy?

6.11. A scatterometer operating at a center frequency of 13.4 GHz is flown on a spacecraft at an altitude of 800 km. In order to cover a wide continuous swath, a fixed-angle conical scanning approach is selected. The desired swath width is 1800 km. Calculate

(a) The off-nadir angle at which the antenna beam has to be pointed

(b) The ground resolution if the antenna is a dish with 1 m diameter

(c) The signal bandwidth if the compressed pulse must have a ground resolution of 5 km

(d) The required antenna spin rate in revolutions per minute

(e) The required pulse repetition frequency if we want to integrate 2 pulses for each antenna footprint

Table 6-2. Measured Values for a Forested Area in the Shasta Trinity National Forest

Parameter	P band	L band	C band
η	0.3301	0.3485	0.2416
ζ	0.6529	0.7122	0.4685
ρ	0.2803 + i0.0167	0.3950 + i0.0582	0.3669 + i0.0016

(f) The data rate assuming that the samples are digitized to 8 bits, and the data window is 2.5 milliseconds long. (This means we sample at 2.2 × bandwidth for 2.5 milliseconds for each pulse we transmit.)

6.12. A 37 GHz scanning radar altimeter is being developed for the Space Shuttle to operate at an altitude of 250 km with a swath width of 75 km. The required surface resolution is 150 meters and the height resolution is desired to be 5 meters. The PRF is 50 kHz. Calculate:

(a) Antenna size along and across track

(d) Scanning rate

(c) The peak power required, assuming a surface backscatter of –17 dB and an SNR of 10 dB

REFERENCES AND FURTHER READING

Allan, T. D. *Satellite Microwave Remote Sensing.* Halsted Press, New York, 1983.

Amelung, F., D. L. Galloway, J. W. Bell, H. A. Zebker, and R. J. Laczniak. Sensing the ups and downs of Las Vegas: InSAR reveals structural control of land subsidence and aquifer-system deformation. *Geology,* **27,** 6, 483–486, 1999.

Amelung, F., S. Jonsson, H. A. Zebker, and P. Segall. Widespread uplift and "trapdoor" faulting on Galapagos volcanoes observed with radar interferometry. *Nature,* **407,** 6807, 993–996, 2000a.

Amelung, F., C. Oppenheimer, P. Segall, and H. A. Zebker, Ground deformation near Gada 'Ale volcano, Afar, observed by radar interferometry. *Geophysical Research Letters,* **27,** 19, 3093–3096, 2000b.

Bamler, R., and M. Eineder. ScanSAR processing using standard high-precision SAR processing algorithms. *IEEE Transactions on Geoscience and Remote Sensing,* **34,** 212–218, 1996.

Barrick, D. E., and C. T. Swift. The Seasat microwave instruments in historical perspective. *IEEE Journal of Oceanic Engineering,* **OE-5,** 74–80, 1980.

Beaudoin, A., T. LeToan, S. Goze, A. Nezry, and A. Lopez. Retrieval of Forest Biomass form SAR Data. *International Journal of Remote Sensing,* **15,** 2777–2796, 1994.

Bindschadler, R., P. Nornberger, D. Blankenship, T. Scambos, and R. Jacobel. Surface velocity and mass balance of ice streams D and E, West Antarctica. *Journal of Glaciology,* **42,** 461–475, 1996.

Blom, R., and C. Elachi. Spaceborne and airborne imaging radar observation of sand dunes. *Journal of Geophysics Research,* **86,** 3061–3070, 1981.

Blom, R., and M. Daily. Radar image processing for rock type discrimination. *IEEE Transactions on Geoscience and Remote Sensing,* **20,** 343–351, 1982.

Blom, R. G., R. J. Crippen, and C. Elachi. Detection of subsurface features in Seasat radar images of Means Valley, Mojave Desert, California. *Geology,* **12,** 346, 1983.

Born, G., J. A. Dunne, and D. B. Lowe. SEASAT mission overview. *Science,* **204,** 1405–1406, 1979.

Bracalante, E. M., et al. The SASS scattering coefficient algorithm. *IEEE Journal of Oceanic Engingeering,* **OE-5,** 166–156, 1980.

Brown, G. S. The average impulse response of a rough surface and its applications. *IEEE Transactions on Antennas and Propagation,* **AP-25,** pp. 67–74, 1977.

Brown, W. E., Jr., C. Elachi, and T. W. Thompson. Radar imaging of ocean surface patterns. *Journal of Geophysics Research,* **31,** 2657–2667, 1976.

Brown, W. M., G. G. Houser, and R. G. Jenkins. Synthetic aperture processing with limited storage and presuming. *IEEE Transactions on Aerosp. Electron. Syst.,* **AES-9,** 166–176, 1973.

Brown, W. M., and L. J. Porcello. An introduction to synthetic aperture radar. *IEEE Spectrum,* **6,** 52–62, 1969.

Carande, R. E. Estimating Ocean Coherence Time Using Dual-Baseline Interferometric Synthetic Aperture Radar. *IEEE Transactions on Geoscience and Remote Sensing,* **32,** 846–854, 1994

Chauhan, N., and R. Lang. Radar Modeling of a Boreal Forest. *IEEE Transactions on Geoscience and Remote Sensing,* **29,** 627–638, 1991.

Chelton, D. B., E. J. Walsh, and J. L. MacArthur. Pulse compression and sea level tracking in satellite altimetry. *Journal of Atmospheric Oceanic Technology,* **6,** 407–438, 1989.

Chi, C. Y., and F. K. Li. A comparative study of several wind estimation algorithms for spaceborne scatterometers. *IEEE Transactions on Geoscience and Remote Sensing,* **26,** 115–121, 1988.

Chuah, T. S. Design and development of a coherent radar depth sounder for measurement of Greenland ice sheet thickness. Radar Systems and Remote Sensing Laboratory RSL Technical Report 10470-5, University of Kansas Center for Research, Inc., 1997.

Cloude, S.R., and K. P. Papathanassiou. Polarimetric SAR Interferometry. *IEEE Transactions on Geoscience and Remote Sensing,* **36,** 1551–1565, 1998.

Colwell, R. N. (Ed). *Manual of Remote Sensing,* Chapters 9, 10, and 13. American Society of Photogrammetry, Falls Church, VA, 1983.

Cook, C., and M. Bernfeld. *Radar Signals: An Introduction to Theory and Application.* Academic Press, New York, 1967.

Curlander, J. C. Location of spaceborne SAR imagery. *IEEE Transactions on Geoscience and Remote Sensing,* **20,** 359–364, 1982.

Curlander, J. C., and R. N. McDonough. *Synthetic Aperture Radar Systems and Signal Processing.* Wiley, New York, 1991.

Cutrona, L. J. Synthetic aperture radar. In M. I. Skolnik (Ed.). *Radar Handbook.* McGraw-Hill, New York, 1970.

Daniels, D. J., D. J. Gunton, and H. F. Scott. Introduction to subsurface radar. *IEE Proc.,* **135 F,** 278–320, 1988.

Dobson, M. C., F. T. Ulaby, M. T. Hallikainen, and M. A. El-Rayes. Microwave dielectric behavior of wet soil—Part II: Dielectric mixing models. *IEEE Transactions on Geoscience and Remote Sensing,* **23,** 35–46, 1985.

Dobson, M. C., F. T. Ulaby, T. LeToan, A. Beaudoin, E. S. Kasischke, and N. Christensen. Dependence of Radar Backscatter on Conifer Forest Biomass. *IEEE Transactions on Geoscience and Remote Sensing,* **30,** 412–415, 1992.

Dobson, M. C., F. T. Ulaby, L. E. Pierce, T. L. Sharik, K. M. Bergen, J. Kellndorfer, J. R. Kendra, E. Li, Y. C. Lin, A. Nashashibi, K. Sarabandi, and P. Siqueira. Estimation of Forest Biophysical Characteristics in Northern Michigan with SIR-C/X-SAR. *IEEE Transactions on Geoscience and Remote Sensing,* **33,** 877–895, 1995.

Donelan, M. A., and W. J. Pierson. Radar scattering and equilibrium ranges in wind-generated waves with application to scatterometry. *Journal of Geophysics Research,* **92,** 4971–5029, 1987.

Dubois, P. C., J. J. van Zyl, and T. Engman. Measuring Soil Moisture with Imaging Radars. *IEEE Transactions on Geoscience and Remote Sensing,* **33,** 915–926, 1995.

Durden, S. L., J. J. van Zyl, and H. A. Zebker. Modeling and Observation of the Radar Polarization Signature of Forested Areas. *IEEE Transactions on Geoscience and Remote Sensing,* **27,** 290–301, 1989.

Durden, S. L., J. J. van Zyl, and H. A. Zebker. The unpolarized component in polarimetric radar observations of forested areas. *IEEE Transactions on Geoscience and Remote Sensing,* **28,** 268–271, 1990.

Elachi, C. Wave patterns across the North Atlantic on September 28, 1974 from airborne radar imagery. *Journal of Geophysics Research,* **81,** 2655–2656, 1976.

Elachi, C. Radar imaging of the ocean surface. *Boundary Layer Meteorology,* **13,** 154–173, 1977.

Elachi, C. Spaceborne imaging radar: Geologic and oceanographic applications. *Science,* **209,** 1073–1082, 1980.

Elachi, C. Radar images of the Earth from space. *Scientific American,* December, 1982.

Elachi, C., and J. Apel. Internal wave observations made with an airborne synthetic aperture imaging radar. *Geophysics Research Letters,* **3,** 647–650, 1976.

Elachi, C., and W. E. Brown. Models of radar imaging of the ocean surface waves. *IEEE Transactions on Antennas and Propagation,* **AP-25,** 84–95, Jan. 1977; also *IEEE Journal of Oceanic Engingeering,* **OE-2,** 84–95, January 1977.

Elachi, C., and N. Engheta. Radar scattering from a diffuse vegetation layer. *IEEE Transactions on Geoscience and Remote Sensing,* **20,** 212, 1982.

Elachi, C., T. Bicknell, R. L. Jordan, and C. Wu. Spaceborne imaging synthetic aperture radars: Technique, Technology and Applications. *Proc. IEEE,* **70,** 1174–1209, 1982.

Elachi, C., L. Roth, and G. Schaber. Spaceborne radar subsurface imaging in hyperarid regions. *IEEE Transactions on Geoscience and Remote Sensing,* **22,** p. 383, 1984 (Note: In this article, the SIR-A and Landsat images in Fig. 2 were mistakenly interchanged.)

Elachi, C., et al. Shuttle Imaging Radar experiment. *Science,* **218,** 996, 1982.

Elachi, C. *Spaceborne Radar Remote Sensing: Applications and Techniques.* IEEE Press, New York, 1988.

Elachi, C., E. Im, L. E. Roth, and C. L. Werner. Cassini Titan Radar Mapper. *Proc. IEEE,* **79,** 867–880, 1991.

Faller, N. P., and E. H. Meier. First Results with the Airboren Single-Pass DO-SAR Interferoemeter. *IEEE Transactions on Geoscience and Remote Sensing,* **33,** 1230–1237, 1995.

Farr, T. G. Microtopographic evolution of lava flows at Cima volcanic field, Mojave Desert, California. *Journal of Geophysics Research,* **97,** p. 15171–15179, 1992.

Farr, T. G. and M. Kobrick. Shuttle radar topography mission produces a wealth of data. *Eos,* **81,** 583–585, 2000.

Ford, J. Seasat orbital radar imagery for geologic mapping: Tennessee–Kentucky–Virginia. *American Association of Petroleum Geology Bulletin,* **66,** 2064–2070, 1981.

Ford, J. P. Resolution versus speckle relative to geologic interpretability of spaceborne radar images: A survey of user preferences. *IEEE Transactions on Geoscience and Remote Sensing,* **20,** 434–444, 1982.

Freeman, A., Y. Shen, and C. L. Werner. Polarimetric SAR Calibration Experiment using Active Radar Calibrators. *IEEE Transactions on Geoscience and Remote Sensing,* **28,** 224–240, 1990.

Freeman, A., J. J. van Zyl, J. D. Klein, H. A. Zebker, and Y. Shen. Calibration of Stokes and scattering matrix format polarimetric SAR data. *IEEE Transactions on Geoscience and Remote Sensing,* **30,** 531–539, 1992.

Freeman, A. SAR calibration: An overview. *IEEE Transactions on Geoscience and Remote Sensing,* **30,** 1107–1121, 1992.

Freeman, A., M. Alves, B. Chapman, J. Cruz, Y. Kim, S. Shaffer, J. Sun, E. Turner, and K. Sarabandi. SIR-C data quality and calibration results. *IEEE Transactions on Geoscience and Remote Sensing,* **33,** 848–857, 1995.

Freeman, A., W. T. K. Johnson, B. Huneycutt, R. Jordan, S. Hensley, P. Siqueira, and J. Curlander. The "Myth" of the minimum SAR antenna area constraint. *IEEE Transactions on Geoscience and Remote Sensing,* **38,** 320–324, 2000.

Fu, L-L., and A. Cazenave. *Satellite Altimetry and Earth Sciences: A Handbook of Techniques and Applications.* Academic Press, San Diego, 2001.

Fujiwara, S., P. A. Rosen, M. Tobito, and M. Murakami. Crustal deformation measurements using repeat-pass JERS-1 synthetic aperture radar interferometry near the Izu Peninsula, Japan. *Journal of Geophysical Research,* **103,** 2411-2426, 1998.

Fung, A. K., Z. Li, and K. S. Chen. Backscattering from a randomly rough dielectric surface. *IEEE Transactions on Geoscience and Remote Sensing,* **30,** 356–369, 1992.

Gabriel, A. K., and R. M. Goldstein. Crossed orbit interferometry: Theory and experimental results from SIR-B. *International Journal of Remote Sensing,* **9,** 857–872, 1988.

Gabriel, A. K., R. M. Goldstein, and H. A. Zebker. Mapping small elevation changes over large areas: Differential radar interferometry. *Journal of Geophysics Research,* **94,** 9183–9191, 1989.

Gatelli, F., A. Monti-Guarnieri, F. Parizzi, P. Pasquali, C. Prati, and F. Rocca. The wavenumber shift in SAR Interferometry. *IEEE Transactions on Geoscience and Remote Sensing,* **32,** 855–865, 1994.

Ghiglia, D. C., and L. A. Romero. Direct phase estimation from phase differences using fast elliptic partial differential equation solvers. *Optics Letters,* **15,** 1107–1109, 1989.

Goldstein, R. M., and H. A. Zebker. Interferometric radar measurements of ocean surface currents. *Nature,* **328,** 707–709, 1987.

Goldstein, R. M., H. A. Zebker, and C. Werner. Satellite radar interferometry: Two-dimensional phase unwrapping. *Radio Science,* **23,** 713–720, 1988.

Goldstein, R. M., T. P. Barnett, and H. A. Zebker. Remote snsing of ocean currents. *Science,* **246,** 1282–1285, 1989.

Goldstein, R. M., H. Englehardt, B. Kamb, and R. M. Frolich. Satellite radar interferometry for monitoring ice sheet motion: Application to an Antarctic ice stream. *Science,* **262,** 1525–1530, 1993.

Goldstein, R. M. Atmospheric limitations to repeat-track radar interferometry. *Geophysics Research Letters,* **22,** 2517–1520, 1995.

Graham, L. C. Synthetic Interferometer Radar for Topographic Mapping. *Proc. IEEE,* **62,** 763–768, 1974.

Gray, A. L., P. W. Vachon, C. E. Livingstone, and T. I. Lukowski. Synthetic aperture radar calibration using reference reflectors. *IEEE Transactions on Geoscience and Remote Sensing,* **28,** 374–383, 1990.

Gray, A. L., and P. J. Farris-Manning. Repeat-pass interferometry with airborne synthetic aperture radar. *IEEE Transactions on Geoscience and Remote Sensing,* **31,** 180–191, 1993.

Harger, R. O. *Synthetic Aperture Radar Systems: Theory and Design.* Academic Press, New York, 1970.

Hess, L. L., J. M. Melack, S. Filoso, and Y. Wang. Delineation of inundated area and vegetation along the amazon floodplain with the SIR-C synthetic aperture radar. *IEEE Transactions on Geoscience and Remote Sensing,* **33,** 896–904, 1995.

Hibbs, A. R., and W. S. Wilson. Satellites map the oceans. *IEEE Spectrum,* **20,** 46–53, Oct. 1983.

Hsu, C. C., H. C. Han, R. T. Shin, J. A. Kong, A. Beaudoin, and T. LeToan. radiative transfer theory for polarimetric remote sensing of pine forest at P-band. *International Journal of Remote Sensing,* **14,** 2943–2954, 1994.

Imhoff, M. L. Radar backscatter and biomass saturation: Ramifications for global biomass inventory. *IEEE Transactions on Geoscience and Remote Sensing,* **33,** 511–518, 1995.

Johnson, J. W., et al. Seasat—A satellite scatterometer instrument evaluation. *IEEE Journal of Oceanic Engineering,* **OE-5,** 138–144, 1980.

Johnson, W. T. K. Magellan Imaging Radar Mission to Venus. *Proc. IEEE,* **79,** 777–790, 1991.

Jones, L. et al. Aircraft measurements of the microwave scattering signature of the ocean. *IEEE Transactions on Antennas and Propagation,* **AP-25,** 52–60, 1977.

Jordan, R. L. The Seasat—A synthetic aperture radar. *IEEE Journal of Oceanic Engineering,* **OE-5,** 154–163, 1980.

Jordan, R. L., B. L. Huneycutt, and M. Werner. The SIR-C/X-SAR synthetic aperture radar system. *IEEE Transactions on Geoscience and Remote Sensing,* **33,** 829–839, 1995.

Joughin, I. R. Estimation of ice-sheet topography and motion using interferometric synthetic aperture radar. Ph.D. Thesis, University of Washington, Seattle, 1995.

Joughin, I. R., D. P. Winebrenner, and M. A. Fahnestock. Observations of ice-sheet motion in Greenland using satellite radar interferometry. *Geophysics Research Letters,* 22, 571–574, 1995.

Joughin, I., D. Winebrenner, M. Fahnestock, R. Kwok, and W. Krabill. Measurement of ice sheet topography using satellite-redar interferometry. *Journal of Glaciology,* **42**, 10–22, 1996a.

Joughin, I., S. Tulaczyk, M. Fahnestock, and R. Kwok. A mini-surge on the Ryder Glacier, Greenland, observed by satellite radar interferometry. *Science,* **274**, 228–230, 1996b.

Joughin, I., R. Kwok, and M. Fahnestock. Estimation of ice sheet motion using satellite radar interferometry: Method and error analysis with application to Humboldt Glacier, Greenland. *Journal of Glaciology,* **42**, 564–575, 1996c.

Kasischke, E. S., N. L. Christensen, and L. L. Bourgeau-Chavez. Correlating radar backscatter with components of biomass in loblolly pine forests. *IEEE Transactions on Geoscience and Remote Sensing,* **33**, 643–659, 1995.

Kirk, J. C. A discussion of digital processing for synthetic aperture radar. *IEEE Transactions on Aerospace Electronics Systems,* **AES-11,** 326–337, May 1975.

Klein, J. D., and A. Freeman. Quadpolarisation SAR calibration using target reciprocity. *Journal of Electromagnetic Waves Applications,* 5, 735–751, 1991.

Kong, J. A., A. A. Swartz, H. A. Yueh, L. M. Novak, and R. T. Shin. Identification of Earth terrain cover using the optimum polarimetric classifier. *Journal of Electromagnetic Waves Applications,* 2, 171–194, 1988.

Kozma, A., E. M. Leith, and N. G. Massey. Tilted plane optical processor. *Applied Optics,* **11,** 1766–1777, 1972.

Kovaly, J. J. *Synthetic Aperture Radar.* Artech House, Dedham, MA, 1976.

Lang, R. H., and J. S. Sidhu. Electromagnetic backscattering from a layer of vegetation. *IEEE Transactions on Geoscience and Remote Sensing,* **21,** 62–71, 1983.

Lang, R. H., N. S. Chauhan, J. K. Ranson, and O. Kilic. Modeling P-band SAR returns from a red pine stand. *Remote Sensing of Environment,* 47, 132–141, 1994.

Larson, T. R., L. I. Moskowitz, and J. W. Wright. A note on SAR imagery of the ocean. *IEEE Transactions on Antennas and Propagation,* **AP-24,** 393–394, May 1976.

Leberl, F., M. L. Bryan, C. Elachi, T. Farr, and W. Campbell. Mapping of sea ice measurement of its drift using aircraft synthetic aperture radar images. *Journal of Geophysics Research,* **84,** 1827–1835, Apr. 20, 1979.

Lee, Jong-Sen, M. R. Grunes, E. Pottier, and L. Ferro-Famil. Unsupervised terrain classification preserving polarimetric scattering characteristics. *IEEE Transactions on Geoscience and Remote Sensing,* **42**, 722–731, 2004.

Leith, E. N. Complex spatial filters for image deconvolution. *Proc. IEEE,* **65,** 18–28, 1977.

Leith, E. N. Quasi-holographic techniques in the microwave region. *Proc. IEEE,* **59,** 1305–1318, Sept. 1971.

LeToan, T., A. Beaudoin, J. Riom, and D. Guyon. Relating forest biomass to SAR data. *IEEE Transactions on Geoscience and Remote Sensing,* **30,** 403–411, 1992.

Li, F. K., and R. M. Goldstein. Studies of multibaseline spaceborne interferometric synthetic aperture radars. *IEEE Transactions on Geoscience and Remote Sensing,* **28,** 88–97, 1990.

Lim, H. H., A. A. Swartz, H. A. Yueh, J. A. Kong, R. T. Shin, and J. J. van Zyl. Classification of Earth terrain using polarimetric synthetic aperture radar images. *Journal of Geophysics Research,* **94,** 7049–7057, 1989.

Long, D. G., P. J. Hardin, and P. T. Whiting. Resolution enhancement of spaceborne scatterometer data. *IEEE Transactions on Geoscience and Remote Sensing,* **31,** 700–715, May 1993.

MacDonald, H. C. Geologic evaluation of radar imagery from Darien Province, Panama. *Modern Geology,* **1,** 1–63, 1969.

Madsen, S. N., H. A. Zebker, and J. Martin. Topographic mapping using radar intereferometry: processing techniques. *IEEE Transactions on Geoscience and Remote Sensing,* **31,** 246–256, 1993.

Marom, M., R. M. Goldstein, E. B. Thronton, and L. Shemer. Remote sensing of ocean wave spectra by interferometric synthetic aperture radar. *Nature,* 345, 793–795, 1990.

Marom, M., L. Shemer, and E. B. Thronton. Energy density directional spectra of nearshore wave field measured by interferometric synthetic aperture radar. *Journal of Geophysics Research,* 96, 22125–22134, 1991.

Massonnet, D., M. Rossi, C. Carmona, F. Adragna, G. Peltzer, K. Freigl, and T. Rabaute. The displacement field of the Landers earthquake mapped by radar interferometry. *Nature,* **364,** 138–142, 1993.

Massonnet, D., K. Freigl, M. Rossi, and F. Adragna. Radar interferometric mapping of deformation in the year after the landers earthquake. *Nature,* 369, 227–230, 1994.

Massonnet, D., P. Briole, and A. Arnaud. Deflation of Mount Etna monitored by spaceborne radar interferometry. *Nature,* 375, 567–570, 1995.

Massonnet, D., and K. Freigl. Satellite radar interferometric map of the coseismic deformation field of the M=6.1 Eureka Valley, California Earthquake of May 17,1993. *Geophysics Research Letters,* 22, 1541–1544, 1995.

Masursky, H., et al. Pioneer Venus radar results. *JGR,* **85,** 8232–8260, 1980.

McCauley, J., et al. Subsurface valleys and geoarcheology of the eastern Sahara revealed by Shuttle Radar. *Science,* **218,** 1004, 1982.

Moccia, A., and S. Vetrella. A Tethered interferometric synthetic aperture radar (SAR) for a topographic mission. *IEEE Transactions on Geoscience and Remote Sensing,* **31,** 103–109, 1992.

Moreira, J., M. Schwabish, G. Fornaro, R. Lanari, R. Bamler, D. Just, U. Steinbrecher, H. Breit, M. Eineder, G. Franceschetti, D. Guedtner, and H. Rinkel. X-SAR interferometry: First results. *IEEE Transactions on Geoscience and Remote Sensing,* **33,** 950–956, 1995.

Naderi, F. M., M. H. Freilich, and D. G. Long. Spaceborne radar measurement of wind velocity over the ocean—An overview of the NSCAT scatterometer system. *Proc. IEEE,* **79,** 850–866, 1991.

Oh, Y., K. Sarabandi, and F. T. Ulaby. An Empirical model and an inversion technique for radar scattering from bare soil surfaces. *IEEE Transactions on Geoscience and Remote Sensing,* **30,** 370–381, 1992.

Okada, Y. Surface deformation due to shear and tensile faults in a half space. *Bulletin of the Seismological Society of America,* **75,** 1135–1154, 1985.

Oliver, C., and S. Quegan. *Understanding Synthetic Aperture Radar Images.* Artech House, Norwood, MA, 1998.

Peake, W. H. Interaction of electromagnetic waves with some natural surfaces. *IRE Transactions on Antennas and Propagation,* **AP-7,** 5324–329, 1959.

Peake, W. H., and T. L. Oliver. The response of terrestrial surfaces at microwave frequencies. Columbus, Ohio: Ohio State University, Electroscience Laboratory, 2440-7, Technical Report AFAL-TR-70301, 1971.

Peltzer, G., K. Hudnut, and K. Feigl. Analysis of coseismic displacemnt gradients using radar interferometry: new insights into the Landers earthquake. *Journal of Geophysics Research,* 99, 21971–21981, 1994.

Peltzer, G., and P. Rosen. Surface displacement of the 17 May 1993 Eureka Valley, California, earthquake observed by SAR interferometry. *Science,* 268, 1333–1336, 1995.

Peltzer, G., P. Rosen, F. Rogez, and K. Hudnut. Postseismic rebound in falt step-overs caused by pore fluid flow. *Science,* **273,** 1202–1204, 1996.

Peltzer, G., P. Rosen, F. Rogez, and K. Hudnut. Crustal deformation in southern California using SAR interferometry. *ESA SP-414,* 545–548, 1997.

Peltzer, G., F. Crampé, and G. King. Evidence of nonlinear elasticity of the crust from the Mw7.6 Manyi (Tibet) Earthquake. *Science,* **286,** 272–276, 1999.

Peplinski, N. R., F. T. Ulaby, and M. C. Dobson. Dielectric properties of soils in the 0.3–1.3-GHz range. *IEEE Transactions on Geoscience and Remote Sensing,* **GE-33,** 803–807, 1995.

Pettengill, G., et al. Pioneer Venus radar results: Altimetry and surface properties. *Journal of Geophysics Research,* **85,** 8261, 1980.

Pierce, L. E., F. T. Ulaby, K. Sarabandi, and M. C. Dobson. Knowledge-based classification of polarimetric SAR images. *IEEE Transactions on Geoscience and Remote Sensing,* **32,** 1081–1086, 1994.

Plant, W. J. The variance of the normalized radar cross section of the sea. *Journal of Geophysics Research,* **96,** 20643–20654, 1991.

Porcello, L. J., R. L. Jordan, J. S. Zelenka, G. F. Adams, R. J. Phillips, W. E. Brown, S. H. Ward, and P. L. Jackson. The Apollo lunar sounder radar system. *Proc. IEEE,* **62,** 769–783, 1974.

Prati, C., and F. Rocca. Limits to the resolution of elevation maps from stereo SAR images. *International Journal of Remote Sensing,* 11, 2215–2235, 1990.

Prati, C., F. Rocca, A. Moni Guarnieri, and E. Damonti. Seismic migration for SAR focussing: Interferometrical applications. *IEEE Transactions on Geoscience and Remote Sensing,* **28,** 627–640, 1990.

Prati, C., and F. Rocca. Improving slant range resolution with multiple SAR surveys. *IEEE Transactions on Aerospace and Electronics Systems,* 29, 135–144, 1993.

Raney, R. K. Synthetic aperture imaging radar and moving targets. *IEEE Transactions on Aerospace and Electronics Systems,* **AES-7,** (3), 499–505, May 1971.

Raney R. K., and D. L. Porter. WITTEX: An innovative three-satellite radar altimeter concept. *IEEE Transactions on Geoscience Remote Sensing,* **39,** 2387–2391, 2001.

Ranson, K. J., and G. Sun. Mapping biomass of a northern forest using multifrequency SAR data. *IEEE Transactions on Geoscience and Remote Sensing,* **32,** 388–396, 1994.

Ranson, K. J., S. Saatchi, and G. Sun. Boreal forest ecosystem characterization with SIR-C/XSAR. *IEEE Transactions on Geoscience and Remote Sensing,* **33,** 867–876, 1995.

Rice, S. O. Reflection of electromagnetic waves by slightly rough surfaces. In Kline, M. (Ed.), *The Theory of Electromagnetic Waves,* Wiley, New York, 1951.

Richards, J. A., G. Sun, and D. S. Simonett. L-band radar backscatter modeling of forest stands. *IEEE Transactions on Geoscience and Remote Sensing,* **25,** 487–498, 1987.

Rignot, E. Tidal motion, ice velocity and melt rate of Peterman Gletscher, Greenland, measured from radar interferometry. *Journal of Glaciology,* **42,** 476–485, 1996.

Rignot, E. J. Fast recession of a west Antarctic glacier. *Science,* **281,** 549–551, 1998.

Rignot, E., and R. Chellappa. Segmentation of polarimetric synthetic aperture radar data. *IEEE Transactions on Image Processing,* **1,** 281–300, 1992.

Rignot, E., and R. Chellappa. Maximum *a-posteriori* classification of multifrequency, multilook, synthetic aperture radar intensity data. *Journal of Optics Society of America A,* **10,** 573–582, 1993.

Rignot, E. J. M., C. L. Williams, J. Way, and L. Viereck. Mapping of forest types in Alaskan boreal forests using SAR imagery. *IEEE Transactions on Geoscience and Remote Sensing,* **32,** 1051–1059, 1994.

Rignot, E., J. Way, C. Williams, and L. Viereck. Radar estimates of aboveground biomass in boreal forests of interior Alaska. *IEEE Transactions on Geoscience and Remote Sensing,* **32,** 1117–1124, 1994.

Rignot, E. J., R. Zimmerman, and J. J. van Zyl. Spaceborne applications of P-band imaging radars for measuring forest biomass. *IEEE Transactions on Geoscience and Remote Sensing,* **33,** 1162–1169, 1995a.

Rignot, E., K. Jezek, and H.G. Sohn. Ice flow dynamics of the Greenland Ice Sheet from SAR interferometry. *Geophysics Research Letters,* **22,** 575–578, 1995.

Rignot, E. J., S. P. Gogineni, W. B. Krabill, and S. Ekholm. North and northeast Greenland ice discharge from satellite radar interferometry. *Science,* **276,** 934–937, 1997.

Rihaczek, A. W. *Principles of High-Resolution Radar.* McGraw-Hill, New York, 1969.

Rodriguez, E. Altimetry for non-Gaussian oceans: Height biases and estimation of parameters. *Journal of Geophysics Research,* **93,** 14107–14120, 1988.

Rodriguez, E., and B. Pollard. The measurement capabilities of wide-swath ocean altimeters. In: D. B. Chelton (Ed.), R*eport of the High-Resolution Ocean Topography Science Working Group Meeting,* pp. 190–215, ref. 2001-4, Oregon State University, Corvallis, OR, 2001.

Rosen, P. A., S. Hensley, H. A. Zebker, F. H. Webb, and E. J. Fielding. Surface deformation and coherence measurements of Kilauea volcano, Hawaii, from SIR-C radar interferometry. *Journal of Geophysical Research,* **101,** 23109–23125, 1996.

Rosen, P. A., C. Werner, E. Fielding, S. Hensley, S. Buckley, and P. Vincent. Aseismic creep along the San Andreas fault nortwest of Parkfield, CA measured by radar interferometry. *Geophysical Research Letters,* **25,** 825–828. 1998.

Rosen, P. A., S. Hensley, I. R. Joughin, F. K. Li, S. N. Madsen, E. Rodríguez, and R. M. Goldstein. Synthetic aperture radar interferometry. *Proceedings of the IEEE,* **88,** 333–382, 2000.

Rufenach, C. L., and W. R. Alpers. Imaging ocean waves by SAR with long integration times. *IEEE Transactions on Antennas and Propagation,* **AP-29,** 422–428, 1981.

Rufenach, C. L., J. J. Bates, and S. Tosini. ERS-1 scatterometer measurements—Part I: The relationship between radar cross section and buoy wind in two oceanic regions. *IEEE Transactions on Geoscience and Remote Sensing,* **36,** 603–622, 1998.

Rufenach, C. L. ERS-1 scatterometer measurements—Part I: An algorithm for ocean-surface wind retrieval including light winds. *IEEE Transactions on Geoscience and Remote Sensing,* **36,** 623–635, 1998.

Sabins, F. F. *Remote Sensing: Principles and Interpretation.* Freeman, San Francisco, 1978.

Sabins, F. F., R. Blom, and E. Elachi. Seasat radar image of the San Andreas Fault, California. *American Association of Petroleum Geologists,* **64,** 614, 1980.

Saunders, R. S. Questions for the geologic exploration of Venus. *Journal of British Interplanetary Society,* **37,** 435, 1984.

Schaber, G. G., C. Elachi, and T. Farr. Remote sensing data of SP lava flow and vicinity in North Central Arizona. *Remote Sensing of the Environment,* **9,** 169, 1980.

Shanmugan, K. S., V. Narayanan, V. S. Frost, J. A. Stiles, and J. C. Holtzman. Textural features for radar image analysis. *IEEE Transactions on Geoscience and Remote Sensing,* **GE-19,** 153–156, 1981.

Shemer, L., and M. Marom. Estimates of ocean coherence time by interferometric SAR. *International Journal of Remote Sensing,* 14, 3021–3029, 1993.

Shi, J., and J. Dozier. Inferring snow wetness using C-band data from SIR-C's polarimetric synthetic aperture radar. *IEEE Transactions on Geoscience and Remote Sensing,* **33,** 905–914, 1995.

Shi, J., J. Wang, A. Hsu, P. O'Neill, and E. T. Engman. Estimation of bare surface soil moisture and surface roughness parameters using L-band SAR image data. *IEEE Transactions on Geoscience and Remote Sensing,* **35,** 1254–1266, 1997.

Spencer, M. W., W. Tsai, and D. G. Long. High resolution measurements with a spaceborne pencil-beam scatterometer using combined range/doppler discrimination techniques. *IEEE Transactions on Geoscience Remote Sensing,* **41,** 567–581, 2003.

Stiles, W. H., and F. T. Ulaby. The active and passive microwave response to snow parameters, Part I: Wetness. *Journal of Geophysics Research,* **85,** 1037–1044, 1980.

Stimson, G. W. *Introduction to Airborne Radar.* Hughes Aircraft, El Segundo, CA, 1983.

Stofan, E. R., D. L. Evans, C. Schmullius, B. Holt, J. J. Plaut, J. van Zyl, S. D. Wall, and J. Way. Overview of results of spaceborne imaging radar-C, X-band synthetic aperture radar (SIR-C/X-SAR). *IEEE Transactions on Geoscience and Remote Sensing,* **23,** 817–828, 1995.

Sun, G., D. S. Simonett, and A. H. Strahler. A radar backscatter model for discontinuous coniferous forest canopies. *IEEE Transactions on Geoscience and Remote Sensing,* **29,** 639–650, 1991.

Thompson, D. R., and J. R. Jensen. Synthetic aperture radar interferometry applied to ship-generated internal waves in the 1989 Loch Linnhe experiment. *Journal of Geophysics Research,* **98,** 10259–10269, 1993.

Tobita, M., S. Fujiwara, S. Ozawa, P. A. Rosen, E. J. Fielding, F. L. Werner, M. Murakami, H. Nakagawa, K. Nitta, and M. Murakami. Deformation of the 1995 North Sakhalin eartquake detected by JERS-1/SAR interferometry. *Earth Planets Space,* **50,** 313–325, 1998.

Tomiyasu, K. Conceptual performance of a satellite borne, wide swath SAR. *IEEE Transactions on Geoscience and Remote Sensing,* **19,** 108–116, 1981.

Tomiyasu, K. Tutorial review of synthetic aperture radar with application to imaging of the ocean surface. *Proc IEEE,* **66,** 563–583, 1978.

Townsend, W. F. An initial assessment of the performance achieved with the Seasat radar altimeter. *IEEE Journal of Oceanic Engingeering,* **OE-5,** 80–92, 1980.

Ulaby, F. T. Radar signatures of terrain: Useful monitors of renewable resources. *Proc. IEEE,* **70,** (12), 1410–1428, 1982.

Ulaby, F. T., and K. R. Carver. Passive microwave remote sensing. *Manual of Remote Sensing,* vol. 1, pp. 475–516. American Society of Photogrammetry, Falls Church, VA, 1983.

Ulaby, F. T., R. K. Moore, and A. K. Fung. *Microwave Remote Sensing: Active and Passive,* volume II: Radar Remote Sensing and Surface Scattering and Emission Theory. Artech House, Dedham, MA, 1982.

Ulaby, F. T., R. K. Moore, and A. K. Fung. *Microwave Remote Sensing.* vol. 3. Artech House, Dedham, MA, 1985.

Ulaby, F. T., and W. H. Stiles. The active and passive microwave response to snow parameters, Part II: Water equivalent of dry snow. *Journal of Geophysics Research,* **85,** 1045–1049, 1980.

Ulaby, F. T., and C. E. Elachi (Eds.). *Radar Polarimetry for Geoscience Applications.* Artech House, Norwood, MA, 1990.

Ulaby, F. T., K. Sarabandi, K. McDonald, M. Whitt, and M. C. Dobson. Michigan microwave canopy sacttering model. *International Journal of Remote Sensing,* **11,** 1223–1253, 1990.

Valenzuela, G. R. Depolarization of EM waves by slightly rough surfaces. *IEEE Transactions on Antennas and Propagation,* **AP-15,** 552–557, 1967.

Van de Lindt, W. J. Digital technique for generating synthetic aperture radar image. *IBM Journal of Research and Development,* 415–432, September 1977.

Vant, M. R., R. W. Herring, and E. Shaw. Digital processing techniques for satellite borne synthetic aperture radars. *Canadian Journal of Remote Sensing,* **5,** 1979.

van Zyl, J. J. *On the Importance of Polarization in Radar Scattering Problems.* Ph.D. Thesis, California Institute of Technology, Pasadena, CA, 1985.

van Zyl, J. J. Unsupervised classification of scattering behavior using radar polarimetry data. *IEEE Transactions on Geoscience and Remote Sensing,* **27,** 36–45, 1989.

van Zyl, J. J. A technique to calibrate polarimetric radar images using only image parameters and trihedral corner reflectors. *IEEE Transactions on Geoscience and Remote Sensing,* **28,** 337–348, 1990.

van Zyl, J. J., and C. F. Burnette. Bayesian classification of polarimetric SAR images using adaptive *a-priori* probabilities. *International Journal of Remote Sensing,* 13, 835–840, 1992.

van Zyl, J. J. The effects of topography on the radar scattering from vegetated areas. *IEEE Transactions on Geoscience and Remote Sensing,* **31,** 153–160, 1993.

van Zyl, J. J. Application of Cloude's target decomposition theorem to polarimetric imaging radar data. In *Radar Polarimetry,* H. Mott and W-M. Boerner (Ed.), *Proc. SPIE 1748,* pp. 184–191, 1993.

van Zyl, J.J. and J. Hjelmstadt. Interferometric techniques in remote sensing. In *The Review of Radio Science 1996–1999,* W. Ross Stone (Ed.), Oxford University Press, 1999.

Wang, Y., J. Day, and G. Sun. Santa Barbara microwave backscattering model for woodlands. *International Journal of Remote Sensing,* **14,** 1146–1154, 1993.

Wentz, F. J., S. Peteherych, and L. A. Thomas. A model function for ocean radar cross sections at 14.6 GHz. *Journal of Geophysics Research,* **89,** 3689–3704, 1984.

Wentz, F. J., L. A. Mattox, and S. Peteherych. New algorithms for microwave measurements of ocean winds: Applications to SEASAT and the special sensor microwave imager. *Journal of Geophysics Research,* **91,** 2289–2307, 1986.

Wright, T. J., B. Parsons, P. C. England, and E. J. Fielding. InSAR observations of low slip rates on the Major faults of western Tibet. *Science,* **305,** 236–239, 2004.

Wu, C., B. Barkan, W. Karplus, and D. Caswell. Seasat SAR data reduction using parallel array processors. *IEEE Transactions on Geoscience and Remote Sensing,* **GE-20,** 352–358, July 1982.

Wurtele, M. G., P. M. Woiceshyn, S. Peteherych, M. Borowski, and W. S. Appleby. Wind direction alias removal studies of SEASAT scatterometer-derived wind fields. *Journal of Geophysics Research,* **87,** 3365–3377, 1982.

Yueh, S. H., J. A. Kong, J. K. Rao, R. T. Shin, and T. LeToan. Branching model for vegetation. *IEEE Transactions on Geoscience and Remote Sensing,* **30,** 390–402, 1992.

Zebker, H. and R. Goldstein. Topographic mapping from Interferometric SAR observations. *Journal of Goephys. Research,* **91,** 4993–4999, 1986.

Zebker, H. A., J. J. van Zyl, and D. N. Held. Imaging radar polarimetry from wave synthesis. *Journal of Geophysics Research,* **92,** 683–701, 1987.

Zebker, H. A., and Y. L. Lou. Phase calibration of imaging radar polarimeter Stokes matrices. *IEEE Transactions on Geoscience and Remote Sensing,* **28,** 246–252, 1990.

Zebker, H. A., J. J. van Zyl, S. L. Durden, and L. Norikane. Calibrated imaging radar polarimetry: technique, examples, and applications. *IEEE Transactions on Geoscience and Remote Sensing,* **29,** 942–961, 1991.

Zebker, H. A., and J. Villasenor. Decorrelation in interferometric radar echoes. *IEEE Transactions on Geoscience and Remote Sensing,* **30,** 950–959, 1992.

Zebker, H. A., S. N. Madsen, J. Martin, K. B. Wheeler, T. Miller, Y. Lou, G. Alberti, S. Vetrella, and A. Cucci. The TOPSAR interferometric radar topographic mapping instrument. *IEEE Transactions on Geoscience and Remote Sensing,* **30,** 933–940, 1992.

Zebker, H. A., C. L. Werner, P. A. Rosen, and S. Hensley. Accuracy of topographic maps derived from ERS-1 interferometric radar. *IEEE Transactions on Geoscience and Remote Sensing,* **32,** 823–836, 1994.

Zebker, H. A., P. A. Rosen, R. M. Goldstein, A. Gabriel, and C. L. Werner. On the derivation of co-seismic displacement fields using differential radar interferometry: The Landers earthquake. *Journal of Geophysics Research,* **99,** 19617–19634, 1994.

Zebker, H. A., F. Amelung, and S. Jonsson. Remote sensing of volcano surface and internal processes using radar interferometry. *AGU Geophysical Monograph,* **116,** pp. 179–205, 2000.

Zieger, A. R., D. W. Hancock, G. S. Hayne, and C. L. Purdy. NASA radar altimeter for the TOPEX/POSEIDON project. *Proc. IEEE,* **79,** 810–826, 1991.

7

OCEAN SURFACE SENSING

The oceans cover three quarters of our planet, and the study of their properties is of major importance in understanding our environment. Even though the ocean surface is generally quasi-homogeneous in composition, its dynamic nature makes global remote sensing a critical tool in understanding and monitoring its behavior.

Electromagnetic wave sensing of the ocean is limited to the ocean surface and the immediate subsurface because of the high absorption and scattering losses encountered by these waves. Ocean surface features such as surface waves, solitary (internal) waves, currents, fronts, and eddies, as well as the near-surface wind, modulate the intensity and structure of the surface capillary and short gravity waves. These in turn strongly affect the scattering of incident electromagnetic waves, thus allowing the remote sensing of these surface features. The surface temperature affects the microwave and thermal emission, thus allowing a mechanism to remotely map the surface temperature. Biological materials near the surface affect the surface reflectivity in the visible and near IR, thus allowing the possibility of detecting their presence, identifying their composition, mapping their extent, and monitoring their dynamic behavior.

Microwave sensors have been the main tools for remotely sensing the ocean surface. Altimeters on the GEOS and SEASAT satellites were used to map ocean topography and r.m.s. wave heights. The scatterometer on SEASAT proved that the surface wind field could be measured, a measurement that is now routinely done from space. The SARs on SEASAT, SIR-A, and SIR-B allowed imaging of ocean features and, in the case of SEASAT, polar ice features. The operational SAR satellites RadarSat and ERS-1/2 now acquire these types of images on a regular basis. The imaging microwave radiometers on the Nimbus satellites provided images of polar ice distribution. Imaging infrared radiometers are used to observe sea surface temperature. Visible and near-IR imagers are also used to image surface features, plankton distribution, and sediment plumes.

Introduction to the Physics and Techniques of Remote Sensing. By C. Elachi and J. van Zyl

7-1   PHYSICAL PROPERTIES OF THE OCEAN SURFACE

The ocean surface is a fluid interface subject to the Earth's gravitational and rotational forces that is modified by currents, tides, and atmospheric forcing. In the absence of any atmospheric and tidal effect and assuming that the ocean water moves at the same angular speed as the solid Earth, the ocean surface will follow an equipotential surface. The geoid is the equipotential surface that corresponds to the mean sea level. The geoid corresponds to a biaxial rotational ellipsoid determined by the mass and rotation of the Earth plus undulations that result from inhomogeneities in the Earth's volumetric and surficial mass distribution (Fig. 7-1). These undulations are of the order of +60 to −100 m.

7-1-1   Tides and Currents

Tides affect both the solid land surface and the ocean surface. Land tides can be accurately computed and have an amplitude of about 20 cm. Ocean tides are more complicated to determine as they correspond to nearly resonant fluid motion in the ocean basins, thus leading to an instantaneous surface significantly different than the tidal equipotential. Prior to the launch of TOPEX/Poseidon, tides were most accurately known in the deep ocean. The knowledge was based primarily on hydrodynamic models that were constrained with measurements from a worldwide network of tide gauges and ocean bottom pressure sensors. Since these sensors are not distributed uniformly globally, the quality of the tide models varied geographically. As a result, tide errors exceeded 10 cm in many areas. The problem was most severe in shallow basins, where the tide amplitude can reach many meters.

As mentioned in Chapter 6, the TOPEX/Poseidon orbit was chosen carefully by including considerations of how the tidal signal would alias with the desired oceanography signal. The orbit was selected such that these alias frequencies would be high enough for the six dominant tidal model constituents so that they could be reliably estimated from the TOPEX/Poseidon data over a period of about a year. Chapter 1 of Fu and Casenave (2001) provides a detailed discussion and references to how the orbit parameters were chosen, and how the various tidal components alias with for different altimeter missions. Chapter 6 in the same book discusses in detail the different approaches used to estimate

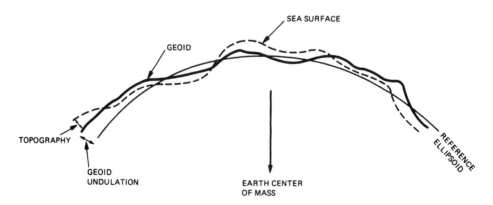

Figure 7-1. Sketch illustrating the geoid, the reference ellipsoid, the geoid undulations (due to variations in the gravity), and the ocean topography (due to currents).

and improve tide models. The conclusion is that in the post-TOPEX/Poseidon era, tides can be estimated globally with an accuracy of 2–3 cm. As a result, tidal aliasing is now a less stringent requirement on the orbit selection for altimeter missions than before the TOPEX/Poseidon launch.

The ocean surface is further displaced from the geoid by ocean circulation and atmospheric effects (wind and pressure). The ocean circulation can be divided into two principal geostrophic components: a large-scale component associated with the quasipermanent mean circulation, which changes on a scale of many months, and a mesoscale component (100 to 1000 km) associated with eddies, which is time dependent. Strong currents can produce height changes of about 1 m over a distance of 100 km.

Due to Earth's rotation, a moving water volume in the northern hemisphere is forced by the Coriolis force off to the right of its initial trajectory. In the southern hemisphere, the water is forced off to the left. If we require that the Coriolis force balance the gravity force, then we have a special kind of motion called geostrophic motion. In this situation, the water tends to flow around areas of high or low water level instead of flowing from highs to lows. Geostrophic motion is characteristic of geophysical fluids on the large scale. It means that fluids tend to move along lines of constant pressure instead of from high to low pressure.

Winds blowing across shallow areas push the water and tend to pile it against the coast, leading to a height change of few meters. In addition, winds generate waves that undulate the mean surface on a horizontal scale of up to a few hundred meters. Pressure changes cause the sea surface to react like an inverted barometer, leading to changes of few centimeters.

All the above effects lead to surface undulation, which is superimposed on the time-independent geoid. The magnitude of this undulation ranges from a few centimeters to many meters (see Table 7-1). Thus, accurate measurement of the ocean topography requires a sensor capable of range accuracy of a few centimeters.

7-1-2 Surface Waves

Ocean surface waves are governed by the Navier–Stokes nonlinear differential equations for incompressible fluids, which express the balance between inertial, convective, and restorative forces of the wave motion.

The main restorative force for surface waves longer than 1.7 cm (called gravity waves) is Earth's gravity. For waves shorter than 1.7 cm (called capillary waves), the restorative force is surface tension.

In the linear approximation for small slopes, the surface displacement $H(x, t)$ of the wave is sinusoidal:

$$H(x, t) = H \cos(Kx - \Omega t) \tag{7-1}$$

where $K = 2\pi/\Lambda$ = wave number and Ω is the angular frequency. Ω and K are related by the dispersion relation of water waves:

$$\Omega = [gK \tanh (KD)]^{1/2} \tag{7-2}$$

for gravity waves, and

$$\Omega = \sqrt{B}\, K^{3/2} \tag{7-3}$$

TABLE 7-1. Magnitude of Ocean Surface Perturbations

Source	Height perturbation	Time Scale
Tides	~ 1 m in deep ocean, many meters in shallow ocean	Hours
Large-scale currents	~ 1 m	Months
Mesoscale currents	~ 1 m	Days
Wind	~ meters near coast	Hours
Pressure field	~ centimeters	Hours

for capillary waves, where D is the water depth and B is the surface tension to water density ratio ($B = 73$ cm^3/sec^2). In the case of the deep ocean ($KD \gg 1$), the dispersion relation for the gravity waves becomes

$$\Omega = (gK)^{1/2} \tag{7-4}$$

Figure 7-2 shows the phase velocity $V_p = \Omega/K$ as a function of the wavelength Λ, with the longer wavelength waves traveling fastest. The minimum speed of water waves is about 23 cm/sec for 1.7 cm waves.

If we follow the motion of a fluid particle, it describes a closed circular orbit with a radius equal to the wave amplitude H. The orbital velocity is

$$V_0 = \frac{2\pi H}{T} = \Omega H \tag{7-5}$$

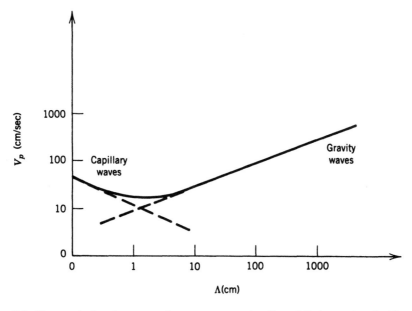

Figure 7-2. Phase velocity of ocean surface waves as a function of their wavelength. The water depth D is assumed to be much larger than Λ.

In the presence of a mean current of velocity U, the angular frequency becomes

$$\Omega = \Omega_0 + \mathbf{K} \cdot \mathbf{U} \tag{7-6}$$

On the ocean, the surface waves are characterized by an infinite spectrum of waves. The Fourier transforms of the spatial and temporal correlation functions of the surface displacement $H(x, t)$ yield the wave number spectrum $F(K)$ and the frequency spectrum $S(\Omega)$, which are interrelated by

$$F(K) \, dK = S(\Omega) \, d\Omega \tag{7-7}$$

where

$$\Omega = (gK + BK^3)^{1/2} + \mathbf{K} \cdot \mathbf{U}$$

The mean square height of the ocean is then obtained by integrating over $\mathbf{K}$ or Λ:

$$H^2 = \int_{-\infty}^{\infty} \int_{-\infty}^{\infty} F(\mathbf{K}) \, d\mathbf{K} = \int_0^{\infty} S(\Omega) \, d\Omega \tag{7-8}$$

Ocean waves are generated through linear resonant coupling of turbulent pressure fluctuations in the air with the water. As the wind increases, the first waves that are generated by this mechanism are 1.7 cm waves, which have the minimum phase speed and travel in the direction of the wind. For higher winds, waves of phase speed V_p are generated at an angle ψ with respect to the wind where

$$\cos \psi = \frac{V_p}{V_w} \tag{7-9}$$

V_w is the convection velocity of the air pressure fluctuations near the water surface. Thus, a whole spectrum of waves will be generated in different directions (see Fig. 7-3). As the wind blows, energy is transferred to the waves, leading to their growth. On the other hand, damping results from viscous losses and wave breaking. In addition, nonlinear resonant interactions lead to energy transfer between different parts of the spectrum.

For fully developed seas (i.e., the wind field blows over a sufficiently long fetch and for a sufficiently long time), it has been found that the Pierson–Morkowitz spectrum gives a good description of the surface:

$$S(\Omega) \simeq 0.1 \, g^2 \Omega^{-5} \exp\left[-\frac{3}{4}\left(\frac{\Omega_m}{\Omega}\right)^4\right] \tag{7-10}$$

where $\Omega_m = g/W$ and W is the wind speed at a height of 19.5 m.

A number of surface and near-surface phenomena, in addition to the surface wind, affect the amplitude of the capillary and small gravity waves. Large surface waves (swells) and internal waves modulate the amplitude of the capillary waves. Surface films and suspended sediments tend to dampen them. Currents affect the growth and damping of small waves depending on their relative velocity vector. Thus, numerous ocean phenomena can be sensed remotely in an indirect way by observing the spatial distribution and temporal

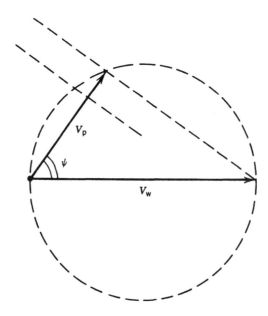

Figure 7-3. A wave with phase speed V_P and direction ψ will be in resonance with wind turbulence moving at speed V_w.

dynamics of the capillary and small gravity waves. This is usually done by microwave scatterometers and imaging radars.

7-2 MAPPING OF THE OCEAN TOPOGRAPHY

High-accuracy global measurement of the ocean surface topography can be accomplished using a spaceborne radar altimeter. This type of sensor provides a measure of the distance between the sensor and the surface area reflecting the signal. Range accuracy of few centimeters is possible using very broad band sensors, as discussed in Chapter 6.

7-2-1 Geoid Measurement

The geoid is the static component of the ocean surface topography. One of the ways in which it is derived is by averaging repetitive measurements over long time periods (Tapley and Kim, 2001). This tends to average out the effects due to temporal phenomena such as tides, currents, and waves. Some of the earliest maps of the geoid were derived from the GEOS-3 and Seasat altimeters (Fig. 7-4). They clearly mimic the major ocean bottom features such as the mid-Atlantic ridge and the Pacific trenches. On a regional basis, altimetric geoid measurements are used to locate and characterize sea mounts and localized density inhomogeneities.

Improved determinations of the geoid were obtained with data from the Geosat mission during its exact repeat mission phase that started in 1986, and again when the data from the ERS 1 mission became available. The accuracy and the resolution of these estimates were limited at the long-wavelength scales by uncertainties in the radial compo-

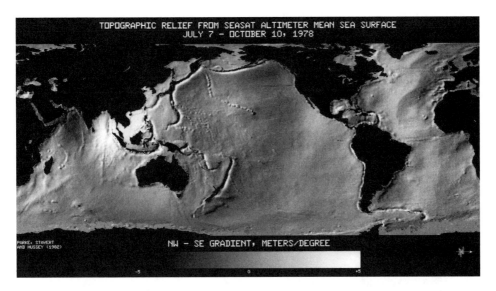

TOPOGRAPHIC RELIEF FROM SEASAT ALTIMETER MEAN SEA SURFACE
JULY 7 - OCTOBER 10, 1978

NW - SE GRADIENT, METERS/DEGREE

Figure 7-4. Global ocean surface average topography derived from the Seasat altimeter.

nent of the spacecraft orbit. At the short wavelengths, the limitation was primarily the spacing between adjacent satellite tracks. This, in turn, is a function of the orbit altitude and repeat period. The TOPEX/Poseidon mission, launched in 1992, was designed specifically to help address some of the issues associated with estimating the geoid, and included specific measurements for precision determination of the satellite orbit. As a result, various different analyses suggest that the radial orbit of this satellite is known to an accuracy of 3–4 cm, nearly an order of magnitude better than previous altimeter satellites.

If the orbit of an altimeter satellite is accurately known and repeats a specific ground track exactly, the altimeter measurements are basically a measurement of the time-varying ocean surface elevation. If a series of such collinear altimetric profiles are acquired over a long period of time (a few months), then the mean topography (i.e., geoid and large-scale, semipermanent current component) and the time-variable perturbations (from mesoscale eddy circulation) can be separated. The mean topography is equal to

$$\overline{H}(x) = \frac{1}{N} \sum_{i}^{N} H_i(x) \qquad (7\text{-}11)$$

and the circulation component is

$$H_c(x) = \overline{H}(x) - H_g \qquad (7\text{-}12)$$

where H_g is the geoid. If the geoid is well known, then $H_c(x)$ can be derived. Figure 7-5 illustrates the effect of the Gulf Stream and its eddies as derived from the Seasat altimeter. It also illustrates the short-term topography variations derived from

$$\Delta H = H_i(x) - \overline{H}(x) \qquad (7\text{-}13)$$

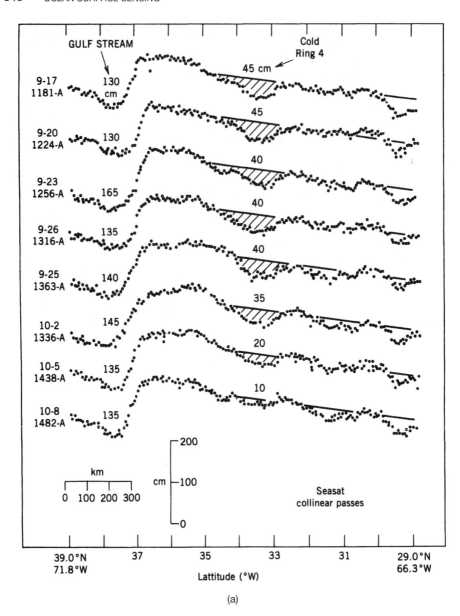

Figure 7-5. Successive Seasat altimetric profiles (a) along the tracks shown in (b), which show the evolution over 21 days of the Gulf Stream and a large, strong cold ring. (From Cheney and Marsh, 1981.)

After observations over neighboring tracks, the eddy circulation velocity ΔV_y in the cross-track direction can then be derived from

$$\Delta V_y = \frac{g}{2\Omega \sin \theta} \frac{\partial \Delta H}{\partial x} \tag{7-14}$$

where $2\Omega \sin \theta$ is the Coriolis parameter, Ω is the rotation rate of the Earth ($\Omega = 7.27 \times 10^{-5}$ rad/sec), θ is the latitude, and g is the local gravitational acceleration.

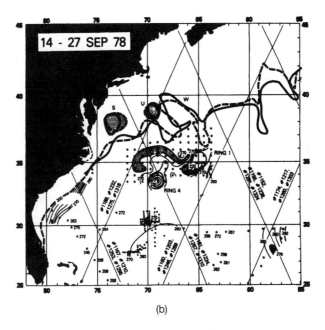

(b)

Figure 7-5. (*continued*)

Estimating the mean sea surface from the measurements of a single altimeter mission is basically a trade-off between temporal and spatial sampling of the sea surface heights. If the orbit has a short repeat period, many individual samples are acquired during a given period of time, and the estimates at those places where measurements are made are improved as a result. A short repeat period, however, means fewer orbits during the repeat cycle, which translates into large spatial gaps between measurements, reducing the spatial resolution of the resulting geoid. The Earth's geoid is typically described in a spherical harmonic expansion. Short repeat cycles, therefore, will limit the maximum harmonic degree that can be estimated reliably to a low value. This situation can be improved by combining the data from several altimeter missions. There were several periods in which as many as three different altimeter missions were flying simultaneously. The combination of the data from all of these missions increases the spatial resolution of the geoid that can be derived. Figure 7-6, from Tapley and Kim (2001), shows the improvement in the spatial resolution of the mean surface height that was achieved over the past two decades using data from multiple satellites. This figure shows two geoid models for a portion of the southern mid-Atlantic ridge system. The spreading center segments are clearly visible in the improved resolution of the geoid shown on the right. By combining data from many satellites, the ocean gravity field has been reconstructed to spatial scales better than 10 km (Tapley and Kim, 2001). Even so, the relatively large errors in the knowledge of the geoid limit the ability of satellite altimeters to improve the models of general ocean circulation.

It is widely recognized that the geoid is not static. Some of the dynamic forces, such as tidal forces, are well known. Not so well known are the driving forces of plate tectonics and the three-dimensional structure of the Earth's interior. Most of the mass redistribution on Earth is due to movement of water between the liquid water in the oceans, the atmospheric water vapor, and the land hydrology in the form of soil moisture, snow cover, and polar ice sheets. Several missions have been launched or are planned for launch early in

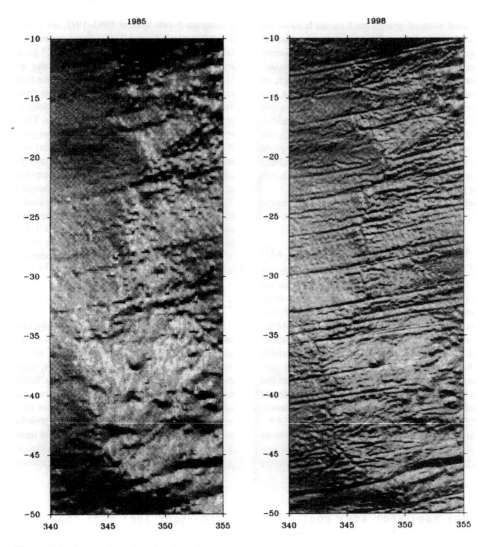

Figure 7-6. Comparison showing the advances in the estimates of the mean sea surface. The result on the left is based on GEOS 3 and Seasat data and is from a model reported by the Goddard Space Flight Center in 1985. The result on the right incorporates data from Geosat, ERS 1, ERS 2, and TOPEX/Poseidon data, and was reported by the University of Texas Center for Space Research in 1998. (Reprinted from Tapley and Kim, 2001. Copyright 2001, with permission from Elsevier.)

the 21st century that would be able to help increase our understanding of these issues. Many of these include next-generation altimeter missions, both for study of the ocean [Jason and the NASA/CNES Ocean Surface Topography Mission (OSTM)] and ice sheets (ICESAT). ICESAT (Ice, Cloud and Land Elevation Satellite), launched in 2003 and carrying a laser altimeter, will extend altimeter measurements north and south to 86° latitudes to the ice-covered regions that still are among the most poorly understood areas of geodesy. Jason, a follow-on to TOPEX/Poseidon, was launched in 2001, and its follow-on, OSTM is scheduled for launch in 2008.

Oceanography in the 21st century will benefit from satellite missions dedicated to gravity recovery studies. Two examples are the CHAMP (A Challenging Micro-Satellite Payload for Geophysical Research and Applications) and the GRACE (Gravity Recovery and Climate Experiment) missions, launched in 2000 and 2002, respectively. CHAMP is a single-satellite mission, whereas the GRACE mission consists of two spacecraft that are flying about 220 km apart on the same orbit. The distance between the two satellites is measured continuously to high accuracy using a microwave link between the two satellites. If the leading satellite approaches an area of increased mass on Earth, the locally increased gravity will speed the satellite up, increasing the distance between the two satellites. As the leading satellite leaves the area of increased mass, it slows down again, while the trailing satellite, now approaching the area of increased mass, will speed up, decreasing the distance between the satellites. By accurately measuring these changes in distance between the satellites, as well as accurately measuring the relative acceleration of each satellite in the vertical direction using some of the world's most sensitive accelerometers, and tracking the satellite positions accurately using GPS satellites, accurate measurements of the Earth's gravity field are made. Figure 7-7 shows two gravity anomaly maps. The one on the left is the best model available prior to the launch of the GRACE satellites, whereas the one on the right shows the new model derived from 363 days of GRACE data. The increased resolution in the GRACE data can be seen clearly in these images.

7-2-2 Surface Wave Effects

Surface wave heights can be several meters. This introduces a potential significant error in the altimetric measurement that must be corrected for.

When the altimeter pulse interacts with the surface, the shape of the reflected echo depends strongly on the surface local topographic variations. In the case of a perfectly smooth surface, the echo is a mirror image of the incident pulse (Fig. 7-8*b*). If the surface has a slight roughness, some return occurs in the backscatter direction, at slight off-vertical angles, as the pulse footprint spreads over the surface (Fig. 7-8*a*). This results in a slow drop-

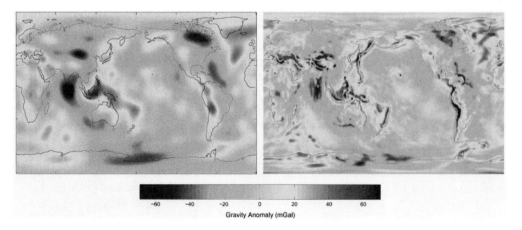

Figure 7-7. Comparison of gravity anomaly models before the GRACE mission (left) and after 363 days of GRACE data (right). The increase in resolution is clearly seen in the image on the right. The unit mGal corresponds to 0.00001 m/s^2. (Courtesy University of Texas Center for Space Research.) See color section.

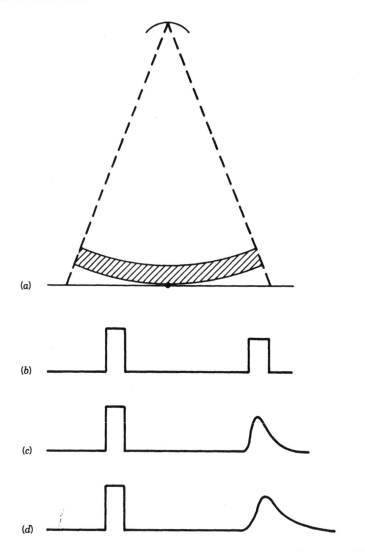

Figure 7-8. (a) Geometry showing the pulse footprint spread on the surface. The shape of the return echo from a perfectly smooth surface (b), a slightly rough surface (c), and a very rough surface (d), respectively.

off of the echo (Fig. 7-8c). If the surface is very rough, then some of the energy is returned when the pulse intercepts the peaks of the ocean waves, and more energy is returned as the pulse intercepts areas across the height distribution of the ocean waves. This leads to a gradual rise in the echo leading edge (Fig. 7-8d). The rise time t_r depends on the root mean square (r.m.s.) height of the waves. Thus, t_r can be used to measure the waves' r.m.s. height and then correct for the mean surface topography. Figure 7-9 shows examples of echo shapes for different cases of wave heights. Figure 6-58 shows an example of standard wave height (SWH) measurements derived from the echo shapes of the Seasat altimeter as it crossed hurricane Fico. Figure 7-10 shows an example of the global measurement of surface wave height derived from the echo shapes of the TOPEX/Poseidon altimeter.

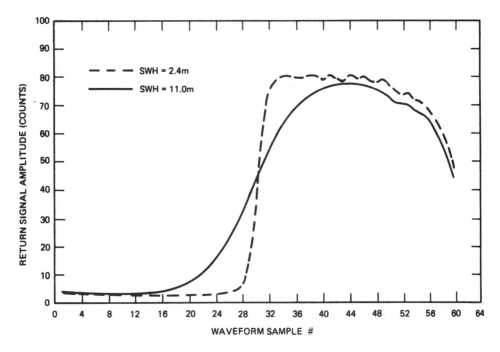

Figure 7-9. Examples of echo shapes from the ocean surface with different wave heights. (From Townsend, 1980.)

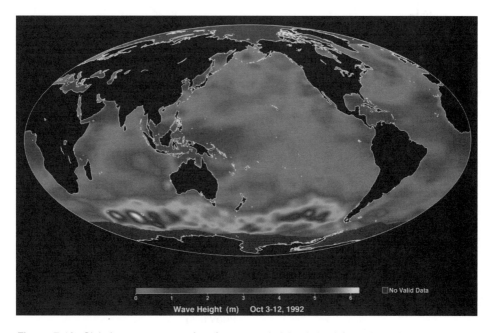

Figure 7-10. Global measurement of surface wave height derived from the echo shape of the TOPEX/Poseidon altimeter. (Courtesy NASA/JPL-Caltech.) See color section.

7-2-3 Surface Wind Effects

As the surface wind increases, the ocean surface becomes rougher, leading to fewer areas in which there is specular return at the nadir and a decrease in the amplitude of the echo. Thus, the altimeter echo amplitude can be used as a measure of the surface wind. Figure 7-11 shows the comparison between wind measurements derived from the altimeter return on GEOS and Seasat and the surface measurement made by ships and buoys. It shows an accuracy of about . Figure 6-58 illustrates how the backscatter return changed as the Seasat altimeter passed over hurricane Fico. Figure 7-12 shows the global distribution of surface wind speed derived from TOPEX/Poseidon data for the same period as the measurement of wave height shown in Figure 7-10. Note the strong correlation between wave height and surface wind speed exhibited by these images.

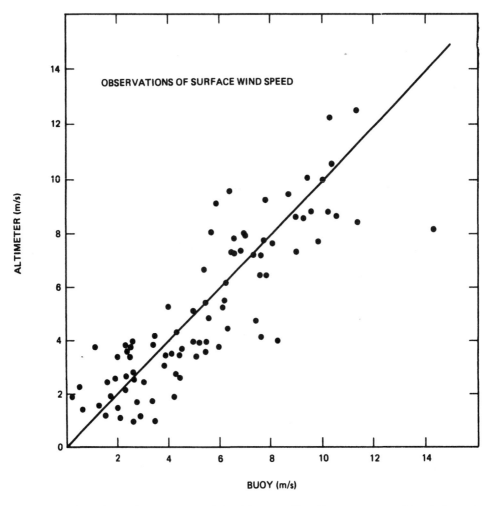

Figure 7-11. Wind-speed measurement derived from satellite altimeter compared to wind-speed measurement at 10 m above the surface made by buoys. (From Fedor and Brown, © 1982 IEEE.)

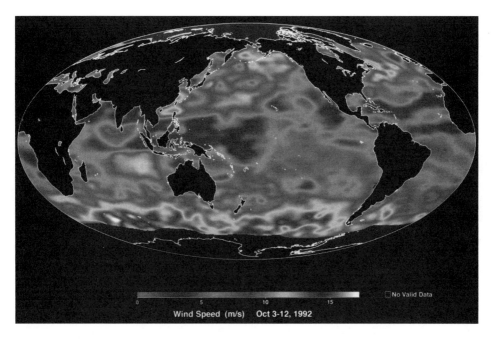

Figure 7-12. Global wind-speed measurement derived from the TOPEX/Poseidon altimeter measurements. (Courtesy NASA/JPL-Caltech.) See color section.

7-2-4 Dynamic Ocean Topography

The real power of operational radar altimeter measurements lies in the derivation of the dynamic topography of the ocean surface. This dynamic component of the ocean topography is basically the deviation of the ocean surface from its mean value, and is sometimes described as the sea surface height anomaly. As mentioned before, the dynamic component of the ocean topography is the result of complex interactions between the ocean surface and the atmosphere. Here, we shall illustrate the measurement of dynamic ocean topography through the well-known El Niño phenomenon that occurs in the Pacific Ocean.

The El Niño phenomenon was named by fishermen who lived on the west coast of Central America to describe the periodic invasion of warmer ocean water into this part of the Pacific, usually around Christmas. During a "normal" year, the trade winds blow from east to west over the Pacific. This causes warmer ocean water to build up in the western Pacific near Australia and the Philippines, causing a local increase in both sea surface temperature and ocean height. As the winds cross the ocean, they pick up moisture, which is released in the form of rain in the western Pacific. At the same time, cold, nutrient-rich water wells up on the eastern part of the Pacific basin. This increase in nutrients fosters the growth of the fish population, particularly anchovy, in the eastern Pacific, leading to excellent fishing off the coast of Central America and the northern part of South America. The left panel in Figure 7-13 shows the dynamic component of the ocean topography for such a "normal" year in December 1996.

The onset of an El Niño event is signaled by weakening trade winds and an eastward spread of the pool of warm water normally associated with the western Pacific basin. The eastward spread of the warm water is accompanied by increased rainfall in the eastern

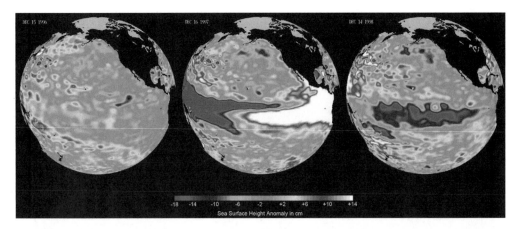

Figure 7-13. This figure shows sea surface height anomalies measured by the TOPEX/Poseidon mission in December 1996 (left), December 1997 (middle), and December 1998 (right). Sea surface height anomalies are the difference between the dynamic and static ocean topography. The left image shows a "normal" year, the middle an El Niño event, and the image on the right a La Niña pattern. (Courtesy NASA/JPL-Caltech.) See color section.

part of the Pacific basin, and an accompanying decrease in precipitation in the western Pacific, leading to severe droughts in Australia and Indonesia. The eastward-spreading warm water reduces the upwelling of the nutrient-rich cold water along the coast of South America, causing fish populations to dwindle, and leading to poor catches for the fishermen. The middle image in Figure 7-13 clearly shows the increased sea surface height in the eastern Pacific in December 1997, and the accompanying decrease in sea surface height near Australia. Figure 4-21 shows the increase in sea surface temperature for the same period in the eastern part of the Pacific.

The opposite of the El Niño weather pattern is the La Niña. In this case, the tradewinds blow stronger than usual, so the effect of piling up of the warmer water in the western Paciific is enhanced. At the same time, the upwelling of nutrient-rich cold water in the eastern Pacific is also enhanced, and spreads to most of the tropical Pacific. The increase in sea surface temperatures and sea surface height in the western Pacific cause increased precipitation in this part of the world. At the same time, the climate is colder and drier off the coast of South America. The right-hand panel in Figure 7-13 shows what is believed to be a La Niña pattern with higher than usual sea surface height north of Australia, and lower than normal sea surface height in the central and western Pacific.

7-2-5 Ancillary Measurements

Altimeter missions such as TOPEX/Poseidon and the Jason mission, the follow-on mission for TOPEX/Poseidon launched in 1999, make additional measurements that are used to correct the pulse travel time to increase the accuracy of the ocean topography measurement. The index of refraction of the ionosphere for high frequencies varies as the square of the ratio of the electromagnetic wave frequency to the plasma frequency of the ionosphere. The plasma frequency of the ionosphere, in turn, is a function of the density of free electrons in the ionosphere. Given that the ionospheric index of refraction varies as the frequency squared, pulses transmitted with different carrier frequencies will be delayed differentially as they propagate through the atmosphere. This differential delay is directly proportional to the integral of the electron density along the propagation path.

The TOPEX/Poseidon and Jason missions use dual-frequency Ku-band and C-band altimeters to measure this variable delay due to the variations in the electron content of the ionosphere. This differential delay is then used to infer the total electron content of the ionosphere, from which the correction to the altimeter range measurement can be inferred. Whereas this quantity is used primarily to correct the delay of the pulses, it is an important measurement in its own right. The highest concentration of charged particles in the earth's atmosphere is found between 250 and 400 km altitude. In this region, ionization is believed to be mostly the result of extreme ultraviolet radiation from the sun. Electron density varies by as much as an order of magnitude diurnally, with peak values occurring typically around 3:00 p.m. local time. Electron densities also vary with latitude, with minimum values typically observed at approximately 15° north and south of the equator. This variation can be as much as a factor two. Figure 7-14 shows an example of the total electron content inferred from the dual-frequency TOPEX/Poseidon measurement. Note the strong change with latitude evident in these measurements.

The propagation speed of the altimeter pulse is also influenced by the atmospheric water vapor. Both the TOPEX/Poseidon and the Jason missions include a three-frequency microwave radiometer operating at frequencies of 18.7, 23.8, and 34 GHz. Three frequencies are used to reduce the effects of clouds and the ocean surface itself on the water vapor estimate. The measurements at 23.8 GHz are used to measure atmospheric water vapor emission. The measurements in the 18.7 GHz channel are used to correct for ocean surface effects, whereas the 34 GHz channel measurements are used to correct for the effects of clouds. Figure 7-15 shows an example of the TOPEX/Poseidon-derived water vapor abundance.

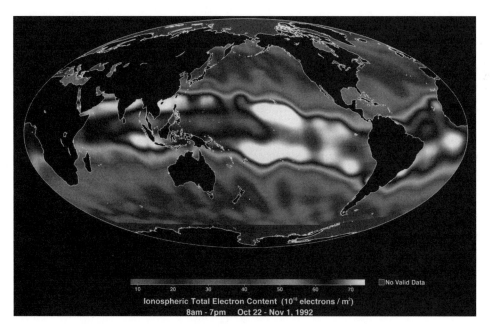

Figure 7-14. Global ionospheric total electron content measurement derived from the TOPEX/Poseidon altimeter measurements. (Courtesy NASA/JPL-Caltech.) See color section.

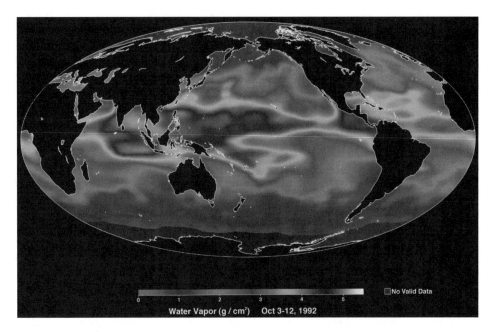

Figure 7-15. Global atmospheric water vapor distribution derived from the TOPEX/Poseidon radiometer measurements. (Courtesy NASA/JPL-Caltech.) See color section.

7-3 SURFACE WIND MAPPING

Based on a number of experimental investigations, an empirical relationship has been derived that relates the radar backscatter, the wind velocity U measured at a reference altitude usually selected to be 19.5 m above the surface, and the angle α between the plane of the incident wave and the wind direction. This relationship is given by

$$\sigma = AU^{\gamma}(1 + a \cos \alpha + b \cos 2\alpha) \qquad (7\text{-}15)$$

where A, a, b, and γ are determined empirically and depend on the incidence angle θ. Figure 7-16 shows the backscatter cross section measured as a function of the wind for a set of fixed incidence angles θ. These curves show that γ varies as a function of θ. Figure 7-17 shows the measured backscatter as a function of α. It is observed that: (1) σ_{VV} is always larger than σ_{HH}; (2) there is a symmetry for $\pm\alpha$ as expected; (3) there is a well-defined 180° periodicity ($b \cos 2\alpha$ term); and (4) there is a slight assymetry between the upwind and downwind observation ($a \cos \alpha$ term). The a and b parameters can be derived from three sets of measurements at different values of α such as

$$\sigma(0) = AU^{\gamma}(1 + a + b) \qquad (7\text{-}16)$$

$$\sigma(90°) = AU^{\gamma}(1 - b) \qquad (7\text{-}17)$$

$$\sigma(180°) = AU^{\gamma}(1 - a + b) \qquad (7\text{-}18)$$

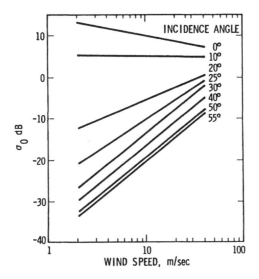

Figure 7-16. Variation of as a function of wind speed for different values of θ. These curves correspond to VV polarization and a signal frequency of 14 GHz (© IEEE).

Once the parameters A, γ, a, and b are known from detailed measurements over a calibration site, Equation 7-15 will provide the algorithm that could be used to derive the wind velocity vector from σ on a global basis.

7-3-1 Observations Required

Because σ is a function of U and α, a single measurement of σ will not allow the derivation of the wind speed and direction. More than one measurement is necessary. To illus-

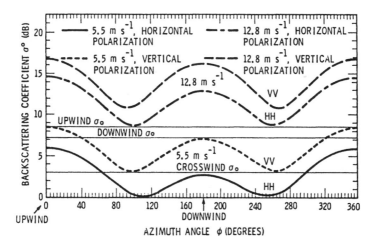

Figure 7-17. Variation of σ with azimuth at constant wind speed.

trate in a simple fashion, let us neglect the $a \cos \sigma$ term, which is usually small. Let σ_1 and σ_2 be two measurements of σ acquired in two orthogonal directions, then

$$\sigma_1 = AU^\gamma(1 + b \cos 2\alpha) \tag{7-19}$$

$$\sigma_2 = AU^\gamma \left[1 + b \cos 2\left(\alpha + \frac{\pi}{2} \right) \right] = AU^\gamma(1 - b \cos 2\alpha) \tag{7-20}$$

Then

$$\sigma_1 + \sigma_2 = 2AU^\gamma \tag{7-21}$$

and

$$\frac{\sigma_1 - \sigma_2}{\sigma_1 + \sigma_2} = b \cos 2\alpha \tag{7-22}$$

This allows the derivation of the wind speed U and the wind direction with a fourfold ambiguity (i.e., $+\alpha$, $-\alpha$, $180° - \alpha$, $-180° + \alpha$). In order to reduce the ambiguity, a third measurement is required. If the third measurement is taken at 45° relative to the first two, then

$$\sigma_3 = AU^\gamma \left[1 + b \cos 2\left(\alpha + \frac{\pi}{4} \right) \right] = AU^\gamma(1 + b \sin 2\alpha) \tag{7-23}$$

This will allow us to resolve the ambiguity between α and $-\alpha$ or $180° - \alpha$ and $-180° + \alpha$, thus leaving only an ambiguity of 180°, which can be resolved by meteorological considerations.

Figure 7-18 illustrates the global wind field derived from the Seasat scatterometer. In this case, only two orthogonal observations were acquired, and the fourfold ambiguity had to be resolved by using an interactive algorithm.

Pencil beam scatterometers, such as the SeaWinds instrument on the QuikScat satellite, simplify the measurement of global winds. The main advantage of these systems (see Chapter 6) in terms of the measurement inversion is that they acquire data at a constant incidence angle. This means that the parameters A, a, b, and γ only have to be determined for this one incidence angle and then apply across the entire swath. In addition, conically scanning systems do not suffer from the measurement gap near the nadir line, so that a wider swath is covered. A dual-beam system such as SeaWinds actually makes four measurements of each pixel everywhere in the inner swath—twice by the inner beam looking forward and aft, and again twice by the outer beam looking forward and aft. The accuracy with which the wind direction can be estimated, however, decreases toward the outer edges of the swath. This is because the relative azimuth angle between the forward-looking and aft-looking measurements decreases toward the edge of the swath. These four measurements at a constant incidence angle means that pencil beam scatterometers typically make more accurate measurements of wind speed and direction than their fan beam counterparts. These measurements, however, still suffer from the 180° ambiguity discussed earlier. To resolve this ambiguity, the output from numerical weather-prediction models are used to "nudge" the solution in the right direction. The SeaWinds instrument

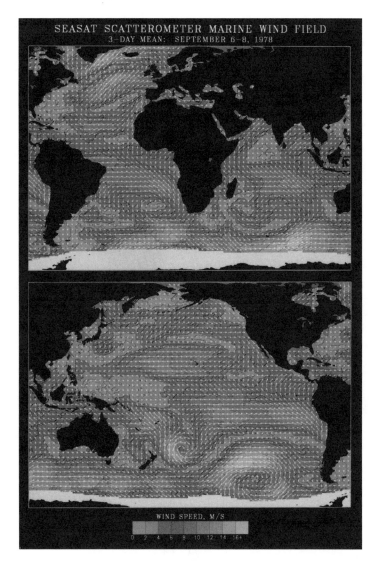

Figure 7-18. Average global winds derived from the Seasat scatterometer for September 6–8, 1978. See color section. (Courtesy of P. Woceishyn, Jet Propulsion Laboratory.)

covers more than 90% of the world's oceans in a day. Figure 6-74 illustrates this by showing the swaths for the SeaWinds instrument for a 24 hour period. QuikScat global measurements of wind speed and direction are shown in Figure 6-77. The relatively high resolution capability of the SeaWinds instrument is illustrated by the measurement of the wind field associated with Hurricane Frances shown in Figure 7-19. This image shows the winds measured on September 4, 2004 as Frances was approaching the east coast of Florida. Frances was responsible for seven deaths, disrupted power to more than 6 million people, and caused damages estimated at several billions of dollars.

Wind stress is the primary driving mechanism for upwelling circulations in many coastal ocean regions. To understand coastal ocean circulation, it is therefore important to

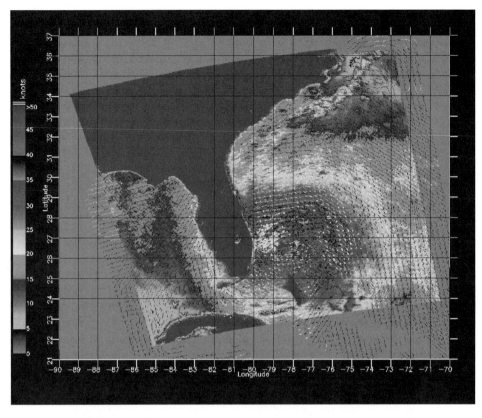

Figure 7-19. SeaWinds measurements of the winds associated with Hurricane Frances approaching the coast of Florida on September 4, 2004. (Courtesy NASA/JPL-Caltech.) See color section.

accurately characterize and understand the coastal ocean wind field. The relationship between the wind stress and the wind vector was reported by Large and Pond (1981) as

$$\tau = \rho_a C_D \boldsymbol{v}_{10} \tag{7-24}$$

where ρ_a is the density of the air, C_D is the drag coefficient, and $\boldsymbol{v}_{10}$ is the wind velocity vector referenced to 10 m above the sea surface. The drag coefficient is generally a function of the wind speed. The availability of vector wind fields from scatterometers therefore makes it possible to estimate the wind stress field. The QuikScat implementation of the Large and Pond algorithm assumes a constant air density at $\rho_a = 1.223$ kg/m^3 and a polynomial parameterization of the wind-speed-dependent drag coefficient for the magnitude of the wind stress as

$$\tau = 0.00270 v_{10} + 0.000142 \nu v_{10} + 0.0000764 v_{10}^3 \tag{7-25}$$

with v_{10} the magnitude of the wind speed at 10 m above sea level. Figure 7-20 shows the global wind stress field for September 3, 2004. Hurricane Frances is visible near Florida, and Typhoon Songda is the large storm in the Pacific approaching Japan. Songda hit Japan on September 7 and 8, and killed 31 people.

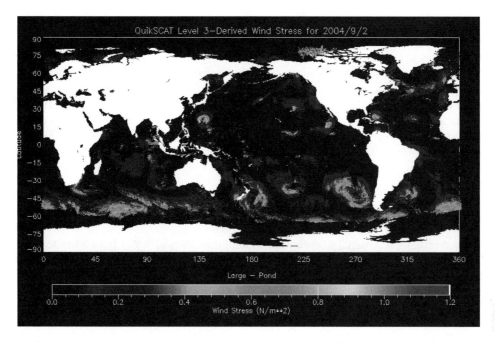

Figure 7-20. SeaWinds measurements of the global wind stress for September 2, 2004. Hurricane Frances is visible near Florida, while Typhoon Songda is approaching Japan. (Courtesy NASA/JPL-Caltech.) See color section.

Polarimetric scatterometers have been proposed (Tsai et al., 2000) to eliminate the need for external information to resolve the ambiguity in wind direction measurement. A polarimetric scatterometer measures the traditional copolarized returns, and also measures the cross-correlation between the copolarized returns and the cross-polarized returns. Models of ocean scattering predict the copolarized returns to be symmetrical with respect to the wind direction, whereas the cross-correlation between the co- and cross-polarized returns is an odd function of the wind direction. This difference in symmetry properties can, in principle, be used to resolve the wind direction ambiguity. This polarimetric principle is also applicable to passive microwave measurements of ocean winds. WIND-SAT is a demonstration project that carries a polarimetric, conically scanning radiometer operating at frequencies of 6.8, 10.7, 18.7, 23.8, and 37 GHz. Three of the channels, (10.7, 18.7, and 37.0 GHz) measure all four Stokes parameters to derive the wind speed and direction. WINDSAT was launched on January 6, 2003, and is expected to operate for three years. Cross-comparisons between the radiometer and scatterometer-derived wind fields are planned.

7-3-2 Nadir Observations

From Figure 7-16, it is apparent that the nadir backscatter is a function of the wind velocity:

$$\sigma(0) \sim \frac{1}{U^x} \qquad (7\text{-}26)$$

This inverse behavior is to be expected. As the surface roughness increases with the wind, there will be fewer and fewer specular areas that will reflect the signal back to the sensor, thus leading to a weaker echo.

Nadir and near-nadir measurements have the limitation that only a narrow strip below the spacecraft is observed. However, if long-term averages are desired, a large number of nadir tracks could be used to develop a global wind map, as was done with Seasat altimetry data. Figure 7-21 shows the near-nadir backscatter return observed across typhoon Carmen. As the wind increases near the center of the typhoon, the backscatter at the nadir decreases rapidly below the return from larger angles in accordance with the specular scattering theory discussed above.

7-4   OCEAN SURFACE IMAGING

Surface imaging is required to observe and monitor features of scales from a few tens of meters to a few kilometers. This includes surface waves, internal waves, current boundaries, plankton density contours, and eddies. Radar and visible sensors are used for high-resolution imaging. Infrared sensors are used for surface temperature mapping, and passive microwave imagers are used for large-scale mapping, particularly ice covers.

7-4-1   Radar Imaging Mechanisms

The tone in a radar image is a direct representation of the backscatter return, which in turn is mainly proportional to the surface roughness at the scale of the radar wavelength. Most imaging radars operate in the spectral region from 1 GHz (30 cm wavelength) to 15 GHz (2 cm wavelength). Thus, the resulting images represent the surficial variation in the intensity of the ocean short gravity waves and capillary waves.

Two additional mechanisms play a role in radar imaging of the ocean surface. The first one is related to the change in the backscatter associated with the change in surface slope. This mechanism is of importance in imaging ocean swell. The second one is related to the Doppler shift due to the surface motion. This has an important effect in the case of synthetic-aperture imaging radars in which the Doppler history is used for image generation.

Surface straining, local wind, surface motion, surface films, nonlinear coupling, and a number of other effects can lead to a variation of the population of the short gravity waves. This leads to a modulation of the population of these waves across a swell, over an internal wave train, or across the boundary of a current or a weather front. From Equation 6-2, it is clearly apparent that the backscatter cross section σ is directly proportional to the surface roughness spectrum $W(K)$.

The surface slope mechanism plays a role in the imaging of swells. To illustrate, let us consider the two-dimensional case in the plane of observation. A small, rough ocean patch is characterized by its energy spectrum $W(K_x)$. Let ψ be the tilt angle of the patch due to the presence of large waves. Thus, the backscatter σ from the patch should be calculated at the local incidence angle $\theta_i = \theta + \psi$. The tilt angle ψ can be written as

$$\psi = \psi_1 + \psi_2 \qquad (7\text{-}27)$$

where ψ_1 is the tilt from all waves of wavelengths less than the resolution of the radar image, and ψ_2 is the tilt due to longer ocean waves. Thus, in computing the backscatter from one resolution element, we have to integrate over ψ_1:

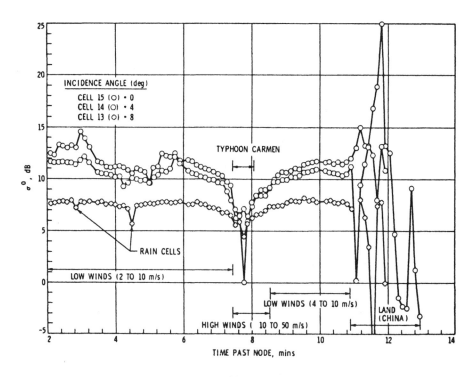

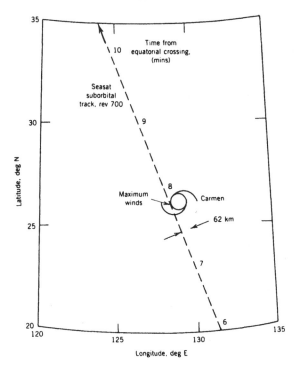

Figure 7-21. Nadir backscatter return measured with the Seasat scatterometer as it flew over typhoon Carmen on August 14, 1978 (From Bracalante et al. 1980).

$$\sigma(\theta, \psi_2) = \int \sigma(\theta_i)p(\tan \psi_1)d(\tan \psi_1) \tag{7-28}$$

where $p(\tan \psi_1)$ is the probability density of slopes for water waves. This can be approximated by a Gaussian distribution:

$$p(\tan \psi_1) = \frac{1}{2\pi m^2} \exp\left(-\frac{\tan^2 \psi_1}{2m^2}\right) \tag{7-29}$$

This allows us then to calculate $\sigma(\theta)$ as a function of θ, m, and $S = \tan \psi_2$. Figure 7-22 shows the change of σ from one side of a swell to the other as a function of S for $\theta = 30°$ and $m = 0.1$. For a slope change of about $\pm 3°$ ($\tan \psi_2 = 0.1$) the backscatter cross section can change by 5 to 10 dB.

The surface motion effect is a result of the mechanism used by synthetic-aperture radars for image generation. Referring to the simplified case illustrated in Figure 7-23, where the surface element velocity $V_p \ll V$, the phase history of the echo is given by [see also Equations (6-105)–(6-107)]

$$\Phi(t) = \frac{4\pi}{\lambda}D(t) \simeq \frac{4\pi R}{\lambda} + \frac{2\pi V^2 t^2}{\lambda R} - \frac{4\pi}{\lambda} VV_p t^2 \tag{7-30}$$

Thus, if the surface velocity V_p varies across a scene, it leads to a modulation of the phase $\Phi(t)$ and of the corresponding image brightness, even if the surface has a uniform scatterer distribution.

7-4-2   Examples of Ocean Features on Radar Images

A wide range of ocean surface phenomena have been imaged with spaceborne imaging radars. Figure 7-24 shows surface swells of a wavelength of about 350 m refracting and

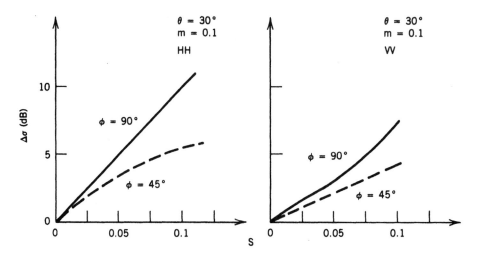

Figure 7-22. Changes of σ from one side of a swell to the other as a function of the maximum swell slope S for $\theta = 30°$ and $m = 0.1$. ϕ is the angle between the observation plane and the swell direction.

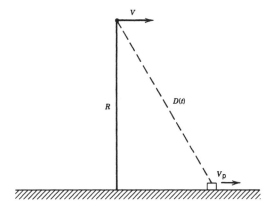

Figure 7-23. The phase history of a point target is determined by the instantaneous distance $D(t)$ between the radar and the target during the formation of the synthetic aperture.

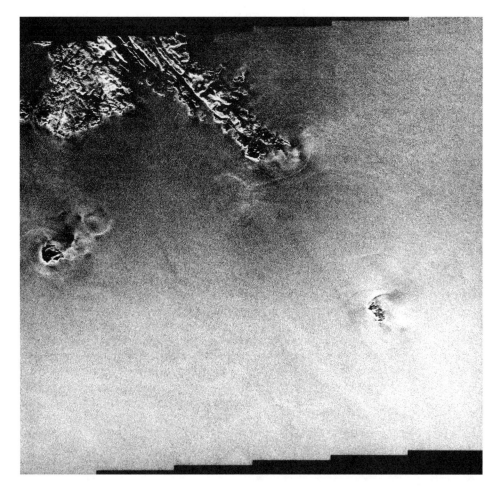

Figure 7-24. Surface waves refracting and defracting around Shetland Island (Scotland). Ocean swell wavelength is about 350 m. Area size is 90 × 90 km. (Courtesy of JPL.)

defracting around the Shetland Islands as a result of change in the shallow bottom topography. Figure 7-25 shows large internal waves in the Gulf of California generated by the interaction of the strong tide with submerged topographic features. Bottom features visible in Figure 7-26 are the result of the modulation of surface current by the varying shallow bathymetry. Figure 7-27 shows the surface expression of tropical rainstorms near the mouth of the Gulf of California and in the western Pacific.

The image on the right in Figure 7-27 combines data from the C-band (HH polarization) in red, L band (HV polarization) in green, and L band (HH polarization) in blue. The white, curved area at the top of the image is a portion of the Ontong Java Atoll, which forms part of the Solomon Islands group. It is believed that the small red cells may be caused by reflection from ice particles in the colder, upper portion of the storm cell. The dark hole in the middle of the large rain cell is the result of very heavy rainfall, which actually smooths out the ocean surface and results in lower radar returns. Also visible in this image are long, thin, dark lines extending across the ocean. These are surface currents. These currents accumulate natural oils deposited by small marine biological organisms. The oils cause the small, wind-generated waves to be damped, which produces a smooth, dark zone on the radar image. Figure 7-28 shows the effect of suspended sediment on the radar backscatter return.

Global wind measurements such as those provided by scatterometry or radiometry are limited in their resolution to tens of kilometers. In coastal regions, however, wind patterns

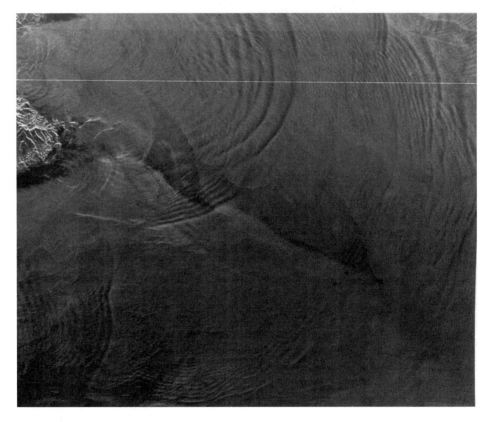

Figure 7-25. Internal waves in the Gulf of Baja near the island of Angel de la Guarda. Area size is 90 × 90 km. (Courtesy of JPL.)

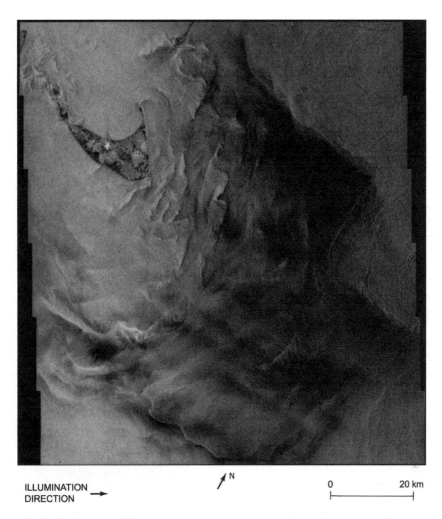

ILLUMINATION
DIRECTION →

N

0   20 km

Figure 7-26. The Nantucket Shoals are shallow-water areas to the south and east of Nantucket Island, south of Cape Cod, and are characterized by ridges and shoals separated by deeper channels.
The surface expressions in this image reflect closely the bathymetric patterns shown on bottom
maps, with the more intense and distinct patterns occurring over areas shallower than 10 fathoms
(18 m). (Courtesy of JPL.)

can change drastically over areas much smaller than these resolutions. Synthetic-aperture
radars, with their much higher resolution, may provide the solution to measuring wind
speed and direction in coastal areas. The problem with using SAR images to estimate
wind speed and direction, however, is that these images are taken with a single look direction. As discussed before, several measurements are needed to decouple the wind speed
and direction. Monaldo et al. (2004) proposed a way to get around this limitation. Their
solution involves using the wind direction predicted by a numerical weather forecast
model together with the measured SAR cross sections in a standard geophysical model of
the form given in Equation (7-15) to estimate the wind speed. They demonstrated that by
using the Naval Operational Global Atmospheric Prediction System (NOGAPS) model
direction to retrieve wind speeds from the RADARSAT-1 SAR, the resulting wind speeds

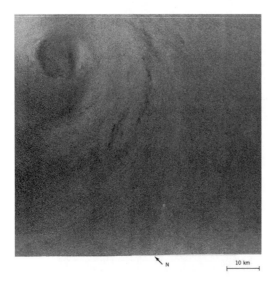
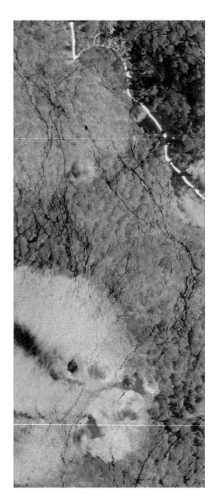

Figure 7-27. Shown on the left is a small-scale, well-organized tropical storm near the mouth of the Gulf of California. Except for its small size, the storm has all the characteristics of a hurricane, for example, the cyclonic spiral arms and a well-defined center of low winds. The image on the right shows an intense rain storm in the western Pacific imaged by the SIR-C radar. The storm is the yellowish area near the bottom of the image. See text for more discussion. (Courtesy NASA/JPL-Caltech.) See color section.

agreed with buoy measurements to a standard deviation of 1.76 m/s. Figure 7-29 shows an example of their results of retrieving wind speed from RADARSAT data, compared to the wind speed and direction measured with the QuikScat scatterometer over part of the Aleutian Islands. The SAR-derived wind fields show much more detail in the coastal wind patterns. Note, for example, the von Karman vortices in the wake of the Pogromni Volcano that are clearly visible in the SAR-derived wind fields.

7-4-3 Imaging of Sea Ice

The Arctic and Antarctic oceans are partially covered with floating ice, the extent of which varies significantly through the year. The extent of the ice cover and its seasonal variation has a strong impact on Earth's climate because the heat flux between the polar

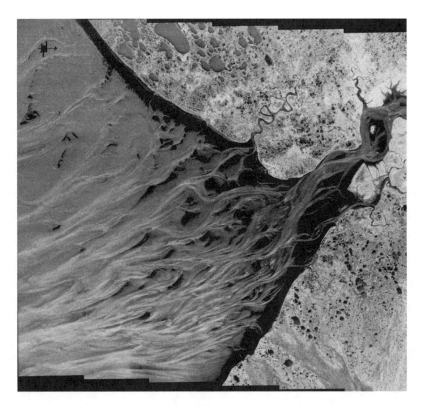

Figure 7-28. The Kuskokwim River in Alaska flows into the southeastern Bering Sea, where it forms large sediment deposits in the Kuskokwim Bay. As this image shows, many of the sediment deposits (dark areas) are exposed throughout the bay and are separated by channels (bright areas) from 7 to 14 m deep. The river flow should be substantial due to the runoff from the seasonal snow melt at this time of year and the additional component from the ebb tidal current of 0.8 m/sec. The linear variations within the channels themselves may be a result of the shear and strain of the current as it is deflected around the sediment deposits. Area size is 90 × 90 km. (Courtesy of JPL.)

oceans and the atmosphere, as well as the mean albedo, are strongly dependent on the ice cover extent. In addition, observation of the ice cover at the onset of local fall gives the extent of the ice that has survived the summer melt. In the winter, this ice will become massive, thick ice, called multiyear ice, which is the greatest environmental hazard for navigation and coastal installations.

Ice cover and dynamics have been monitored with active and passive microwave sensors. Microwave sensors are of particular interest because they show little sensitivity to cloud cover and can operate day and night. Imaging radars are used to acquire high-resolution surface images that delineate ice shape and areal extent. Periodic observations allow accurate determination of the ice motion (Figs. 7-30 and 7-31). Passive microwave imaging radiometers are used to measure, at a lower resolution, ice extent and to separate different ice ages. At 1.55 cm, multiyear ice has emissivity $\varepsilon_m = 0.84$, first-year ice has $\varepsilon_f = 0.92$, and open water has emissivity $\varepsilon_w = 0.5$. Thus if the surface thermodynamic temperature is $T = 273°K$, the surface effective emission temperature $T_e = \varepsilon T$ is 229°K, 251°K, and 137°K for multiyear ice, first-year ice, and open water, respectively. The observed microwave temperature is then given by

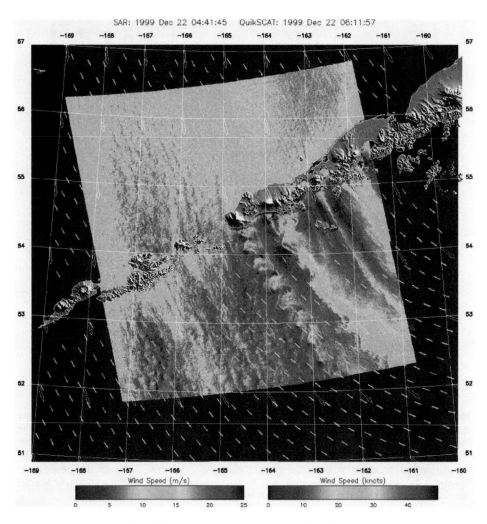

Figure 7-29. Combined QuikSCAT scatterometer and RADARSAT SAR wind-speed products reported by Monaldo et al. (2004). The small arrows represent the color-coded scatterometer wind vectors. The background image is the SAR-retrieved wind-speed field. The large arrows represent the NOGAPS model wind directions used to initialize the SAR wind-speed retrievals. When the QuikSCAT arrows blend into the SAR image, the wind speeds agree. This wind field covers a portion of the Aleutian Islands. See color section. (From Monaldo et al. © 2004 IEEE.)

$$T_{\mathrm{m}} = \varepsilon T + (1 - \varepsilon)T_{\mathrm{s}} \tag{7-31}$$

where T_{s} is the sky temperature. Figure 7-32 shows an example of passive microwave radiometer images of polar sea ice. See also Figures 5-17 and 5-18 for examples of passive microwave images of sea ice concentration.

7-4-4   Ocean Color Mapping

Suspended sediments and phytoplankton change the color of the ocean. Surface radiance measurement in several narrow bands in the visible region of the spectrum can be used to

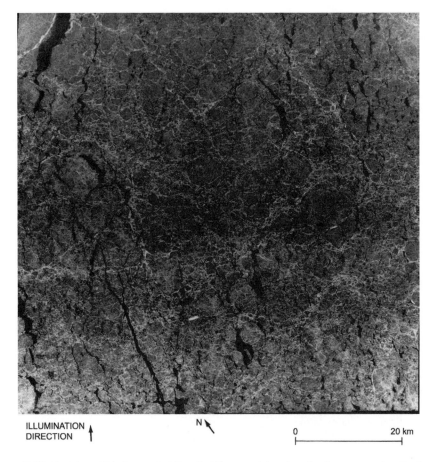

ILLUMINATION
DIRECTION

N

0   20 km

Figure 7-30. Pack ice within the central Beaufort Sea consists primarily of an aggregate of both mul-tiyear and first-year ice floes, some as large as 10 km in diameter. Multiyear floes (3 to 5 m thick) are rounded from extensive grinding, whereas first-year floes (1 to 1.5 m thick) tend to be more angular. Ridging and rafting of floes caused by compressive forces results in bright, linear features that often border the floes. The leads between floes appear as dark linear bands, indicating either a calm ocean surface or recently frozen ice. Of special interest is the bright, anomalous feature entrapped within the ice pack. This feature, with small ice floes cemented to it, is probably an ice island calved from the ice shelf on Ellesmere Island in the Arctic Ocean. (Courtesy of JPL.)

calculate the concentration of chlorophyll-like pigments and suspended sediments in the upper layers of the sea. Phytoplankton photosynthetic pigments are the single most im-portant contributor to ocean color. Chlorophyll-a is the primary photosynthetic pigment found in all plants and is also highly fluorescent. Thus, precise measurement of ocean col-or and chlorophyll fluorescence will give a good estimate of chlorophyll concentration and phytoplankton abundance.

Chlorophyll-a absorbs more blue and red light than green, making plants appear green. As a consequence, the ocean color as viewed from space changes from deep blue to green with increasing phytoplankton concentration. An intuitive measure of the chlorophyll concentration in ocean water would, therefore, be to compare the ratio of reflected light in the green part of the spectrum to that in the blue part of the spectrum.

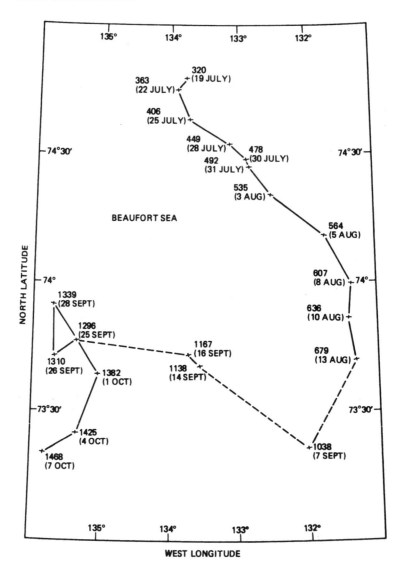

Figure 7-31. The bright feature in Figure 7-30 was imaged by the Seasat SAR on 20 separate passes, many of these over 3-day repeat sequences, providing a unique opportunity for the study of the drift of an ice island within pack ice. During the 80-day period from July 19 to October 7, it traveled circuitously approximately 435 km, an average of 5.4 km per day, as shown on the map. The greatest drift velocity was found to be between September 28 and October 1, when it averaged 12.2 km per day. The crosses indicate the position and orientation of the ice island plus the revolution number and date. The dashed lines indicate a significant gap in the SAR coverage of this feature. (Courtesy of B. Holt, Jet Propulsion Laboratory.)

A large percentage of the upwelling radiance observed from space in the visible spectrum comes from the atmosphere and must be removed in order to determine the ocean surface spectral reflectance. This is usually done by assuming that the ocean does not emit in the red. Thus, all the reflectivity in the red is used as a benchmark for atmospheric contribution. The measured red radiance is then extrapolated to shorter wavelengths using

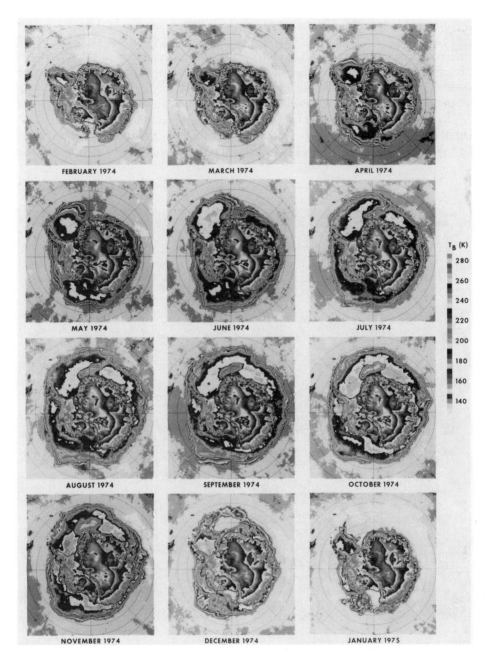

Figure 7-32. Passive microwave imagery of floating ice acquired by a spaceborne sensor. The twelve images show the thermally emitted radiance temperature (color code on the right) at a microwave frequency of 19.35 GHz. The images show the change of sea ice coverage from a minimum extent in January/February to a maximum extent in September. See color section.

known spectral properties of aerosols and Rayleigh scattering of light. The extrapolated radiance is subtracted from the observed radiances, thus providing surface radiance. The surface spectral radiance is then used to determine the concentration of chlorophyll-like pigments. This correction technique must be used with care when observing turbid waters in which suspended sediments might have some reddish hue.

An ocean color sensor was flown on Nimbus 7 in 1978 (Coastal Zone Color Scanner, CZCS). Landsat multispectral data have also been used to study the ocean color. Figure 7-33a illustrates the chlorophyll concentration measured by CZCS off the western coast of the United States. A number of empirical algorithms are used to relate chlorophyll con-

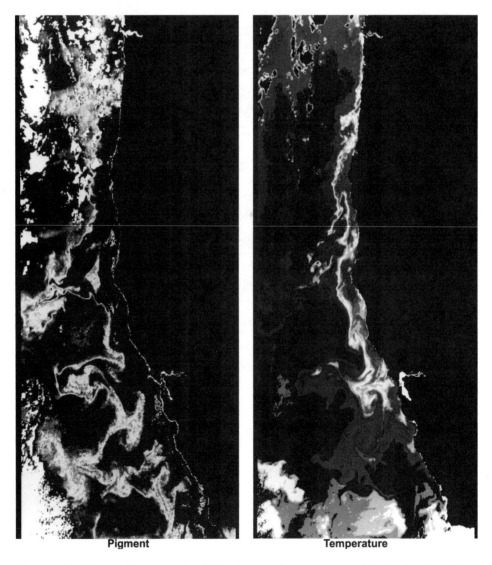

Pigment   **Temperature**

Figure 7-33. Chlorophyll concentration (left) and sea surface temperature (right) derived from Nimbus 7 CZCS along the U.S. west coast acquired July 7, 1981. The correlation between the parameters of the two images is clearly apparent.

centration to ocean reflectivity at a multiplicity of wavelengths in the 0.43 to 0.8 μm region (see Stewart, 1985 for a detailed discussion). O'Reilly and coworkers (1998) give a summary of more recent algorithms and a comparison of their performance. Ocean color is now monitored routinely from space with various instruments, including the Sea-viewing Wide Field-of-view Sensor (SeaWiFS), Medium Resolution Imaging Spectrometer (MERIS), Moderate Resolution Imaging Spectroradiometer (MODIS), Ocean Colour and Temperature Scanner (OCTS), and POLarization and Directionality of the Earth's Reflectances (POLDER). The characteristics of these sensors are compared in O'Reilly et al. (1998). Figure 7-34 shows the chlorophyll concentration associated with a phytoplankton bloom in the Gulf of Alaska acquired with the MODIS instrument on the NASA Aqua spacecraft on April 11, 2005. Blooms of phytoplankton often occur in these latitudes during the spring when the days are getting longer and the angle of the sun above the horizon increases. This means that more sunlight becomes available for photosynthesis and for heating of the water near the ocean surface. The warmer water allows the phytoplankton to remain close to the surface where they use the sunlight to produce the complex hydrocarbons. These blooms disappear once the nutrients are consumed by the phytoplankton.

7-4-5 Ocean Surface Temperature Mapping

Thermal emission from the ocean surface covers the whole spectrum from the near infrared to the microwave and peaks near 10 μm. Both infrared and passive microwave sensors have been used to measure and map the ocean surface temperature.

In the infrared region, the most appropriate spectral regions correspond to the 3–4 μm and 10–12 μm atmospheric windows. Even though the emitted radiation at 10 μm is stronger than at 3 μm, the sensitivity to surface temperature variation is significantly larger at the shorter wavelength. The relative sensitivity is given by the ratio of the derivative of Planck's spectral brightness relative to temperature ($\partial S/\partial T$ from Equation 4-1) over S:

$$\frac{1}{S}\frac{\partial S}{\partial T} = \frac{c_2}{\lambda T^2}\frac{\exp(c_2/\lambda T)}{\exp(c_2/\lambda T) - 1}$$

In most of the near infrared region $c_2 \gg \lambda T$; then,

$$\frac{1}{S}\frac{\partial S}{\partial T} \simeq \frac{c_2}{\lambda T^2}$$

This shows that the relative sensitivity at 3 μm is about 3.3 times higher than the sensitivity at 10 μm. In addition, the error due to water vapor in the 3–4 μm region is significantly lower than in the 10 μm region. On the other hand, the error due to aerosol scattering is significantly higher at the shorter wavelengths (Rayleigh scattering). The above has led to the use of multispectral observations to correct for humidity and aerosol effects. Usually observations are made at a minimum of three wavelengths (3.7, 10.5, and 11 μm) and sometimes at up to 24 bands to measure ocean temperature in the presence of clouds to an accuracy of 1°C.

High-resolution infrared radiometers have been flown on a number of NOAA satellites in polar orbits since 1966. Some of them are summarized in Table 7-2. They usually operate in the visible (~0.65 μm), near infrared (~3.5 μm), and thermal infrared (~10.5 μm) regions, with spatial resolutions of one to a few kilometers and noise equivalent ΔT of down to 0.1°K. Figure 7-33b shows an example of the data acquired with the AVHRR.

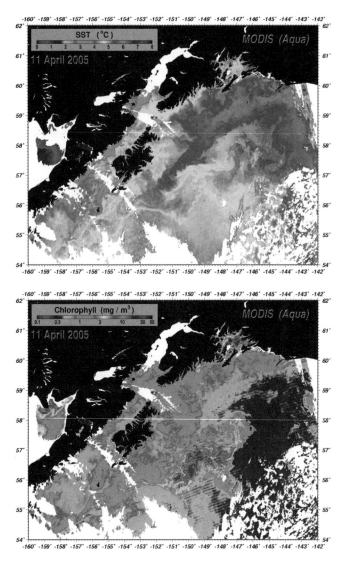

Figure 7-34. Measurements of sea surface temperature (top) and chlorophyll concentration (bottom) off the coast of Alaska made with the MODIS instrument on the NASA Aqua spacecraft. High concentrations of chlorophyll are shown in red. The images were acquired on 11 April 2005. (Courtesy NASA/Goddard Space Flight Center.) See color section.

In the microwave region, the effect of clouds is usually negligible. However, the surface emissivity is only 0.4, leading to an effective temperature of about 120°K and a corresponding reduction in the contrast due to temperature variation. The surface effective emissivity is influenced by the surface roughness (which varies with wind), foam cover, and salinity (which affects the dielectric constant). This leads to the need for multifrequency observations in order to separate the effect due to the surface temperature. Additional channels are also used to separate atmospheric effects.

A scanning multichannel microwave radiometer was flown on Seasat and Nimbus 7 in 1978. It operated at five frequencies (wavelengths 0.8, 1.4, 1.7, 2.8, and 4.5 cm) and two

TABLE 7-2. Characteristics of Some of the Spaceborne Infrared and Microwave Radiometers Used for Surface Temperature Mapping

Satellite	Instrument	Wavelength	Resolution (km)	Swath (km)	Sensitivity (K)
Nimbus 1, 2, 3 (1966–1972)	High Resolution Infrared Radiometer (HRIR)	3.4–4.2 μm	8	2700	1–4
NOAA 2, 3, 4, 5 (1972–1979)	Very High Resolution Radiometer (VHRR)	0.6–0.7 μm 10.5–12.5 μm	1	2900	1–4
NOAA 7, 8 (1981–)	Advanced Very High Resolution Radiometer (AVHRR)	0.58–0.68 μm; 0.7–21.1 μm; 3.55–3.93 μm; 10.3–11.3 μm; 11.5–12.5 μm	1 and 4	2700	0.12
GOES-5 (1981)	Visible Infrared Spin Scan Radiometer/ Atmospheric Sounder (VAS)	0.55–0.7 μm 3.2–4.2 μm 9–13 μm 10.5–12.6 μm	1 8	13,000 (full disc)	0.3
Nimbus 6 (1975)	Electronic Scanning Microwave Radiometer (ESMR)	0.8 cm	20	1270	1.5
Nimbus 7, Seasat (1978)	Scanning Multichannel Microwave Radiometer (SMMR)	0.8 cm 1.4 cm 1.7 cm 3 cm 4.5 cm	21 38 44 74 121	600	0.9–1.5
DMSP (1987-)	Special Sensor Microwave Imager (SSM/I)	0.35 0.81 1.36 1.58	16 × 14 38 × 30 60 × 40 70 × 45	1392	0.3–0.7

polarizations. The measurements were used to derive surface temperature, wind speed, and vertical column water vapor. The accuracy achieved was better than 1.2°C, 2 m/sec, and 0.4 g/cm^2, respectively. Table 7-2 lists the characteristics of some microwave radiometers used for surface temperature mapping. See also Chapter 5 for a discussion of some of these sensors.

EXERCISES

7.1 The backscatter cross section of the ocean surface in the K-band frequency region as a function of wind velocity is given by

$$\sigma = CU^{\gamma}(1 + a \cos 2\alpha)$$

where C, γ, and a are constants, and α is the angle between the observation plane and the wind vector. In order to measure the wind velocity and direction unambiguously, measurements in more than two directions are required. Let us assume that the first two measurements are taken in orthogonal directions:

(a) Show that if a third measurement is taken, the best observation direction will be at 45° relative to the first two directions

(b) If we take more than three measurements, can we eliminate the α, $\alpha + \pi$ ambiguity?

7.2. The Pierson–Moskowitz frequency spectrum for deep ocean waves is given by

$$S(\Omega) = \frac{0.0081 g^2}{\Omega^5} \exp\left[-\frac{3}{4}\left(\frac{\Omega_m}{\Omega}\right)^4\right]$$

where $\Omega_m = g/W$, with W being the wind speed at a height of 19.5 m above the sea surface (which was the height of the anemometers on the weather ships used by Pierson and Moskowitz in their research).

(a) Derive the corresponding wave number spectrum where $\Omega = \sqrt{gK}$. Write it in the form

$$F(K) = f(K) \exp\left[-\left(\frac{K_m}{K}\right)^2\right]$$

and give the expression for K_m.

(b) Derive the expression for the radar backscatter cross section using the small perturbation model expressions fur this spectrum.

(c) Assuming that $\varepsilon = 81$, plot the behavior of the backscattering cross section as a function of the angle of incidence for two different radar wavelengths ($\lambda = 1$ m and $\lambda = 0.25$ m) for a wind speed of $W = 5$ m/sec.

REFERENCES AND FURTHER READING

Akitomo, K., M. Ooi, T. Awaji, and K. Kutsuwada. Interannual variability of the Kuroshio transport in response to the wind stress field over the North Pacific: Its relation to the path variation south of Japan. *Journal Geophysics Research,* **101** (C6), 14,057–14,072, 1996.

Alpers, W., U. Pahl, and G. Gross. Katabatic wind fields in coastal areas studied by ERS-1 synthetic aperture radar imagery and numerical modeling. *Journal Geophysics Research,* **103** (C4), 7875–7886, 1998.

Alpers, W. R., D. B. Ross, and C. L. Rufenach. On the detectability of ocean surface waves by real and synthetic aperture radar. *Journal Geophysics Research,* **86** (C7), 6481–6498, 1981.

Alpers, W. R., and C. L. Rufenach. The effects of orbital motions on synthetic aperture imagery of ocean waves. *IEEE Transactions on Antennas and Propagation,* **AP-27**, 685–690, 1979.

Asanuma, I., K. Matsumoto, H. Okano, T. Kawano, N. Hendiarti, and S. I. Sachoemar. Spatial distribution of phytoplankton along the Sunda Islands: The monsoon anomaly in 1998. *Journal Geophysics Research,* **108**(C6), 3202, doi:10.1029/1999JC000139, 2003.

Babin, S. M., J. A. Carton, T. D. Dickey, and J. D. Wiggert. Satellite evidence of hurricane-induced phytoplankton blooms in an oceanic desert. *Journal Geophysics Research,* **109**, C03043, doi:10.1029/2003JC001938, 2004.

Barrick, D. E. Rough surface scattering based on the specular point theory. *IEEE Transactions on Antennas and Propagation,* **AP-16**, 449–454, 1968.

Barrick, D. E. First-order theory and analysis of MF/HF/UHF scatter from the sea. *IEEE Transactions on Antennas and Propagation,* **AP-20,** 2–10, 1972.

Barrick, D. E. The use of skywave radar for remote sensing of sea states. *Marine Technical Society Journal,* **7** (1), 29–33, 1973.

Barrick, D. E., M. W. Evans, and B. L. Weber. Ocean surface currents mapped by radar. *Science,* **198** (4313), 138–144, 1977.

Beal, R. C. Spaceborne imaging radar: Monitoring of ocean waves. *Science,* **208** (4450), 1373–1375, 1980.

Beal, R. C. Spatial evolution of ocean wave spectra. In R. C. Beal, P. S. DeLeonibus, and I. Katz (Eds.). *Spaceborne Synthetic Aperture Radar for Oceanography,* pp. 110–127. The Johns Hopkins University Press, Baltimore, 1981.

Beal, R. C., P. S. DeLeonibus, and I. Katz (eds.). *Spaceborne Synthetic Aperture Radar for Oceanography.* The Johns Hopkins University Press, Baltimore, 1981.

Beckmann, P., and A. Spizzichino. *The Scattering of Electromagnetic Waves from Rough Surfaces.* Macmillian, New York, 1963.

Bernstein, R. L. Sea surface temperature estimation using the NOAA 6 satellite advanced very high resolution radiometer. *Journal Geophysics Research,* **87** (C12), 9455–9465, 1982.

Bernstein, R. L., G. H. Born, and R. H. Whritner. Seasat altimeter determination of ocean current variability. *Journal Geophysics Research,* **87** (C5), 3261–3268, 1982.

Bernstein, R. L., L. Breaker, and R. Whritner. California current eddy formation: Ship, air, and satellite results. *Science,* **195,** 353–359, 1977.

Bernstein, R. L., and W. B. White. Zonal variability of eddy energy in the midlatitude north Pacific Ocean. *Journal Physical Oceanography,* **7** (1), 123–126, 1977.

Born, G. H., M. A. Richards, and G. W. Rosborough. An empirical determination of the effects of sea state bias on Seasat altimetry. *Journal Geophysics Research,* **87** (C5), 3221–3226, 1982.

Boulanger, J.-P., S. Cravatte, and C. Menkes. Reflected and locally wind-forced interannual equatorial Kelvin waves in the western Pacific Ocean. *Journal Geophysics Research,* **108** (C10), 3311, doi:10.1029/2002JC001760, 2003.

Bracalante, E. M., et al. The SASS Scattering Coefficient Algorithm. *IEEE Journal of Oceanic Engineering,* **OE-5,** 145–154, 1980.

Brandt, P., W. Alpers, and J. O. Backhaus. Study of the generation and propagation of internal waves in the Strait of Gibraltar using a numerical model and synthetic aperture radar images of the European ERS 1 satellite. *Journal Geophysics Research,* **101**(C6), 14,237–14,252, 1996.

Brown, W. L., C. Elachi, and T. W. Thompson. Radar imaging of ocean surface patterns. *Journal Geophysics Research,* **81** (15), 2657–2667, 1976.

Brüning, C., R. Schmidt, and W. Alpers. Estimation of the ocean wave–radar modulation transfer function from synthetic aperture radar imagery. *Journal Geophysics Research,* **99** (C5), 9803–9816, 1994.

Carr, M.-E., P. T. Strub, A. C. Thomas, and J. L. Blanco. Evolution of 1996–1999 La Niña and El Niño conditions off the western coast of South America: A remote sensing perspective. *Journal Geophysics Research,* **107**(C12), 3236, doi:10.1029/2001JC001183, 2002.

Carr, M.-E. Simulation of carbon pathways in the planktonic ecosystem off Peru during the 1997–1998 El Niño and La Niña. *Journal Geophysics Research,* **108** (C12), 3380, doi:10.1029/1999JC000064, 2003.

Chase, Z., A. van Geen, P. M. Kosro, J. Marra, and P. A. Wheeler. Iron, nutrient, and phytoplankton distributions in Oregon coastal waters. *Journal Geophysics Research,* **107** (C10), 3174, doi:10.1029/2001JC000987, 2002.

Chelton, D. B., K. J. Hussey, and M. E. Park. Global satellite measurements of water vapor, wind speed, and wave height. *Nature,* **234,** 529–532, 1981.

Chen, G., B. Chapron, R. Ezraty, and D. Vandemark. A dual-frequency approach for retrieving sea surface wind speed from TOPEX altimetry. *Journal Geophysics Research,* **107**(C12), 3226, doi:10.1029/2001JC001098, 2002.

Cheney, R. E., and J. G. Marsh. Seasat altimeter observations of dynamic topography in the Gulf Stream. *Journal Geophysics Research,* **86** (1), 473–483, 1981.

Chu, P. C., J. M. Veneziano, C. Fan, M. J. Carron, and W. T. Liu. Response of the South China Sea to Tropical Cyclone Ernie 1996. *Journal Geophysics Research,* **105** (C6), 13,991–14,009, 2000.

Chubb, S. R., F. Askari, T. F. Donato, R. Romeiser, S. Ufermann, A. L. Cooper, W. Alpers, S. A. Mango, and J.-S. Lee. Study of Gulf Stream Features with a Multifrequency Polarimetric SAR from the Space Shuttle. *IEEE Transactions on Geoscience and Remote Sensing,* **37**, 2495–2507, 1999.

Colton, M. C., and G. A. Poe. Intersensor Calibration of DMSP SSM/I's: F-8 to F-14, 1987–1997. *IEEE Transactions on Geoscience and Remote Sensing,* **37**, 418–439, 1999.

Doney, S. C., D. M. Glover, S. J. McCue, and M. Fuentes. Mesoscale variability of Sea-viewing Wide Field-of-view Sensor (SeaWiFS) satellite ocean color: Global patterns and spatial scales. *Journal Geophysics Research,* **108** (C2), 3024, doi:10.1029/2001JC000843, 2003.

Draper, D. W., and D. G. Long. An assessment of SeaWinds on QuikSCAT wind retrieval. *Journal Geophysics Research,* **107** (C12), 3212, doi:10.1029/2002JC001330, 2002.

Elachi, C., and W. E. Brown. Models of radar imaging of ocean surface waves. *IEEE Transactions on Antennas and Propagation,* **AP-25**, v84–95, 1977.

Elachi, C. Radar imaging of the ocean surface. *Boundary Layer Meteorology,* **13**, 165–179, 1978.

Fedor, L. S., and G. S. Brown. Waveheight and wind speed measurement from the Seasat altimeter. *Journal Geophysics Research,* **87** (C5), 3254–3620, 1982.

Fu, L. L. On the wavenumber of oceanic mesoscale variability observed by the Seasat altimeter. *Journal Geophysics Research,* **88** (C7), 4331–4341, 1983.

Fu, L. L., and B. Holt. *Seasat Views Oceans and Sea Ice with Synthetic-Aperture Radar.* Pasadena: NASA Jet Propulsion Laboratory Publication 81-1D, 1981.

Fu, L. L., and B. Holt. Some examples of detection of oceanic mesoscale eddies by the Seasat Synthetic-Aperture Radar. *Journal Geophysics Research,* **88** (C3), 1844–1852, 1983.

Fu, L-L., and A. Cazenave. *Satellite Altimetry and Earth Sciences: A Handbook of Techniques and Applications.* Academic Press, San Diego, 2001.

Hasselmann, K. Feynman diagrams and interaction rules of wave-wave scattering processes. *Review of Geophysics,* **4** (1), 1–32, 1966.

Hasselmann, K. A simple algorithm for the direct extraction of the two-dimensional surface image spectrum from the return signal of a synthetic aperture radar. *International Journal Remote Sensing,* **1**, 219–240, 1980.

Jain, A. Determination of ocean wave heights from synthetic aperture radar imagery. *Applied Physics,* **13**, v371–382, 1977.

Jain, A. Focusing effects in synthetic aperture radar imaging of ocean waves. *Applied Physics,* **15**, 323–333, 1978.

Jain, A. SAR imaging of ocean waves: Theory. IEEE Journal Oceanic Engineering, **OE-6** (4), 130–139, 1981.

Jain, A., G. Medlin, and C. Wu. Ocean wave height measurement with Seasat SAR using speckle diversity. *IEEE Journal Oceanic Engineering,* **OE-7** (2), 103–107, 1982.

Jones, W. L., D. H. Boggs, E. M. Bracalente, R. A. Brown, T. H. Guymer, D. Chelton, and L. C. Schroeder. Evaluation of the Seasat wind scatterometer. *Nature,* **294** (5843), 704–707, 1981.

Jones, W. L., V. E. Delnore, and E. M. Bracalente. The study of mesoscale ocean winds. In R. C. Beal, P. S. DeLeonibus, and I. Katz (Eds.). *Spaceborne Synthetic Aperture Radar for Oceanography,* pp. 87–94. The Johns Hopkins University Press, Baltimore, 1981.

Jones, W. L., and L. C. Schroeder. Radar backscatter from the ocean: Dependence on surface friction velocity. *Boundary-Layer Meteorology,* **13,** 133–149, 1978.

Jones, W. L., L. C. Schroeder, D. H. Boggs, E. M. Bracalante, R. A. Brown, G. J. Dome, W. J. Pierson, and F. J. Wentz. The Seasat-A satellite scatterometer: The geophysical evaluation of remotely sensed wind vectors over the ocean. *Journal Geophysics Research,* **87** (C5), 3297–3317, 1982.

Jones, W. L., L. C. Schroeder, and J. L. Mitchell. Aircraft measurements of the microwave scattering signature of the ocean. *IEEE Transactions on Antennas and Propagation,* **AP-25** (1), 52–61, 1977.

Large, W.G. and S. Pond. Open ocean momentum flux measurements in moderate to strong winds. *Journal of Physical Oceanography,* **11,** 324–336, 1981.

Legeckis, R. A Survey of world wide sea surface temperature fronts detected by environmental satellites. *Journal Geophysics Research,* **83,** 4501–4522, 1978.

Liu, W. T., and X. Xie. Double intertropical convergence zones-a new look using scatterometer. *Geophysics Research Lett.,* **29** (22), 2072, doi:10.1029/2002GL015431, 2002.

Monaldo, F. M., D. R. Thompson, W. G. Pichel, and P. Clemente-Colón. A Systematic Comparison of QuikSCAT and SAR Ocean Surface Wind Speeds. *IEEE Transactions on Geoscience and Remote Sensing,* **42,** 283–291, 2004.

Murtugudde, R. G., S. R. Signorini, J. R. Christian, A. J. Busalacchi, C. R. McClain, and J. Picaut. Ocean color variability of the tropical Indo-Pacific basin observed by SeaWiFS during 1997–1998. *Journal Geophysics Research,* **104** (C8), 18,351–18,366, 1999.

O'Reilly, J. E., S. Maritorena, B. G. Mitchell, D. A. Siegel, K. L. Carder, S. A. Garver, M. Kahru, and C. McClain. Ocean color chlorophyll algorithms for Sea WiFS. *Journal Geophysics Research,* **103**(C11), 24,937–24,953, 1998.

Phillips, O. M. *The Dynamics of the Upper Ocean.* Cambridge University Press, Cambridge, 1966.

Risien, C. M., C. J. C. Reason, F. A. Shillington, and D. B. Chelton. Variability in satellite winds over the Benguela upwelling system during 1999–2000. *Journal Geophysics Research,* **109,** C03010, doi:10.1029/2003JC001880, 2004.

Samelson, R., P. Barbour, J. Barth, S. Bielli, T. Boyd, D. Chelton, P. Kosro, M. Levine, E. Skyllingstad, and J. Wilczak. Wind stress forcing of the Oregon coastal ocean during the 1999 upwelling season. *Journal Geophysics Research,* **107**(C5), 3034, doi:10.1029/2001JC000900, 2002.

Schlax, M.G., and D. B. Chelton. Detecting aliased tidal errors in altimeter height measurments. *Journal Geophysics Research,* **99** (C6), 12,603–12,612, 1994.

Schlax, M.G., and D. B. Chelton. Aliased tidal error in TOPEX/POSEIDON sea surface height data. *Journal Geophysics Research,* **99** (C12), 24,761–24,775, 1994.

Seasat Special Issue. Geophysical evaluation. *Journal Geophysics Research,* **87,** 1982.

Shuchman, R. A., and J. S. Zelenka. Processing of ocean wave data from a synthetic aperture radar. *Boundary-Layer Meteorology,* **13,** 181–191, 1978.

Special Joint Issue on Radio Oceanography. *IEEE Journal Oceanic Eng.,* **OE-2,** 1–59, 1977.

Stewart, R. H. *Methods of Satellite Oceanography.* University of California Press, Berkeley, 1985.

Stewart, R. H. Satellite oceanography: The instruments. *Oceanus,* **24** (3), 66–74, 1982.

Tang, D. L., H. Kawamura, H. Doan-Nhu, and W. Takahashi. Remote sensing oceanography of a harmful algal bloom off the coast of southeastern Vietnam. *Journal Geophysics Research,* **109,** C03014, doi:10.1029/2003JC002045, 2004.

Tapley, B. D., and M.-C. Kim. Applications to Geodesy. In L.-L. Fu and A. Cazenave (Eds.), *Satellite Altimetry and Earth Sciences: A Handbook of Techniques and Applications.* Academic Press, San Diego, 2001.

Townsend, W. F. The initial assessment of the performance achieved by the Seasat altimeter. *IEEE Journal Oceanic Eng.,* **OE-5,** 80–92, 1980.

Tsai, W.-Y., S. V. Nghiem, J. N. Huddleston, M. W. Spencer, B. W. Stiles, and R. D. West. Polarimetric Scatterometry: A Promising Technique for Improving Ocean Surface Wind Measurements from Space. *IEEE Transactions on Geosci. Remote Sens.,* **38,** 1903–1921, 2000.

Valenzuela, G. R. Theories for the interaction of electromagnetic and oceanic waves—A review. *Boundary-Layer Meteorology,* **13,** 61–85, 1978.

Valenzuela, G. R. An asymptotic formulation for SAR images of the dynamical ocean surface. *Radio Science,* **15,** 105–114, 1980.

Vesecky, J. F., and R. H. Stewart. The observation of ocean surface phenomena using imagery from the Seasat synthetic aperture radar: An assessment. *Journal Geophysics Research,* **87** (C3), 3397–3430, 1982.

Wilson, C., and D. Adarnec. Correlations between surface chlorophyll and sea surface height in the tropical Pacific during the 1997–1999 El Niño—Southern oscillation event. *Journal Geophysics Research,* **106** (C12), 31,175–31,188, 2001.

Yu, L., R. A. Weller, and W. T. Liu. Case analysis of a role of ENSO in regulating the generation of westerly wind bursts in the Western Equatorial Pacific. *Journal Geophysics Research,* **108** (C4), 3128, doi:10.1029/2002JC001498, 2003.

Zwally, H. J., and P. Gloersen. Passive microwave images of the polar regions and research applications. *Polar Record,* VII-44, **18** (116), 431–450, 1977.

8

BASIC PRINCIPLES OF ATMOSPHERIC SENSING AND RADIATIVE TRANSFER

The interactions of electromagnetic waves with planetary atmospheres are governed by the characteristics of the propagating wave (mainly its wavelength), the physical characteristics of the atmosphere (pressure, temperature, and suspended particulates), and its constituents. These interaction mechanisms are relatively complex to model because of the three-dimensional nature of the propagation medium and the multiplicity of the interaction mechanisms: scattering, absorption, emission, and refraction.

In this chapter, we give a brief overview of the physical and chemical properties of planetary atmospheres that are relevant to remote sensing. The basic concepts of temperature, composition, pressure, density, and wind sensing will be established as an introduction to the more detailed analysis in the following chapters.

8-1 PHYSICAL PROPERTIES OF THE ATMOSPHERE

The atmospheric density decreases with altitude. In the case of hydrostatic equilibrium, the pressure $p(z)$ and density $\rho(z)$ are related by the following relationship:

$$dp(z) = -g\rho(z)dz \qquad (8\text{-}1)$$

where g is the planet's gravity, assumed to be constant through the thin atmospheric layer, and p is measured vertically upward from the surface. The above expression basically states that the difference of pressure between levels z and $z + dz$ is equal to the weight of the atmosphere between these two levels.

Introduction to the Physics and Techniques of Remote Sensing. By C. Elachi and J. van Zyl

The equation of state for a perfect gas relates the pressure and density by

$$\rho(z) = M M_0 \frac{p(z)}{kT(z)}$$

or

$$N(z) = \frac{p(z)}{kT(z)} \tag{8-2}$$

where M is the average molecular weight of the atmosphere ($M = 28.97$ for the Earth's atmosphere), M_0 is the atomic mass unit ($M_0 = 1.66 \times 10^{-27}$ kg), k is Boltzmann's constant ($k = 1.38 \times 10^{-23}$ JK^{-1}), N is the number density (molecules/m^3), and T is the temperature. Combining Equations 8-1 and 8-2 we get

$$\frac{dp}{p} = -g \frac{M M_0}{kT(z)} dz = -\frac{dz}{H(z)} \tag{8-3}$$

where

$$H(z) = \frac{kT(z)}{g M M_0} \tag{8-4}$$

H is known as the scale height. The solution for Equation 8-3 is simply

$$p(z) = p(0) \exp\left[-\int_0^z \frac{d\zeta}{H(\zeta)}\right] \tag{8-5}$$

where $p(0)$ is the surface pressure. In the case of an isothermal atmosphere ($T =$ constant),

$$p(z) = p(0) \exp \frac{-z}{H} \tag{8-6}$$

which shows that the pressure decreases exponentially with altitude. The same is true for the density

$$\rho(z) = \rho(0) \exp \frac{-z}{H} \tag{8-7}$$

where

$$\rho(z) = \frac{M M_0 p(0)}{kT} = \frac{p(0)}{gH} \tag{8-8}$$

Similarly, the number density N (molecules/m^3) is given by

$$N(z) = N(0) \exp \frac{-z}{H} \tag{8-9}$$

By integrating Equation 8-7, we get the total atmospheric mass M_T in a column of unit area:

$$M_T = \int_0^\infty \rho(z)dz = \rho(0)H = \frac{p(0)}{g} \tag{8-10}$$

This shows that the scale height corresponds to the thickness of a homogeneous atmospheric layer of density $\rho(0)$ and mass equal to the atmospheric mass.

To illustrate, in the case of the Earth's atmosphere, we have $M = 28.97$, $p(0) = 1$ atmosphere $= 10^5$ Newton/m^2, $g = 9.81$ m/sec^2, and $T = 288$ K. This gives:

$$H = 8.4 \text{ km} \qquad \text{(from Equation 8-4)}$$

$$\rho(0) = 1.21 \text{ kg/m}^3 \qquad \text{(from Equation 8-8)}$$

$$M_T = 10.200 \text{ kg/m}^2 \qquad \text{(from Equation 8-10)}$$

Table 8-1 gives some of the parameters for the atmosphere of Venus, Mars, and Titan.

Most of the atmospheric mass is in a very thin layer above the surface relative to the planet's radius. In the case of the Earth, 99% of the atmospheric mass is below 32 km. Thus, in most models, the atmosphere is considered as a locally plane parallel slab, and the planetary curvature is neglected except in the case of limb sounding and occultation experiments.

The above illustrations assumed an isothermal atmosphere. Usually, this is not exactly true. In the case of the Earth, the temperature near the surface decreases as a function of altitude up to an altitude of 11 km, then remains constant up to the 25 km level. The change of the atmospheric temperature can be derived for a simplified case in which it is assumed that the atmosphere is transparent to all radiation and contains no liquid particles. If we assume a unit mass moving upward in an atmosphere in hydrostatic equilibrium, the first law of thermodynamics gives

$$C_v dT + p \, dV = dq = 0 \tag{8-11}$$

where C_v is the specific heat at constant volume, dV is the change in volume, dT is the change in temperature, and dq is the heat input, which is equal to zero. Differentiating the equation of state (Equation 8-2) and remembering that $\rho = 1/V$, we get

$$p \, dV + V \, dp = \frac{k}{MM_0} dT \tag{8-12}$$

TABLE 8-1. Some Properties of Planetary Atmospheres

	Venus	Earth	Mars	Titan
Surface pressure (bar)	90	1	0.02	1-3
Surface temperature (K)	880	290	~250	~94
Scale height (km)	15	8.4	~10	~20
Main constituents	97% CO_2	78% N_2	CO_2	CH_4
	N_2	21% O_2	N_2	
		1% Ar	Ar	
		H_2O, CO_2, O_3		

For a perfect gas we have $k/MM_0 = C_p - C_v$, where C_p is the specific heat at constant pressure. Combining Equations 8-11 and 8-12 we get:

$$V\, dp = C_p dT$$

Replacing dp by its value from Equation 8-1, we get

$$\frac{dT}{dz} = -\frac{g}{C_p} = -\Gamma_a \qquad (8\text{-}13)$$

where Γ_a is known as the adiabatic lapse rate. For the Earth's atmosphere, $C_p \simeq 1000$ J/kg K. This gives

$$\Gamma_a = 9.81 \text{ K/km}$$

If the temperature at the bottom of the atmosphere is equal to $T(0)$, then the solution for Equation 8-12 gives the temperature profile as

$$T(z) = T(0) - \Gamma_a z \qquad (8\text{-}14)$$

In reality, the temperature in the Earth's lower atmosphere does decrease linearly with altitude but at a rate of 6.5 K/km, somewhat lower than Γ_a. At higher altitudes, the temperature profile is significantly more complex, as shown in Figure 8-1.

8-2 ATMOSPHERIC COMPOSITION

The composition of the atmosphere varies significantly among planets, as shown in Table 8-1. In the case of the Earth, carbon dioxide, water vapor, and ozone dominate the interaction with electromagnetic radiation. Carbon dioxide is substantially uniformly mixed up to about 100 km. The interaction of the CO_2 molecule with electromagnetic waves is strongest in the infrared region, near 4.3 μm and 15 μm. It plays a dominant role in the energy budget of the mesosphere (50 to 90 km), which is cooled as a result of CO_2 radiative emission in the 15 μm band.

Water vapor plays an important role in the energy budget of the troposphere (below about 15 km) because of its role in cloud formation, precipitation, and energy transfer in the form of latent heat. Water vapor concentration is highly variable in space and time. At sea level, it varies from 10^{-2} g/m^3 in a very cold, dry climate up to 30 g/m^3 in hot, humid regions. The density decreases exponentially as a function of altitude with a scale height of about 2.5 km. For an average surface density of 10 g/m^3 the total columnar mass per unit area is 25 kg/m^2, which is about 0.25% of the total atmospheric mass. However, it is particularly important in remote sensing because of the strong water vapor absorption lines in the infrared (Fig. 8-2) and microwave regions.

Ozone strongly absorbs ultraviolet radiation and causes a shortwave cutoff of the Earth's transmission at 0.3 μm and shorter. It is mostly concentrated between 20 and 50 km altitude, and its distribution is also highly variable. The formation and dissociation of ozone involves a series of complex chemical catalytic reactions among nitrogen, hydrogen, and chlorine compounds, some of which arise from human activities.

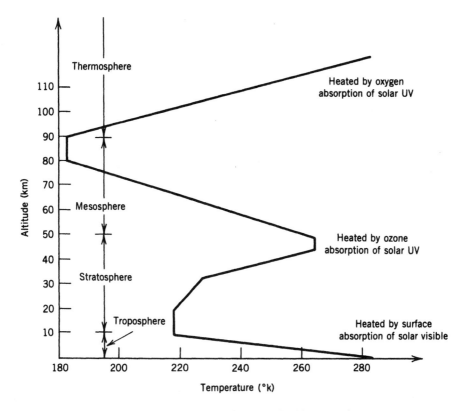

Figure 8-1. Temperature profile of the Earth's atmosphere.

In addition to CO_2, H_2O, and O_3 a number of other minor constituents play a role in the Earth's upper atmosphere chemistry. The distribution of some of these constituents is shown in Figure 8-3.

8-3 PARTICULATES AND CLOUDS

Particulates are atmospheric dust particles of radii between 0.1 and 10 μm. Clouds of liquid and solid water have particles of size varying between 1 and 100 μm. Particulates are usually concentrated in the lowest few kilometers and tend to scatter radiation in the visible and near- to mid-IR regions. Particulates can be ejected directly into the atmosphere or formed in the visible atmosphere from trace gases.

The height distribution of the number density of large particles can also be approximated by an exponential decay similar to Equation 8-9:

$$N(z) = N(0) \exp \frac{-z}{H_p} \tag{8-15}$$

where H_p is the particles' scale height. In the case of large particles in the Earth's atmosphere, H_p is about 1 km. The other important factor is the size distribution of the particles.

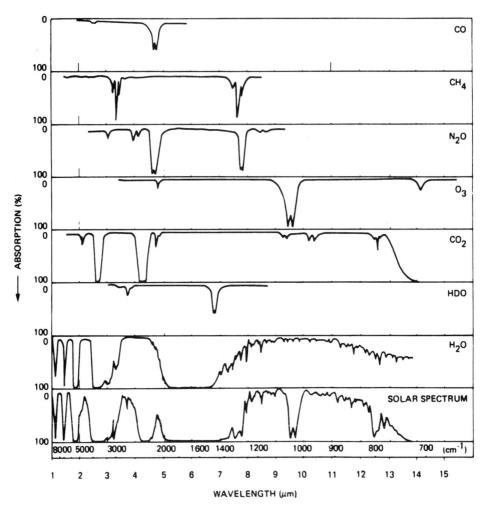

Figure 8-2. Absorption along a vertical atmosphere path by a variety of constituents in the spectral region from 1.0 to 16μm. (From J. H. Shaw, 1970.)

A number of models have been proposed to describe this distribution, and different ones are used to describe clouds, haze, or continental particulates.

Clouds completely obscure the surface in the visible and near-IR regions up to 3 μm because of scattering and, to some extent, absorption. At longer wavelengths, their effect is considerable up to the millimeter region; then they become gradually more transparent.

Cloud coverage varies considerably as a function of location and season. On the average, 40% of the Earth is cloud covered at any time. It is appreciably higher than this value in the tropical regions and appreciably lower over desert regions at midlatitude. In contrast, Venus is completely cloud covered all the time. The Venusian clouds are thought to consist of sulfuric acid droplets. They completely eliminate the use of visible or infrared sensors for imaging the surface. The same is true in the case of Titan, where a global layer of methane clouds/haze covers the surface and blocks its visibility in the visible and infrared regions.

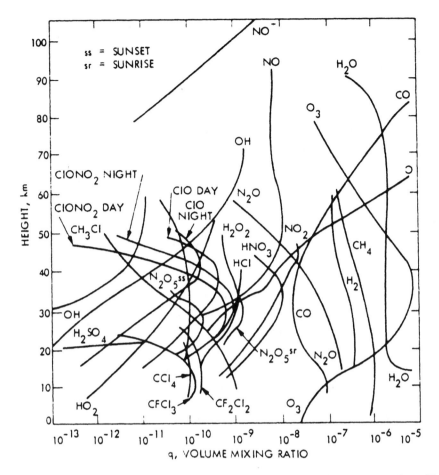

Figure 8-3. Distribution of some chemically active constituents in the Earth's atmosphere. (From Waters, 1984.)

8-4 WAVE INTERACTION MECHANISMS IN PLANETARY ATMOSPHERES

The interaction of electromagnetic waves with planetary atmospheres involves resonant interactions corresponding to molecular and atomic energy levels, nonresonant interactions, and refraction. In addition, suspended particles scatter the light waves and have their own radiative properties. These different interaction mechanisms are discussed in this section.

8-4-1 Resonant Interactions

When an electromagnetic wave interacts with a gaseous molecule, it may excite the molecule to a higher energy level and in the process transfer all or part of its energy to the molecule. A molecule in an excited state may drop to a state with a lower energy level and in the process emit energy in the form of an electromagnetic wave. The energy levels of gaseous molecules are well defined and discrete. Thus, the related interaction occurs at a very specific frequency leading to spectral lines. The environmental factors, specifically temperature and pressure, lead to the broadening of these lines to narrow bands.

The basic mechanisms behind the energy states of gaseous molecules are similar to what was discussed in the case of solids (Chapters 2 and 3). Electronic energy levels result from the transfer of electrons between different orbits. Vibrational energy levels result from the different vibrational modes of the molecule. In addition, gaseous molecules have rotational energy states that correspond to rotation of the molecule around different axes. On the other hand, the interaction mechanisms that were related to the crystalline structure in solids, such as crystal field effects, semiconducting bands, and color centers, do not exist in the case of gases.

The lowest energy levels correspond to the rotational states. These levels depend on the three principal moments of inertia of the molecule. Four types of rotating molecules exist, some of which are shown in Figure 8-4:

1. All three principal moments of inertia are different. This is the case for H_2O. The molecule is called asymmetric top.
2. Two of the moments are equal. This is a symmetric top molecule.
3. All three moments are equal. This is the case for CH_4. The molecule is called spherical top.
4. Two of the moments are equal and the third one is negligible. This is the case of a linear molecule such as CO_2.

No energy transitions are allowed in the pure rotation spectrum for a molecule that possesses no permanent dipole moment. The symmetric linear CO_2 molecule is such an example.

To illustrate, let us consider the case of the water molecule and oxygen molecule; Figure 8-5 shows their rotational modes. The lowest spectral line for water vapor occurs at

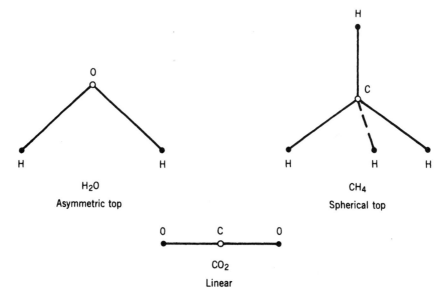

Figure 8-4. Three different molecules representing three different types of rotational effects.

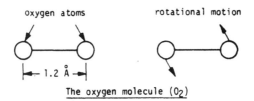

The oxygen molecule (O₂)

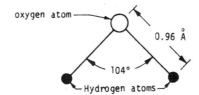

The water vapor molecule (H₂O)

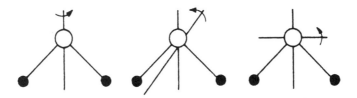

Figure 8-5. Rotational modes of the water and oxygen molecules.

22.2 GHz, the next-lowest one is at 183.3 GHz. Figure 8-6 shows the absorption spectrum of water vapor from 3 to 300 GHz at two pressure levels. An interesting observation is that as the pressure decreases, the absorption at the center of the line increases because the absorption from the same amount of water vapor is now limited to a narrower band.

The oxygen molecule has no permanent electrical dipole moment but has a permanent magnetic dipole moment resulting from the fact that two of the orbital electrons are unpaired. The two lowest spectral lines occur at 60 GHz (which is a band consisting of multiple lines) and 118.75 GHz (Fig. 8-7).

The absorption due to atmospheric oxygen and water vapor dominates the resonant interaction of waves in the Earth's lower atmosphere across the microwave spectrum from 20 GHz to 180 GHz.

In the upper atmosphere, the trace molecules play a more tangible role, and their spectral signature is usually very rich in spectral lines, as shown in Figure 8-8.

In the more simple model of a diatomic molecule and elementary calculation of the energy levels, the molecule can be regarded as a dumbbell consisting of two masses at a fixed distance. Let I be the moment of inertia relative to the axis perpendicular to the line connecting the two masses and intersecting it at the center of mass. In quantum mechanics, it can be shown that the corresponding rotational energy levels are given by

$$E = \frac{h}{8\pi^2 I} j(j+1) \tag{8-16}$$

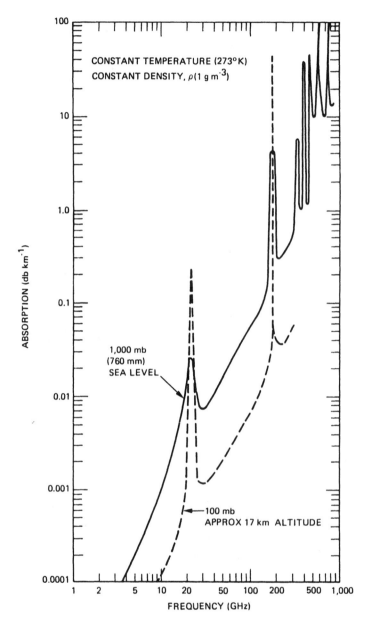

Figure 8-6. Absorption spectrum of water vapor at two pressures—1 bar and 0.1 bar—at tempera-ture 273 K and for a water vapor density of 1 g/m³ (from Chahine et al., 1983).

where j is the rotational quantum number ($j = 0, 1, 2, \ldots$). The selection rule requires that transitions can occur only between adjacent energy levels (i.e., $\Delta j = \pm 1$). Thus, the energy transfer for the transition from level j to $j - 1$ is

$$\Delta E_j = \frac{h}{8\pi^2 I}[j(j+1) - (j-1)j] = \frac{h}{4\pi^2 I}\,j \qquad (8\text{-}17)$$

which indicates that the spectrum consists of a series of equidistant lines.

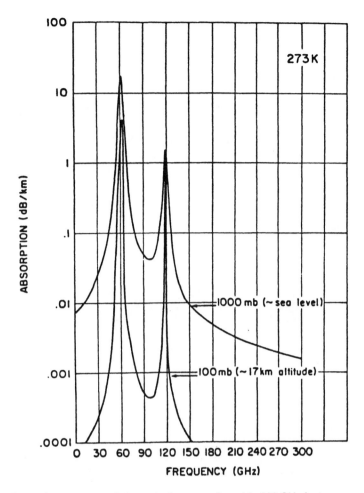

Figure 8-7. Absorption spectrum of atmospheric oxygen from 1 to 300 GHz for two pressures (1 bar and 0.1 bar) and at a constant temperature of 273 K (from Chahine et al., 1983).

The vibrational transitions correspond to significantly higher energy than the rotational transitions. Thus, the corresponding spectral lines appear mainly in the infrared region. One of the most important transitions used in atmospheric remote sensing is the 15 μm line of CO_2, which corresponds to the bending of the linear molecule.

As a molecule vibrates, its effective moment of inertia varies. Thus, we would expect the presence of a family of spectral lines around a vibrational line, which corresponds to rotation–vibration interactions. The rotational transitions appear as fine structure near the vibrational lines. This is clearly seen in the spectrum of the CO_2 molecule near the 15 μm band (wave number of 667 cm^{-1}).

The electronic transitions correspond to the highest energies, and the corresponding spectral lines are usually in the visible and ultraviolet part of the spectrum.

8-4-2 Spectral Line Shape

The spectral lines are not infinitely narrow. Line broadening results mainly from three factors: excited state lifetime, pressure-induced collisions, and thermal motion.

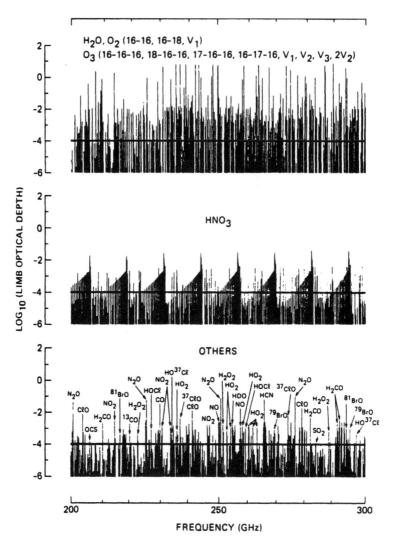

Figure 8-8. Spectral lines of a variety of atmospheric molecules in the Earth's upper atmosphere (from Waters et al., 1984).

If an excited state has a lifetime τ, the spectral line will have a frequency width Δv at least equal to:

$$\Delta v = \frac{1}{2\pi\tau} \tag{8-18}$$

This can be derived by assuming the emission from an excited state to be a pulse with exponentially decaying amplitude as a function of time. The Fourier transform of such a pulse will have a spectral width given by Equation 8-18.

The thermal motion of a molecule during emission or absorption of radiation gives rise to a Doppler shift. Considering that the thermal motion is random and the molecule's velocity distribution is related to the temperature, it is normal to expect that the thermal motion induces a broadening of the spectral line. The resultant shape is given by

$$k(v) = \frac{S}{v_D \sqrt{\pi}} \exp\left[-\left(\frac{v - v_0}{v_D}\right)^2\right] \tag{8-19}$$

where S is a constant representing the total strength of the line $[S = \int_0^\infty k(v)dv]$, v_0 is the center frequency, and v_D is the Doppler line width. v_D is related to the gas temperature by

$$v_D = \frac{v_0}{c} \sqrt{\frac{2\pi RT}{M}} \tag{8-20}$$

where

R = universal gas constant = k/M_0 = 8314 JK^{-1} kg^{-1}

M = molecular weight of the gas

T = temperature of the gas

The Doppler broadening dominates at high altitudes.

Another mechanism of line broadening results from pressure-induced collisions. This is given by a Lorentz shape function:

$$k(v) = \frac{S}{\pi} \frac{v_L}{(v - v_0)^2 + v_L^2} \tag{8-21}$$

where v_L is the Lorentz width. It is related to the mean time between collisions t_c by

$$v_L = \frac{1}{2\pi t_c} \tag{8-22}$$

The collision time in turn is inversely proportional to the pressure. Thus v_L is linearly proportional to the pressure:

$$\frac{v_L}{P} = \frac{v_L(P_0)}{P_0} = \text{constant} \tag{8-23}$$

The Lorentz broadening dominates at high pressure and thus is particularly important at low altitudes. As shown in Figure 8-9a, the Lorentz broadening is slowly decaying and tends to have broad wings in comparison to Doppler broadening.

To illustrate the relative variation of the Doppler broadening and the pressure broadening, let us consider the lower 50 km of the Earth's atmosphere. In this layer, the pressure varies by three orders of magnitude, from 1 bar to 1 millibar, leading to a three orders of magnitude variation in the line broadening due to pressure. In comparison, the temperature varies by only 30%, between 300 K and 210 K. This leads to a change in the line broadening due to temperature of only about 14%. The drastic change in the pressure broadening of spectral lines allows the use of this effect for sounding of the atmospheric properties as a function of height.

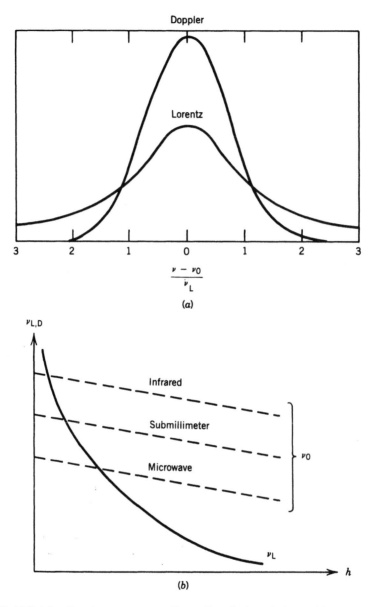

Figure 8-9. (a) Relative line shapes corresponding to Doppler broadening and Lorentz broadening. Both lines have the same strength and $v_L = v_D$. (b) The behavior of v_L and v_L as a function of altitude.

It should be noted that the Doppler broadening is proportional to v_0. Thus, the Doppler effect will become particularly significant in the visible and infrared region of the spectrum (Fig. 8-9b).

8-4-3 Nonresonant Absorption

As seen in Figure 8-6, there is significant absorption that still occurs between spectral bands. The most common cause of absorption is a result of the slowly decaying wings of distant spectral lines. The wings of the absorption lines, far away from the line center, fall off even slower than the Lorentz profile. However, the absorption does decrease significantly with pressure.

The two most-used atmospheric transparent windows in the infrared are at 3.5 μm to 4.1 μm and 10.5 μm to 12.5 μm. The latter one is particularly important because it is centered near the peak emission of a blackbody at typical Earth surface temperature. The main sources of absorption in the 11 μm window are the wings of the CO_2 lines at 15, 10.4, and 9.4 μm, and the wings of the neighboring water vapor lines. The effect of the CO_2 gas can be accurately modeled because it is uniformly mixed and has a relatively constant distribution. The water vapor contribution is hard to model because of its variability in space and time.

In the microwave region, continuous absorption is a result mainly of the water vapor (Fig. 8-6) and to a lesser extent oxygen (Fig. 8-7). It is clearly apparent that most of the absorption occurs in the lower atmosphere as a result of the high pressure and corresponding increase in the absorption coefficient.

In the case of Venus, the continuous absorption in the microwave region is mainly due to carbon dioxide and nitrogen (nitrogen absorption is also dominant for the Titan atmospheric), for which the absorption coefficient is given by (Ho et al., 1966)

$$\alpha_a = 2.610^7 [f^2_{CO_2} + 0.25 f_{CO_2} f_{N_2} + 0.005 f^2_{N_2}] \frac{v^2 P^2}{T^2} \tag{8-24}$$

where α_a is in km^{-1}, f is the volume mixing ratio, P is in atmosphere, v in GHz, and T in K. Figure 8-10 shows the total microwave absorption in the atmospheres of Venus and Titan.

If we consider a layer of gas of thickness dz and density ρ, the absorption through this layer is

$$\alpha_a\, dz = \rho \alpha\, dz \tag{8-25}$$

and the amount of intensity reduction encountered by an electromagnetic wave of intensity I is

$$dI = -\alpha_a\, dz\, I = -\rho \alpha\, dz\, I \tag{8-26}$$

α_a is called the absorption extinction coefficient, and α is called the absorption coefficient.

8-4-4 Nonresonant Emission

If we consider an atmospheric layer of thickness dz and temperature T, thermal radiation will be generated following Planck's law:

$$dI = \psi_t\, dz = \alpha_a B(v, T) dz \tag{8-27}$$

where $B(v, T)$ is the Planck function:

$$B(v, T) = \frac{2hv^3}{c^2} \frac{1}{\exp(hv/kT) - 1} \tag{8-28}$$

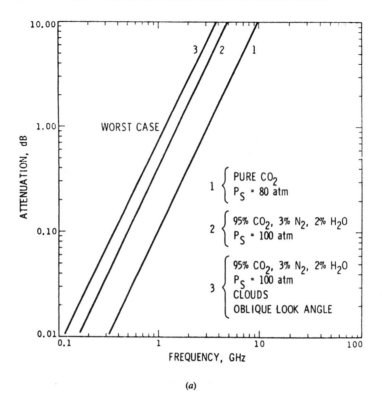

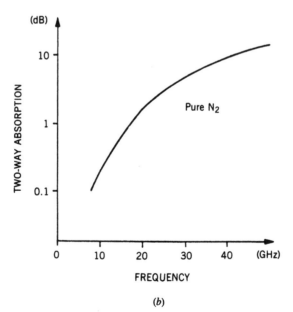

Figure 8-10. Total microwave absorption in the atmospheres of (a) Venus and (b) Titan over the frequency range from 1 to 30 GHz.

ψ_t is called the thermal source term. The reader should be careful when comparing the above equation to the expressions of the Planck function in Equations (2-30) and (4-1), where it is expressed as function of wavelength λ. The transformation is given by

$$B(v)dv = B(\lambda)d\lambda \Rightarrow B(v) = \frac{\lambda^2}{c}B(\lambda)$$

and the spectral radiance B is related to the spectral emittance S by

$$S = \pi B$$

In the case of planetary atmospheres, the temperature and constituent density are a function of the altitude z. Thus, the thermal source term is expressed as

$$\psi_t(z) = \alpha_a(z)B[v, T(z)] \tag{8-29}$$

8-4-5 Wave Particle Interaction and Scattering

In the most simple case, atmospheric particles can be considered as small spheres. The rigorous solution for the scattering of a plane monochromatic wave by a spherical dielectric particle with a complex index of refraction n was derived by Mie. The expressions for the scattering cross section and total extinction cross section are given by

$$\sigma_s = \frac{\lambda^2}{2\pi} \sum_{m=1}^{\infty} (2m + 1)[|a_m(n, q)|^2 + |b_m(n, q)|^2] \tag{8-30}$$

$$\sigma_E = \frac{\lambda^2}{2\pi} \sum_{m=1}^{\infty} (2m + 1)Re[a_m(n, q) + b_m(n, q)] \tag{8-31}$$

where the coefficients a_m and b_m are coefficients in the scattered field representing contributions from the induced magnetic and electric dipoles, quadrupoles, and so on. These coefficients can be expressed in terms of spherical Bessel functions with arguments $q = \pi D/\lambda$. D is the diameter of the sphere.

In the case of a small particle relative to the wavelength ($q \gg 1$), the above scattering cross sections reduce to

$$\sigma_s = \frac{2}{3} \frac{\pi^5 D^6}{\lambda^4} |K|^2 \tag{8-32}$$

$$\sigma_E = \frac{\pi^2 D^3}{\lambda} Im(-K) + \frac{2}{3} \frac{\pi^5 D^6}{\lambda^4} |K|^2 \tag{8-33}$$

where $K = (n^2 - 1)/(n^2 + 2)$. The absorption cross section σ_a is given by

$$\sigma_a = \sigma_E - \sigma_s \tag{8-34}$$

If the atmospheric unit volume contains N particles of equal size, the medium is then characterized by a particle extinction coefficient α_E given by:

$$\alpha_E = \sigma_E N \tag{8-35}$$

If the scattering particles have a size distribution given by , then

$$\alpha_E = \int \sigma_E(r)N'(r)dr \tag{8-36}$$

Similar expressions can be written for the scattering coefficient α_s and the absorption coefficient α_a. In the above expressions, multiple scattering has been neglected.

When the wave scatters from a particle, the scattered intensity has an angular pattern called the phase function $p(\theta_i, \theta_s, \phi_i, \phi_s)$ where θ_i and ϕ_i are the angles of the incident wave and θ_s and ϕ_s are the angles of the scattered wave. In simple terms, the phase function can be thought of as (1) the percentage of energy scattered per unit solid angle in a certain direction, (2) an angular weighting function for the scattered radiation, or (3) the equivalent of an antenna radiation pattern.

8-4-6   Wave Refraction

As a wave propagates in a medium, its speed differs from the wave speed in vacuum due to interaction with the medium constituents. This is characterized by the medium index of refraction n, which equals the ratio of the wave speed in vacuum to the wave speed in the medium. Thus, n is always greater than unity.

In the case of a gas mixture of oxygen, water vapor, and carbon dioxide similar to the Earth's atmosphere, the index of refraction for frequencies less than 200 GHz is given by

$$N = (n - 1)\,10^6 = 0.0776\frac{P}{T} + 373\frac{e}{T^2} \tag{8-37}$$

where e is the partial pressure of water vapor in bar, P is the total pressure in bar, and N is called the refractivity. Under mean conditions, P, e, and T decrease with height such that N decreases roughly exponentially with height. Thus, the atmosphere can be considered as a spherically inhomogeneous shell. A wave incident obliquely on the top of the atmosphere will refract and therefore has a curved trajectory as it propagates through the atmosphere. This effect is used in occultation experiments to derive information about planetary atmospheres.

8-5   OPTICAL THICKNESS

Let us consider an atmospheric thin slab of thickness D. An incident monochromatic wave will be partially scattered and absorbed as it propagates through the slab. If α is the total extinction coefficient (including absorption and scattering), the intensity loss is

$$dI = -\alpha I\,dz \Rightarrow I(z) = I_0 e^{-\alpha z} \tag{8-38}$$

At the output of the slab the intensity is

$$I(D) = I_0 e^{-\alpha D} \tag{8-39}$$

The slab is thus characterized by a transmission coefficient,

$$T = e^{-\alpha D} \tag{8-40}$$

The term αD is usually called the slab optical thickness τ. If the slab is inhomogeneous (i.e., α varies with altitude z) and covers the altitudes from z_1 to z_2, then the slab optical thickness is

$$\tau(v, z_1, z_2) = \int_{z_1}^{z_2} \alpha(v, z)dz \tag{8-41}$$

and the slab transmission coefficient is given by

$$T(v) = e^{-\tau} \tag{8-42}$$

The total normal optical thickness of a planet's atmosphere is given by

$$\tau(v) = \int_0^\infty \alpha(v, z)dz \tag{8-43}$$

The optical thickness at an altitude z is defined as

$$\tau(v, z) = \int_z^\infty \alpha(v, \zeta)d\zeta \tag{8-44}$$

In the case where the wave is oblique, at an incidence angle θ, then

$$\tau(\theta, v, z) = \frac{\tau(v, z)}{\cos\theta} \tag{8-45}$$

If the extinction coefficient α decreases exponentially with altitude,

$$\alpha(v, z) = \alpha(v, 0) \exp\frac{-z}{H} \tag{8-46}$$

the optical thickness is then

$$\tau(v, z) = \int_z^\infty \alpha(v, 0) \exp\frac{-\zeta}{H}d\zeta = H\alpha(v, 0) \exp\frac{-z}{H} \tag{8-47}$$

and

$$\tau(v, 0) = H\alpha(v, 0) \tag{8-48}$$

If an atmosphere consists of a mixture of gases and particles, the total optical thickness is equal to the sum of the individual optical thicknesses:

$$\tau = \sum_g \tau_g + \sum_p \tau_p \tag{8-49}$$

The "optical thickness" is a parameter that is very widely used to characterize atmospheres.

8-6 RADIATIVE TRANSFER EQUATION

The radiative transfer equation is the fundamental equation describing the propagation of electromagnetic radiation in a scattering and absorbing medium. At a given point in the medium, the change in the intensity $I(z, \theta, \phi)$ as the wave traverses a distance dz in the direction (θ, ϕ) consists of the following elements (see Fig. 8-11).

The wave is attenuated due to absorption in the gas and the suspended particles. This corresponds to change of the wave energy to heat. The corresponding intensity loss is given by

$$\frac{dI}{dz} = -\alpha_a I \tag{8-50}$$

where α_a is the sum of the absorption coefficient of all the gases and particles in the medium.

Some of the wave energy is scattered by the particles, which results in a loss of intensity in the (θ, ϕ) direction, even though the total wave energy is conserved. The corresponding intensity loss is given by

$$\frac{dI}{dz} = -\alpha_s I \tag{8-51}$$

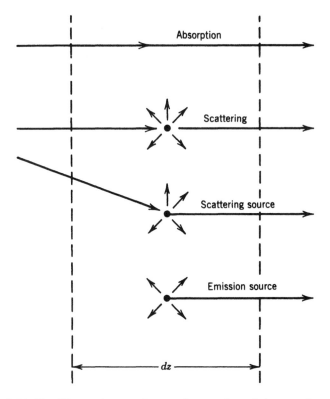

Figure 8-11. The different elements that contribute to the radiation transfer equation.

Some energy is added to the wave as a result of thermal emission from the medium. This source term is given by

$$\frac{dI}{dz} = \psi_t(z) = +\alpha_a B(v, T) \tag{8-52}$$

Some energy is added to the wave in the (θ, ϕ) direction as a result of scattering of waves incident from other directions. This source term is given by

$$\frac{dI}{dz} = \psi_s = \alpha_s(z) J(\theta, \phi) \tag{8-53}$$

where

$$J(\theta, \phi) = \frac{1}{4\pi} \int_0^{4\pi} I(\theta', \phi') p(\theta', \theta, \phi', \phi) d\Omega'$$

In the above expression, p is the scattering phase function that describes the angular distribution of the scattered field. Therefore, the radiative transfer equation can be written as

$$\frac{dI}{dz} = -\alpha_a I - \alpha_s I + \alpha_a B + \alpha_s J \tag{8-54}$$

or, in a more simplified form,

$$\frac{dI}{dz} = -\alpha(z) I + \psi(z, \theta, \phi) \tag{8-55}$$

where the loss terms and the source terms are combined.

The radiative equation is often written using the medium elementary optical thickness $d\tau = -\alpha\, dz$ as a variable. From Equation 8-54 we can write

$$\frac{dI}{d\tau} = +I - \frac{\alpha_a}{\alpha} B - \frac{\alpha_s}{\alpha} J \tag{8-56}$$

or

$$\frac{dI}{d\tau} = +I - (1 - \omega) B - \omega J \tag{8-57}$$

where

$$\omega = \frac{\alpha_s}{\alpha}$$

The solution of the radiative transfer equation is very involved in the general case, and usually simplifying assumptions and numerical methods are used. The reader is referred to the book by Chandrasekhar, *Radiative Transfer,* for a good study of the solution methods.

The radiative equation solutions become even more involved when the polarization of the wave is also included. In this case, three additional parameters have to be accounted for in addition to the wave intensity. These are degree of polarization, polarization plane, and ellipticity.

Most of the techniques used in practical remote sensing situations use numerical methods. One such technique is the Monte Carlo technique, in which photons are followed through the atmosphere as they get absorbed and scattered in a probabilistic sense. This method can be used with homogeneous as well as inhomogeneous atmospheres.

8-7   CASE OF A NONSCATTERING PLANE PARALLEL ATMOSPHERE

To illustrate, let us consider the case of a nonscattering atmosphere in which the geometry is plane parallel (see Fig. 8-12). Let us consider the case of a wave propagating at an angle θ relative to the vertical. We assume that the semiinfinite atmosphere is bounded at $z = 0$, which is also the origin for the optical thickness variable ($\tau = 0$). The radiative transfer equation can be written as

$$\cos\theta \frac{dI}{d\tau} = I(\tau, 0) - B$$

or

$$\mu \frac{dI}{d\tau} = I(\tau, \mu) - \psi(\tau, \mu) \tag{8-58}$$

where $\mu = \cos\theta$ and ψ is the source term ($\psi = \psi_r/\alpha_a$). The solution of the above equation is given by

$$I(\tau, \mu) = Ae^{\tau/\mu} + \frac{1}{\mu}\int_{\tau}^{\tau_0} \psi(\eta, \mu)e^{(\tau-\eta)/\mu}\,d\eta \tag{8-59}$$

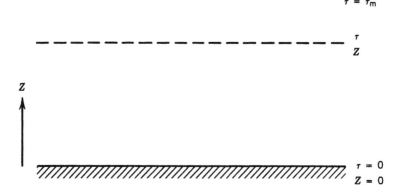

Figure 8-12. Geometry for the case of a plane parallel atmosphere.

where A and τ_0 are constants. τ varies from $\tau = 0$ to $\tau = \tau_m$, where τ_m is the optical thickness of the total atmospheric layer.

In the case of inward radiation from the atmosphere toward the surface, the total intensity at level τ is

$$I(\tau, \mu) = \frac{1}{\mu} \int_{\tau}^{\infty} \psi(\eta, \mu) e^{(\tau-\eta)/\mu} d\eta \qquad (8\text{-}60)$$

In the case of outward radiation toward space,

$$I(\tau, -\mu) = I(0, -\mu) e^{-\tau/\mu} + \frac{1}{\mu} \int_{0}^{\tau} \psi(\eta, -\mu) e^{-(\tau-\eta)/\mu} d\eta \qquad (8\text{-}61)$$

and if the observing point is above the atmosphere, then

$$I(\tau_m, -\mu) = I(0, -\mu) e^{-\tau m/\mu} + \frac{1}{\mu} \int_{0}^{\tau_m} \psi(\eta, -\mu) e^{-(\tau-\eta)/\mu} d\mu \qquad (8\text{-}62)$$

where $I(0, -\mu)$ is the intensity of the upward radiation in the direction μ at the surface, and τ_m is the total optical thickness of the atmosphere.

8-8 BASIC CONCEPTS OF ATMOSPHERIC REMOTE SOUNDING

The techniques of atmospheric sounding can be divided into three general categories: occultation, scattering, and emission.

In the case of occultation techniques, the approach is to measure the changes that the atmosphere imparts on a signal of known characteristics as this signal propagates through a portion of the atmosphere. The signal source can be the sun, a star, or a man-made source such as a radio or radar transmitter. The geometry corresponds usually to limb sounding. Figure 8-13a shows the case in which the sun is used as a source, whereas Figure 8-13b shows the case in which a spacecraft radio transmitter is the source. This latter case is commonly used in planetary occultation. A special configuration is shown in Figure 8-13c in which the surface is used as a mirror to allow the signal to pass through the total atmosphere twice.

In the case of scattering, the approach is to measure the characteristics of the scattered waves in a direction or directions away from the incident wave direction. The source can be the sun, as in Figure 8-13d, or man-made, as in Figure 8-13e.

In the case of emission, the radiation source is the atmosphere itself (Figs. 8-13f and g), and the sensor measures the spectral characteristics and intensity of the emitted radiation.

The different atmospheric sounding techniques aim at measuring the spatial and temporal variations of the atmospheric properties, specifically temperature profile, constituents nature and concentration, pressure, wind, and density profile. In the rest of this section, the techniques used for measuring these properties are briefly described in order to give an overview of the following three chapters.

8-8-1 Basic Concept of Temperature Sounding

If a sensor measures the radiation emitted from gases whose distribution is well known, such as carbon dioxide or molecular oxygen in the Earth's atmosphere, then the radiance

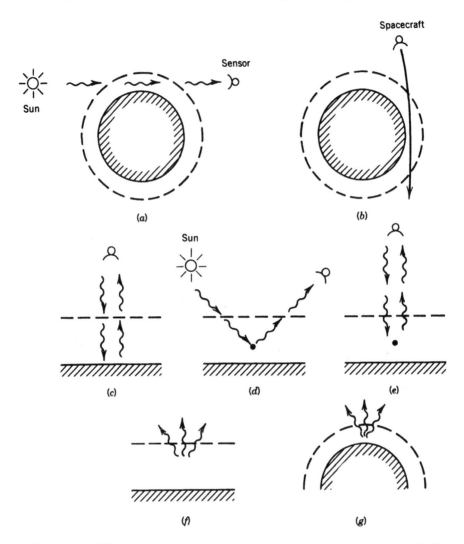

Figure 8-13. Different configurations for atmospheric sounding (see text for explanation).

can be used to derive the temperature (Equation 8-29). At first glance, it seems that only the mean temperature can be derived because the radiation detected at any instant of time is a composite of waves emitted from all the different layers in the atmosphere. However, if we can measure the radiance variation as a function of frequency near a spectral line, the temperature vertical profile can be derived. This can be explained as follows.

The contribution from the layers at the top of the atmosphere is very small because the density (i.e., number of radiating molecules) is low. As we go deeper in the atmosphere, the contribution increases because of the higher atmospheric density. However, for the deep layers near the surface, even though the radiation source is the largest, the emitted radiation has to traverse the whole atmosphere, where it gets absorbed. Thus, its net contribution to the total radiance is small. This implies that, for a certain atmospheric optical thickness, there is an optimum altitude layer for which the combination of gas density (i.e., source strength) and attenuation above it are such that this layer contributes most to the total radiance. If the optical thickness changes, then the altitude of the peak contribution changes. Thus, if we observe the radiance at a number of neighboring frequencies for which the optical thickness varies over a wide range (this occurs when we look around an absorption spectral band), the altitude of the contribution peak will vary, thus allowing temperature measurement at different altitudes. This technique will be formulated quantitatively and discussed in detail in Chapter 9.

In order to get an accurate temperature profile, the absorption band used should have the following properties:

1. The emitting constituent should have a known mixing ratio and preferably be uniformly mixed in the atmosphere. This is the case for molecular oxygen and carbon dioxide in the Earth's atmosphere up to 100 km. The most commonly used bands are the 60 GHz band for oxygen and the 15 μm and 3.4 μm infrared bands for carbon dioxide. In the case of the Martian and Venusian atmospheres, CO_2 infrared bands can be used. In the case of Jupiter, methane is uniformly mixed and its 7.7 μm infrared line can be used.

2. The absorption band involved should not be overlapped by bands from other atmospheric constituents. The 60 GHz oxygen line in the Earth's atmosphere, the 15 μm CO_2 line in the Earth, Mars, and Venus atmospheres, and the 7.7 μm CH_4 line in the Jovian atmosphere satisfy this requirement.

3. Local thermodynamic equilibrium should apply so that the Planck emission law is appropriate. This is usually the case in the lower 80 km of the Earth's atmosphere.

4. The wavelength should be long enough such that the scattered solar radiation is insignificant compared to the thermal emission. This is always the case in the microwave, millimeter, and thermal infrared part of the spectrum.

8-8-2 Basic Concept of Composition Sounding

The identification of atmospheric constituents is usually based on detecting the presence of a spectral line or lines associated with a certain molecule. The spectral signature is in effect the "fingerprint" of a gaseous constituent.

In order to determine the abundance of a constituent, a more detailed analysis of the spectral signature is required. The line strength is usually related to the number density of molecules. This usually requires knowledge of the local pressure and temperature. Once the temperature is derived using the radiance from a homogeneously mixed constituent as discussed earlier, the corresponding abundance profiles can be derived by measuring the spectral radiance around other spectral lines. The "sounding" spectral lines should satisfy the same properties discussed in the previous section except for the first one, which is relevant only to temperature sounding.

8-8-3 Basic Concept of Pressure Sounding

The total columnar absorption is strongly related to the columnar mass of a constituent in the atmosphere, particularly near a resonant line of the constituent. If the constituent is homogeneously mixed in the atmosphere, its total mass is then directly proportional to the surface pressure. Thus, surface pressure sounding can be achieved by devising a technique to measure the total columnar absorption of a homogeneously mixed gas such as oxygen in the Earth's atmosphere.

8-8-4   Basic Concept of Density Measurement

The atmospheric refractivity N is directly proportional to the atmospheric density. Thus, one approach is to derive the refractivity profile as a function of altitude. This is done to derive the density profile of planetary atmospheres using the refraction of the radio communication signal as orbiting or flyby spacecraft are occulted by the atmosphere. The radio occultation technique is discussed in more detail in Chapter 9.

8-8-5   Basic Concept of Wind Measurement

The simplest technique for wind measurement is to take a time series of cloud photographs, which would allow derivation of the wind field at cloud level. In order to get the wind field at any other altitude, the Doppler effect is used. The Multi-angle Imaging SpectroRadiometer (MISR) instrument that flies on the Terra spacecraft uses a series of images taken forward and aft of the spacecraft at different look angles and image matching algorithms to measure vector winds at the cloud levels.

Any molecule in motion will have its spectral line shifted by the Doppler effect. The Doppler shift is equal to

$$\Delta v = \frac{V}{\lambda} \cos \theta$$

where λ is the line wavelength and $V \cos \theta$ is the molecule velocity along the line of observation. Thus, by accurately measuring the line center for a known atmospheric constituent and comparing it to the frequency of the line for the same constituent in a static case, one of the velocity components can be derived. To illustrate, if a carbon dioxide molecule is moving at a line of sight velocity of 1 m/sec, the 15 μm line will have a frequency shift of

$$\Delta v = \frac{1}{15 \times 10^{-6}} \simeq 66.7 \text{ kHz}$$

which is small but measurable. If we use the oxygen line at 60 GHz, then

$$\Delta v = \frac{1}{5 \times 10^{-3}} = 200 \text{ Hz}$$

Another technique for wind measurement, also based on the Doppler effect, uses the scattered wave from an illuminating laser or radar beam. The incident wave is scattered by moving particles and the returned signal is shifted by a frequency Δv given by

$$\Delta v = \frac{2V}{\lambda} \cos \theta$$

The returned signal is then mixed with a reference identical to the transmitted signal to derive the Doppler shift.

EXERCISES

8.1. A planetary atmosphere is composed of two gases, G_1 and G_2. The molecular weight of gas G_2 is half that of gas G_1. At a height of 10 km, the number densities of the two gases were measured to be equal. Assuming that the scale height of gas G_1 is 15 km and that the temperature is constant in altitude, then:

(a) Calculate the scale height of gas G_2.

(b) Calculate the ratio of number densities of gas G_1 to G_2 at the surface.

(c) Calculate the ratio at altitude of 30 km.

(d) Calculate the ratio of the total number of atoms of gas G_1 above 30 km to the total number of atoms of gas G_2 above 30 km.

(e) Calculate the ratio of the total mass of the two gases.

(f) Calculate the altitude at which the total mass is 1% of the mass at the surface.

8.2. A planetary atmosphere is characterized by a scale height H and a surface number density $N(0)$. A photon flux F_0 at a certain wavelength strikes the top of the atmosphere. If σ is the absorption cross section in square meters of the atmospheric gas, calculate:

(a) The flux at altitudes $2H$, H, $H/2$, and the surface.

(b) The ratio of the flux at a certain height h to the flux at height $h + H$.

8.3. Let us assume that the scale height is a linear function of altitude:

$$H = H_0 + ah$$

where H_0 and a are constants. Give the expression of the number density and the pressure as a function of altitude. Compare the two cases in which $H_0 = 10$ km, $a = 0.5$, and $H_0 = 1$ km, $a = 0$

8.4. Derive the analytical expression, then calculate the ratio of σ_s/σ_a over the wavelength region 0.1 μm $\leq \lambda \leq 10$ μm for the case of particles characterized by the following radius and index of refraction:

(a) $r = 0.5$ μm, $n = 1.5 + i0.001$

(b) $r = 0.5$ μm, $n = 1.5 + i0.01$

(c) $r = 2.0$ μm, $n = 1.5 + i0.001$

(d) $r = 2.0$ μm, $n = 1.5 + i0.01$

Assume that the imaginary part of the index of refraction is much smaller than the real part.

8.5. Repeat the previous exercise for the wavelength region 1 cm $\leq \lambda \leq 10$ cm and for

(a) $r = 1$ mm, $n = 7 + i1$

(b) $r = 1$ mm, $n = 2 + i0.01$

8.6. Using Equations (8-19) and (8-21) for the Doppler and Lorentz line-shape broadening, calculate the pressure in the atmosphere at which the half amplitude width due to collision broadening is equal to that due to Doppler broadening for the following spectral lines:

(a) Carbon dioxide line at 15 μm

(b) Water vapor line at 6.25 μm

(c) Water vapor line at 0.1 mm

Assume that all lines have collision-broadening half width of $\Delta v = 3$ GHz at the surface pressure of 1 bar. Also, assume that $M = 29$ and $T = 273$ K.

REFERENCES AND FURTHER READING

Caulson, K., J. Dave, and Z. Sekera. *Tables Related to Radiation Emerging from a Planetary Atmosphere with Rayleigh Scattering.* University of California Press, Los Angeles, 1960.

Chahine, M. T., et al. Interaction mechanisms within the atmosphere. In *Manual of Remote Sensing.* American Society of Photogrammetry, Falls Church, VA, 1983.

Chandrasekhar, S. *Radiative Transfer.* Oxford University Press, 1950.

Chen, M., R. B. Rood, and W. G. Read. Seasonal variations of upper tropospheric water vapor and high clouds observed from satellites. *Journal of Geophysics Research,* **104,** 6193–6197, 1999.

Dessler, A. E., M. D. Burrage, J.-U. Grooss, J. R. Holton, J. L. Lean, S. T. Massie, M. R. Schoeberl, A. R. Douglass, and C. H. Jackman. Selected science highlights from the first 5 years of the Upper Atmosphere Research Satellite (UARS) program. *Review of Geophysics,* **36,** 183–210, 1998.

Diner, D. J., J. C. Beckert, T. H. Reilly, C. J. Bruegge, J. E. Conel, R. Kahn, J. V. Martonchik, T. P. Ackerman, R. Davies, S. A. W. Gerstl, H. R. Gordon, J-P. Muller, R. Myneni, R. J. Sellers, B. Pinty, and M. M. Verstraete. Multi-angle Imaging SpectroRadiometer (MISR) description and experiment overview. *IEEE Transactions on Geoscience and Remote Sensing,* **36,** 1072–1087, 1998.

Drouin, B. J., J. Fisher, and R. R. Gamache. Temperature dependent pressure induced lineshape of O_3 rotational transitions in air. *Journal of Quantitative Spectroscopy and Radiative Transfer,* **83,** 63–81, 2004.

Hansen, J. E., and L. D. Travis. *Space Science Review,* **16,** 527, 1974.

Herman, B. M., S. R. Bowning, and R. J. Curran. The effects of atmospheric aerosols on scattered sunlight. *Journal of Atmospheric Science,* **28,** 419–428, 1971.

Ho, W., et al. Laboratory measurements of microwave absorption in models of the atmosphere of Venus. *Journal of Geophysics Research,* **21,** 5091–5108, 1966.

Liou, K. N. *An Introduction to Atmospheric Radiation* (2nd ed.). Academic Press, San Diego, 2002.

Oh J. J., and E. A. Cohen. Pressure broadening of ozone lines near 184 and 206 GHz by nitrogen and oxygen. *Journal of Quantitative Spectroscopy and Radiative Transfer,* **48,** 405–408, 1992.

Pickett, H. M., D. E. Brinza and E. A. Cohen. Pressure broadening of ClO by nitrogen. *Journal of Geophysics Research,* **86,** 7279–7282, 1981.

Rodgers, C. D. *Inverse Methods for Atmospheric Sounding: Theory and Practice.* World Scientific Publishing Co. Ltd., Singapore, 2000.

Shaw, J. H. Determination of the Earth's surface temperature from remote spectral radiance observation near 2600 cm^{-1}. *Journal of Atmospheric Science,* **27,** 950, 1970.

Thomas, G. E., and K. Stamnes. *Radiative Transfer in the Atmosphere and Ocean.* Cambridge University Press, Cambridge, 1999.

Van de Hulst, H. C. *Light Scattering by Small Particles.* Wiley, New York, 1957.

Waters, J., et al. A balloon borne microwave limb sounder for stratospheric measurements. *Journal of Quantitative Spectroscopy and Radiative Transfer,* **32,** 407–433, 1984.

Waters, J. W., et al. The UARS and EOS Microwave Limb Sounder Experiments. *Journal of Atmospheric Science,* **56,** 194–218, 1999.

9

ATMOSPHERIC REMOTE SENSING IN THE MICROWAVE REGION

Microwave sensors are used in a wide variety of configurations to sound atmospheric properties. Passive microwave sensors are used to measure the emitted thermal radiation and derive from it the atmospheric temperature profile and the distribution of some atmospheric constituents such as water vapor. Active microwave sensors are used to map precipitation and to measure surface pressure. Occultation techniques are used to derive the density profile of planetary atmospheres. One of the key advantages of using the microwave part of the spectrum is the fact that it is usually insensitive to the presence of clouds, particularly at the lower part of the spectral region, thus allowing the sounding of cloudy atmospheres. In the upper part of the microwave spectral region the effect of the clouds becomes significant.

9-1 MICROWAVE INTERACTIONS WITH ATMOSPHERIC GASES

The interaction of microwave radiation with atmospheric gases can be divided into two major types: (1) resonant interactions that are spectrally localized to narrow frequency bands that correspond mainly to the rotational energy levels, and in some cases vibrational levels, of the constituent molecules; and (2) nonresonant interactions that depend on the bulk properties of the gas and result in the wave propagating at a speed less than that in vacuum. This in turn results in wave refraction. In addition, continuous absorption takes place.

A number of atmospheric molecules have microwave spectral lines, the most prominent ones being water vapor (H_2O) and Oxygen (O_2). Water vapor has rotational spectral lines at 22.235 GHz and 183.3 GHz (see Fig. 8-6). Oxygen has a series of rotational spectral lines that combine into a band with a peak near 60 GHz and a single line at 118.75 GHz (see Fig. 8-7). Because oxygen is homogeneously mixed in the Earth's atmosphere, the emission near its spectral lines is used to sound the temperature profile. In addition, ozone has a number of absorption lines below 100 GHz, but they are weak compared to the H_2O and O_2 lines.

Introduction to the Physics and Techniques of Remote Sensing. By C. Elachi and J. van Zyl
Copyright © 2006 John Wiley & Sons, Inc.

In the regions between the spectral lines, absorption of microwave radiation results from the far wings of the water vapor and oxygen spectral lines as well as nonresonant interactions that transfer wave energy to heat as a result of collisions. Figure 9-1 shows the opacity for a "standard" Earth atmosphere in the tropics (300°K surface temperature and 4 g/cm² of precipitable water) and the arctic (249°K surface temperature and 0.2 g/cm² of precipitable water). It is apparent that even far away from the spectral lines, significant absorption still takes place. In the case of planetary atmospheres, the continuum absorption plays a significant role. This is shown in Figure 8-10 for Venus and Titan, where nitrogen is the main absorbing molecule.

The microwave emission from the atmospheric gases follows the Rayleigh–Jeans approximation of Planck's law ($hv \ll kT$). This implies that the spectral radiance for a blackbody can be written as

$$B(v, T) = \frac{2kTv^2}{c^2} \tag{9-1}$$

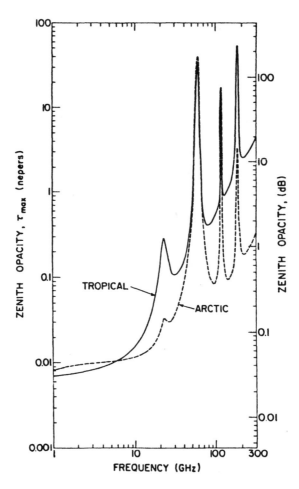

Figure 9-1. Total Earth atmosphere opacity across the microwave spectrum for two standard atmospheres corresponding to the Arctic and tropics. (From Chahine et al., 1983.)

9-2 BASIC CONCEPT OF DOWNLOOKING SENSORS

Let us consider a sensor looking down on a scatter-free plane-parallel atmosphere (see Fig. 9-2). An elementary layer of thickness dz at altitude z will have its spectral radiance given by

$$\Delta B = \alpha(v, z)B(v, T)dz \qquad (9\text{-}2)$$

As the emitted wave propagates upward toward the sensor, it gets partially absorbed by the atmospheric layers at higher altitude. Thus, the effective spectral radiance observed by the sensor is

$$\Delta B_e = \Delta B e^{-\tau(v,z)} \qquad (9\text{-}3)$$

where $\tau(v, z)$ is the optical thickness of the atmosphere above z and is given by

$$\tau(v, z) = \int_z^\infty \alpha(v, \zeta)d\zeta \qquad (9\text{-}4)$$

Thus, the total spectral radiance of the atmosphere is given by

$$B_a(v) = \int_0^\infty \alpha(v, z)B[v, T(z)]e^{-\tau(v,z)}dz \qquad (9\text{-}5)$$

The sensor observes a spectral radiance $B_t(v)$ equal to

$$B_t(v) = B_a(v) + B_s(v)e^{-\tau_m(v)} \qquad (9\text{-}6)$$

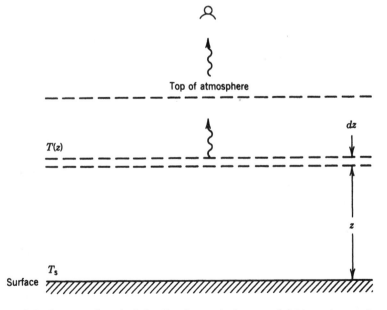

Figure 9-2. Geometry for calculating the observed microwave brightness temperature.

where $B_s(v)$ is the surface radiance and τ_m is the total atmosphere optical thickness at the frequency of observation.

In the case of the Rayleigh–Jeans approximation, which is valid across the microwave spectrum, Equations 9-5 and 9-6 can be expressed in terms of the effective brightness temperature $T_t(v)$ observed by the sensor, which is defined as

$$B_t(v) = \frac{2kv^2}{c^2} T_t(v) \tag{9-7}$$

Replacing $B[v, T(z)]$ in Equation 9-5 by its expression from Equation 9-1 we get

$$T_t(v) = T_a(v) + T_s(v)e^{-\tau_m(v)}$$

and

$$T_a(v) = \int_0^\infty \alpha(v, z)T(z)\exp\left[-\int_z^\infty \alpha(v, \zeta)\, d\zeta\right]dz \tag{9-8}$$

which can be written as

$$T_a(v) = \int_0^\infty W(v, z)T(z)dz \tag{9-9}$$

where $T_s(v)$ is the surface brightness temperature (i.e., physical temperature multiplied by the surface emissivity), $T(z)$ is the atmosphere temperature profile, and $W(v, z)$ is a weighting function given by

$$W(v, z) = \alpha(v, z)\exp\left[-\int_z^\infty \alpha(v, \zeta)\, d\zeta\right]$$

The sensor measures $T_t(v)$ as a function of frequency. The objective is to derive from these measurements information about the atmosphere temperature profile $T(z)$ and the constituents' density profile, which is directly related to $\alpha(v, z)$.

9-2-1 Temperature Sounding

In order to be able to derive the temperature profile in the lower atmosphere, the brightness temperature is observed around the emission line or band of an atmospheric constituent, which is homogeneously mixed in the atmosphere. Oxygen is such a constituent in the Earth's atmosphere.

The absorption coefficient for a pressure-broadened line is related to the constituent density ρ_c and the observation frequency in a nonstraightforward manner. It is proportional to the product of the density and the line shape (Equation 8-21), which in turn is a function of the atmospheric density ρ_z (because the Lorentz line width v_L is related to the pressure). Near the line center ($v = v_0$), we have

$$\alpha \sim \frac{\rho_c}{v_L} \sim \frac{\rho_c}{\rho_a}$$

For a homogeneously mixed constituent such as oxygen, this ratio is independent of height. Far away from the line center $(v - v_0 \gg v_L)$ we have

$$\alpha \sim \rho_c v_L \sim \rho_c \rho_a$$

For a homogeneously mixed constituent this product is proportional to ρ_c^2 or ρ_a^2.

To illustrate the basic principle of atmospheric temperature sounding in a simple analytical fashion, let us consider the simplified case where α is proportional to ρ_a. Let us first take the case of an exponentially decaying atmospheric density as a function of altitude. Then,

$$\alpha(v, z) = \alpha_0(v) e^{-z/H} \tag{9-10}$$

Thus, from Equation (9-9), we have

$$T_a(v) = \alpha_0(v) \int_0^\infty T(z) e^{-z/H} \exp\left[-\alpha_0(v) \int_z^\infty e^{-\zeta/H} d\zeta\right] dz$$

$$\rightarrow T_a(v) = \alpha_0(v) \int_0^\infty T(z) \exp\left[-\frac{z}{H} - \alpha_0(v) H e^{-z/H}\right] dz \tag{9-11}$$

which can be written as

$$T_a(v) = \int_0^\infty W(v, z) T(z)\, dz \tag{9-12}$$

where

$$W(v, z) = \alpha_0(v) \exp\left[-\frac{z}{H} - \tau_m(v) e^{-z/H}\right] \tag{9-13}$$

and

$$\tau_m(v) = \alpha_0(v) H$$

Equation 9-12 shows that the observed temperature is equal to the integral of the atmospheric temperature multiplied by a weighting function $W(v, z)$, which is a function of the observing frequency. Figures 9-3 and 9-4 show the general behavior of $W(v, z)$, which is discussed in more detail in Appendix C. The following observations can be made:

1. Most of the contribution is from a limited altitude region. This effect is to be expected, as explained in Section 8-1 of Chapter 8.
2. The altitude location of the region of maximum contribution varies with $\tau_m(v)$ and is given by

$$z_M(v) = H \log \tau_m \tag{9-14}$$

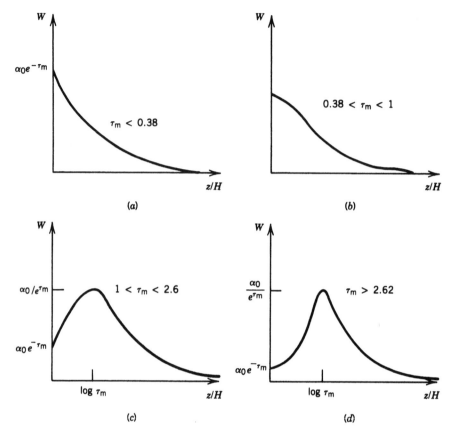

Figure 9-3. Behavior of the weighting function $W(v, z)$ corresponding to an exponentially decaying atmosphere for different values of the τ_m for a downlooking sensor.

Thus, by measuring the observed brightness at a wide range of τ_ms, the peaks of the corresponding weighting functions will cover a wide range of altitudes, allowing one to retrieve the temperature profile with altitude. The τ_ms can be changed by using different observing frequencies on the wings of a spectral line. Since we are operating on the wings of the line, changing the frequency changes the amount of absorption and, therefore, changes τ_m.

In the case of oxygen, the 60 GHz band is commonly used. This band extends over about a 10 GHz region. At high altitudes (>50 km) Zeeman splitting of the O_2 lines becomes dominant, complicating matters significantly. At low altitudes and for observation frequencies far on the wings of the band the absorption coefficient is proportional to the square of the atmospheric density, as discussed earlier. In the case of the exponential atmosphere, α is thus proportional to $\exp(-2z/H)$ and the above derivation is still valid with H replaced by $H/2$.

If we assume a linearly decaying atmosphere and an absorption coefficient proportional to the density such that

$$\alpha(v, z) = \alpha_0(v)\left(1 - \frac{z}{H}\right) \quad \text{for } z < H$$

$$\alpha(v, z) = 0 \qquad\qquad \text{for } z > H$$

$$(9\text{-}15)$$

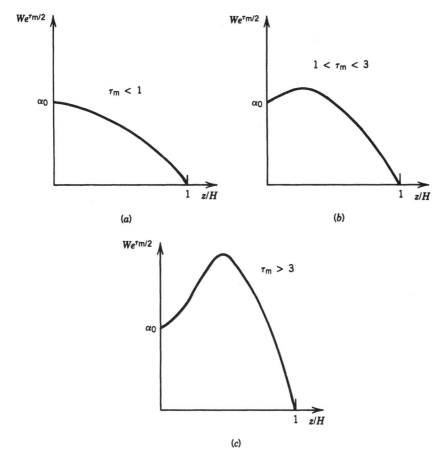

Figure 9-4. Behavior of $W(v, z)$ corresponding to a linearly decaying atmosphere for different values of τ_m for a downlooking sensor.

the weighting function will have the same general behavior as illustrated in Figure 9-4. By using the actual profile of the Earth's atmosphere, and the correct expression for the absorption coefficient, including the line shape, the weighting function corresponding to oxygen emission is illustrated in Figure 9-5, which shows a similar bell-shaped behavior.

9-2-2 Constituent Density Profile: Case of Water Vapor

Let us consider the situation in which the temperature profile is known and the variable looked for is the density profile of one of the atmospheric constituents. This is usually the case for water vapor measurement in the Earth's atmosphere. In this case, brightness temperature measurements are acquired near the water vapor line at 22 GHz and in neighboring spectral regions where the atmosphere is semitransparent.

The absorption coefficient of water vapor near the 22 GHz line is given by (Waters, 1976):

$$\alpha = \rho f[T, P, v] \tag{9-16}$$

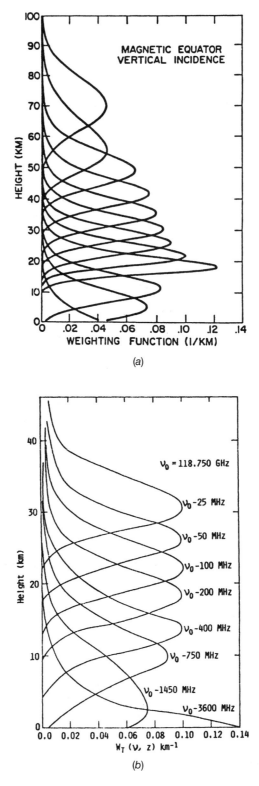

Figure 9-5. Temperature weighting functions as a function of altitude above the surface for observations from space that view the nadir. The curves correspond to the emission by oxygen. (a) around 60 GHz (from Lenoir, 1968). (b) Around 118 GHz (from Wilheit et al., 1977).

where

$$f = \frac{v^2 \Delta v}{T^{3/2}} \left[\frac{718 \, e^{-644/T}}{T} \frac{1}{(v - v_0)^2 + 4v^2 \Delta v^2} + 2.7 \times 10^{-6} \right]$$

ρ = water vapor density in g/m^3

v_0 = 22.235 GHz

P = total pressure in bar

T = temperature in °K

v = frequency in GHz

$\Delta v = 2.6P(318/T)^{0.6}$ in GHz

α = absorption coefficient in m^{-1}

The total atmospheric opacity or optical thickness is given by

$$\tau(v) = \int_0^\infty \rho(z) f[T(z), P(z), v] \, dz \tag{9-17}$$

Thus, $f(z, v)$ acts like a weighting function that applies for both uplooking and downlooking sensors. The total opacity can be determined by measuring the attenuation from a known source, such as the sun, or looking at the emission from a surface with known or easily measurable temperature, such as the ocean. Figure 9-6 shows the weighting functions for water vapor. Their general behavior can be derived by looking at the variation in the shape of pressure-broadened spectral lines as a function of pressure (e.g., altitude).

Figure 9-7 shows in a simplified form the shape of the spectral line as a function of pressure. The following can be observed:

- Very close to the center of the line (location A), the absorption coefficient increases as the pressure decreases for the same concentration of water vapor. Thus, the weighting function increases as a function of altitude. This is consistent with the 22.23 GHz curve in Figure 9-6.
- Far out on the line wing (location C), the absorption coefficient decreases with the pressure, leading to a decreasing weighting function as a function of altitude. This is consistent with the 19 GHz curve in Figure 9-6.
- At the intermediate location (B), the absorption coefficient first increases then decreases as the pressure (altitude) decreases. This corresponds to the 21.9 GHz line in Figure 9-6.

This behavior can be visualized quantitatively by considering the change of the absorption coefficient as a function of the Lorentz width v_L. From Equation (8-21) we have

$$k = \frac{S}{\pi} \frac{v_L}{(v - v_0)^2 + v_L^2} \tag{9-18}$$

For a fixed observing frequency v, the behavior of k as a function of v_L (i.e., pressure or altitude) is shown in Figure 9-8. The three curves correspond to the three frequency lines

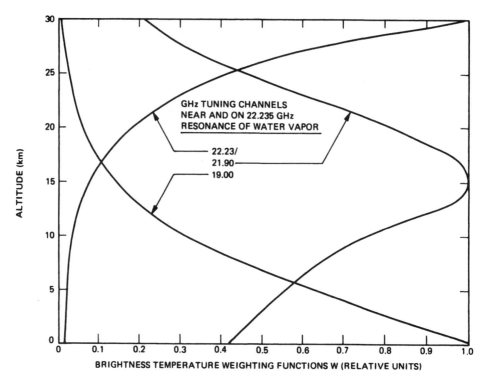

Figure 9-6. Normalized weighting function curves for water-vapor density in the atmosphere at three representative frequencies near and on the 22.235 GHz resonance of water vapor. The curves are derived for brightness temperature measurements from the surface of the Earth. (From Staelin, 1969.)

shown in Figure 9-7 and the three brightness weighting functions in Figure 9-6. An important fact is that for $v - v_0$ at intermediate values, the absorption function has a bell-shaped curve, indicating that most of the absorption (or emission) at a certain frequency is localized in a well-defined atmospheric layer centered around the altitude where

$$v_L = v - v_0 \rightarrow P = P_0 \frac{v - v_0}{v_L(P_0)} \tag{9-19}$$

9-3   BASIC CONCEPT FOR UPLOOKING SENSORS

Considering the geometry shown in Figure 9-2, the spectral emittence observed by an uplooking sensor will have the same expression as in Equation (9-5), except in this case

$$\tau(v, z) = \int_0^z \alpha(v, \zeta)\, d\zeta \tag{9-20}$$

because the emitted wave is absorbed by the layers below the elementary source layer. Thus, the expression for the brightness temperature becomes [from Equation (9-4)]

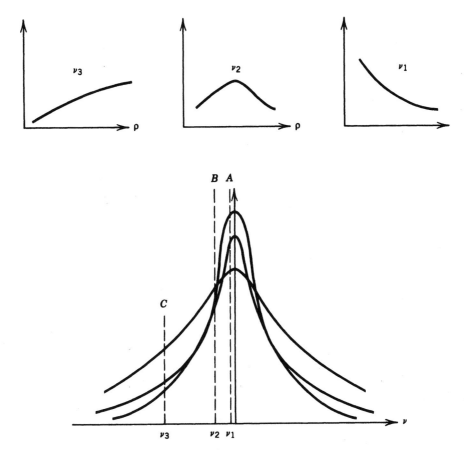

Figure 9-7. Behavior of a pressure-broadened spectral line as the pressure changes. The upper curves show the change of the absorption as a function of pressure for three different frequencies.

$$T_a(v) = \int_0^\infty \alpha(v, z) T(z) \exp\left[-\int_0^z \alpha(v, \zeta) d\zeta\right] dz \qquad (9\text{-}21)$$

Taking again the simple case of an exponentially decaying atmosphere and an absorption coefficient proportional to the density, the temperature weighting function is given by:

$$W(v, z) = \alpha_0(v) \exp[-\tau_m(v) - z/H + \tau_m(v)e^{-z/H}] \qquad (9\text{-}22)$$

Figure 9-9 shows the behavior of $W(v, z)$. It is consistent with the fact that the lower layers provide the strongest emission, which is also the least attenuated, whereas the higher layers provide low emission, which, in addition, is highly attenuated by the lower layers before it reaches the sensor. Similar behavior for $W(v, z)$ is derived in the case of a linear atmosphere.

Figure 9-10 shows the uplooking weighting functions of oxygen emission near the 60 GHz band using the correct dependence of α on the density.

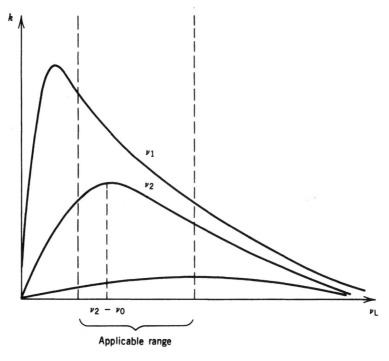

Figure 9-8. Behavior of k as a function of ν_L for fixed value of $\nu - \nu_0$.

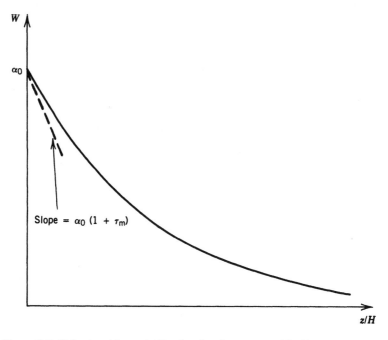

Figure 9-9. Behavior of the weighting function for an upward-looking sensor.

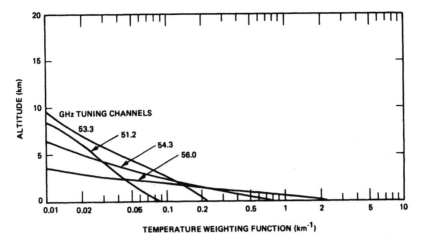

Figure 9-10. Unnormalized weighting functions for temperature as a function of height above the surface for observations from the surface looking at the zenith. The curves correspond to the emission by oxygen near the 60 GHz region (from Pearson and Kreiss, 1968).

9-4 BASIC CONCEPT FOR LIMBLOOKING SENSORS

Limblooking geometry is used in sensing the tenuous upper atmosphere. In this geometry, the sensor is looking through a large volume extending many hundreds of kilometers horizontally (Fig. 9-11), and the total mass of the upper atmosphere within the observation field of view is significantly larger than in the case of the downlooking sensor. This allows much better sensitivity to the emission from trace constituents such as CO, NO, N_2O, and ClO.

The weighting functions for a limblooking sensor are derived in a similar fashion as a downlooking sensor, taking into account the different geometry. Assuming a spherical

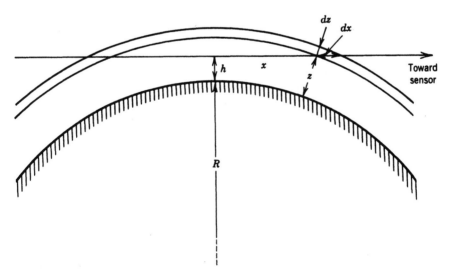

Figure 9-11. Geometry for a limb sounder.

geometry, if we consider a spherical layer (shell) of thickness dz, its elementary spectral emittence is given by (see Fig. 9-11)

$$\Delta B = \alpha(v, z)B(v, T)dx \tag{9-23}$$

and

$$dx = [(z + R)/\sqrt{(z + R)^2 - (h + R)^2}]dz = g(z, h)dz \tag{9-24}$$

$$\rightarrow \Delta B = \alpha(v, z)B(v, T)g(z, h)dz \tag{9-25}$$

h is called the tangent altitude, and R is the planet radius. The effective spectral emittence observed by the sensor is

$$\Delta B_e = \Delta B[e^{-\tau_1(v,z)} + e^{-\tau_2(v,z)}] \tag{9-26}$$

where

$$\tau_1(v, z) = \int_z^\infty \alpha(v, \zeta)g(\zeta, h)d\zeta \tag{9-27}$$

$$\tau_2(v, z) = \int_h^z \alpha(v, \zeta)g(\zeta, h) \, d\zeta + \int_h^\infty \alpha(v, \zeta)g(\zeta, h)d\zeta \tag{9-28}$$

The first term corresponds to the emission from the layer section nearer to the sensor, whereas the second term corresponds to the layer section away from the sensor (see Fig. 9-11), which is seen through a much thicker atmosphere. The total spectral radiance is then given by [see Equation (9-5)]

$$B_a(v) = \int_h^\infty \alpha(v, z)B[v, T(z)]g(z, h)[e^{-\tau_1(v,z)} + e^{-\tau_2(v,z)}]dz \tag{9-29}$$

and the corresponding brightness temperature is

$$T_a(v, h) = \int_h^\infty \alpha(v, z)T(z)g(z, h)[e^{-\tau_1(v,z)} + e^{-\tau_2(v,z)}]dz \tag{9-30}$$

which can be written as

$$T_a(v, h) = \int_0^\infty W(v, h, z)T(z)dz \tag{9-31}$$

where the weighting function is given by

$$W(v, h, z) = \alpha(v, z)g(z, h)[e^{-\tau_1(v,z)} + e^{-\tau_2}(v, z)] \quad \text{for } z > h$$
$$W(v, h, z) = 0 \quad\quad\quad\quad\quad\quad\quad\quad\quad\quad \text{for } z < h \tag{9-32}$$

Figure 9-12*a* shows the temperature weighting functions for an infinitesimal pencil-beam sensor looking at a tangent altitude of 12 km for the emission near the 118 GHz

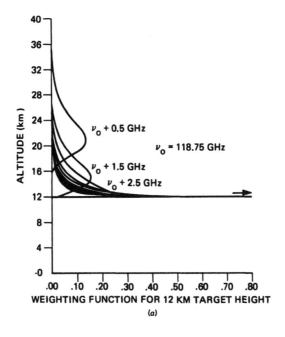

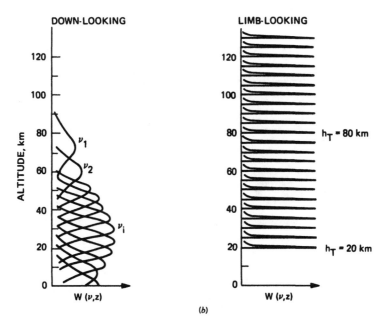

Figure 9-12. (a) Temperature weighting functions for an infinitesimal pencil beam tangent to the terrestrial atmosphere at 12 km altitude; the frequencies are on the wings of the strong 118 GHz oxygen resonance (from Chahine et al., 1983). (b) Downward- and limb-looking single-ray weighting functions for a species with constant volume mixing ratios (from Waters and Wofsy, 1978).

oxygen line. Very close to the center of the line, it is apparent that most of the contribution comes from the upper layers. The contribution from the deep layers is minimal because of the absorption in the higher ones. As the observation frequency moves away from the line center, the main contribution comes from deeper and deeper layers, similar to the case of downlooking sensors. In actuality, observations are conducted by operating in the optically thin region and scanning the sensor field of view. This allows a good height resolution. Figure 9-12*b* compares the downlooking and limblooking weighting functions for a species with constant volume mixing ratio.

The two Microwave Limb Sounding (MLS) instruments flown on the Upper Atmosphere Research Satellite (UARS) and the Aura Satellite are examples of limb sounders that operate in the microwave region. The UARS was launched in 1991, and Aura was launched in 2004. The UARS MLS instrument was the first application of the microwave limb sounding technique from a long-term satellite in space. This instrument uses double-sideband heterodyne radiometers that operate near 63 GHz (they measure stratospheric pressure, but also provide stratospheric temperature), 205 GHz (to measure stratospheric ClO, O_3 and H_2O_2), and 183 GHz (to measure stratospheric and mesospheric H_2O and O_3). Limb scans are performed every 65 seconds along track. Figure 9-13 shows examples of ClO and O_3 measurements from the UARS MLS instrument.

Advances in microwave technology, as well as the experience gained from the UARS MLS measurements, allowed several improvements to be incorporated in the MLS instrument on the AURA satellite. This instrument operates at frequencies near 118, 190, 240, 640, and 2500 GHz, and provides more measurements at better precision than the UARS

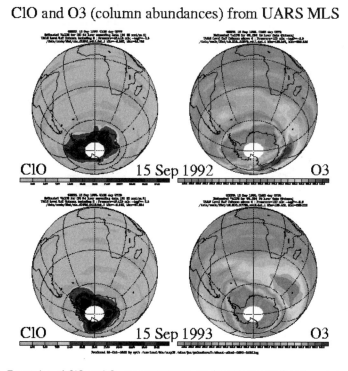

Figure 9-13. Examples of ClO and O_3 measurements made with the MLS instrument on the UARS satellite. (Courtesy NASA/JPL-Caltech.) See color section.

MLS. The improved technology that allows measurements at the higher frequencies is particularly important because more molecules have spectral lines at these frequencies, and spectral lines are stronger and provide greater instantaneous spectral bandwidth for measurements at lower altitudes. In addition, the polar orbit of the Aura satellite allows near-global coverage of these measurements. The vertical resolution of the EOS MLS instrument is generally in the 2 to 3 km range. Limb scans are taken approximately every 165 km along track, which is about three times more often than in the case of the UARS MLS instrument. The nominal scan range is expected to be from about 2 to 60 km above the surface.

9-5 INVERSION CONCEPTS

A passive microwave sensor measures the total emission radiated by the different atmospheric layers with a certain weighting function. This leads to the integral equations 9-5, 9-9, and 9-12. Thus, in order to derive vertical profiles of temperature or constituent density from the sensor spectral measurements, inversion techniques (also called retrieval techniques) have to be used. The detailed study of these techniques is beyond the scope of this text. Only the basic concept is discussed here in its most simple form. The reader is referred to the books by Houghton et al. (1984) and Ulaby et al. (1986) for a comprehensive discussion of retrieval theory.

Let us use the case of Equation (9-12) for illustration. Let us divide the atmosphere into N layers, each of constant temperature and of equal thickness Δz. The nth layer is centered at z_n and has a temperature T_n. In this case, Equation (9-12) becomes

$$T_a(v) = \sum_{n=1}^{N} W(v, z_n) T_n \Delta z \tag{9-33}$$

If the brightness temperature is measured at M different frequencies v_m, then we can write

$$T_{am} = \sum_{n=1}^{N} W_{nm} T_n \tag{9-34}$$

where $T_{am} = T_a(v_m)$ and $W_{nm} = W(v_m, z_n)\Delta z$. The above equation can be written in a matrix form as

$$\mathbf{T_a} = \mathbf{WT} \tag{9-35}$$

where $\mathbf{T_a}$ is a column vector of M elements containing the M measured brightness temperatures, $\mathbf{W}$ is a matrix of $M \times N$ elements containing the weighting function values for frequency v_m and altitude z_n, and $\mathbf{T}$ is an unknown N elements vector. In order not to have an underdefined situation, M must be equal to or larger than N, thus implying that the "granularity" of the temperature profile is driven by the number of measurement frequencies. The solution for $\mathbf{T}$ is

$$\mathbf{T} = (\mathbf{\tilde{W}W})^{-1}\mathbf{\tilde{W}T_a} \tag{9-36}$$

where $\mathbf{\tilde{W}}$ is the transpose of $\mathbf{W}$ (Franklin 1968).

More generally, the temperature profile can be written as a series expansion:

$$T(z) = \sum_{j=1}^{J} a_j F_j(z) \tag{9-37}$$

where the F_js are known polynomials or sines and cosines, and the a_js are the unknown coefficients. Equation (9-12) can then be written as

$$T_a(v) = \sum_{j=1}^{J} a_j \int_0^\infty W(v, z) F_j(z) dz = \sum_{j=1}^{J} a_j H_j(v) \tag{9-38}$$

Again, by making measurements at a number of frequencies and using the same matrix formulation as above, the a_js can be derived.

Unfortunately, the above simple methods are not fully satisfactory in actual situations in which the number of observations is limited and experimental errors exist. This has led to the use of a number of techniques that rely on some predetermined knowledge of the mean behavior of the atmospheric profile. The reader is referred to the reference list at the end of the chapter for further study of these techniques.

9-6 BASIC ELEMENTS OF PASSIVE MICROWAVE SENSORS

The basic function of a passive microwave sensor is to measure the intensity (and in some cases the polarization) of the microwave radiation emanating from the atmosphere within the beam of the sensor's antenna and within a certain frequency band or bands. Figure 9-14 shows in a simplified form the main generic functional elements of a heterodyne multispectral passive microwave radiometer (e.g., spectrometer). Figure 9-15 shows the signal spectrum at different stages in the sensor.

Usually, a single antenna reflector is used with a single or multiple feed network, depending on the total spectral coverage of the sensor. A switch allows the measurement of a calibration signal on a periodic basis in order to compensate for any sensor drift. Multiple calibrators of different temperatures can be used to bracket the range of measured amplitudes. The signal is then band-pass filtered and mixed with the signal from a local oscillator at frequency f_0. The next filter passes only the difference frequency. At the output of this filter the band of interest is in a fairly low frequency region (see Fig. 9-15), which is easier to measure than when the frequencies are higher. It should be emphasized that the parameter of interest, that is, incident signal amplitude, is conserved within a known factor. If high spectral resolution is desired, the low-frequency band signal is then inputted to the spectrometer, which can be visualized as a bank of filters. The detected output of the filters gives the signal power at specific frequencies.

If the polarization is an important parameter, a dual-polarized feed is used. The rest of the sensor is the same as in Figure 9-14.

9-7 SURFACE PRESSURE SENSING

Global measurement of surface pressure is one of the most basic elements required for modeling the dynamics of the atmosphere. A change in the surface pressure corresponds

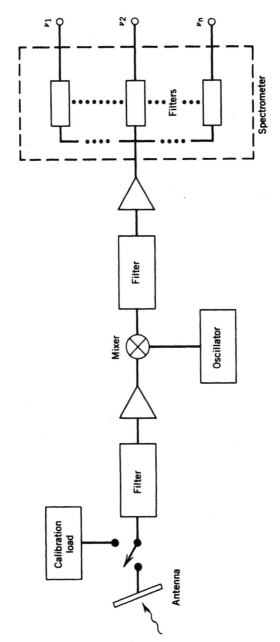

Figure 9-14. Main functional elements of a passive microwave spectrometer.

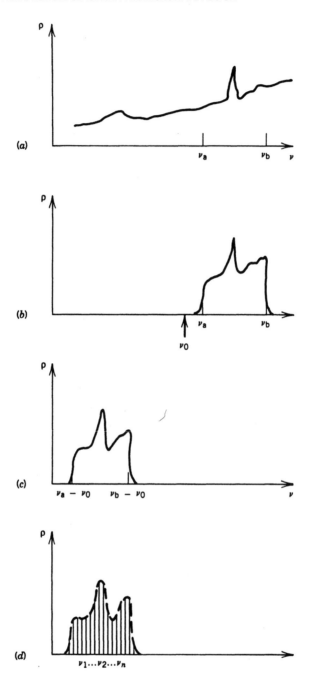

Figure 9-15. Signal spectra at different stages in the spectrometer shown in Figure 9-14. (a) Signal incident on the antenna. (b) Signal at the output of the front band-pass filter (band from ν_a to ν_b). (c) Signal at the output of the mixer and the low-pass filter. ν_0 is oscillator frequency. (d) Output of the different filter elements in the spectrometer.

to a change in the total columnar mass of the atmosphere. Thus, surface pressure can be measured if a remote sensing technique can be devised to determine the total columnar mass of the atmosphere or of a gas that is homogeneously mixed in the atmosphere, such as oxygen in the case of the Earth.

Let us consider a nadir-looking radar sensor operating at a pair of frequencies v_1 and v_2 in the wing of the 60 GHz oxygen band. One frequency v_1 is sufficiently close to the band center that the corresponding absorption coefficient is significantly affected by changes in the total mass of oxygen in the propagation path. The second frequency v_2 is far out on the wing and the corresponding absorption coefficient is little affected by the changes in the oxygen mass. Thus, if the transmitted signals at the two frequencies are equal, the power ratio of the received echoes will be proportional to:

$$\frac{P(v_1)}{P(v_2)} \sim \frac{R(v_1)T_0(v_1)T_w(v_1)}{R(v_2)T_w(v_2)} \qquad (9\text{-}39)$$

where R is the surface reflectivity, T_0 is the round-trip transmittivity due to oxygen absorption, and T_w is the transmittivity due to absorption by all other atmospheric elements, particularly water vapor. If v_1 and v_2 are not very different, then we can assume that $R(v_1) \simeq R(v_2)$. T_w can be derived by using a number of other frequencies away from the oxygen line. Thus, the above ratio will allow the derivation of $T_0(v_1)$, which is related to the oxygen absorption coefficient α_0 by

$$T_0(v_1) = 2\int_0^\infty \alpha_0(v, P, T)\rho(z)dz \qquad (9\text{-}40)$$

which in turn is directly related to the surface pressure. For more detail on this technique, the reader is referred to the paper by Peckham et al. (1983).

9-8 ATMOSPHERIC SOUNDING BY OCCULTATION

Occultation techniques have been widely used to sound planetary atmospheres from orbiting or flyby spacecrafts. As the telecommunication link radio signal is occulted by a planet's atmosphere, refraction occurs because of the increase in the medium index of refraction (see Fig. 9-16). By measuring the extent of refraction as the signal traverses deeper layers, the profile of the index of refraction as a function of altitude can be derived. In turn, this can be used to determine the atmospheric density profile (assuming a certain atmospheric composition) and from it, information about the atmospheric pressure and temperature using the hydrostatic equation and the equation of state.

Figure 9-16 shows the geometry of occultation. The planet location, the spacecraft location, and its velocity are usually known as a function of time from the spacecraft ephemeris and radio tracking just before occultation. The radio signal, which is received on Earth at a certain time t, emanates from the spacecraft at an angle α relative to the spacecraft–planet line. This angle can be derived by measuring the Doppler shift of the received signal. From the amount of ray bending through the atmosphere and the use of Snell's law in spherical geometry, the index of refraction profile can be derived. Figure 9-17 shows the refractivity profile and pressure–temperature profiles of the Venusian atmosphere derived from occultation data.

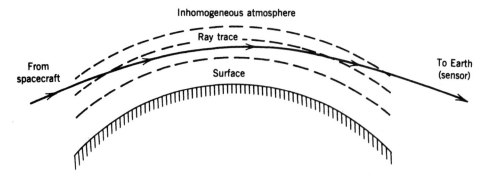

Figure 9-16. The geometry of radio occultation by a planetary atmosphere.

The availability of the Global Positioning Satellite (GPS) constellation provides an excellent opportunity for occultation sounding of the Earth's atmosphere. The basic measurement uses a GPS satellite as the transmitting source and a satellite in low earth orbit as the receiver, as shown in Figure 9-18. The GPS constellation consists of 24 satellites orbiting in six different planes around the earth. This means that several GPS satellites are typically in view of any satellite in low earth orbit, and every time a GPS satellite is rising or setting relative to the receiving satellite, an occultation measurement can be made. A single receiving satellite in the proper orbit can have as many as 500 occultation opportunities per day (Kursinski et al., 1997).

The GPS occultation measurements derive the bending angle α from a measurement of the Doppler shift of the GPS signal at the receiving spacecraft. The Doppler shift, in turn, is measured as the time derivative of the phase of the GPS signals. The bending angle is related to the index of refraction, n, of the Earth's atmosphere. The index of refraction contains contributions from the dry neutral atmosphere, water vapor, free electrons in the ionosphere, and particulates, primarily in the form of liquid water. To first order, the refractivity, N, for the Earth's atmosphere can be written as (Kursinski et al., 1997)

$$N = (n - 1)10^6 = 77.6 \frac{P}{T} + 3.73 \times 10^5 \frac{e}{T^2} + 4.03 \times 10^7 \frac{n_e}{f^2} + 1.4W \qquad (9\text{-}41)$$

where e is the partial pressure of water vapor in millibars, P is the total pressure in millibars, T is the temperature in Kelvin, n_e is the electron density expressed as number of electrons per cubic meter, f is the frequency in Hertz, and W is the liquid water content in grams per cubic meter.

Each GPS satellite broadcasts a dual-frequency signal at 1575.42 MHz and 1227.60 MHz, respectively. By measuring the differential phase delay between these two signals, one can estimate the number density of the electrons, n_e, and then the contribution of the ionosphere to the bending angle can be removed. Care must be taken when doing this correction, however, because of a subtle but important fact implied by the third term in Equation (9-41). Because the contribution to the index of refraction by the free electrons is a function of frequency, the bending angle, and hence the actual path through the ionosphere, will be different for the two signals. The only way the two signals can then originate at the same point in space (the GPS transmitter) and arrive at the same point in space (the receiver) is for these two signals to propagate through slightly different paths also before

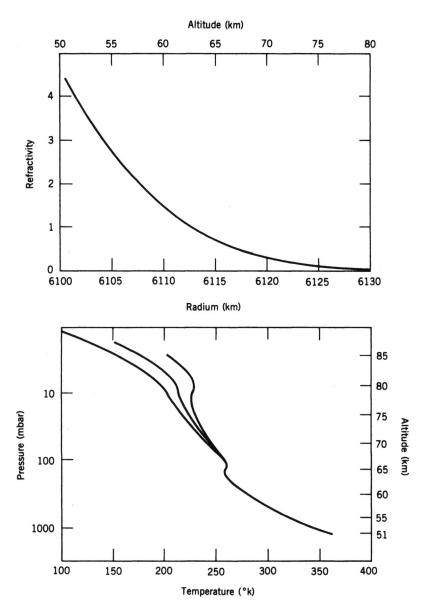

Figure 9-17. Refractivity profile and pressure–temperature profiles of the Venusian atmosphere derived from Mariner 10 occultation data. (From Nicholson and Muhleman, 1982.)

they enter the ionosphere, and also after they leave the ionosphere, as shown in Figure 9-18. Vorob'ev and Krasil'nikova (1993) introduced a method that recognizes this subtlety to calibrate the contribution of the free electrons in the ionosphere to the index of refraction.

Once the contribution from the ionosphere has been removed, only the dry and wet atmospheric contributions to the refractivity remain. If the atmosphere is dry enough, which is usually the case above 5–7 km altitude, the contribution of the wet atmosphere can be neglected, and only the first term in Equation (9-41) remains. Combining this term with

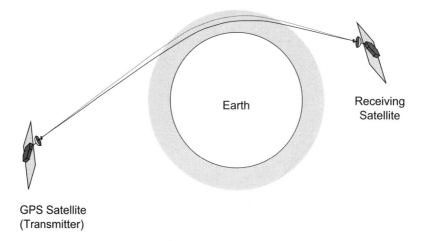

Earth

Receiving
Satellite

GPS Satellite
(Transmitter)

Figure 9-18. Geometry for a dual-frequency occultation measurement using a GPS satellite as a transmitter and a receiver on a different satellite in low earth orbit.

the equation of state of a perfect gas (see Chapter 8), one can then write the density of the atmosphere as a function of the refractivity. The pressure in the atmosphere can then be found by integrating the density between the altitude of observation and the top of the atmosphere [see Equation (8-1)]. Once the pressure profile is known, the temperature profile is calculated using the relationship in the first term in Equation (9-41). If the atmosphere is moist, however, the second term in Equation (9-41) must also be taken into account. In this case, one solution described by Kursinski et al. (1997) uses an estimate of the humidity in an initial density profile. Pressure and density profiles are then derived using this estimate, and these results are used to update the density profile. This iteration is repeated until the derived temperature profile converges.

9-9 MICROWAVE SCATTERING BY ATMOSPHERIC PARTICLES

When an electromagnetic wave encounters a particle, some of the energy is scattered and some of it is absorbed. The scattering and absorption cross sections, σ_s and σ_a respectively, are strongly dependent on the size of the particle relative to the incident wave wavelength. An exact expression for the scattering and absorption cross sections requires the general treatment of the interaction of a plane wave with a sphere. This was done by Mie in 1908, and the cross sections are expressed in terms of a complicated multipole expansion (see Equations 8-30 and 8-31).

In the case of microwave scattering, the particles' radius a is usually much smaller than the radiation wavelength, thus allowing significant simplification in the expressions of the scattering cross section. In this limit where $a \ll \lambda$, which is called the Rayleigh limit, the cross sections are given by [from Equations (8-32) and (8-33)]

$$\frac{\sigma_s}{\pi a^2} = \frac{128}{3} \pi^4 \left(\frac{a}{\lambda} \right)^4 |K|^2 \tag{9-42}$$

$$\frac{\sigma_a}{\pi a^2} = 8\pi \frac{a}{\lambda} \, \text{Imag} \, (-K) \tag{9-43}$$

where $K = (n^2 - 1)/(n^2 + 2)$, and n is the complex index of refraction of the scattering particle. For a low-loss dielectric such as ice, $\sigma_a \ll \sigma_s$. For a high-loss dielectric such as water, $\sigma_a \gg \sigma_s$.

One particularly valuable parameter is the backscatter cross section , which is given by

$$\frac{\sigma}{\pi a^2} = 64\pi^4 \left(\frac{a}{\lambda}\right)^4 |K|^2 \tag{9-44}$$

If the wave is incident on a volume containing suspended particles, the total cross section can be assumed to be equal to the sum of all the individual cross sections. This neglects the effects of multiple scattering.

One important observation is that the absorption cross section is proportional to a^3 and, hence, the volume of the particle. In the case of a cloud of particles, the total extinction would then be proportional to the total volume (or mass) of the ensemble of suspended particles.

In order to obtain the attenuation coefficient α, we should take the total loss due to all particles within a volume of unit cross section and unit length. Thus,

$$\alpha = \sum_j (\sigma_{aj} + \sigma_{sj}) = \frac{8\pi^2}{\lambda} \sum_j a_j^3 \left[\mathrm{Imag}(-K) + \frac{16}{3} |K|^2 \left(\frac{\pi a_j}{\lambda}\right)^3 \right] \tag{9-45}$$

9-10 RADAR SOUNDING OF RAIN

The use of radar sensors for rain detection, sounding, and tracking began in the early 1940s. As a radar pulse interacts with a rain region, echoes are returned that are proportional to the radar backscatter of the rain particles. As the pulse propagates deeper in the rain cell, additional echoes are returned at later times. Thus, a time analysis of the returned echo provides a profile of the rain intensity along the radar line of sight.

Three radar configurations are used to map a rain cell. The plan position indicator (PPI), the range height indicator (RHI), and the downlooking profiler (see Fig. 9-19). The PPI configuration corresponds to a horizontal cut around the sensor acquired by rotating the radar beam continuously, deriving an image of precipitation cells on a conical surface. The RHI gives a vertical cut in one plane acquired by scanning the beam up and down at a fixed azimuth angle. A combined PPI and RHI system would allow the acquisition of a three-dimensional image of the rain cells in the volume surrounding the sensor.

The downlooking profiler corresponds to an airborne or spaceborne sensor, and it provides a vertical cut of the precipitating region along the flight line. Figure 9-20 gives an example of such data acquired at the X band with an airborne radar sensor. It clearly shows the top profile of the precipitation region and the bright band that corresponds to the melting level. If the radar beam of a downlooking profiler is scanned back and forth across the track, then a wide volume of the atmosphere is mapped as the platform (aircraft or spacecraft) moves by.

If we consider the backscatter cross section of a volume of small particles, then [from Equation (9-43)]

$$\sigma = \frac{\pi^5}{\lambda^4} |K|^2 \sum_j D_j^6 \tag{9-46}$$

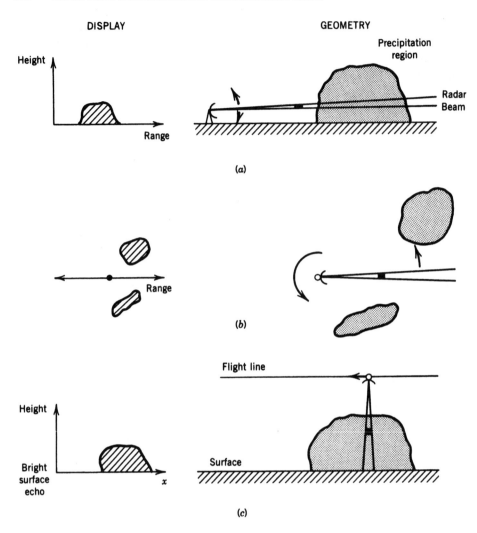

Figure 9-19. Three rain radar configurations. (a) Range height indicator, (b) plan position indicator, and (c) downlooking profile.

where $D_j = 2a_j$ is the diameter of the jth particle. The summation term is commonly called the reflectivity factor Z:

$$Z = \sum_j D_j^6 \tag{9-47}$$

and is expressed in mm^6/m^3 or in m^3. In the case of a distribution of particles,

$$Z = \int N(D) D^6 \, dD \tag{9-48}$$

where $N(D)$ is the particles' size distribution. Numerous research activities have been conducted to relate the parameter Z to the rain rate R expressed in mm/hr and the liquid water content M expressed in mg/m^3. The liquid water content is given by

4 km

Figure 9-20. Airborne radar data of rain region above the ocean surface off the Alaskan coast. The image represents rain distribution in a vertical plane below the aircraft.

$$M = \int_0^\infty \rho \frac{\pi D^3}{6} N(D)\, dD = \frac{\pi \rho}{6} \int_0^\infty D^3 N(D)\, dD \qquad (9\text{-}49)$$

where ρ is the water density. In the case of an exponential drop distribution,

$$N(D) = N_0 e^{-D/\overline{D}} \qquad (9\text{-}50)$$

and, from Equations (9-48) and (9-49),

$$Z = 720 N_0 (\overline{D})^7 \qquad (9\text{-}51)$$

$$M = \pi \rho N_0 (\overline{D})^4 \qquad (9\text{-}52)$$

Thus,

$$Z = \frac{720}{(\pi \rho)^{7/4} N_0^{3/4}} M^{7/4} \sim M^{1.75} \qquad (9\text{-}53)$$

If all the particles were of the same size, then

$$Z = ND^6 \qquad (9\text{-}54)$$

$$M = \frac{\pi \rho}{6} ND^3 \qquad (9\text{-}55)$$

$$\rightarrow Z = \frac{36}{(\pi \rho)^2 N} M^2 \sim M^2 \qquad (9\text{-}56)$$

Experimental investigations showed that for clouds, snow, and rain,

$$Z = CM^a \qquad (9\text{-}57)$$

where $1.8 < \alpha < 2.2$, and C is a constant given by (when M is in mg/m^3 and Z in mm^6/m^3):

$C \simeq 5 \times 10^{-8}$ for clouds

$C \simeq 0.01$ for snow

$C \simeq 0.08$ for rain

If we assume that all the drops fall at the same velocity V, then the rain rate R is related to the liquid water content by

$$R = \frac{VM}{\rho} \tag{9-58}$$

Statistical measurements showed that $R \sim M^{0.9}$ and that

$$Z = BR^b \tag{9-59}$$

where $1.6 < b < 2$ and B is a constant given by (when R is mm/hr):

$B = 200$ for rain

$B = 2000$ for snow

9-11 RADAR EQUATION FOR PRECIPITATION MEASUREMENT

From Equations (9-46) and (9-47), the radar backscatter of a unit volume of small particles is

$$\sigma = \frac{\pi^5}{\lambda^4} |K|^2 Z \tag{9-60}$$

If a radar sensor radiates a pulse of peak power P_t and length τ through an antenna of area A and gain G, the power density P_i at a distance r is

$$P_i = \frac{P_t G}{4\pi r^2} = \frac{P_t A}{\lambda^2 r^2} \tag{9-61}$$

The backscattered power P_s at any instant of time corresponds to the return from the particles in a volume of depth $c\tau/2$ and area S equal to the beam footprint. Thus,

$$P_a = P_i \sigma \frac{Sc\tau}{2} \tag{9-62}$$

where

$$S = \frac{\lambda^2 r^2}{A}$$

$$\rightarrow P_s = \frac{P_i \sigma \lambda^2 r^2 c\tau}{2A}$$

The received power P_r is equal to the power collected by the antenna:

$$P_r = \frac{P_s}{4\pi r^2}A = \frac{P_i\lambda^2 c\tau}{8\pi}$$

$$\rightarrow P_r = \frac{\pi^4}{8}Z|K^2|\frac{P_t Ac\tau}{\lambda^4 r^2}$$

(9-63)

In addition, P_r should be reduced by the loss factor due to absorption and scattering along the path of length r.

In the case of a coherent radar sensor, the received echo is mixed with the appropriate signal from the reference local oscillator in order to derive any shift in the echo's frequency relative to the transmitted pulse. This corresponds to the Doppler shift resulting from the motion of the droplets relative to the sensor.

9-12   THE TROPICAL RAINFALL MEASURING MISSION (TRMM)

The Tropical Rainfall Measuring Mission, launched in 1997, is a joint mission between the United States and Japan, and it is the first satellite Earth observation mission to monitor tropical rainfall. The TRMM satellite carries a total of five instruments, including the Precipitation Radar (PR), the TRMM Microwave Imager (TMI), the Visible Infrared Scanner (VIRS), the Clouds and the Earth's Radiant Energy System (CERES), and the Lightning Imaging Sensor (LIS). Japan provided the Precipitation Radar and the launch of the TRMM satellite, and the United States provided the spacecraft bus and the remaining four sensors, and also operates the TRMM satellite.

The TRMM satellite orbits the Earth at an altitude of 350 km and an inclination of 35°. Approximately 210 minutes of data can be stored on bulk memory cards in the command and data handling system on the spacecraft before downlinking through the Tracking and Data Relay Satellite network.

The precipitation radar is the primary instrument on the TRMM satellite. The radar operates at 13.8 GHz and utilizes an active phased-array antenna to scan the radar beam from side to side to image a swath of 220 km. The range resolution is about 250 m, and the spatial resolution is 4 km. This imaging geometry provides a three-dimensional view of the rainfall along the spacecraft orbit track.

Figure 9-21 shows an example of the TRMM measurements. This image shows heavy rains associated with springtime showers and thunderstorms in Texas and Oklahoma in the United States as measured by TRMM at 6:30 UTC on April 24, 2004. The heaviest rains fell on the evening of April 23 and the morning of April 24. The image on the left shows the horizontal distribution of rain intensity as seen by the TRMM satellite. The measurements in the inner swath are from the precipitation radar, and those from the wider, outer swath are from the TMI. Several mesoscale convective systems are clearly visible: one over south-central Texas, one over north-central Texas and one along the Oklahoma–Arkansas border. The image on the right shows the vertical structure of the rainfall as measured by the precipitation radar through the storm system in north-central Texas. The numerous towers in the vertical direction are indicative of intense convection.

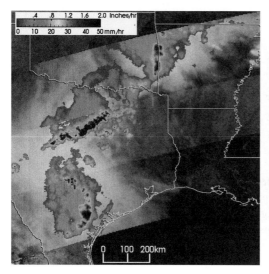

Figure 9-21. Examples of rainfall measurements from the TRMM satellite over Texas and Oklahma. The image on the left shows the horizontal distribution of rainfall, whereas the image on the right shows the vertical structure in the rainfall for one of the three storms visible in the left image. See color section. (Courtesy NASA/Goddard Space Flight Center.)

EXERCISES

9-1. Let us assume an atmosphere such that its opacity in the observation spectral region is equal to

$$\alpha(z) = \begin{cases} \alpha_0 & \text{for } 0 \leq z \leq H \\ 0 & \text{for } z > H \end{cases}$$

Calculate the corresponding weighting functions for upward- and downward-looking sensors and describe their behavior.

9-2. Let us assume an atmosphere is such that its opacity is given by

$$\alpha(z) = \begin{cases} \alpha_0 - \alpha_0 z/H & \text{for } 0 \leq z \leq H \\ 0 & \text{for } z > H \end{cases}$$

Derive the corresponding weighting functions for downlooking sensors and describe their behaviors.

9-3. Let us assume that the temperature in the atmosphere changes in a linear fashion:

$$T(z) = \begin{cases} T_0 - az & \text{for } 0 \leq z \leq T_0/a \\ 0 & \text{for } z > T_0/a \end{cases}$$

Give the expression for the brightness temperature for a nadir-looking sensor. How many spectral measurements are required to fully describe the temperature profile?

9-4. A nadir-looking microwave sensor is making multispectral emissivity measurements around an emission line with opacity given by

$$\tau_m(\nu) = \frac{\tau_0 \nu_L^2}{(\nu - \nu_0)^2 + \nu_L^2}$$

where $\nu_0 = 60$ GHz, $\nu_L = 5$ GHz, and $\tau_0 = 2$. Calculate and plot the weighting function at 50 GHz, 55 GHz, 58 GHz, 60 GHz, 62 GHz, 65 GHz, and 70 GHz. Assume an exponentially decaying atmosphere.

9-5. Let us assume a spherical atmosphere where the density is given by

$$\rho(z) = \rho_0 e^{-z/H}$$

where z is the altitude and ρ_0 and H are constants. Consider a limb sounding sensor with an infinitesimal beam looking at a tangent altitude h. Calculate the total atmospheric mass along the sensor line of sight and compare it to the corresponding mass above h for a downlooking sensor. Illustrate numerically by taking $h = 12$ km, $H = 6$ km, and the radius of the planet $R = 6000$ km.

REFERENCES AND FURTHER READING

Alcala, C. M., and A. E. Dessler. Observations of deep convection in the tropics using the Tropical Rainfall Measuring Mission (TRMM) precipitation radar. *Journal of Geophysical Research,* **107**(D24), 4792, doi:10.1029/2002JD002457, 2002.

Ali, D. S., P. W. Rosenkranz, and D. H. Staelin. Atmospheric sounding near 118 GHz. *Journal of Applied Meteorology,* **19**, 1234–1238, 1980.

Allen, K. C., and H. J. Liebe. Tropospheric absorption and dispersion of millimeter and submillimeter waves. *IEEE Transactions on Antennas and Propagation,* **AP-31**, 221–223, 1983.

Atlas, D. Radar meteorology. *Encyclopedia Britannica,* Vol. 18, p. 873, 1963.

Atlas, D., C. Elachi, and W. E. Brown. Precipitation mapping with an airborne SAR. *Journal of Geophysical Research,* **82**, 3445–3451, 1977.

Azeem, S. M. I., S. E. Palo, D. L. Wu, and L. Froidevaux. Observations of the 2-day wave in UARS MLS temperature and ozone measurements. *Geophysical Research Letters,* **28**(16), 3147–3150, 2001.

Barath, F. T., et al. The Upper Atmosphere Research Satellite microwave limb sounder instrument. *Journal of Geophysical Research,* **98**(D6), 10,751–10,762, 1993.

Barrett, A., and V. Chung. A method for the determination of high altitude water vapor abundance from ground-based microwave observations. *Journal of Geophysical Research,* **67**, 4259–4266, 1962.

Barros, A. P., M. Joshi, J. Putkonen, and D. W. Burbank. A study of the 1999 monsoon rainfall in a mountainous region in central Nepal using TRMM products and rain gauge observations. *Geophysical Research Letters,* **27**(22), 3683–3686, 2000.

Basharinov, A. E., A. S. Gurvich, S. T. Yegorov, A. A. Kurskaya, D. T. Matvyev, and A. M. Shutko. *The results of Microwave Xounding of the Earth's Surface According to Experimental Data from the Satellite Cosmos 243.* Space Research, Vol. 11. Akademie-Verlag, Berlin, 1971.

Basharinov, A. E., A. S. Gurvich, and S. T. Yegorov. *Radiation of the Earth as a planet.* Akademii Nauk SSSR, Institute of Atmospheric Physics, Scientific Publishing House, Moscow, USSR, 1974.

Basili, P., P. Ciotti, and D. Solimini. Inversion of ground-based radiometric data by Kalman filtering. *Radio Science,* **16,** 83–91, 1981.

Battan, L. J. *Radar Meteorology.* University of Chicago Press, Chicago, 1959.

Battan, L. J. *Radar Observes the Weather.* Doubleday, New York, 1962.

Bean, B. R., and E. J. Dutton. *Radio Meteorology.* Dover, New York, 1968.

Bent, A. E. Radar echoes from atmospheric phenomena. MIT Radiation Laboratory Report #173, 1943.

Carr, E. S., et al. Tropical stratospheric water vapor measured by the Microwave Limb Sounder (MLS). *Geophysical Research Letters,* **22,** 691–694, 1995.

Chahine, M., et al. Interaction mechanisms within the atmosphere. Chapter 5 in *Manual of Remote Sensing.* American Society of Photogrammetry, Falls Church, VA, 1983.

Chipperfield, M. P., M. L. Santee, L. Froidevaux, G. L. Manney, W. G. Read, J. W. Waters, A. E. Roche, and J. M. Russell. Analysis of UARS data in the southern polar vortex in September 1992 using a chemical transport model. *Journal of Geophysical Research,* **101**(D13), 18,861–18,882, 1996.

Clark, H. L., R. S. Harwood, P. W. Mote, and W. G. Read. Variability of water vapor in the tropical upper troposphere as measured by the Microwave Limb Sounder on UARS. *Journal of Geophysical Research,* **103**(D24), 31,695–31,708, 1998.

Crewell, S., R. Fabian, K. Künzi, H. Nett, T. Wehr, W. Read, and J. Waters. Comparison of ClO measurements by airborne and spaceborne microwave radiometers in the Arctic winter stratosphere 1993. *Geophysical Research Letters,* **22**(12), 1489–1492, 1995.

Danilin, M. Y., et al. Trajectory hunting as an effective technique to validate multiplatform measurements: Analysis of the MLS, HALOE, SAGE-II, ILAS, and POAM-II data in October–November 1996. *Journal of Geophysical Research,* **107**(D20), 4420, doi:10.1029/2001JD002012, 2002.

Deepak, A. Inversion methods. In A. Deepak (Ed.), *Atmospheric Remote Sounding.* Academic Press, New York, 1977.

Deepak, A. (Ed.). *Remote Sensing of Atmospheres and Oceans.* Academic Press, New York, 1980.

Dessler, A. E., M. D. Burrage, J.-U. Grooss, J. R. Holton, J. L. Lean, S. T. Massie, M. R. Schoeberl, A. R. Douglass, and C. H. Jackman. Selected science highlights from the first 5 years of the Upper Atmosphere Research Satellite (UARS) program. *Rev. Geophysical,* **36**(2), 183–210, 1998.

Fjeldbo, G., and V. R. Eshleman. The atmosphere of Mars analyzed by integral inversion of the Mariner IV occultation data. *Planetary Space Science,* **16,** 1035–1059, 1968.

Fjeldbo, G., A. J. Kliore, and V. R. Eshleman. The neutral atmosphere of Venus as studied with the Mariner V radio occultation experiments. *Astronomy Journal,* **76,** 123–140, 1971.

Franklin, J. N. *Matrix Theory.* Prentice-Hall, Englewood Cliffs, NJ, 1968.

Franklin, J. N. Well-posed stochastic extensions of ill-posed linear problems. *Journal of Mathematical Analysis Applications,* **31,** 682–716, 1970.

Grody, N. C. Remote sensing of atmospheric water content from satellites using microwave radiometry. *IEEE Transactions on Antennas and Propagation,* **AP-24,** 155–162, 1976.

Grody, N. C., and P. P. Pellegrino. Synoptic-scale studies using Nimbus-6 scanning microwave spectrometer. *Journal of Applied Meteorology,* **16,** 816–826, 1977.

Haddad, Z. S., J. P. Meagher, R. F. Adler, E. A. Smith, E. Im, and S. L. Durden. Global variability of precipitation according to the Tropical Rainfall Measuring Mission. *Journal of Geophysical Research,* **109,** D17103, doi:10.1029/2004JD004607, 2004.

Harada, C., A. Sumi, and H. Ohmori. Seasonal and year-to-year variations of rainfall in the Sahara desert region based on TRMM PR data. *Geophysical Research Letters,* **30**(6), 1288, doi:10.1029/2002GL016695, 2003.

Haroules, G. G., and W. E. Brown III. The simultaneous investigation of attenuation and emission by the Earth's atmosphere at wavelengths from 4 cm to 8 mm. *Journal of Geophysical Research,* **76,** 4453–4471, 1969.

Herman, B. M., and L. J. Battan. Calculation of Mie backscattering from melting ice spheres. *Journal of Meteorology,* **18,** 468–478, 1961.

Houghton, J. T., et al. *Remote Sensing of Atmospheres.* Cambridge University Press, 1984.

Jiang, J. H., B. Wang, K. Goya, K. Hocke, S. D. Eckermann, J. Ma, D. L. Wu, and W. G. Read. Geographical distribution and interseasonal variability of tropical deep convection: UARS MLS observations and analyses. *Journal of Geophysical Research,* **109,** D03111, doi:10.1029/2003JD003756, 2004.

Kerr, D. *Propagation of Short Radio Waves.* MIT Rad. Lab. Series #13, McGraw-Hill, New York, 1951.

Kliore, A. et al. The atmosphere of Mars from Mariner 9 radio occultation experiments. *Icarus,* **17,** 484–516, 1972.

Kliore, A., and I. R. Patel. Vertical structure of the atmosphere of Venus from Pioneer Venus radio occultation. *Journal of Geophysical Research,* **85,** 7957–7962, 1980.

Kummerow, C., W. Barnes, T. Kozu, J. Shiue, and J. Simpson. The Tropical Rainfall Measuring Mission (TRMM) Sensor Package. *Journal of Atmospheric and Oceanic Technology,* **15**(3), 809–817, 1998.

Kursinski, E. R., G. A. Hajj, J. T. Schofield, R. P. Linfield, and K. R. Hardy. Observing Earth's atmosphere with radio occultation measurements using the Global Positioning System. *Journal of Geophysical Research,* **102**(D19), 23,429–23,466, 1997.

Kursinski, E. R., G. A. Hajj, S. S. Leroy, and B. Herman. The GPS radio occultation technique. *Terrestrial Atmospheric and Oceanic Science,* **11**(1), 235–272, 2000.

Ledsham, W. H., and D. H. Staelin. An extended Kalman filter for atmospheric temperature profile retrieval with passive microwave sounder. *Journal of Applied Meteorology,* **17,** 1023–1033, 1978.

Lenoir, W. G. Microwave spectrum of molecular oxygen in the mesosphere. *Journal of Geophysical Research,* **73,** 361.

Lindal, G. F. The atmosphere of Neptune: An analysis of radio occultation data acquired with Voyager. *Astronomical Journal,* **103,** 967–982, 1992.

Ligda, M. Radar storm observation. *Comparative Meteorology,* 1265–1282, 1951.

Manney, G. L., L. Froidevaux, M. L. Santee, N. J. Livesey, J. L. Sabutis, and J. W. Waters. Variability of ozone loss during Arctic winter (1991–2000) estimated from UARS Microwave Limb Sounder measurements. *Journal of Geophysical Research,* **108**(D4), 4149, doi:10.1029/2002JD002634, 2003.

Mariner Stanford Group. Venus-Ionosphere and atmosphere as measured by dual-frequency radio occultation of Mariner 5. *Science,* **158,** 1678–1683, 1979.

Meeks, M. L., and A. E. Lilley. The microwave spectrum of oxygen in the Earth's atmosphere. *Journal of Geophysical Research,* **68,** 1683, 1963.

Mie, G. Beitrage zur Optik Truber Medien, Speziell Kolloidaler Metallsugen. *Ann. Phys.,* **XXV,** 377, 1908.

Njoku, E. G., J. M. Stacey, and F. T. Barath. The Seasat Scanning Multichannel Microwave Radiometer (SMMR): Instrument description and performance. *IEEE Journal of Oceanic Engineering,* **OE-5,** 100–115, 1980.

Njoku, E. G., E. J. Christensen, and R. E. Cofield. The Seasat Scanning Multichannel Microwave Radiometer (SMMR): Antenna pattern corrections—development and implementation. *IEEE Journal of Oceanic Engineering,* **OE-5,** 125–137, 1980.

Njoku, E. G. Passive microwave remote sensing of the Earth from space—A review. *Proceedings of IEEE,* **70,** 728–749, 1982.

Oki, R., K. Furukawa, S. Shimizu, Y. Suzuki, S. Satoh, H. Hanado, K. Okamoto, and K. Nakamura. Preliminary results of TRMM: Part I, A comparison of PR with ground observations. *Marine . Technology Society Journal,* **32,** 13–23, 1998.

Pearson, M., and W. Kreiss. Ground based microwave radiometry for recovery of average temperature profiles of the atmosphere. Boeing Scientific Research Laboratory Technical Report, D1-82-0781, 1968.

Peckham G., et al. Optimizing a remote sensing instrument to measure atmospheric surface pressure. *International Journal of Remote Sensing,* **14,** 465–478, 1983.

Read, W. G., D. L. Wu, J. W. Waters, and H. C. Pumphrey. Dehydration in the tropical tropopause layer: Implications from the UARS Microwave Limb Sounder. *Journal of Geophysical Research,* **109,** D06110, doi:10.1029/2003JD004056, 2004.

Read, W. G., D. L. Wu, J. W. Waters, and H. C. Pumphrey. A new 147–56 hPa water vapor product from the UARS Microwave Limb Sounder. *Journal of Geophysical Research,* **109,** D06111, doi:10.1029/2003JD004366, 2004.

Rius, A., G. Ruffini, and A. Romeo. Analysis of ionospheric electron-density distribution from GPS/MET occultations. *IEEE Transactions on Geoscience and Remote Sensing* **36**(2), 383–394, 1998.

Rodgers, C. D. Retrieval of atmospheric temperature and composition from remote measurements of thermal radiation. *Review of Geophysical and Space Physics,* **14,** 609–624, 1976.

Rodgers, C. D. Statistical principles of inversion theory. In A. Deepak (Ed.), *Inversion Methods in Atmospheric Remote Sounding,* pp. 117–138. Academic Press, New York, 1977.

Rosenkranz, P. W., et al. Microwave radiometric measurements of atmospheric temperature and water from an aircraft. *Journal of Geophysical Research,* **77,** 5833–5844, 1972.

Rosenkranz, P. W. Shape of the 5 mm oxygen band in the atmosphere. *IEEE Transactions on Antennas and Propagation,* **AP-23,** 498–506, 1975.

Rosenkranz, P. W. Inversion of data from diffraction-limited multiwavelength remote sensors, 1, Linear case. *Radio Science,* **13,** 1003–1010, 1978.

Rosenkranz, P. W., D. H. Staelin, and N. C. Grody. Typhoon June (1975) viewed by a scanning microwave spectrometer. *Journal of Geophysical Research,* **83,** 1857–1868, 1978.

Rosenkranz, P. W., and W. T. Baumann. Inversion of multiwavelength radiometer measurements by three-dimensional filtering. In A. Deepak (Ed.), *Remote Sensing of Atmospheres and Oceans,* pp. 277–311. Academic Press, New York, 1980.

Rosenkranz, P. W. Inversion of data from diffraction-limited multiwavelength remote sensors, 2, Nonlinear dependence of observables on the geophysical parameters. *Radio Science,* **17,** 245–256, 1982.

Rosenkranz, P. W. Inversion of data from diffraction-limited multiwavelength remote sensors, 3, Scanning multichannel microwave radiometer data. *Radio Science,* **17,** 257–267, 1982.

Rosenkranz, P. W., M. J. Komichak, and D. H. Staelin. A method for estimation of atmospheric water vapor profiles by microwave radiometry. *Journal of Applied Meteorology,* **21,** 1364–1370, 1982.

Santee, M. L., G. L. Manney, J. W. Waters, and N. J. Livesey. Variations and climatology of ClO in the polar lower stratosphere from UARS Microwave Limb Sounder measurements. *Journal of Geophysical Research,* **108**(D15), 4454, doi:10.1029/2002JD003335, 2003.

Santee, M. L., G. L. Manney, N. J. Livesey, and W. G. Read. Three-dimensional structure and evolution of stratospheric HNO3 based on UARS Microwave Limb Sounder measurements. *Journal of Geophysical Research,* **109,** D15306, doi:10.1029/2004JD004578, 2004.

Schoeberl, M. R., et al. Earth observing system missions benefit atmospheric research. *Eos Transactions on AGU,* **85**(18), 177, 2004.

Schreiner, W. S., S. S. Sokolovskiy, C. Rochen, and R. H. Ware. Analysis and validation of GPS/MET radio occultation data in the ionosphere. *Radio Science,* **34,** 949–966, 1999.

Sealy, A., G. S. Jenkins, and S. C. Walford. Seasonal/regional comparisons of rain rates and rain characteristics in West Africa using TRMM observations. *Journal of Geophysical Research,* **108**(D10), 4306, doi:10.1029/2002JD002667, 2003.

Sokolovskiy, S. V. Inversions of radio occultation amplitude data. *Radio Science,* **35,** 97–105, 2000.

Sokolovskiy, S. V. Modeling and inverting radio occultation signals in the moist troposphere. *Radio Science,* **36,** 441–458, 2001.

Sokolovskiy, S. V. Tracking tropospheric radio occultation signals from low Earth orbit. *Radio Science,* **36,** 483–498, 2001.

Solheim, F. S., J. Vivekanandan, R. H. Ware, and C. Rocken. Propagation delays induced in GPS signals by dry air, water vapor, hydrometeros, and other particulates. *Journal of Geophysical Research,* **104**(D8), 9663–9670, 1999.

Staelin, D. H. Measurements and interpretations of the microwave spectrum of the terrestrial atmosphere near 1-cm wavelength. *Journal of Geophysical Research,* **71,** 2975, 1966.

Staelin, D. H. Passive microwave remote sensing. *Proceedings of IEEE,* **57,** 427, 1969.

Staelin, D. H. Inversion of passive microwave remote sensing data from satellites. In A. Deepak (Ed.), *Inversion Methods in Atmospheric Remote Sounding,* pp. 361–394. Academic Press, New York, 1977.

Staelin, D. H., A. L. Cassel, K. F. Künzi, R. L. Pettyjohn, R. K. L. Poon, P. W. Rosenkranz, and J. W. Waters. Microwave atmospheric temperature sounding: Effects of clouds on the Nimbus-5 satellite data. *Journal of Atmospheric Science,* **32,** 1970–1976, 1975.

Staelin, D. H., K. F. Künzi, R. L. Pettyjohn, R. K. L. Poon, R. W. Wilcox, and J. W. Waters. Remote sensing of atmospheric water vapor and liquid water with the Nimbus 5 microwave spectrometer. *Journal of Applied Meteorology,* **15,** 1204–1214, 1976.

Staelin, D. H. Progress in passive microwave remote sensing: Nonlinear retrieval techniques. In A. Deepak (Ed.), *Remote Sensing of Atmospheres and Oceans,* pp. 259–276. Academic Press, New York, 1980.

Staelin, D. H. Passive microwave techniques for geophysical sensing of the Earth from satellites. *IEEE Transactions on Antennas and Propagation,* **AP-29,** 683–687, 1981.

Stratton, J. A. The effects of rain and fog upon the propagation of very short radio waves. *Proceedings of Institute of Electrical Engineers,* **18,** 1064–1075, 1930.

Tomiyasu, K. Remote sensing of the Earth by microwaves. *Proceedings of IEEE,* **62,** 86–92, 1974.

Tsang, L., J. A. Kong, E. Njoku, D. H. Staelin, and J. W. Waters. Theory for microwave thermal emission from a layer of cloud or rain. *IEEE Transactions on Antennas and Propagation,* **AP-25,** 650–657, 1977.

Ulaby, F. T. Passive microwave remote sensing of the Earth's surface. *IEEE Transactions on Antennas and Propagation,* **AP-24,** 112–115, 1976.

Ulaby F., et al. *Microwave Remote Sensing,* Vol. III. Artech House, Dedham, MA, 1986.

Vorob'ev V. V., and T. G. Krasil'nikova. Estimation of the accuracy of the atmospheric refractive index recovery from Doppler shift measurements at frequencies used in the NAVSTAR system. *Phys. Atmos. Ocean,* **29,** 602–609, 1993.

Waters, J., and S. Wofsy. Applications of high resolution passive microwave satellite systems to the stratosphere, mesophere and lower thermosphere. In *High Resolution Passive Microwave Satellites,* pp. 7-1 to 7-69. MIT Res. Lab. of Electronics, Staelin and Rasenkranz (Eds.), 1978.

Waters, J. W., K. F. Könzi, R. L. Pettyjohn, R. K. L. Poon, and D. H. Staelin. Remote sensing of atmospheric temperature profiles with the Nimbus 5 microwave spectrometer. *Journal of Atmospheric Science,* **32,** 1953–1969, 1975.

Waters, J. W. Absorption and emission of microwave radiation by atmospheric gases. In M. L. Meeks (Ed.), *Methods of Experimental Physics,* Ch. 12, Part B, Radio Astronomy. Academic Press, New York, 1976.

Waters, J. W., W. J. Wilson, and F. I. Shimabukuro. Microwave measurement of mesospheric carbon monoxide. *Science,* March, 1174–1175, 1976.

Waters, J. W., et al. The UARS and EOS Microwave Limb Sounder (MLS) Experiments. *Journal of the Atmospheric Sciences,* **56,** 194–218, 1999.

Westwater E. R., J. B. Snider, and A. V. Carlson. Experimental determination of temperature profiles by ground-based microwave radiometry. *Journal of Applied Meteorology,* **14,** 524–539, 1975.

Westwater, E. R., and M. T. Decker. Application of statistical inversion to ground-based microwave remote sensing of temperature and water vapor profiles. In A. Deepak (Ed.), *Inversion Methods in Atmospheric Remote Sounding,* pp. 395–428. Academic Press, New York, 1977.

Westwater, E. R. The accuracy of water vapor and cloud liquid determination by dual-frequency ground-based microwave radiometry. *Radio Science,* **13,** 677–685, 1978.

Westwater, E. R., and N. C. Grody. Combined surface- and satellite-based microwave temperature profile retrieval. *Journal of Applied Meteorology,* **19,** 1438–1444, 1980.

Westwater, E. R., and F. O. Guiraud. Ground-based microwave radiometric retrieval of precipitable water vapor in the presence of clouds with high liquid content. *Radio Science,* **15,** 947–957, 1980.

Wickert, J., C. Reigber, G. Beyerle, R. Konig, C. Marquardt, T. Schmidt, L. Grunwaldt, R. Galas, T. K. Meehan, W. G. Melbourne, and K. Hocke. Atmosphere sounding by GPS radio occultation: First results from CHAMP. *Geophysical Research Letters,* **28,** 3263–3266, 2001.

Wilheit, T., et al. A satellite technique for quantitatively mapping rainfall rates over the ocean. *Journal of Applied Meteorology,* **16,** 551–560, 1977.

Wilheit, T. T., R. F. Adler, R. Burpee, R. Sheets, W. E. Shenk, and P. W. Rosenkranz. Monitoring of severe storms. In D. Staelin and P. Rosenkranz (Eds.), *High Resolution Passive Microwave Satellites.* Applications Review Panel Final Report, Research Laboratory of Electronics, Massachusetts Institute of Technology, Cambridge, MA, 1978.

Yuan, L. L., R. A. Anthes, R. H. Ware, C. Rocken, W. D. Bonner, M. G. Bevis, and S. Businger. Sensing climate change using the global positioning system. *Journal of Geophysical Research,* **98,** 14,925–14,937, 1993.

10

MILLIMETER AND SUBMILLIMETER SENSING OF ATMOSPHERES

The millimeter and submillimeter region (about 100 GHz to a few THz) of the electro-magnetic spectrum is rich in spectral signatures of atmospheric and interstellar gaseous constituents. A large number of rotational transitions occur in this spectral region. In addition, the temperatures of dense interstellar clouds range from about 6°K to 100°K, which correspond to thermal emission maxima at 500 to 30 μm (frequency of 600 GHz to 10 THz) and a significant amount of emitted radiation in the rest of the millimeter and submillimeter region. In addition, because of its relatively longer wavelength, submillimeter radiation is less affected by interplanetary dust than infrared or visible radiation.

10-1 INTERACTION WITH ATMOSPHERIC CONSTITUENTS

In order to illustrate the richness of the millimeter and submillimeter region in spectral transitions of gases, let us consider the case of the water molecule. Figure 10-1 illustrates the energy level diagram of H_2O. It is apparent that a very large number of transitions are possible. The familiar 22 GHz water vapor line corresponds to one of the $J = 6 \rightarrow 5$ transitions. A large number of transitions have emission in the 100 GHz to a few THz spectral region. Table 10-1 shows the water transitions below 800 GHz. This includes the ground state transition at 557 GHz. The $H_2{}^{18}O$ isotope has a transition at 547.7 GHz.

A very important molecule in the interstellar medium is carbon monoxide, CO, which is the second most abundant known molecule after molecular hydrogen. It has a number of transitions in the millimeter and submillimeter region, such as the $J = 5 \rightarrow 4$ transition at 576 GHz and $J = 6 \rightarrow 5$ transition at 691 GHz.

Numerous other molecules of importance in astrophysics and planetary atmospheres have transitions in the millimeter and submillimeter region. Figure 8-8 shows some of the spectral lines in the 200 to 300 GHz regions for molecules in the Earth's upper atmosphere. Figure 10-2 shows the abundance of spectral lines in a 400 MHz band around the 265.75 GHz frequency. Table 10-2 shows some of the important spectral lines in the 530 GHz to 695 GHz.

Introduction to the Physics and Techniques of Remote Sensing. By C. Elachi and J. van Zyl
Copyright © 2006 John Wiley & Sons, Inc.

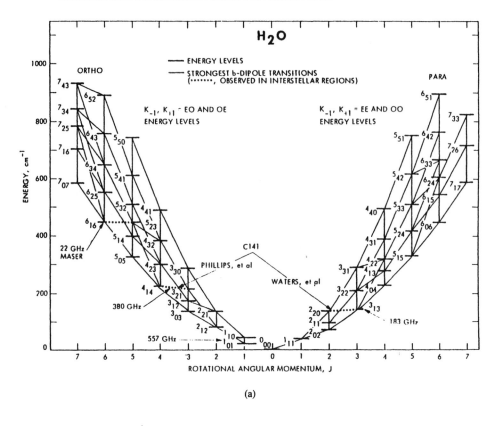

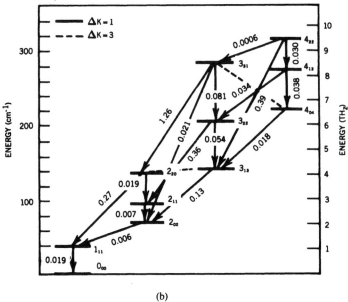

Figure 10-1. (a) Energy level diagram of H_2O. (b) Detail of the lower part of the para-H_2O rotational energy levels, showing the allowed transitions that connect the levels. The Einstein coefficients (probability of spontaneous decay per second) are also shown (from Kuiper et al., 1984).

Spectral lines in the submillimeter region are broadened mainly due to pressure. Typically, molecular lines are pressure broadened by about 3 GHz/bar. At 1 millibar pressure, this corresponds to a 3 MHz line width. In comparison, the Doppler broadening leads to a line width v_D, given by [from Equation (8-20)]

$$v_D = 0.76 \times 10^{-6} v \sqrt{\frac{T}{M}} \quad (10\text{-}1)$$

TABLE 10-1. Transitions of Water below 800 GHz

	Transition frequency (GHz)
Ortho ladder	
$1_{10}\text{--}1_{01}$	556.9
$4_{14}\text{--}3_{21}$	380.2
$4_{23}\text{--}3_{30}$	448.0
$5_{33}\text{--}4_{40}$	474.7
$6_{16}\text{--}5_{23}$	22.2
$6_{43}\text{--}5_{50}$	439.2
$7_{52}\text{--}6_{61}$	443.0
$10_{29}\text{--}9_{36}$	321.2
Para ladder	
$2_{11}\text{--}2_{02}$	752.0
$3_{13}\text{--}2_{20}$	183.3
$5_{15}\text{--}4_{22}$	325.2
$5_{32}\text{--}4_{41}$	620.7
$6_{42}\text{--}5_{51}$	470.9
$6_{24}\text{--}7_{17}$	488.5
$7_{53}\text{--}6_{60}$	437.3

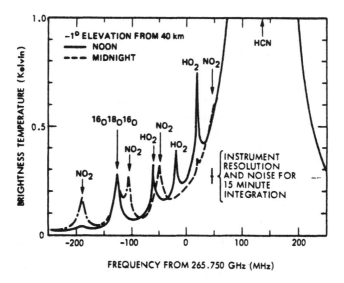

Figure 10-2. Spectral lines within a ±200 MHz band around 265.75 GHz, corresponding to H_2O, NO_2, HCN, and $^{16}O^{18}O^{16}O$ in the Earth's upper atmosphere. (From Waters et al., 1984.)

TABLE 10-2. Molecular Transitions of Interest and Importance to Astrophysics and Planetary Atmosphere Studies between 530 GHz and 695 GHz

Molecule	Frequency (GHz)
PH_3	534
$H_2{}^{18}O$	547.7
	537.3
	554.8
^{13}CO	551 and 661
$H_2{}^{16}O$	556.936 (ortho)
	620.7 (ortho)
HDO	559.8 and 559.9
NH_3	572.5
CO	576 and 691
$HC^{18}O$	603.025
$H^{13}CN$	604.2
SO	611.5
NH_2D	613
HCN	620.3
$H^{37}Cl$	624
HCO^+	625
HCl	626
SO_2	632.2
CS	636.5
DCO^+	648.192
HDS	650.3
DCN	651.5
SiH	682
H_2S	712
	687.3

For a temperature of 200°K and a mass $M = 28$ (i.e., CO), the Doppler broadening at 576 GHz is 1.15 MHz. In the infrared region, the Doppler broadening will be 6.1 MHz at 100 μm and 61 MHz at 10 μm. Thus, for this illustrative case, pressure broadening will dominate down to a pressure of about 1 millibar in the millimeter region, down to few millibars in the very far infrared, and few tens of millibars in the thermal infrared. This shows that submillimeter spectroscopy, which uses the pressure broadening effect to derive vertical profiles, permits the derivation of the mixing ratio of molecules over a larger altitude range than infrared spectroscopy. Figure 10-3 shows the spectral linewidths in the terrestrial atmosphere. It is clear that pressure broadening dominates up to about 60 km altitude for the 1 THz region.

10-2 DOWNLOOKING SOUNDING

Downlooking sounders in the millimeter spectral region are sometimes used to derive the temperature profile and detect the presence of and measure the mixing ratio profile of constituents in planetary upper atmospheres, including those of Earth. The measurement approach is based on the fact that the spectral line shape is strongly dependent on the local pressure up to very high altitudes (e.g., low pressure levels). To illustrate, let us consider

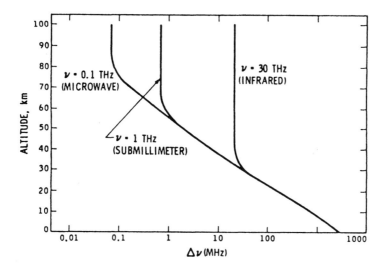

Figure 10-3. Spectral linewidths in the terrestrial atmosphere. Typical values are shown. Actual values will vary by about a factor of two from species to species. (Courtesy of J. Waters, Jet Propulsion Laboratory.)

the optical thickness τ of an atmospheric layer near one spectral line of its constituent under study. Using the downlooking configuration,

$$\tau(v) = \int_{\text{layer}} \rho(z)k(v, z)\, dz \tag{10-2}$$

where

$$k(v, z) = \frac{S}{\pi} \frac{v_L}{(v - v_0)^2 + v_L^2} \tag{10-3}$$

$$v_L(P) = v_L(P_0)\frac{P(z)}{P_0} \tag{10-4}$$

v_0 = center frequency of spectral line

P_0 = pressure at a reference level

If we assume that the pressure is exponentially decaying in altitude with a scale height H, then

$$k(v, z) = A\frac{e^{-z/H}}{a^2 + e^{-2z/H}}$$

$$A = \frac{S}{\pi v_L(P_0)} \tag{10-5}$$

$$a = \frac{|v - v_0|}{v_L(P_0)}$$

$k(v, z)$ can be considered as a weighting function for measuring the mixing ratio profile [note that $v_L(P_0) \sim$ total density, thus $A \sim 1$/total density]. Figure 10-4 shows the behavior of $k(v, z)$ as a function of z for different values of a (i.e., operating frequency v). It is clear that by measuring the optical thickness at a number of frequencies along the wing of the spectral line, the mixing ratio profile can be determined assuming the temperature profile is known. This technique has been exploited on Venus and Mars using the ground state and first excited rotational transition of CO, which occur at 115 GHz and 230 GHz. This technique can be used to get the mixing ratio of many trace molecules in the planetary upper atmospheres, such as HCl (625.9 GHz), PH_3 (534 GHz), HCN (620.3 GHz), SO_2 (632.2 GHz), and H_2S (687.3 GHz), and parent molecules outgassed by comets, such as H_2O, NH_3, HCN, and CO, all of which have strong rotational transitions in the 500–700 GHz region.

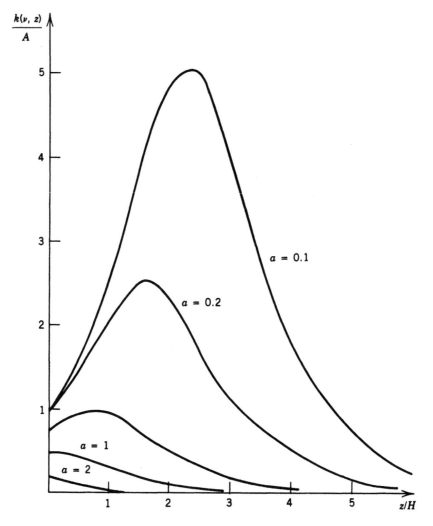

Figure 10-4. Behavior of $k(v, z)$ as a function of z for different values of $a = |v - v_0|/v_L(P_0)$.

10-3 LIMB SOUNDING

In the limb sounding geometry, the optical path in the atmosphere is much longer than in the downlooking geometry. This allows the detection and study of trace constituents in the upper atmosphere with very low mixing ratios.

Let us consider the simple case of an optically thin atmosphere. In this case, the absorption can be neglected and Equation (9-30) reduces to

$$T_a(v, h) = \int_h^\infty \alpha(v, z) T(z) g(z, h) \, dz \qquad (10\text{-}6)$$

Taking into account that the planet radius is usually much larger than z or h, $g(z, h)$ can be written as [from Equation (9-24)]

$$g(z, h) \simeq \sqrt{\frac{R}{2(z - h)}} \qquad (10\text{-}7)$$

Taking the additional simplifying assumption that $T(z)$ is constant, then the observed temperature can be written as

$$T_a(v, h) = T \sqrt{\frac{R}{2}} \int_h^\infty \rho(z) \frac{k(v, z)}{\sqrt{z - h}} \, dz = \int_h^\infty \rho(z) W(v, h, z) \, dz \qquad (10\text{-}8)$$

where $W(v, h, z)$ is the weighting function. To illustrate, in the case of an exponentially decaying pressure, W can be written as [from Equations (10-3), (10-4), and (10-5)]

$$W(v, h, z) = \frac{ST}{\pi} \sqrt{\frac{R}{2}} \frac{v_L}{(v - v_0)^2 + v_L^2} \frac{1}{\sqrt{z - h}}$$

$$= A' \frac{e^{-z/h}}{(a^2 + e^{-2z/h}) \left(\dfrac{z - h}{H} \right)^{1/2}}$$

$$A' = \frac{ST}{\pi v_L(P_0)} \sqrt{\frac{R}{2H}} \qquad (10\text{-}9)$$

$$a = \frac{|v - v_0|}{v_L(P_0)}$$

Figure 10-5 shows the behavior of $W(v, h, z)$ for different values of a (i.e., frequency displacement from the line center).

Let us consider the observed brightness temperature T_a far out on the wings of the spectra line ($v - v_0 \gg v_L$). The weighting function will reduce to

$$W(v, h, z) = \frac{A'}{a^2} \frac{e^{-z/h}}{\left(\dfrac{z - h}{H} \right)^{1/2}} \qquad (10\text{-}10)$$

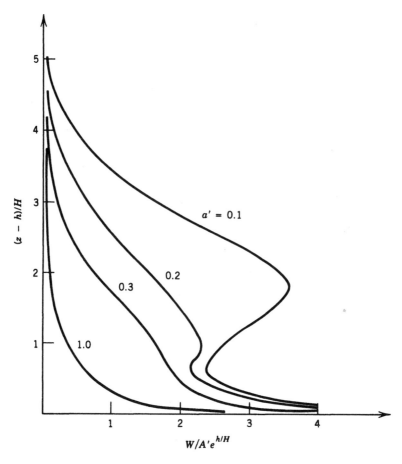

Figure 10-5. Behavior of the weighting function W as a function of altitude for different values of $a' = a \exp(h/H)$.

which is a fast-decaying function of z. Thus, most of the contribution to T_a results from a thin layer of atmosphere just above the tangent height h. This can be illustrated by considering the ratio

$$C(\Delta h) = \frac{\int_h^{h+\Delta h} W(v, h, z)\, dz}{\int_h^\infty W(v, h, z)\, dz} = \frac{\int_0^{\Delta h/H} \dfrac{e^{-x}}{\sqrt{x}}\, dx}{\int_0^\infty \dfrac{e^{-x}}{\sqrt{x}}\, dx} \tag{10-11}$$

as a function of $\Delta h/H$. With the appropriate variable transformation ($x = y^2$), this ratio can be written as a function of the error function as

$$C(\Delta h) = \mathrm{erf}\left(\sqrt{\frac{\Delta h}{H}}\right) \tag{10-12}$$

where

$$\mathrm{erf}(\xi) = \frac{2}{\sqrt{\pi}} \int_0^\xi e^{-y^2} \, dy$$

This function is shown in Figure 10-6a. For $\Delta h/H = 0.23$, $C(\Delta h) = 0.5$. This shows that half the contribution is from a layer about equal to one-quarter of the scale height H. For a typical H of 7 km, more than half the contribution is from a layer of thickness 1.75 km.

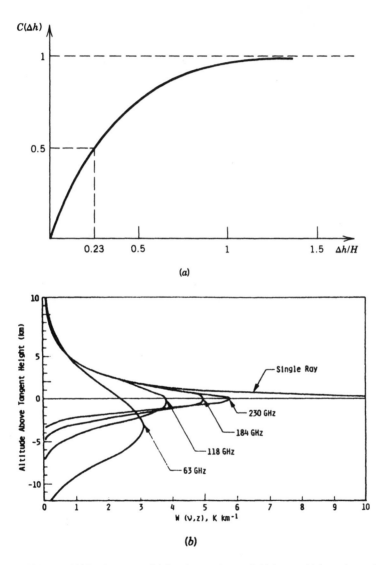

(a)

(b)

Figure 10-6. (a) Ratio $C(\Delta h)$ of the contribution from a layer of thickness Δh just above the tangent height h to the total contribution of the atmosphere in a limb sounding geometry. (b) Limb weighting functions for a 1 m antenna at a tangent height of 50 km in a 250 km atmosphere. (Courtesy of J. Waters, Jet Propulsion Laboratory.)

Thus, if the limb sounder scans the upper atmospheric region vertically (i.e., change in h), a fairly high accuracy profile of the constituent can be derived in a straightforward manner.

In reality, the receiving antenna pattern plays a significant role. The ratio in Equation (10-11) must be convolved with the antenna pattern, which can be expressed as a function of Δh for a fixed scan position (i.e., fixed h). This usually leads to a significant broadening of the weighting function, as illustrated in Figure 10-6b.

10-4 ELEMENTS OF A MILLIMETER SOUNDER

The millimeter sounding sensor consists of a collecting aperture and a receiver/spectrometer that measures the intensity of the collected signal as a function of frequency. This is usually done using a heterodyne receiver with which the signal frequency band is translated down to a lower frequency band using a nonlinear mixer. This translation process preserves the relative position of the received frequencies and their relative amplitudes. This allows the high-resolution spectral analysis of the collected signal to be done in a lower-frequency region in which high-resolution digital and analog spectrometers are available.

Figure 10-7 shows the basic elements of a millimeter heterodyne receiver. Figure 10-8 is a block diagram of a balloon-borne cooled millimeter spectrometer (Waters et al., 1984) used to sound ClO (204.352 GHz line), H_2O_2 (204.575 GHz line), and O_3 (206.132 GHz line). The filter bank consists of 35 filters with spectral resolution of 2 MHz, 8 MHz, and 32 MHz, as indicated in Figure 10-9. After the two-step mixing, the spectral range for each one of the three spectral lines under study is the same, thus allowing the use of identical filter banks to derive the line shape. The heterodyning arrangement is shown in Figure 10-10.

For submillimeter sensors, the basic elements are still the same, except that the main challenge is the development of high-efficiency mixers and high-frequency stable local

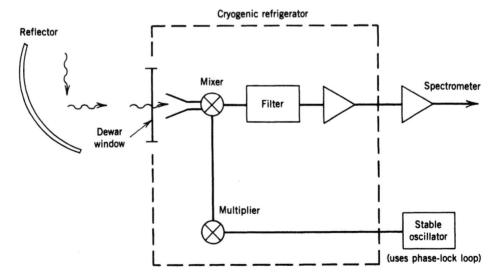

Figure 10-7. Basic elements of a millimeter heterodyne spectrometer.

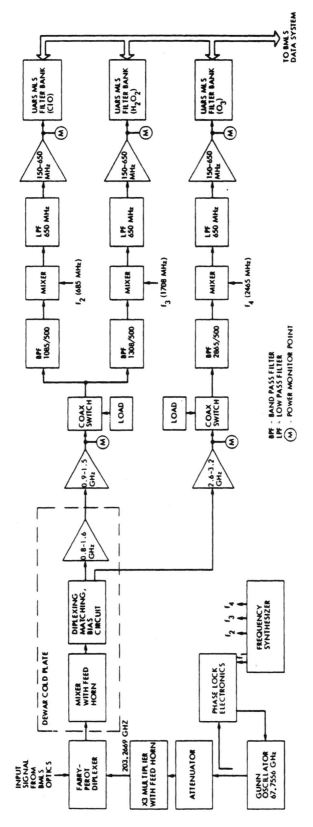

Figure 10-8. Block diagram of a cooled 205 GHz spectrometer (from Waters et al., 1984).

459

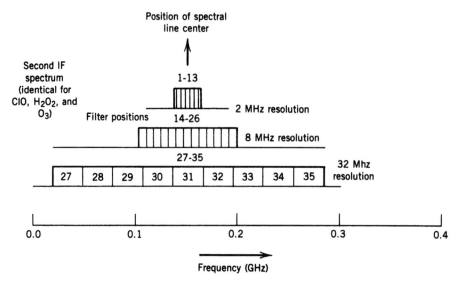

Figure 10-9. Filter bank arrangement for the spectrometer in Figure 10-8.

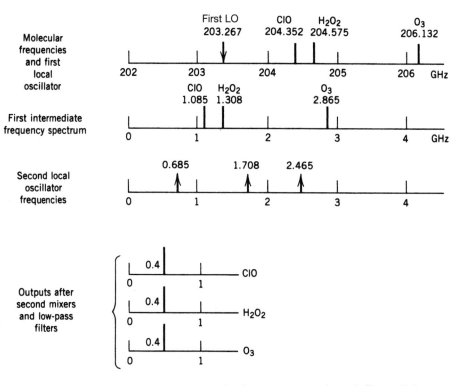

Figure 10-10. Heterodynic arrangement for the spectrometer shown in Figure 10-8.

oscillators. To illustrate, two types of mixers are briefly described: the Schottky mixer and the SIS mixer.

A Schottky diode consists of a metal contact on a lightly doped semiconducting layer epitaxially grown on a heavily doped conducting substrate. The epitaxial layer is conductive except near the metal contact, where an insulating depletion layer is formed. The corresponding current–voltage relationship has the form

$$I \sim e^{(aV)}$$

This nonlinear transfer function allows the generation of mixed frequencies when two voltages at frequencies v_1 and v_2 are input to the device. Schottky diodes have been used both at room temperature and at low temperatures.

SIS (superconductor–insulator–superconductor) devices use the tunneling process to provide a sharp nonlinearity. The barrier is formed by the insulating oxide layer. As illustrated in Figure 10-11, it is apparent that when the applied voltage is equal to or exceeds the band gap in the superconducting materials, a sharp increase in the current will occur.

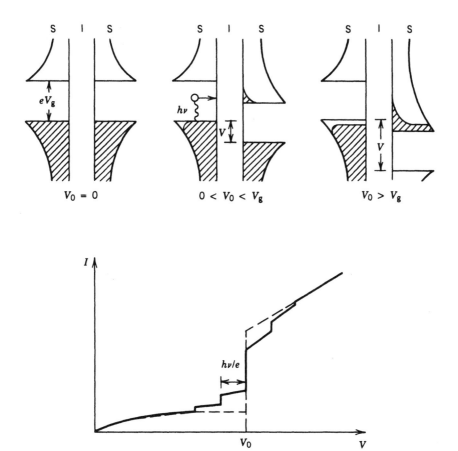

Figure 10-11. Schematic diagram showing the relative positions of the energy density states in an SIS as a function of the applied voltage.

In actuality, the increase in the current will occur in steps due to the energy of the incident photon. This nonlinearity in the I–V transfer function allows the generation of the mixed frequencies.

Schottky and SIS mixers will operate at the lower end of the submillimeter region, and work is ongoing to improve their capability at higher frequencies. At the other end of the spectrum, gallium-doped germanium photoconductive mixers have been used down to a frequency region of 2 to 3 THz.

The stable local oscillator is usually a Gunn oscillator or a carcinotron. Gallium arsenide Gunn oscillators have been used up to 100 GHz. Indium phosphide oscillators have operated up to 180 GHz. By using multipliers, reference frequencies of many hundreds of GHz can be generated. At the upper frequency end, multifrequency lasers have been used where two neighboring spectral lines are mixed down to generate submillimeter signals.

Most of the elements in a submillimeter sensor use quasioptic techniques, such as dichroic filters (plates with an array of appropriately sized holes), comb filters, Fabry–Perot diplexers, polarizing grids, and array detectors.

The Earth Observing System Microwave Limb Sounder (EOS MLS) instrument is an example of a state-of-the-art atmospheric sounder that includes measurements in the millimeter and submillimeter parts of the electromagnetic spectrum. EOS MLS was launched as one of four atmospheric sensors on NASA's Aura satellite in 2004. Figure 10-12 shows a block diagram of the EOS MLS instrument, which includes a channel at 640 GHz and one at 2.5 THz. The 640 GHz channel includes spectral lines of H_2O, N_2O, O_3, ClO, HCl, HOCl, BrO, HO_2, and SO_2, whereas the 2.5 THz channel includes a spectral line of stratospheric OH. The EOS MLS measurements of stratospheric OH on a global scale will fill a gap that currently exists in atmospheric chemistry. Current models of atmospheric chemistry assume that OH is a well-behaved constituent controlling, among other things, the conversion of CH_4 to H_2O and the rate of oxidation of sulfur gases (SO_2 and OCS) to sulfate aerosols. The EOS MLS measurements will allow this assumption to be tested on a global scale.

The EOS MLS radiometers use heterodyne receivers to record the incoming radiation. The incoming radiation for the four lower frequency channels (118, 190, 240, and 640 GHz) is collected by a three-reflector system that is used to scan the limb vertically. Radiometric calibration is provided by periodically switching to calibration targets or to space. This is accomplished using a switching mirror following the antenna system. The signal from the switching mirror is separated into different paths by an optical multiplexer consisting of dichroic plates and polarization grids, and fed to the inputs of the four radiometers. The incoming signals are translated to intermediate frequencies using local oscillator signals and advanced planar-technology mixers. The solid-state local oscillators drive the mixers in the subharmonic pumped mode. The intermediate frequency outputs are fed to spectrometers via a switch network. Digitized data from the spectrometers are passed to the command- and data-handling system for transmission to the ground through the spacecraft data-handling system.

The atmospheric and calibration signals for the 2.5 THz radiometer are obtained via a dedicated telescope and scanning mirror whose operation is synchronized with that of the GHz antenna and the GHz switching mirror. The 2.5 THz measurements are performed simultaneously at both polarizations to provide increased signal to noise for the important OH measurement. The THz radiometer also operates in the heterodyne mode. In this case, the local oscillator is a CO_2-pumped methanol (CH_3OH) gas laser. All radiometers operate at ambient temperature.

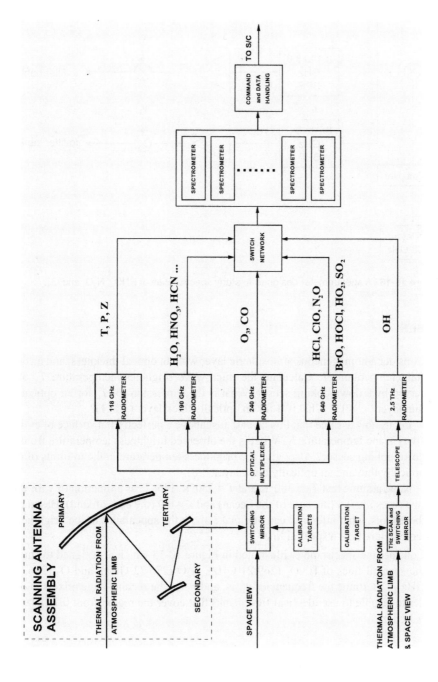

Figure 10-12. Block diagram of the EOS MLS instrument. (From Waters et al., 1993. Copyright © 1993 Wiley & Sons, Inc. Reprinted with permission of John Wiley & Sons, Inc.)

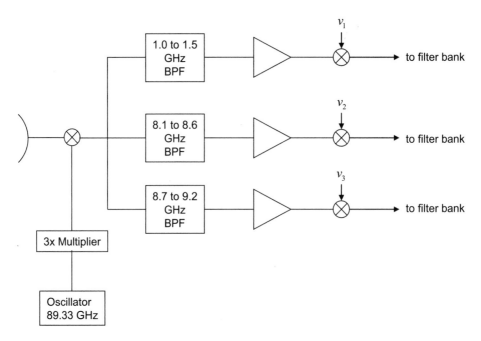

Figure 10-13. A spectrometer designed to study spectral lines of HNO_3, N_2O, and O_3.

EXERCISES

10-1. Consider a homogeneous atmospheric layer with an optical thickness and a constant temperature T. Calculate the microwave brightness temperature T_a observed by a downlooking sensor. What is the expression for T_a for an optically thin layer ($\tau \ll 1$)? What is it for an optically thick layer ($\tau \gg 1$)?

Let us now assume that this layer is just above a perfectly flat surface of emissivity ε and temperature T_g. What is the observed brightness temperature from a downlooking sensor? Also, give the brightness temperature in the to limits of an optically thin and an optically thick layer.

Now, assume that $T_g = 300°K$ and $T = 250°K$. Plot T_a as a function of τ for the two cases $\varepsilon = 0.4$ (the case of the ocean) and $\varepsilon = 0.9$ (the case of land). Based on these plots, what inference can you make about the appearance of the brightness temperature near a spectral line?

10-2. Consider the spectrometer illustrated in Figure 10-13 which is designed to study the spectral lines of HNO_3 (269.21 GHz), N_2O (276.22 GHz), and O_3 (276.92 GHz). Determine the frequencies v_1, v_2, and v_3 for the second-stage mixers in order to be able to use identical filter banks that cover the region from 0.1 GHz to 0.3 GHz.

REFERENCES AND FURTHER READING

Allen, D. R., R. M. Bevilacqua, G. E.Nedoluha, C. E. Randall, and G. L. Manney. Unusual Stratospheric transport and mixing during the 2002 Antarctic winter. *Geophysical Research Letters,* **30,** 10.1029/2003GL017117, 2003.

Chandra, S., J. R. Ziemke, and R. V. Martin. Tropospheric ozone at tropical and middle latitudes derived from TOMS/MLS residual: Comparison with a global model. *Journal of Geophysical Research,* **108**(9), doi:10.1029/2002JD002912, 2003.

Clark, H. L., and R. S. Harwood. Upper-tropospheric humidity from MLS and ECMWF reanalysis. *Monthly Weather Review,* **131**(3), 542-555, 2003.

Kuiper T. B., et al. 183 GHz water line variations: An energetic outburst in Orion KL. *Astrophysical Journal,* **283**, 106–116, 1984.

Livesey, N. J., M. D. Fromm, J. W. Waters, G. L. Manney, M. L. Santee, and W. G. Read. Enhancements in lower stratospheric CH_3CN observed by UARS MLS fillowing boreal forest fires, *Journal of Geophysical Research,* **109**, (D06308), doi:10.1029/2003JD004055, 2004.

Oh, J .J., and E. A. Cohen. Pressure broadening of ClO by N_2 and O_2 near 204 and 649 GHz and new frequency measurements between 632 and 725 GHz. *Journal of Quantitative Spectroscopy and Radiatative Transfer,* **54**, 151–156, 1994.

Peckham G. E., et al. Optimizing a remote sensing instrument to measure atmospheric surface pressure. *International Journal of Remote Sensing,* **4**, 465–478, 1983.

Penfield, H., M. M. Litvak, C. A. Gottlieb, and A. E. Lilley. Mesospheric ozone measured from ground-based millimeter-wave observations. *Journal of Geophysical Research,* **81**, 6115–6120, 1976.

Phillips, T. J., and D. P. Woody. Millimeter and submillimeter wave receivers. *Annual Review of Astronomy and Astrophysics,* **10**, 285–321, 1982.

Poynter, R. L., and H. M. Pickett. Submillimeter, millimeter, and microwave spectral line catalogue. JPL Pub. 80-23, Rev. 1, NASA Jet Propulsion Laboratory, Pasadena, CA., 1981.

Read, W. G., L. Froidevaux, and J. W. Waters. Microwave Limb Sounder (MLS) measurements of SO_2 from Mt. Pinatubo volcano. *Geophysical Research Letters,* **20**, 1299, 1993.

Santee, M. L., G. L. Manney, N. J. Livesey, and W. G. Read. Three-dimensional structure and evolution of stratospheric HNO_3 based on UARS Microwave Limb Sounder measurements. *Journal of Geophysical Research,* **109**, D15306, doi:10.1029/2004JD004578, 2004.

Schoeberl, M. R., R. S. Stolarski, A. R. Douglass, P. A. Newman, L. R. Lait, and J. W. Waters. MLS ClO observations and arctic polar vortex temperatures. *Geophysical Research Letters,* **20**, 2861–2854, 1993.

Schoeberl, M. R., et al. The EOS Aura Mission. *EOS, Transactions of the American Geophysical Union,* **85**(18), 2004.

Siegel, P. H., I. Mehdi, R. J. Dengler, J. E. Oswald, A. Pease, T. W. Crowe, W. Bishop, Y. Li, R. J. Mattauch, S. Weinreb, J. East, and T. Lee. Heterodyne radiometer development for the Earth Observing System Microwave Limb Sounder. In *Infrared and Millimeter-Wave Engineering, SPIE 1874*, 124, 1993.

Smith, E. K. Centimeter and millimeter wave attenuation and brightness temperature due to atmospheric oxygen and water vapor. *Radio Science,* **17**, 1455–1464, 1982.

Stachnik, R. A., J. C. Hardy, J. A. Tarsala, J. W. Waters, and N. R. Erickson. Submillimeter wave heterodyne measurements of stratospheric ClO, HCl, O3, and HO2: First results. *Geophysical Research Letters,* **19**, 1931, 1992.

Waters, J. W. Ground-based measurement of millimeter-wavelength emission by upper stratospheric O_2. *Nature,* **242**, 506–508, 1973.

Waters J. Absorption and emission by atmospheric gases. Chapter 2.3 in *Methods of Experimental Physics,* Vol. 12: Astrophysics, Part B. Academic Press, New York, 1976.

Waters J. W., et al. A balloon-borne microwave limb sounder for stratospheric measurements. *Journal of Quantitative Spectroscopy and Radiatative Transfer,* **32**, 407–433, 1984.

Waters, J. W., J. J. Gustincic, R. K. Kakar, H. K. Roscoe, P. N. Swanson, T. G. Phillips, T. De Graauw, A. R. Kerr, and R. J. Mattauch. Aircraft search for millimeter-wavelength emission by stratospheric ClO. *Journal of Geophysical Research,* **84**, 7034–7040, 1979.

Waters, J. W., J. J. Gustincic, P. N. Swanson, and A. R. Kerr. Measurements of upper atmospheric H₂O emission at 183 GHz. In *Atmospheric Water Vapor,* pp. 229–241. Academic Press, New York, 1980.

Waters, J. W., J. C. Hardy, R. F. Jarnot, and H. M. Pickett. Chlorine monoxide radical, ozone, and hydrogen peroxide: Stratospheric measurements by microwave limb sounding. *Science,* **214,** pp. 61–64, 1981.

Waters, J. W., J. C. Hardy, R. F. Jarnot, H. M. Pickett, and P. Zimmerman. A balloon-borne microwave limb sounder for stratospheric measurements. *Journal of Quantitative Spectroscopy and Radiatative Transfer,* **32,** 407–433, 1984.

Waters, J.W. Microwave Limb Sounding. In M. A. Janssen (Ed.), *Atmospheric Remote Sensing by Microwave Radiometry.* John Wiley, New York, 1993.

Waters, J. W., et al. The UARS and EOS Microwave Limb Sounder (MLS) Experiments. *Journal of the Atmospheric Sciences,* **56,** 194–218, 1999.

Waters, J. W. Observations for chemistry (remote sensing): Microwave. In . J. Holton, J. Curry, and J. Pyle (Eds.), *Encyclopedia of Atmospheric Sciences,* pp. 1516–1528. Academic Press, 2003.

Wu, D. L., et al. Mesospheric temperature from UARS MLS: Retrieval and validation. *Journal of Atmospheric Solar-Terrestial Physics,* **65,** Issue 2, 245–267, 2003.

11

ATMOSPHERIC REMOTE SENSING IN THE VISIBLE AND INFRARED REGIONS

Sensors in the visible and infrared part of the spectrum are used to measure the characteristics of solar radiation after it interacts with the atmosphere through scattering and/or absorption of the emitted radiation that originates from atmospheric molecules. From these characteristics, information can be derived about atmospheric temperature, composition, and dynamics as a function of altitude and lateral position. Because of the relatively short wavelength, high spatial resolution is easier to achieve in comparison to microwave sensors. Imaging spectrometry can be achieved by using two-dimensional detector arrays. On the other hand, the strong scattering by suspended particles significantly limits vertical sounding through cloud layers.

11-1 INTERACTION OF VISIBLE AND INFRARED RADIATION WITH THE ATMOSPHERE

In the visible and near infrared, the sun is the main source of radiation. The maximum solar irradiance occurs at 0.47 μm, with about half of the total energy in the visible band (0.4 to 0.76 μm). In the case of the Earth, about 35% of the sunlight is reflected back into space, about 18% is absorbed in the atmosphere, and 47% is absorbed at the surface.

The thermal infrared radiation is generated mainly by the surface and the atmosphere. It becomes dominant somewhere between 4 and 18 μm, depending on the planet's temperature and albedo.

Atmospheric constituents interact with electromagnetic radiation all across the visible and infrared region as a result of vibrational and rotational processes, thus impinging their "fingerprints" on the spectral signature of the radiation emitted or scattered toward space.

Introduction to the Physics and Techniques of Remote Sensing. By C. Elachi and J. van Zyl
Copyright © 2006 John Wiley & Sons, Inc.

11-1-1 Visible and Near-Infrared Radiation

As the sunlight passes through an atmospheric layer, some of the energy is absorbed, some scattered, and some transmitted (Fig. 11-1). At the surface, the total spectral irradiance E_g contains two components—the direct sunlight E_s and the diffuse skylight E_d:

$$E_g = E_s + E_d \tag{11-1}$$

The direct sunlight spectral irradiance is given by

$$E_s = S \cos \theta \exp(-\tau/\cos \theta) \tag{11-2}$$

where τ is the total optical thickness of the atmosphere, S is the solar spectral irradiance at the top of the atmosphere, and θ is the angle of incidence (or observation). The diffuse skylight consists of all the components that result from single or multiple scattering of the direct sunlight as well as the reflected sunlight. Figure 11-2 shows the ratio of the direct sunlight E_s to the total irradiance at the surface [i.e., $E_s/(E_s + E_d)$] as a function of the optical thickness $\tau/\cos \theta$. It shows that the direct sunlight is dominant for optical thickness less than 1.1. For a representative optical thickness of $\tau = 0.5$, the direct sunlight is dominant for solar zenith angles θ less than 63° (cos $\theta > 0.5/1.1$).

The energy scattered upward by the surface enters the base of the atmosphere, and some of it is scattered back toward the surface. Figure 11-3 shows the reflectance of a

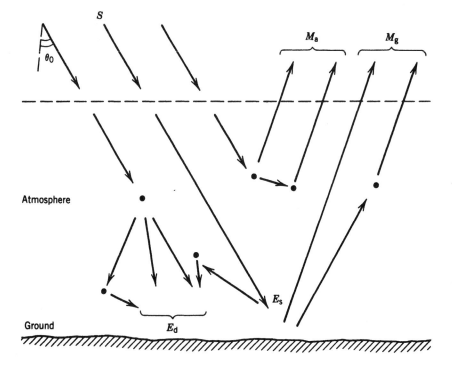

Figure 11-1. Interaction of sunlight with the surface and the atmosphere. E_s is direct sunlight at the surface. E_d is diffuse skylight at the surface. M_a is scattered sunlight that never reaches the surface. M_g is upwelling that has interacted with the surface.

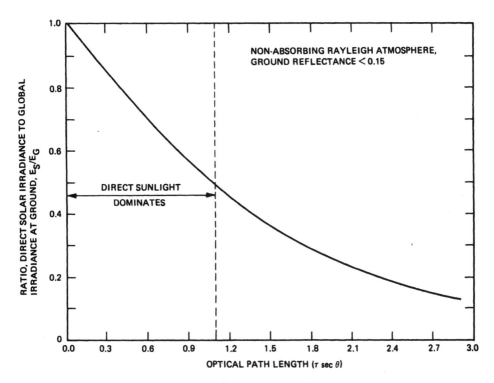

Figure 11-2. Ratio of the direct solar irradiance E_s to the total irradiance E_g as a function of the optical thickness for a nonabsorbing Rayleigh atmosphere and a ground reflectance less than 15% (from Chahine et al., 1983).

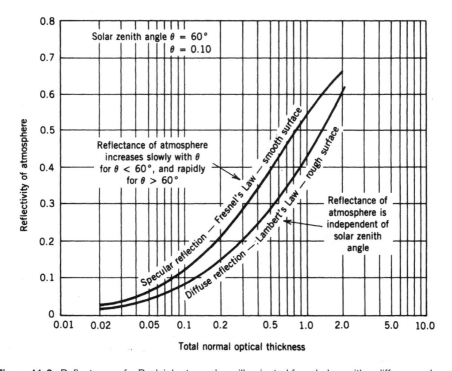

Figure 11-3. Reflectance of a Rayleigh atmosphere illuminated from below with a diffuse rough surface (lower curve) and a specular Fresnel surface.

nonabsorbing Rayleigh scattering atmosphere. It shows that the reflectance is relatively insensitive to extreme variations of the character of ground reflectance.

At the top of the atmosphere, the total spectral irradiance M emitted toward space consists of the upwelling radiation due to scattering in the atmosphere M_a and the upwelling radiation that has interacted at least once with the surface M_g:

$$M = M_a + M_g \tag{11-3}$$

The relative values of these two components are shown in Figure 11-4. As the optical thickness increases, more scattering occurs in the atmosphere, leading to a larger M_a. The transmittance to the surface decreases, leading to a smaller M_s.

In the case of limb sounders in the upper part of the atmosphere, the radiation is mainly a result of the direct sunlight scattering L_s and, in the case of occultation, of the direct sunlight L_d (see Fig. 11-5):

$$L = L_s + L_d \tag{11-4}$$

where

$$L_d = S \exp(-\tau')$$

and τ' is the optical thickness along the line of sight between the sun and the sensor.

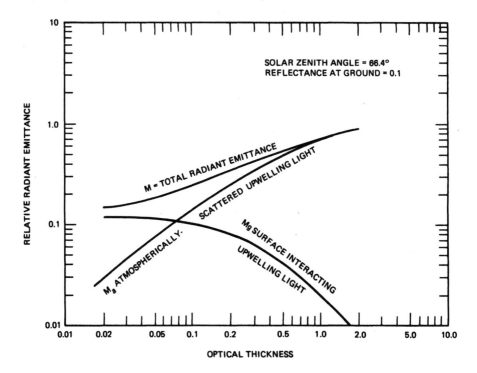

Figure 11-4. Radiant emittance M, M_f, and M_g at the top of a Rayleigh atmosphere as a function of optical thickness. Surface reflectance is 0.1 and solar zenith angle is 66° (from Fraser, 1964).

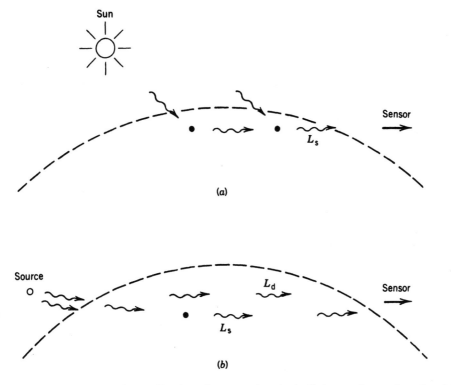

Figure 11-5. Component of outwelling from the atmosphere in the limb scanning configuration: (*a*) Nonoccultation configuration, and (*b*) occulation configuration.

11-1-2 Thermal Infrared Radiation

The total thermal spectral radiance I observed at a frequency v and zenith angle θ consists of (see Fig. 11-6)

$$I = I_s + I_a + I_d + I_h \tag{11-5}$$

where:

I_s = surface emission

I_a = atmospheric emission upward

I_d = reflected atmospheric emission downward

I_h = reflected solar radiance

The surface emitted radiance reaching the sensor can be written as

$$I_s = (v,\ \theta) = \varepsilon_s(v,\ \theta)B(v,\ T_s)\exp[-\tau(v,\ \theta)] \tag{11-6}$$

where ε_s is the surface emissivity, τ is the atmospheric spectral optical thickness, and $B(v,\ T_s)$ is the Planck function given by Equation (8-28).

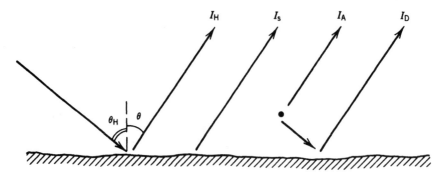

Figure 11-6. Components of the thermal irradiance in the atmosphere.

The atmospheric emission for a plane, parallel atmosphere in local thermodynamic equilibrium is given by

$$I_a = \int_0^\infty \psi_t(v, z) \exp[-\tau(v, z)] \, dz \qquad (11\text{-}7)$$

where

$\psi_t(v, z)$ = source term = $\alpha_a B[v, T_a(z)]$

$T_a(z)$ = atmospheric temperature

$\tau(v, z)$ = optical thickness above $z = \dfrac{1}{\cos \theta} \displaystyle\int_z^\infty \alpha_a(v, \xi) \, d\xi$

The reflected atmospheric downward emission I_d originates in the atmosphere above the surface and is reflected by the surface. The radiant energy reflected into the solid angle $d\Omega_r$ in the direction (θ_r, ϕ_r) comes from all directions above the surface. Thus we can write

$$I_d(v, \theta_r, \phi_r) = \exp[-\tau(v, \theta_r, \phi_r)] \int dA \int d\Omega_i \, I'(v, \theta_i, \phi_i) \rho(v, \theta_i, \phi_i, \theta_r, \phi_r) \cos \theta_i \qquad (11\text{-}8)$$

where I' is the atmospheric downward radiation at the surface, ρ is the surface reflectivity, $d\Omega_i$ is the solid angle of the incident radiation = $\sin \theta_i \, d\theta_i \, d\phi_i$, the second integral is over a solid angle of 2π, and the first integral is over the area viewed by the sensor.

The reflected solar radiance term is given by:

$$I_h(v, \theta) = S(v) \cos \theta_h \exp[-\tau(v, \theta_h)] \rho(v, \theta_h, \theta) \exp[-\tau(v, \theta)] \qquad (11\text{-}9)$$

where θ_h is the sun angle and θ is the observation angle. Of course, $I_h = 0$ during the nighttime.

11-1-3 Resonant Interactions

Resonant interactions with the atmosphere in the visible and infrared regions correspond mainly to the vibrational and vibrational–rotational energy levels of the atmospheric constituents. Figure 8-1 shows the absorption due to some of the main constituents in the Earth's atmosphere. CO_2 is also the main constituent of the atmospheres of Venus and Mars. CH_4 is an important element in the atmospheres of Jupiter, Saturn, and Titan. Figure 11-7 shows the rich, fine structure of the water molecule in the 26–31 μm region and of the carbon dioxide molecule in the 13–15 μm region. Note that the spectrum is usually plotted as a function of wave number, which is the inverse of the wavelength.

Carbon dioxide is substantially uniformly mixed in the Earth's atmosphere and is by far the dominant constituent of the atmospheres of Venus and Mars. This allows the use of the strong CO_2 bands at 4.3 and 15 μm for temperature sounding. Methane is uniformly mixed in the atmospheres of Jupiter and Saturn. Its spectral band at 7.7 μm can then be used for temperature sounding of the giant planet's atmospheres.

11-1-4 Effects of Scattering by Particulates

Particulates play a major role in the visible and infrared region of the spectrum because they usually are of similar size as the wavelength. In order to formulate quantitatively the effect of scattering, the angular distribution of the scattered radiation must be specified. This is described by the scattering phase function $p(\cos \alpha)$, where α is the angle between the incident and scattered radiation. Usually, $p(\cos \alpha)$ is normalized such that

$$\int p(\cos \alpha) \frac{d\Omega}{4\pi} = \omega \tag{11-10}$$

For the case of no absorption loss, $\omega = 1$. Otherwise ω is a constant smaller than one. The simplest case is when the scattering is isotropic. Then,

$$p(\cos \alpha) = \omega = \text{constant} \tag{11-11}$$

One important case is the Rayleigh phase function:

$$p(\cos \alpha) = \tfrac{3}{4}(1 + \cos^2 \alpha) \tag{11-12}$$

Another important function that has been used in planetary studies is

$$p(\cos \alpha) = \omega(1 + \chi \cos \alpha) \tag{11-13}$$

where χ is a constant between -1 and $+1$. In general, the phase function can be written as a series of Legendre polynomials:

$$p(\cos \alpha) = \sum_{n=0}^{\infty} \omega_n P_n(\cos \alpha) \tag{11-14}$$

Figure 11-8 shows the phase functions for atmospheres of increasing turbidity or decreasing visibility, hence, increasing influence of particles. Curve 1 corresponds to the theoretical Rayleigh phase function. It is almost identical to Curve 1a, which corresponds to the measured phase function of dry air at a mountain observatory with a 220 km visibility (extremely clear atmosphere). As the particulates influence increases, the phase function becomes more asymmetrical with enhanced forward scattering. Curve 10 corresponds to

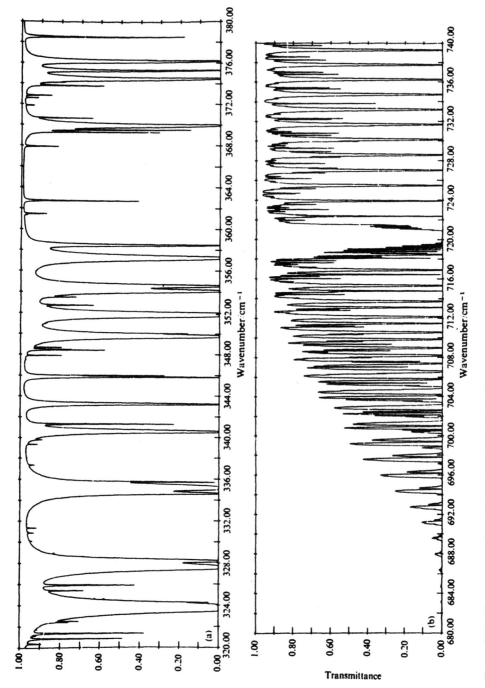

Figure 11-7. Fine structure of molecular absorption bands is illustrated in the transmittance of a 10 km path at 12 km altitude as computed by McClatchy and Selby (1972). The lines in the 320–380 cm^{-1} (31–25 μm) region are due to water vapor. The lines in the 680–740 cm^{-1} (15–13 μm) region are mainly due to carbon dioxide.

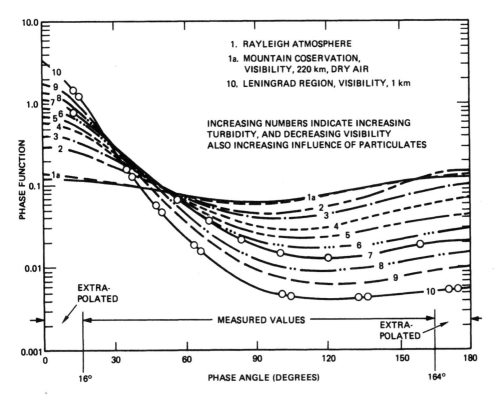

Figure 11-8. Average measured phase functions of different states of the atmosphere. (From Chahine et al., 1983.)

the case of an atmosphere with 1 km visibility. In this case, the forward scattering is about 600 times stronger than the backward scattering.

In the case in which the atmosphere contains clouds or aerosols, the scattering source term in the radiative transfer equation becomes important. This term is expressed as

$$J(\theta, \phi) = \frac{1}{4\pi} \int_0^{4\pi} I(\theta, \phi) p(\cos \alpha) \, d\Omega \qquad (11\text{-}15)$$

and

$$\cos \alpha = \cos \theta_i \cos \theta_s + \sin \theta_i \sin \theta_s \cos(\phi_s - \phi_i) \qquad (11\text{-}16)$$

where θ_i, ϕ_i and θ_s, ϕ_s are the spherical angular coordinates of the incident and scattered beams, respectively. The radiation transfer equation in the case of a plane parallel scattering nonabsorbing atmosphere can then be written as [from Equation (8-56)]

$$\mu \frac{dI(\tau, \mu, \phi)}{d\tau} = I(\tau, \mu, \phi) - \frac{1}{4\pi} \int_{-1}^{+1} \int_0^{2\pi} p(\mu, \phi, \mu_i, \phi_i) I(\tau, \mu_i, \phi_i) d\mu_i \, d\phi_i \qquad (11\text{-}17)$$

where $\mu = \cos \theta$ and $\mu_i = \cos \theta_i$. This differential integral equation cannot be usually solved analytically. In simple cases, it is solved by solving for the following related integrals:

$$\text{Flux integral:} \quad F = \frac{1}{\pi} \iint \mu I \, d\mu \, d\phi \tag{11-18}$$

$$K \text{ integral:} \quad K = \frac{1}{4\pi} \iint {}^2 I \, d\mu \, d\phi \tag{11-19}$$

11-2 DOWNLOOKING SOUNDING

The upwelling radiation carries information about the atmosphere temperature profile (emitted component), suspended particles (scattered component), and constituents. By measuring the spectral characteristics of the upwelling radiation in certain bands of the spectrum, the above geophysical parameters can be measured. In this section, we discuss the use of infrared and visible upwelling radiation to measure temperature profiles and constituent concentration.

11-2-1 General Formulation for Emitted Radiation

Considering a plane parallel nonscattering atmosphere, a layer of thickness dz at altitude z will have an emitted spectral radiance $\Delta B(z)$ given by

$$\Delta B(z) = \alpha(v, \xi) B[v, T(z)] dz \tag{11-20}$$

where $B[v, T]$ is the Planck's function and α is the absorption coefficient, which is a function of the local constituent and atmospheric pressure. The contribution of the layer dz to the total spectral radiance emanating from the top of the atmosphere is

$$\Delta B_a = \Delta B(z) e^{-\tau(v, z)} \tag{11-21}$$

where

$$\tau(v, z) = \int_z^{\infty} \alpha(v, \xi) d\xi \tag{11-22}$$

Thus,

$$\Delta B_a = \alpha(v, z) B[v, T(z)] \exp\left(-\int_z^{\infty} \alpha(v, \xi) d\xi\right) dz \tag{11-23}$$

$$= B[v, T(z)] d\beta$$

where

$$\beta(v, z) = \exp[-\tau(v, z)] \tag{11-24}$$

$\beta(v, z)$ is the transmittance of the atmosphere above level z. The total spectral radiance observed by the sensor is

$$B_t(v) = B_s(v)\beta_m(v) + \int_0^{\infty} B[v, T(z)] \frac{d\beta}{dz} dz = B_s(v)\beta_m(v) + B_a(v) \tag{11-25}$$

where $B_s(v)$ is the surface spectral radiance, $\beta_m(v)$ is the total atmosphere transmittance, and $B_a(v)$ is the atmospheric spectral radiance.

11-2-2 Temperature Profile Sounding

The atmospheric spectral radiance can be written as

$$B_a(v) = \int_0^\infty B[v, T(z)]W(v, z)\, dz \tag{11-26}$$

where $W(v, z) = d\beta/dz$ acts as a weighting function. Sometimes, it is convenient to use instead of z another altitude-dependent variable such as the pressure p or $y = -\ln p$. In this case, we will have

$$B_a(v) = \int_{p_s}^0 B[v, T(p)]\frac{d\beta}{dp}\, dp \tag{11-27}$$

or

$$B_a(v) = \int_{-\ln p_s}^\infty B[v, T(y)]\frac{d\beta}{dy}\, dy \tag{11-28}$$

Let us consider the case of a homogeneously mixed constituent (such as CO_2 in the atmosphere of Earth, Venus, or Mars) and the emission around a single collision broadened spectral line centered at v_0 (see Equation 8-21). In the wing of the line where $v - v_0 \gg v_L$,

$$\alpha = \rho\frac{S}{\pi}\frac{v_L}{(v - v_0)^2} = \rho\frac{S}{\pi}\frac{v_L(p_0)}{(v - v_0)^2}\frac{p}{p_0} \tag{11-29}$$

where ρ is the constituent concentration. In the case of a homogeneously mixed constituent, the concentration ρ is proportional to $-dp/dz$. Thus,

$$\alpha = -bp\frac{dp}{dz} \tag{11-30}$$

and

$$\tau = +\frac{b}{2}p^2 \Rightarrow \beta = \exp\left(-\frac{b}{2}p^2\right) \tag{11-31}$$

where b is a constant which depends on the line strength and width, the concentration of the absorber, and the observation frequency. This leads to a weighting function in Equation (11-28):

$$W(v_j, y) = \frac{d\beta}{dp}\frac{dp}{dy} = bp^2\exp\left(-\frac{b}{2}p^2\right) \tag{11-32}$$

This weighting function is plotted in Figure 11-9. It shows that a peak occurs at $p_m = \sqrt{2/b}$, which depends on the observation frequency. Thus, by changing the frequency, the

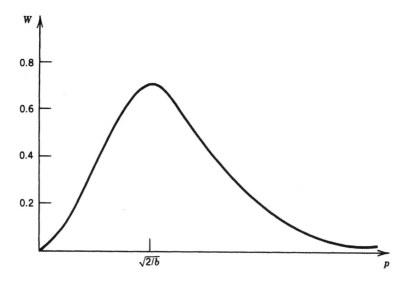

Figure 11-9. Behavior of the weighting function $W(v, y)$ as a function of the pressure p.

peak of the weighting function can be moved to different pressure levels (therefore, different altitude levels).

The sensor measures the collected spectral radiance B_t at a number of frequencies v_j. These measured values can be written as [from Equation (11-26)]:

$$B_t(v_j) = \int_0^\infty B[v_j, T(z)]W(v_j, z)\, dz \qquad (11\text{-}33)$$

where for simplicity we did not include the surface component. The object is to derive the temperature profile from the values of $B_t(v_j)$. Equation (11-33) can be viewed as a nonlinear transformation from $T(z)$ to $B(v)$ as in

$$B(v) = F[T(z)] \qquad (11\text{-}34)$$

and the temperature profile is derived by performing an inverse transformation:

$$T(z) = F^{-1}[B(v)] \qquad (11\text{-}35)$$

This is done in an interative process that uses the fact that the weighting function has a maximum that depends on v and that variations of $T(z)$ around the maximum affect strongly $I(v)$, whereas the effect is small away from the location of the peak. The theory associated with the inversion techniques (also called retrieval techniques) is beyond the scope of this textbook. The reader is referred to the work of Chahine (1970) and the texts by Twomey (1977) and Chahine et al. (1983) for the details of the transformation techniques.

11-2-3 Simple Case Weighting Functions

Let us assume that the temperature varies slowly and monotonously as a function of z. The Planck function can then be approximated by

$$B(v, T) = B(v, \overline{T}) + \frac{\partial B}{\partial T}(T - \overline{T}) \qquad (11\text{-}36)$$

and Equation (11-26) can be written as

$$B_a(v) = B(v, \overline{T}) \int W(v, y) dy + \int \frac{\partial B}{\partial T} W(v, y)[T - \overline{T}] dy \qquad (11\text{-}37)$$

This shows that the product $(\partial B/\partial T)W(v, y)$ is a direct weighting function of the temperature. Figure 11-10 shows this product weighting function for some of the channels on the High Resolution Infrared Radiation Sounder (HIRS) flown on Nimbus 6 and TIROS-N. The characteristics of some of the illustrated channels are given in Table 11-1. The CO_2 bands at 15 and 4.3 μm are commonly used for temperature sounding of the terrestrial planets' atmospheres.

In order to illustrate the concept in a simple analytical form, let us consider the simplified case in which k is assumed to be weakly dependent or even independent of temperature and pressure. Then we can write

$$\alpha(v, z) = k(v)\rho(z) \qquad (11\text{-}38)$$

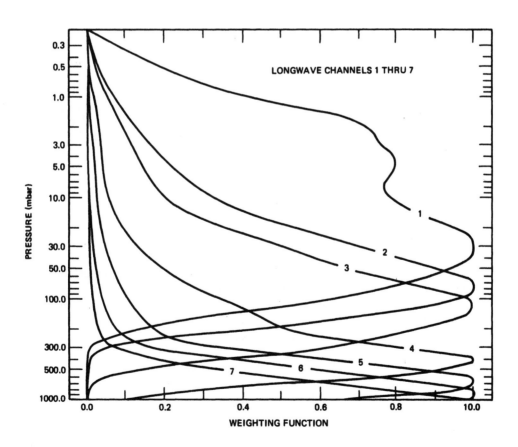

Figure 11-10. Curves of the product $(\partial B/\partial T)W$ as a function of pressure for some of the channels on the HIRS around the 15 μm CO_2 band (from Smith et al., 1975). See Table 11-1 for more description.

TABLE 11-1. Characteristics of Some of the HIRS Channels

Channel no.	Central wavelength (μm)	Principal absorbing constituent	Peak level (mb)
1	15.0	CO_2	30
2	14.7	CO_2	60
3	14.4	CO_2	100
4	14.2	CO_2	250
5	14.0	CO_2	500
6	13.6	CO_2/H_2	750
7	13.4	CO_2/H_2O	900

and

$$\tau(v, z) = k(v) \int_z^\infty \rho(\xi)\, d\xi \tag{11-39}$$

Assuming an exponential behavior for ρ with a scale height H, then we get:

$$\tau(v, z) = k(v)\rho_0 H e^{-z/H} \tag{11-40}$$

and

$$W(v, z) = k(v)\rho_0 \exp\left[-\frac{z}{H} - k(v)\rho_0 H e^{-z/H}\right] \tag{11-41}$$

which is identical to the microwave weighting function in Equation (9-13) and illustrated in Figure 9-3, except that the weighting function here corresponds to the Planck function, which is a function of the temperature instead of the temperature itself. The weighting function peak occurs at an altitude z_m given by

$$z_m = H \log[\rho_0 H k(v)] \tag{11-42}$$

11-2-4   Weighting Functions for Off-Nadir Observations

Off-nadir observations lead to changes in altitude of the peak of the weighting function, thus allowing the derivation of a temperature profile using multiple angle observations at a single frequency. This can be qualitatively explained by the fact that, at an oblique angle, the radiation emitted from a layer dz toward the sensor will propagate through more of the atmosphere than in the normal observation case. Thus, the atmosphere will have a higher effective optical thickness, which implies that the region of maximum contribution has to be at a higher altitude.

Quantitatively, if θ is the observation angle relative to vertical, and assuming the case of an exponential atmosphere with k independent of the pressure and temperature, the weighting function in Equation (11-41) will become

$$W(v, z, \theta) = \frac{k(v)\rho_0}{\cos\theta} \exp\left[-\frac{z}{H} - \frac{k(v)\rho_0 H}{\cos\theta} e^{-z/H}\right] \tag{11-43}$$

which has a peak at altitude

$$z_m(\theta) = H\left[\log\frac{\rho_0 Hk(v)}{\cos\theta}\right] = z_m(0) + H\log\left(\frac{1}{\cos\theta}\right) \quad (11\text{-}44)$$

The location of the peak will move upward by one scale height H for an observation angle change from the nadir to $68°$. One major limitation of this approach is that the atmosphere must be assumed to be laterally homogeneous across the range of angular observations. This is usually not exactly true. This technique was used by Orton et al. (1975) for temperature sounding of the Jovian atmosphere using the Pioneer 10 infrared radiometer.

11-2-5 Composition Profile Sounding

The constituent concentration profile appears explicitly in the expression of $\beta(v, p)$:

$$\beta(v, p) = \exp\left[-\int_p^\infty k(v, p')\rho(p')\,dp'\right] \quad (11\text{-}45)$$

Thus, a mapping transformation can be applied to derive $\rho(p)$, $\rho(y)$, or $\rho(z)$ from a set of measurements $S_a(v_j)$ using Equation (11-26) or (11-27). This is done by iterative techniques as discussed by Chahine (1972). To illustrate, let us consider the simple case of an optically thin atmosphere at the frequency of observation. Then,

$$\frac{d\beta}{dp} = k(v, p)\rho(p)\exp\left[-\int_p^\infty k(v, p')dp'\right] \simeq k(v, p)\rho(p) \quad (11\text{-}46)$$

and Equation (11-26) becomes:

$$S_a(v) = \int_0^\infty \rho(p)W'(v, p)\,dp \quad (11\text{-}47)$$

where

$$W'(v, p) = k(v, p)\beta[v, T(p)] \quad (11\text{-}48)$$

If $T(p)$ is known or derived from emission measurements at other wavelengths, then the weighting function $W'(v, p)$ is a known function of v and p, and $\rho(p)$ can be derived by the inverse transformation of Equation (11-47).

Downlooking sensors operating in the blue and ultraviolet spectral region can be used to derive composition profiles by measuring the backscatter solar radiation. This can be used specifically to sound the ozone layer. Taking the simplified case in which the sun is straight behind the sensor and the atmosphere is infinitely deep so that we can neglect the ground effect, we have

$$I(v, p) = I(v)\exp[-k(v)n(p)] \quad (11\text{-}49)$$

where $I(v, p)$ is the incident radiation at the pressure level p, $I(v)$ is the incident solar radiation at the top of the atmosphere, $k(v)$ is the absorption coefficient per molecule of ozone, and $n(p)$ is the number of molecules above the level of pressure p.

The amount of radiation backscattered upward from the molecules in a layer in which the pressure changes by dp around level p will be proportional to $I(v, p)$ and dp:

$$dI'(v, p) = aI(v, p)\, dp \tag{11-50}$$

where a is the proportionality factor. This radiation will pass again through the atmosphere above level p. The resulting upwelling is then

$$I'(v) = \int_0^\infty dI'(v, p)\, \exp[-k(v)n(p)] = aI(v) \int_0^\infty \exp[-2k(v)n(p)]\, dp \tag{11-51}$$

$$\Rightarrow \frac{I'(v)}{I(v)} = a \int_0^\infty p \exp[-2k(v)n(p)]\, d(\ln p) \tag{11-52}$$

Thus, by measuring the ratio $I'(v)/I(v)$ at a number of frequencies v_j that give different values of k, we can invert the above integral relation and derive the profile $n(p)$.

11-3   LIMB SOUNDING

Limb-looking geometry is used to sound the atmosphere by emission or occultation. In the emission case, the spectral radiance is given by a similar expression as in the downlooking geometry except that the integration has to be carried along the line of sight and there is no contribution from the ground. In the occultation case, the sun (or in some cases a star) is used as the source and the atmospheric absorption has to be calculated along the line of sight from the source to the sensor. In both cases, the measurement accuracy is significantly enhanced by the fact that the optical thickness of a finite layer is significantly larger in limb-sounding geometry than in downlooking geometry. This allows the detection and study of minor constituents with extremely low concentrations.

11-3-1   Limb Sounding by Emission

The total spectral radiance emanating from the atmosphere in a limb-sounding geometry (Fig. 11-11) is expressed in a similar way as a downlooking sensor [Equation (11-25)] except that the integration is along the line of sight:

$$B_t(v, h) = \int_{-\infty}^{+\infty} B[v, T(x)] \frac{d\beta(v, x, h)}{dx}\, dx \tag{11-53}$$

where h is the tangent height, and

$$\beta(v, z, h) = \exp\left[-\int_x^{+\infty} \alpha(v\,\xi, h)\, d\xi\right] \tag{11-54}$$

The coordinate x is related to the altitude z by

$$(R + h)^2 + x^2 = (R + z)^2$$

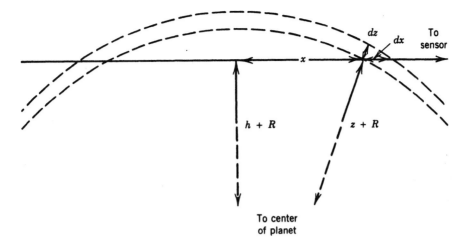

Figure 11-11. Geometry for limb sounding.

Usually, h and z are much smaller than R, thus

$$x = \pm\sqrt{2R(z-h)} \qquad (11\text{-}55)$$

By making the x to z transformation, Equation (11-53) can be written as

$$B_t(v, h) = \int_0^\infty B[v, T(z)]W(v, h, z)\, dz \qquad (11\text{-}56)$$

where $W(v, h, z)$ is a weighting function. If we are observing the emission from a gas with known distribution, such as CO_2 in the Earth's upper atmosphere, then W is known and the measurements allow the derivation of $B(z)$ and from it $T(z)$. Figure 11-12 shows the weighting functions for the CO_2 band around 15 μm for different tangent altitudes. It is apparent that most of the contribution usually is from a small layer just above the tangent altitude.

Once the temperature profile is known, observations in other spectral bands will allow the derivation of the corresponding $d\beta(v, z)/dz$ and from it the constituent profile.

As seen in Figure 11-12, most of the contribution to the emission is from altitudes close to h. Thus, we can write Equation (11-53) as

$$B_t(v, h) \simeq B[v, T(h)]\int_{-\infty}^{+\infty} \frac{d\beta(v, x, h)}{dx}\, dx$$

$$= B[v, T(h)][\beta(v, +\infty, h) - \beta(v, -\infty, h)] \qquad (11\text{-}57)$$

$$= B[v, T(h)][1 - \beta_t(v, h)] = B[v, T(h)]\varepsilon_t(v, h)$$

where $\beta_t(v, h)$ is the total atmospheric transmittance along the path and $\varepsilon_t(v, h)$ is the total emittance. Thus, if we know $T(h)$, we can derive $\varepsilon_t(v, h)$ and, hence, the total absorber amount $m(h)$ along the limb path.

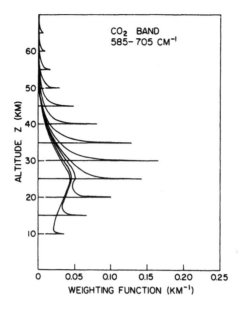

Figure 11-12. A set of weighting functions for a limb sounder with a narrow field of view, observing the emission from the CO_2 band near 15 μm. The different weighting functions correspond to different tangent heights. (From Gille and House, 1971.)

For a uniformly mixed gas, we may write for the total constituent mass $m(h)$:

$$m(h) = c \int \rho(x)\, dx = c \int_h^\infty \rho(z) \sqrt{\frac{2R}{z-h}}\, dz \qquad (11\text{-}58)$$

where c is the mixing ratio and $\rho(z)$ is the atmospheric density.

The measurement of $\beta_t(v, h)$ as a function of h is acquired by scanning the field of view of the sensor in the vertical plane or by using a sensor with a wide vertical field of view and a linear array of detectors in the sensor's focal plane.

11-3-2 Limb Sounding by Absorption

In the case of absorption, the sensor observes the spectral emission from a source (usually the sun) as the line of sight is occulted by the atmosphere. This allows the measurement of the total transmittance $\beta_t(v, h)$ along the line of observation, and, hence, the total absorber mass $m(h)$. To a first order, the measurement does not depend on knowing the temperature profile, as is the case with the emission mode. On the other hand, absorption measurements can be done only during source occultation as viewed from the sensor. In the case of the sun occultation viewed from an orbiting platform, this corresponds to twice in each orbit.

11-3-3 Illustrative Example: Pressure Modulator Radiometer

To illustrate one of the techniques used in limb sounding by emission, we briefly describe here the concept of a pressure modulator radiometer (PMR). The PMR uses gas correlation spectroscopy to measure radiation from the emission lines of specific radiatively active atmospheric constituents with a very high spectral resolution (about 0.001 cm^{-1} in comparison with 10 cm^{-1} for filters and ~1 cm^{-1} with standard spectrometers). This allows observation of radiation from very close to the center of the lines, thus allowing the sounding of high-altitude layers.

Figure 11-13 shows a simplified diagram of the PMR. The collected radiation is passed through a cell that contains the same gas as the one being observed. The pressure in the cell is modulated at an angular frequency ω:

$$p(t) = p_0 + \Delta p \cos(\omega t)$$

This leads to a modulation of the cell transmission T. The depth of modulation is a function of the optical frequency:

$$T(v, t) = T_0(v) + \Delta T(v) \cos(\omega t)$$

The shape of $\Delta T(v)$ depends on the pressure in the cell p, the pressure modulation Δp, and the specific spectral line. This is shown in Figure 11-14. It is clear that by varying Δp and p it is possible to vary the main portion of $\Delta T(v)$ all along the wings of the spectral line, thus allowing measurement of the incident radiation intensity as a function of v around the line center. One of the nice features of this technique is that the modulation affects in the same manner all the neighboring spectral lines of the gas in the cell, thus allowing the measurement of the incoming radiation in all the emission lines of the same gas in the atmosphere over a reasonably wide spectral band.

11-3-4 Illustrative Example: Fourier Transform Spectroscopy

High-resolution infrared Fourier spectroscopy is used to measure the spectrum of the detected radiation over a wide spectral range almost simultaneously. This allows the simultaneous detection of a large number of trace species. This technique has been used to measure the spectrum of solar radiation as it is occulted by the upper atmosphere. The high spectral resolution is achieved by using an interferometer such as a Michelson interferometer. The concept is illustrated in Figure 11-15. The radiation collected by the sensor is divided by the beamsplitter into two beams. Each one of them is reflected by a mir-

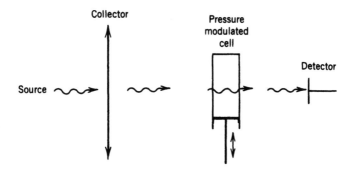

Figure 11-13. Simplified sketch of a pressure modulator radiometer.

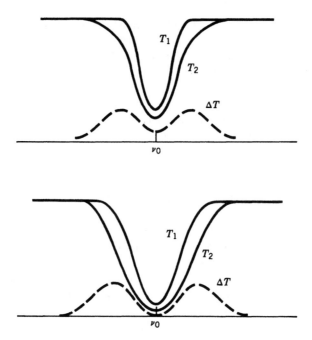

Figure 11-14. Transmission modulation $\Delta T(v)$ peaks at different wavelengths depending on the pressure and pressure modulation.

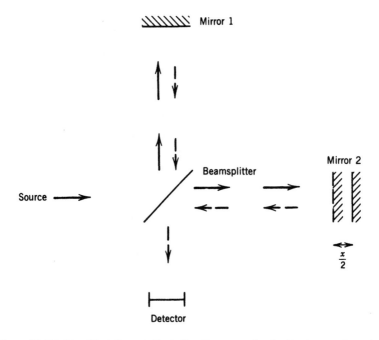

Figure 11-15. Simplified diagram illustrating the concepts of a Fourier spectrometer.

ror and passed again through the beamsplitter. The overlapping beams are directed toward the detector.

Let us first assume that the source is monochromatic, emitting radiation at frequency v_0. If one of the mirrors is moved by a distance $x/2$ from the zero point position (the position where both mirrors are equidistant from the beamsplitter), then the beam energy $D(x)$ incident on the detector is given by

$$D(x) = \left[\text{envelope of} \left(\sqrt{B} \cos v_0 t + \sqrt{B} \cos\left(v_0 t + \frac{2\pi v_0}{c} x \right) \right) \right]^2$$

$$\Rightarrow D(x) = 4B \cos^2\left(\frac{\pi v_0}{c} x \right) = 2B \left[1 + \cos\left(\frac{2\pi}{c} v_0 x \right) \right]$$

where $\sqrt{B}$ corresponds to the amplitude of the signal from each of the beams when the other one is blocked. $D(x)$ goes through maxima and minima as the moving mirror is displaced.

Let us now consider the case in which the incident radiation has a continuous spectrum. In this case, B is a function of v and the detected signal is given by

$$D(x) = 2 \int B(v) \left[1 + \cos\left(\frac{2\pi v}{c} x \right) \right] dv$$

$$= 2 \int B(v) dv + 2 \int B(v) \cos\left(\frac{2\pi v}{c} x \right) dv$$

For $x = 0$,

$$D(0) = 4 \int B(v) \, dv$$

Thus,

$$I(x) = D(x) - \frac{1}{2} D(0) = 2 \int B(v) \cos\left(\frac{2\pi v}{c} x \right) d v$$

$I(x)$ is called the interferogram function. It is measured by the sensor as the mirror is moved. The function $B(v)$ can then be derived from the inverse Fourier transform:

$$B(v) = \frac{1}{2} \int I(x) \cos\left(\frac{2\pi v}{c} x \right) dx$$

The measurement spectral resolution is limited by the realizable range of x. If x is limited to a $\pm L$ range, it can be shown that the sensor spectral resolution is given by

$$\Delta v = \frac{c}{2L}$$

Thus, it is desirable to have L as large as possible. A number of techniques are used to optimize the interferometer response and reduce its sidelobes.

The Fourier transform spectroscopy technique was used in the ATMOS instrument, which was flown on the Space Shuttle to study the trace molecules in the Earth's upper atmosphere. The ATMOS sensor continuously monitored the solar radiation spectrum as the sun was occulted by the atmosphere. The measured spectrum contained the absorption features of the atmosphere's constituents. By continuously monitoring during occulation, a vertical profile of the constituent concentration was measured. Figure 11-16 shows an example of the ATMOS data for illustration. Table 11-2 gives a summary of the key sensor characteristics.

A more recent example of a spaceborne Fourier transform spectrometer is the Tropospheric Emission Spectrometer (TES), one of four atmospheric instruments launched on NASA's AURA spacecraft in 2004. The primary task of the TES instrument is to measure the global three-dimensional distribution of tropospheric ozone and the chemical species

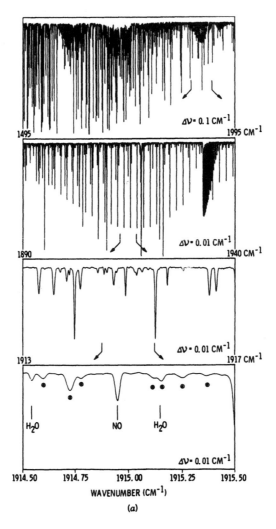

Figure 11-16. (*a*) Example of ATMOS data. The top trace shows a 50 cm⁻¹ region of the spectrum. The second trace is an expansion of a 50 cm⁻¹ region. The third trace is an expansion of a 4 cm⁻¹ region. The bottom trace is an expansion of a 2 cm⁻¹ region centered at 1915 cm⁻¹, which shows a NO line. (Courtesy of C. Farmer, Jet Propulsion Laboratory.)

involved in its formation and destruction. Because many of these species exist only in very low concentrations (on the order of 1 part in 10^9 by volume), both limb and nadir sounding are used. Since the spectral lines of the species to be observed are spread over a wide region of the infrared spectrum, the TES instrument was designed to cover a wide spectral region from approximately 2.35 to 15.4 micron wavelengths. On the other hand, in order to resolve the spectral lines, a high resolution in frequency space is required. Furthermore, the line widths, to first order, are independent of species and frequency, so a constant resolution in frequency space is required. These are exactly the properties of a Fourier transform spectrometer. The TES FTS design is a four-port Connes (Connes and Connes, 1966) configuration, which allows further subdivision of the spectral range into four subbands, for which the detectors can then individually be optimized for performance. The detectors, all photovoltaic mercury cadmium telluride optimized for the individual spectral ranges, are operated at 65°K using active pulse-tube coolers. The optical bench with the back-to-back translating cube-corner reflectors is passively cooled to 180°K. See Beer et al. (2001) for more details on the TES instrument.

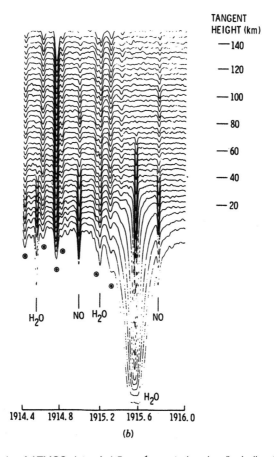

Figure 11-16. (*b*) Example of ATMOS data. A 1.5 cm^{-1} spectral region (including the region in the bottom of *a*) is shown as observed for different tangent heights. It clearly shows the change in the constituent concentrations as a function of altitude. (Courtesy of C. Farmer, Jet Propulsion Laboratory.)

TABLE 11-2. Key Characteristics of the ATMOS Sensor

Spectral interval covered	550–4800 cm^{-1}
	2.1–18 μm
Spectral resolution	0.01 cm^{-1}
Spectral precision	0.001 cm^{-1}
Spatial resolution	1–2.5 km
Aperture size	7.5 cm
Scan time	1 sec
Detector	HgCdTe cooled to 77°K
Data rate	15.8 Mbps

11-4 SOUNDING OF ATMOSPHERIC MOTION

The interaction of an electromagnetic wave with a moving gas leads to a frequency shift as a result of the Doppler effect (see Chapter 2). If a gaseous medium has an emission spectral line at frequency v_0, the same gas moving at a speed along the observation line (see Fig. 11-17a) will emit at a frequency v:

$$v = v_0 + v_D \tag{11-59}$$

where

$$v_D = \frac{v}{c}v_0 = \frac{v}{\lambda_0}$$

Similarly, if a wave passes through a layer of moving gas (Fig. 11-17b) with an absorption line at v_0, the absorption line is shifted by the frequency v_D.

A third situation occurs when a wave of frequency v_0 is scattered by a moving medium. In this case, the scattered wave is Doppler shifted by (see Fig. 11-17c)

$$v_D = \frac{v}{\lambda_0}(\cos \theta_i + \cos \theta_s) \tag{11-60}$$

where θ_i and θ_s are the incident and scattered waves angles relative to the velocity vector v. By using the Doppler effect, a number of techniques can measure and map atmospheric motion. These techniques are discussed in this section.

11-4-1 Passive Techniques

Some passive techniques basically rely on accurate measurement of the frequency shift of known emission (or absorption) spectral lines as a result of the motion of the corresponding atmospheric constituent. The shift is usually measured by comparing the spectral signature of the received radiation to the spectral properties of an appropriately selected static gas within the sensor. This can best be illustrated by the sensor shown in Figure 11-18, which is designed for measuring wind velocities in the upper atmosphere.

The radiation emitted from the upper atmospheric region under observation is passed through an electrooptic modulator, where the index of refraction n is changed in a periodic fashion:

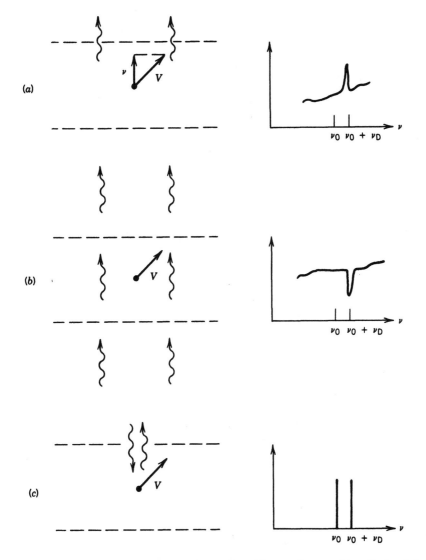

Figure 11-17. Three configurations of wave interaction with a moving gas: (a) emitted radiation, (b) transmitted radiation, (c) scattered radiation.

$$n(t) = n_0 + \Delta n \cos \omega_n t \tag{11-61}$$

where Δn is the modulation term (usually $\Delta n \ll n_0$) and ω_n is the modulation angular frequency, which is controllable. At the output of the modulator, the collected radiation spectrum will consist of a number of spectral lines displaced by integers of ω_n away from any incident radiation line ω_0 (see Figure 11-18b). This can be easily understood by looking at the effect of the electrooptic modulator on a monochromatic input signal given by

$$I(t) = I_0 \cos(\omega_0 t + \omega_D t) \tag{11-62}$$

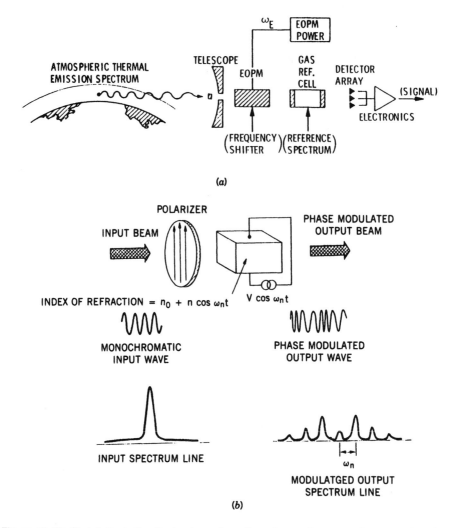

Figure 11-18. Sketch illustrating the basic configuration of a passive sensor for measuring atmospheric motion. (*a*) Basic element of the sensor. (*b*) Effect of the electrooptical modulator on the line spectrum of the incoming radiation. (Courtesy of D. McCleese, Jet Propulsion Laboratory.)

The modulator adds a phase delay ϕ, given by

$$\phi = \frac{2\pi}{\lambda_0} Ln(t) = \phi_0 + \Delta\phi \cos \omega_n t \tag{11-63}$$

where L is the modulator length and

$$\phi_0 = \frac{2\pi L}{\lambda} n_0$$

$$\Delta\phi_0 = \frac{2\pi L}{\lambda} \Delta n$$

The output signal is then

$$O(t) \sim \cos(\omega_0 t + \omega_D t + \phi_0 + \Delta\phi \cos \omega_n t) \tag{11-64}$$

The spectrum of this signal contains components at frequencies $\omega_D + \omega_0 \pm j\omega_n$, with $j = 0$, 1, 2, . . . (see Fig. 11-18). It is given by (McCleese and Margolis, 1983)

$$O(t) \sim J_0(\delta)\cos(\omega_0 + \omega_D)t + \sum_{j=1}^{\infty} J_j(\delta)\cos(\omega_0 + \omega_D + j\omega_n)t + \sum_{j=1}^{\infty}(-1)^j J_j(\delta)\cos(\omega_0 + \omega_D - j\omega_n)t$$

where J_j are ordinary Bessel functions of order j, and δ is a modulation index that depends on the crystal used and the intensity of the modulating field. By changing δ, the amount of energy in the sidebands relative to the central band can be controlled.

The modulated signal is then passed through a cell containing a reference gas selected to be the same as one of the known gases in the atmospheric region under observation. This gas would then have an absorption line at ω_0. If the incident radiation has no Doppler shift (Fig. 11-19a), then the modulated spectral lines are displaced away from the absorption lines, leading to little or no absorption in the cell. If the incident radiation is Doppler shifted, the modulation frequency can be adjusted by quick scanning such that one of the sidebands matches the absorption line in the reference cell, leading to significant absorption and decrease in the output signal (Fig. 11-19b). It is readily apparent that as the modulation frequency is scanned, the minimum output from the reference gas cell will occur when the modulation frequency ω_n is equal to the Doppler frequency ω_D. This allows accurate measurement of ω_D.

It should be pointed out that if the emission gas (and the reference gas) spectrum consists of a large number of neighboring lines, the described technique is not affected because all the lines will be Doppler shifted by approximately the same amount and the electrooptic modulator will displace all the lines by the same approximate amount.

The sensor described above allows the measurement of the wind velocity along the sensor line of sight. In order to get the wind vector, a second measurement is required from a different location to get two vector components. From satellites, this can be easily done by observing the same atmospheric region from two locations (i.e., at two times along the orbit) using two identical sensors with different pointing angles or the same sensor with a pointing mirror (see Figure 11-20).

In the case of the Earth's atmosphere, the emission lines of CO_2 around 15 μm can be used to measure the wind velocity in the upper atmosphere (60 to 120 km), and the emission lines of N_2O around 8 μm can be used to measure the wind velocity at medium altitudes (25 to 50 km).

The accuracy of the measurement depends on the strength of the emission line and its spectral width. Narrow, strong lines would provide the most accurate measurement. Pressure-broadened lines will degrade the measurement accuracy, thus limiting the applicability of this technique to the upper atmosphere. In addition the spacecraft velocity must be known very accurately because it has to be subtracted from the measured Doppler shift in order to derive the component due to the wind.

11-4-2 Passive Imaging of Velocity Field: Helioseismology

One of the most imaginative and promising applications of passive spectral line shift measurement techniques is in the field of helioseismology.

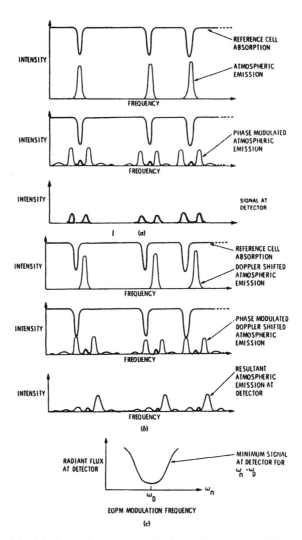

Figure 11-19. Principle of the line shift measurement using a reference gas cell. (a) Case of no Doppler shift (no wind). (b) Case of a Doppler shift that is exactly compensated by the modulator frequency. (c) Output signal as a function of the modulation frequency illustrating the measurement technique.

Continuous wave motions that are similar to seismic waves in the Earth agitate the interior of the sun. In the gaseous solar atmosphere, two types of waves exist:

1. pressure (acoustic) waves, in which the medium is alternately compressed and dilated along the direction of travel; such waves propagate throughout the solar interior in the same way seismic acoustic waves propagate throughout the Earth's interior.

2. Gravity waves, which correspond to oscillations of parcels of fluid around their equilibrium position as a result of the buoyancy restoring force (pressure is the restoring force for acoustic waves).

Both types of waves establish resonant modes similar to acoustic or electromagnetic waves in a spherical inhomogeneous cavity. The spatial and temporal structure of these

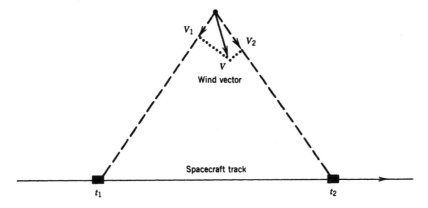

Figure 11-20. Successive observations of the same atmospheric region would allow the derivation of two components of the wind vector.

modes depends on the properties of the solar interior (composition, pressure, and temperature).

On reaching the surface of the sun, the waves cause the surface gases to move up and down, resulting in a Doppler shift of the spectral lines in the solar emitted light. The line displacements are directly proportional to the gas velocity. Thus, a two-dimensional image of the sun that allows the exact measurement of the line displacement in each image pixel can be used to derive the velocity field on the sun surface. This velocity field can then be inverted to derive the wave field in the solar interior, which sheds light on the interior composition, pressure, and temperature distributions. Figure 11-21 shows some examples of the sun's surface oscillations based on computer models. One of the strongest solar oscillations has a period of five minutes, with a spatial distribution scale of few tens of thousands of kilometers and maximum radial velocity of many hundreds of meters per second.

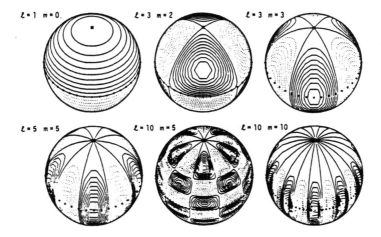

Figure 11-21. Computer-generated models of the sun's surface velocity fields, which correspond to six of the acoustic resonance modes. Many thousands of these modes can exist.

One possible method to map the spectral line displacements on a pixel-by-pixel basis is shown in Figure 11-22. A telescope images the sun onto a CCD array through a very narrow band tunable filter. The filter is first tuned to pass the energy in one of the wings of a solar absorption line. After sufficient integration time, the filter is tuned to pass the energy in the other wing. The two measured intensities are subsequently compared on a pixel-by-pixel basis to derive the displacement of the line in each pixel and from it the velocity field. The spectral line displacement measurement approach is illustrated in Figure 11-23. Let I_1 be the intensity measured by one element in the CCD array when the filter is centered to the left of the line center at v. Let I_2 be the corresponding intensity when the filter is centered at the right of line center. The ratio I_1/I_2 is directly related to the line displacement frequency v_D.

A number of tunable filter types can be used, such as a magneto-optical filter (Agnelli et al. 1975), a Fabry–Perot filter, a birefringent filter, or an acoustooptic filter. The spectral line used can be any of the solar Fraunhofer lines, such as the sodium lines at 16,956 cm^{-1} and 16,973 cm^{-1}.

11-4-3 Multiangle Imaging SpectroRadiometer (MISR)

MISR is one of five instruments that were launched on the NASA spacecraft Terra in December 1999. It employs nine discrete cameras pointed at fixed angles to make measurements of the land surface and atmospheric aerosols and clouds. Of the nine cameras, one points in the nadir (vertically downward) direction and four each are viewing the forward and aft directions along the spacecraft ground track. The off-nadir cameras view the earth at angles of 26.1, 45.6, 60.0, and 70.5° forward and aft of the local vertical. The fore and aft camera angles are the same, that is, the cameras are arranged symmetrically about the nadir. In general, large viewing angles provide enhanced sensitivity to atmospheric aerosol effects and to cloud reflectance effects, whereas more modest angles are required for land surface viewing.

The nadir-viewing camera provides imagery that is less distorted by surface topographic effects than any of the other cameras, and provides a comparison with the other cameras for determining the way that imagery changes appearance with angle of view (called bidirectional reflectance measurement). It also offers an opportunity to cross-compare observations with the more traditional single-view-angle sensors.

The cameras pointing at 26.1° are used for stereoscopic image mapping to determine topographic heights and cloud heights. A base/height ratio near unity is considered optimal for stereo work. This quantity is the linear distance between points observed by two

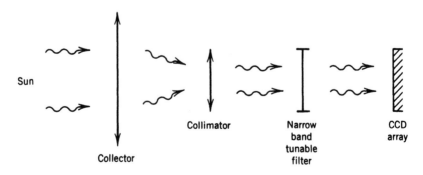

Figure 11-22. Simple diagram for a solar oscillation imager.

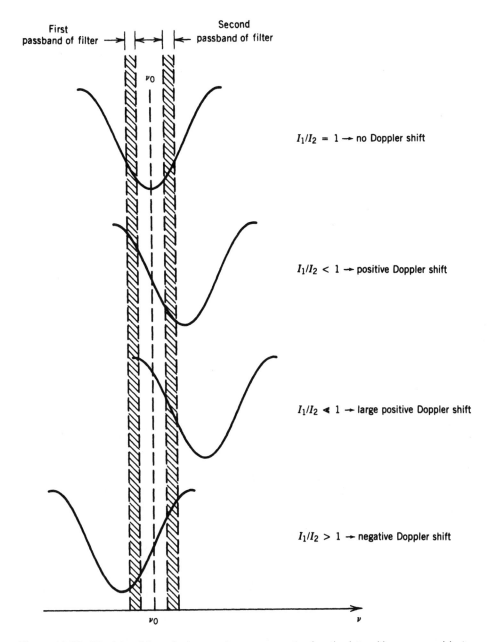

Figure 11-23. Principle of line displacement measurement using the intensities measured in two band-pass regions placed symmetrically on both sides of the line center.

cameras divided by the height of the cameras above Earth. The angular separation between the forward- and aft-pointing 26.1° cameras provides this optimal base/height ratio from the Terra orbit at 705 km altitude.

Sensitivity to aerosol properties is increased when the Earth's surface is viewed at an angle instead of straight down, thereby looking through a larger amount of atmosphere. The effect obviously increases with increasing viewing angle. The 45.6° cameras are positioned to exploit this aerosol sensitivity. At a view angle of 60°, the path length through the atmosphere is twice that of the nadir view. Theoretical studies of the transfer of radiation to and from clouds also suggest that directionally oriented reflectance variations among many different types of clouds are minimized at this angle. Finally, the 70.5° view angle cameras provide the highest possible view angle within practical limitations.

One of the key applications of the MISR data is the measurement of cloud heights and the associated wind fields at these heights. The MISR operational cloud height retrieval algorithm is based on a stereophotogrammetric technique (Maroney et al., 2002) that exploits the multiangle capability of the instrument. Stereo matching algorithms estimate the height of an object based on the apparent distance between similar features in a given image pair (Muller et al., 2002). Winds at the cloud elevation cause advection of the clouds, which changes the appearance and position of the clouds between successive image pairs. The wind field can, therefore, be estimated using these changes between different pairs of MISR images.

Figure 11-24 shows an example of a cloud height measurement from the MISR instrument for hurricanes Ivan and Frances. This same technique can also be applied to clouds of dust and smoke. Figure 11-25 shows an example of the height of smoke clouds due to fires in the taiga forests of eastern Siberia in 2003. The left and center panels are natural-color views from MISR's vertical-viewing (nadir) and 70° forward-viewing cameras, respectively. The steeply looking camera enhances the appearance of smoke, such that many areas, including Lake Baikal (the long dark water body along the left-hand image

Figure 11-24. MISR measurements of cloud heights associated with hurricanes Frances and Ivan. The panel on the left in each case is a natural color image from the nadir-looking camera. (Courtesy NASA/JPL-Caltech.) See color section.

Figure 11-25. MISR measurements of the extent and height of smoke from numerous fires in the Lake Baikal region on June 11, 2003 are shown in the panel on the right. Areas where heights could not be retrieved are shown as dark gray. See text for detailed discussion. (Courtesy NASA/JPL-Caltech.) See color section.

edge), are almost completely obscured by smoke at the oblique viewing angle. On the right is a map of stereoscopically retrieved heights for features exhibiting sufficient spatial contrast. The heights correspond to elevations above sea level. Taking into account the surface elevation, the smoke plumes range from about 2–5 km above the surface. Larger heights are mostly associated with clouds.

11-4-4 Active Techniques

In the case of active techniques, an active source with a well-known narrow-band spectral signature (i.e., a laser) is used to radiate the region of interest. The backscattered light is Doppler shifted by the motion of the scattering particles and molecules. The collected light is then spectrally compared with the emitted light in order to derive the Doppler shift and the corresponding atmospheric wind velocity.

One of the advantages of active techniques is that by measuring the time difference between the emitted and received signals, the location of the measurement region can be accurately determined. This allows the direct derivation of a wind velocity profile. In addition, because of the round-trip effect, the Doppler shift is double the amount observed in passive techniques, thus making it easier to measure. On the other hand, active techniques require the use of high-power laser sources, which are heavy and require a lot of input power.

11-5 ATMOSPHERIC SENSING AT VERY SHORT WAVELENGTHS

Most of the spectral lines that correspond to electronic energy levels are in the short visible and ultraviolet (UV) regions. Table 11-3 gives some of the UV lines of gases encountered in planetary atmospheres. Ultraviolet spectrometers have been flown on numerous planetary missions to measure the intensity of Lyman α emission and to measure the UV airglow in order to determine the distribution of constituents such as CO_2, CO_2^+, CO, O, H, and C. The UV sensors usually are photomultipliers or channel electron multiplier detectors placed after a spectrometer or a set of filters. Typically, the UV spectrometers are of the Ebert scanning type, and they cover the region from 0.02 to 0.4 μm with resolution of about 10 to 20 Å.

EXERCISES

11.1. Consider a homogeneous atmospheric layer of optical thickness τ_1 and constant temperature T_1. Give the expression for the upward spectral radiance above the

TABLE 11-3. Ultraviolet Spectral Lines of Common Atoms and Molecules in Planetary Atmospheres

Atom/molecule	Wavelength (Å)
Atoms	
Argon	1048
Nitrogen	1200
Hydrogen	1216 (Lyman α)
Oxygen	1302, 1304, 1306
Carbon	1657
Molecules	
Oxygen	2026, 2885
Nitrogen	986, 1450, 2010, 3370
Hydroxyl	3064
Nitric acid	1909, 2198, 2262
Carbon monoxide	1544, 1804, 2063
Ions	
Oxygen	2610
Nitrogen	1549, 3911
Nitric oxide	1368
Carbon monoxide	2191, 4900

layer. Plot the effective emissivity ε_1 of the layer as a function of its optical thickness.

Let us now assume that the layer is directly on top of a surface of temperature T_g and emissivity ε_g. Give the expression for the upward spectral radiance above the layer. Write the expression in the form

$$S = S_0 + S_1 e^{-\tau} + S_2 e^{-2\tau}$$

Explain in physical terms the behavior of S when:

(a) $\varepsilon_g = 0$

(b) $\varepsilon_g = 1$

(c) τ small

Let us take the case of small τ. Give the conditions under which an absorption line will show

(a) as a dip

(b) as a peak in the spectral radiance

11-2. Consider two homogeneous layers of optical depth τ_1 and τ_2, and of constant temperatures T_1 and T_2, respectively. Layer 1 is on top of layer 2.

(a) Give the expression for the upward spectral radiance above the top layer.

(b) Under what condition does a spectral line in layer 1 shows as a spectral peak or trough?

(c) How does a spectral line in layer 2 show in the emission spectrum?

11-3. Assuming an exponential decay of the gas density with a scale height H and that k does not depend strongly on temperature and pressure, plot the behavior of the location of the weighting function peak as a function of the observing frequency shift from the center of the absorption line. Use the expression

$$z = H\left[\log A - \log\left\{1 + \left(\frac{v - v_0}{v_L}\right)^2\right\}\right]$$

Consider the following cases

(a) $A = 1$

(b) $A = 2$

(c) $A = 5$

Describe what happens for $A < 1$.

11-4. Plot the behavior of the scattering phase function as a function of α for the following cases:

(a) Rayleigh phase function

(b) For $p = \omega(1 + \chi \cos \alpha)$ and

   a. $\omega = 1, \chi = 1$

   b. $\omega = 1, \chi = 0.5$

   c. $\omega = 0.75, \chi = 1$

   d. $\omega = 0.75, \chi = 0.5$

11-5. Let us consider the line displacement scheme illustrated in Figure 11-24. Assume two bandpass regions of width v_D centered at $v_0 + \Delta v$ and $v_0 - \Delta v$ and that the width of the bands is small enough that the line intensity inside the band can be considered constant. Derive the ratio of the two intensities I_+/I_- as a function of v_D for the following line shapes:

(a) $A(v) = A_0[1 + \alpha e^{-(v-v_0)^2/v_L^2}]$

(b) $A(v) = A_0 \left[1 + \alpha \dfrac{v_L^2}{(v - v_0)^2 + v_L^2} \right]$

In both cases, plot the ratio I_+/I_- as a function of v_D/v_L for $\alpha = 0.8$ and $\Delta v = 0.5 v_L$.

REFERENCES AND FURTHER READING

Agnelli, G., A. Cacciani, and M. Fofi. The magneto-optical filter. *Solar Physics,* **40,** 509–518, 1975.

Barnett, J. J., et al. The first year of the selective chopper radiometer on Nimbus 4. *Quarterly Journal of the Royal Meterological Society,* **98,** 17–37, 1972.

Barnett, J. J., and C. D. Walshaw. Temperature measurements from a satellite: Applications and achievements. In Barrett and Curtis (Eds.), *Environmental Remote Sensing,* p. 213. Arnold, London, 1972.

Barnett, J. J. Analysis of stratospheric measurements by the Nimbus IV and V selective chopper radiometer. *Quarterly Journal of the Royal Meterological Society,* **99,** 173–188, 1973.

Barnett, J. J., et al. Stratospheric observations from Nimbus 5. *Nature,* **245,** 141–143, 1973.

Barnett, J. J. The mean meridional temperature behaviour of the stratosphere from November 1970 to November 1971 derived from measurements by the selective chopper radiometer on Nimbus IV. *Quarterly Journal of the Royal Meterological Society,* **100,** 505–530, 1974.

Barnett, T. L. Application of a nonlinear least square method to atmospheric temperature sounding. *Journal of Atmospheric Science,* **26,** 457, 1969.

Barteneva, O. D. Scattering functions of light in the atmospheric boundary layer. *Iz. Geophys. Ser.,* pp. 1237–1244, 1960. (English translation)

Beer, R., *Remote Sensing by Fourier Transform Spectroscopy.* Wiley, New York, 1992.

Beer, R., T. A. Glavich, and D. M. Rider. Tropospheric emission spectrometer for the Earth Observing System's Aura satellite. *Applied Optics,* **40,** 2356–2367, 2001.

Bender, M. L., et al. Infrared radiometer for the Pioneer 10 and 11 missions to Jupiter. *Applied Optics,* **13,** 2623–2628, 1976.

Chahine, M. T. Determination of the temperature profile in an atmosphere from its outgoing radiance. *Journal of Optical Society of America,* **58,** 1634, 1968.

Chahine, M. Inverse problems in radiative transfer: Determination of atmospheric parameters. *Journal of Atmospheric Science,* **27,** 960–967, 1970.

Chahine, M. Remote sounding of cloudy atmospheres. *Journal of Atmospheric Science,* **34,** 744, 1977.

Chahine, M. A general relaxation method for inverse solution of the full radiation transfer equation. *Journal of Atmospheric Science,* **29,** 741–747, 1972.

Chahine, M. T. Remote soundings of cloudy atmospheres, 1, The single cloud layer. *Journal of Atmospheric Science,* **31,** 233–243, 1976.

Chahine, M., et al. Interaction mechanisms within the atmosphere. Chapter 5 in *Manual of Remote Sensing.* American Society of Photogrammetry, Falls Church, VA, 1983.

Chapman, W., et al. A spectral analysis of global atmospheric temperature fields observed by selective chopper radiometer on the Nimbus 4 satellite during the year 1970–1. *Proceedings of Royal Society London,* **A338,** 57–76, 1974.

Christensen–Dalsgaard, J., D. O. Gough, and J. Toomre. Seismology of the sun. *Science,* **229,** 923–931, 1985.

Coffey, M. T. Water vapour absorption in the 10–12 μm atmospheric window. *Quarterly Journal of the Royal Meterological Society,* **103,** 685–692, 1977.

Connes, J., and P. Connes. Near-infrared planetary spectra by Fourier spectroscopy. I Instruments and Results. *Journal of Atmospheric Science,* **56,** 896–910, 1966.

Conrath, B. J. On the estimation of relative humidity profiles from medium resolution infrared spectra obtained from a satellite. *Journal of Geophysical Research,* **74,** 3347, 1969.

Conrath, B. J., R. A. Hanel, V. G. Kunde, and C. Prabhakara. The infrared interferometer experiment on Nimbus 3. *Journal of Geophysical Research,* **75,** 5831–5857, 1970.

Conrath, B. J. Vertical resolution of temperature profiles obtained from remote radiation measurements. *Journal of Atmospheric Science,* **29,** 1262–1271, 1972.

Conrath, B., et al. Atmospheric and surface properties of Mars obtained by infrared spectroscopy on Mariner 9. *Journal of Geophysical Research,* **78,** 4267–4278, 1973.

Curtis, P. D., et al. A pressure modulator radiometer for high altitude temperature sounding. *Proceedings of Royal Society London,* **A336,** 1, 1973.

Curtis, P. D., J. T. Houghton, G. D. Peskett, and C. D. Rodgers. Remote sounding of atmospheric temperature from satellites V. The pressure modulator radiometer for Nimbus F. *Proceedings of Royal Society London,* **A337,** 135–150, 1974.

Curtis, P. D., and J. Houghton. Un radiometer satellitaire pour determineries temperature de la stratosphere. *La Meteorologie,* Societe Meteorologique de France, Nov. 1975, 105–118, 1975.

Diner, D., et al. MISR: A multiangle imaging spectroradiometer for geophysical and climatological research from Eos. *IEEE Transactions on Geoscience and Remote Sensing,* **27,** 200–214, 1989.

Diner, D., et al. Multi-angle Imaging SpectroRadiometer (MISR) instrument description and experiment overview. *IEEE Transactions on Geoscience and Remote Sensing,* **36,** 1072–1087, 1998.

Fraser, R. S. *Theoretical Investigation, The Scattering of Light by Planetary Atmosphere.* TRW Space Technology Laboratory, Redondo Beach, CA, 1964.

Gille, J. C., and P. L. Bailey. In A. L. Fymat and V. E. Zuev (Eds.), *Remote Sensing of the Atmosphere: Inversion Methods and Applications,* pp. 101–113. Elsevier, Amsterdam, 1978.

Gille, J. C., and F. B. House. On the inversion of limb radiance measurements: Temperature and thickness. *Journal of Atmospheric Science,* **29,** 1427–1442, 1971.

Hanel, R. A., and B. J. Conrath. Interferometer experiment on Nimbus 3: Preliminary results. *Science,* **165,** 1258–1260, 1969.

Hanel, R. A. Recent advances in satellite radiation measurements. In *Advances in Geophysics,* Vol. 14, p. 359. Academic Press, New York, 1970.

Hanel, R. A., B. Schlachman, F. D. Clark, C. H. Prokesh, J. B. Taylor, W. M. Wilson, and L. Chaney. The Nimbus 3 Michelson interferometer. *Applied Optics,* **9,** 1967–1970, 1970.

Hanel, R. A., B. Schlachman, D. Rogers, and D. Vanous. Nimbus-4 Michelson Interferometer. *Applied Optics,* **10,** 1376–1382, 1971.

Hanel, R., et al. Investigation of the Martian environment by infrared spectroscopy on Mariner 9. *Icarus,* **17,** 423–442, 1972.

Hanel, R., et al. Infrared observations of the Jovian system from Voyager 1. *Science,* **204,** 972–976, 1979.

Hanel, R., et al. Infrared spectrometer for Voyager. *Applied Optics,* **19,** 1391–1400, 1980.

Hanel, R., et al. Infrared observations of the Saturnian system for Voyager 1. *Science,* **212,** 192–200, 1981.

Hansen, J. E., and J. W. Hovenier. Interpretation of the polarization of Venus. *Journal of Atmospheric Science,* **31,** 1137–1160, 1974.

Heath, D. F., C. L. Mateer, and A. J. Krueger. The Nimbus-4 backscatter ultraviolet (BUV) atmosphere ozone experiment—Two year's operation. *Pure and Applied Geophysics,* **106–108,** 1239–1253, 1973.

Horváth, Á., and R. Davies. Feasibility and error analysis of cloud motion wind extraction from near-simultaneous multiangle MISR measurements. *Journal of Atmospheric and Oceanic Technology,* **18,** 591–608, 2001.

Horváth, Á., and R. Davies. Simultaneous retrieval of cloud motion and height from polar-orbiter multiangle measurements. *Geophysical Research Letters,* **28,** 2915–2918, 2001.

Houghton, J. T. Stratospheric temperature measurements from satellites. *Journal of British Interplanetary Society,* **19,** 382–386, 1963.

Houghton, J. T., and S. D. Smith. *Infra-red Physics.* Oxford University Press, London, 1966.

Houghton, J. T. Absorption and emission by carbon dioxide in the mesosphere. *Quarterly Journal of the Royal Meterological Society,* **95,** 1–20, 1969.

Houghton, J. T., and S. D. Smith. Remote sounding of atmospheric temperature from satellites. *Proceedings of Royal Society London,* **A320,** 23–33, 1970.

Houghton, J. T., and G. E. Hunt. The detection of ice clouds from remote measurements of their emission in the far infrared. *Quarterly Journal of the Royal Meterological Society,* **97,** 1–17, 1971.

Houghton, J. T. The selective chopper radiometer on Nimbus 4. *Bulletin of the American Meterological Society,* **53,** 27–28, 1972.

Houghton, J. T., and A. C. L. Lee. Atmospheric transmission in the 10–12 μm window. *Nature,* **238,** 117–118, 1972.

Houghton, J. T., and F. W. Taylor. Remote sounding from artificial satellites and space probes of the atmosphere of the Earth and the planets. *Reports on Progress in Physics,* **36,** 827–919, 1973.

Houghton, J. T. Calibration of infrared instruments for the remote sounding of atmospheric temperature. *Applied Optics,* **16,** 319–321, 1977.

Houghton, J. T. The stratosphere and mesosphere. *Quarterly Journal of the Royal Meterological Society,* **104,** 1–29, 1978.

Houghton, J. T. The future role of observations from meteorological satellites. *Quarterly Journal of the Royal Meterological Society,* **105,** 1–28, 1979.

Houghton, J. T. Remote sounding of the atmosphere and ocean for climate research. *Proceedings of IEE,* **128,** 442–448, 1981.

Leibacher, J. W., R. W. Noyes, J. Toomre, and R. K. Ulrich. Helioseismology. *Scientific American,* **253,** 48–59, Sept. 1985.

Martonchik, J. V., D. J. Diner, R. Kahn, T. P. Ackerman, M. M. Verstraete, B. Pinty, and H. R. Gordon. Techniques for the retrieval of aerosol properties over land and ocean using multi-angle imaging. *IEEE Transactions on Geoscience and Remote Sensing,* **36,** 1212–1227, 1998.

Martonchik, J. V., D. J. Diner, K. A. Crean, and M. A. Bull. Regional aerosol retrieval results from MISR. *IEEE Transactions on Geoscience and Remote Sensing,* **40,** 1520–1531, 2002.

McClatchey, R. A., and J. E. A. Selby. Atmospheric transmittance at 7–30 μm. Air Force Cambridge Research Laboratory Environmental Research Paper #419, 1972.

McCleese, D., and J. Margolis. Remote sensing of stratospheric and mesospheric winds by gas correlation electrooptic phase modulation spectroscopy. *Applied Optics,* **22,** 2528–2534, 1983.

Moroney, C., R. Davies, and J.-P. Muller. Operational retrieval of cloud-top heights using MISR data. *IEEE Transactions on Geoscience and Remote Sensing,* **40,** 1532–1540, 2002.

Muller, J.-P., A. Mandanayake, C. Moroney, R. Davies, D. J. Diner, and S. Paradise. MISR stereoscopic image matchers: Techniques and results. *IEEE Transactions on Geoscience and Remote Sensing,* **40,** 1547–1559, 2002.

Orton, G., et al. The thermal structure of Jupiter. *Icarus,* **26,** 125–158, 1975.

Reigler, G. R., J. F. Drake, S. C. Liu, and R. J. Circerone. Stellar occulation measurements of atmospheric ozone and chlorine from OAO 2. *Journal of Geophysical Research,* **81,** 4997–5001, 1976.

Rodgers, C. D. Remote sounding of the atmospheric temperature profile in the present of cloud. *Quarterly Journal of the Royal Meterological Society,* **96,** 654–666, 1970.

Rodgers, C. D. Some theoretical aspects of remote sounding in the earth's atmosphere. *J. Quant. Spectrosc. Radiat. Transfer,* **11,** 767–777, 1971.

Rodgers, C. D. Retrieval of atmospheric temperature and compositions from remote measurements of thermal radiation. *Rev. Geophys. Space. Phys.,* **14,** 609–624, 1976.

Rodgers, C. D. In Deepak (Ed.), *Inversion Methods in Atmospheric Remote Sounding,* pp. 117–158. Academic Press, New York; 1976.

Smith, S. D., M. J. Collis, and G. Peckham. The measurement of surface pressure from a satellite. *Quarterly Journal of the Royal Meterological Society,* **98,** 431–433, 1972.

Smith, W. L. An iterative method for deducing tropospheric temperature and moisture profiles from satellite radiation measurements. *Monthly Weather Review,* **95,** 363, 1967.

Smith, W. L. An improved method for calculating tropospheric temperature and moisture from satellite radiometer measurements. *Monthly Weather Review,* **96,** 387–396, 1968.

Smith, W. L. Iterative solution of the radiation transfer equation for the temperature and absorbing gas profile of an atmosphere. *Applied Optics,* **9,** 1993, 1970.

Smith, W. L. *Bulletin of the American Meterological Society,* **53,** 1074, 1972.

Smith, W. L., and H. M. Woolf. The use of eigenvectors of statistical covariance matrices for interpreting satellite sounding radiometer observations, *Journal of Atmospheric Science,* **33,** 1127–1140, 1976.

Smith, W. L., et al. The High Resolution Infrared Radiation Sounder (HIRS) experiment. Nimbus 6 User's Guide, 37-58 NASA (Goddard Space Flight Center), 1975.

Smith, W. L. et al. Nimbus 6 earth radiation budget experiment. *Applied Optics,* **16,** 306–318, 1977.

Staelin, D. Passive remote sensing at microwave wavelengths. *Proceedings of IEEE,* **57,** 427, 1969.

Taylor, F. W. Temperature sounding experiments for the Jovian planets. *Journal of Atmospheric Science,* **29,** 950–958, 1972.

Taylor, F. W., et al. Radiometer for remote sounding of the upper atmosphere. *Applied Optics,* **11,** 135–141, 1972.

Taylor, F. W. Remote temperature sounding in the presence of cloud by zenith scanning. *Applied Optics,* **13,** 1559–1566, 1974.

Taylor, F. W. Interpretation of Mariner 10 infrared observation of Venus. *Journal of Atmospheric Science,* **32,** 1101–1106, 1975.

Taylor, F. W., et al. Polar clearing in the Venus clouds observed from Pioneer Venus Orbiter. *Nature,* **279,** 5714–5716, 1979.

Taylor, F. W., et al. Infrared radiometer for the Pioneer Venus Orbiter: Instrument description. *Applied Optics,* **18,** 3893–3900, 1979.

Taylor, F. W., et al. Structure and meteorology of the middle atmosphere of Venus: Infrared remote sensing from the Pioneer Orbiter. *Journal of Geophysical Research,* **85,** 7963–8006, 1980.

Taylor, F. W., et al. Comparative aspects of Venusian and terrestrial meteorology. *Weather,* **36,** 34–40, 1981.

Taylor, F. W. *The Pressure Modulator Radiometer: Spectrometric Techniques,* vol. 4. Academic Press, New York, 1983.

Twomey, S. On the deduction of the vertical distribution of ozone by ultraviolet spectral measurements from a satellite. *Journal of Geophysical Research,* **66,** 2153–2162, 1961.

Twomey, S. On the numerical solution of Fredholm integral equations of the first by the inversion of the linear system produced by quadrature. *Journal of Ass. Comput. Mach.,* **10,** 97, 1963.

Twomey, S. *Introduction to the Mathematics of Inversion in Remote Sensing and Indirect Measurements.* Elsevier, New York, 1977.

Waters, J., and D. Staelin. Statistical inversion of radiometric data. MIT Research Laboratories Electronics, Quarterly Progress Report No. 39, 1968.

Westwater, E. R., and O. N. Strand. Statistical information content of radiation measurements used in indirect sensing. *Journal of Atmospheric Science,* **25,** 750–8, 1968.

Yamamoto, G., et al. Radiative transfer in water clouds in the infrared region. *Journal of Atmospheric Science,* **27,** 1970.

12

IONOSPHERIC SENSING

The ionosphere is a spherical layer of ionized gas (electrons and ions) that surrounds planets with atmospheres. The ionization results mainly from the interaction of high-energy ultraviolet sun radiation with the upper atmospheric molecules. The energetic radiation frees some of the electrons, leading to a gas of electrons and ions called plasma. Because it consists of charged particles, the ionosphere acts as an electrically conducting medium, and it strongly impacts the propagation of radio waves. This allows the use of radio sensors to sense and study the properties of the ionosphere.

In the case in which the planet has a weak magnetic field or none at all (Venus and Mars are such examples), the propagation of radio waves in the ionosphere is easy to formulate. This case will be used in this chapter to illustrate the techniques of ionospheric sensing. In the case of planets with a strong magnetic field (such as the Earth and Jupiter), the ionosphere becomes anisotropic and the interactions with radio waves become very complex. This is only briefly discussed here. The reader is referred to specialized textbooks such as *Ionospheric Radio Waves* by Davies for a comprehensive analysis.

12-1 PROPERTIES OF PLANETARY IONOSPHERES

All planets with an atmosphere have an ionosphere that covers the outer part of the atmosphere. In the case of the Earth, the ionospheric shell is a thick layer extending from an altitude of 50 km to a few thousand kilometers. The peak density occurs at about 400 to 500 km altitude with a density of about 5×10^5 to 10^6 electrons per cubic centimeter (Fig. 12-1). The peaked shape of the ionospheric density can be simply explained as follows. As the energetic solar radiation penetrates deeper into the atmosphere, it encounters an increasing gas density, which results in an increasing number of ionized particles. However, in the same process the energetic radiation density decreases. Thus, after a certain depth

Introduction to the Physics and Techniques of Remote Sensing. By C. Elachi and J. van Zyl
Copyright © 2006 John Wiley & Sons, Inc.

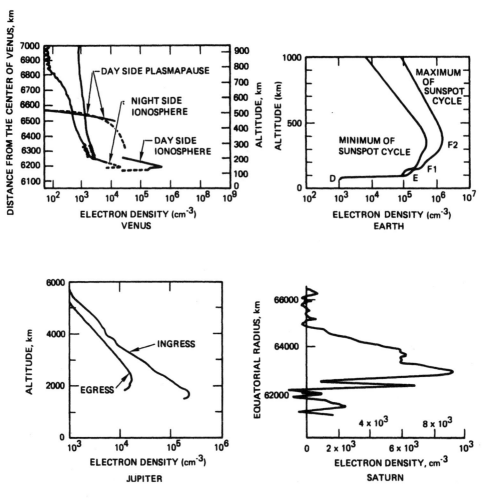

Figure 12-1. Ionospheric electron density profiles for Venus, Earth, Jupiter, and Saturn. (From Gringuaz and Breus, 1970.)

the amount of radiation is so small that fewer particles are ionized, leading to a decrease in the ionospheric density and to a peak at some intermediate altitude.

As electrons are formed by the solar radiation, they are also lost due to recombination (the electron combines with an ion to produce a neutral atom and a photon of radiation) or attachment (the electron is attached to a molecule such as O_2 to form a negative ion O_2^-). These loss phenomena have a finite time constant, which explains the presence of the ionosphere even on the nightside of the planets.

Figure 12-1 shows the ionosphere density profiles for Venus, Earth, Jupiter, and Saturn. In the case of the Earth, the charged particles are formed in the lower part of the ionospheric region and diffuse upward along magnetic lines of force to the upper regions (10,000 to 30,000 km). This results in a thick ionospheric region surrounding the Earth, in contrast to the thin ionosphere surrounding Venus (which has a much thicker atmosphere than the Earth) or Mars (which has a thinner atmosphere). Both Venus and Mars have a very weak magnetic field.

12-2 WAVE PROPAGATION IN IONIZED MEDIA

In its most simple form, an ionosphere can be described as a medium containing N electrons and N ions per unit volume. It is characterized by a natural frequency of oscillation, called the plasma frequency v_p, which occurs when the light electrons are displaced from the heavy ions and then allowed to move freely thereafter. The attraction force between the electrons and ions leads to periodic oscillation of the electrons.

Referring to Figure 12-2, let us assume that all the electrons (charge e and mass m) in the plasma slab are displaced by a distance x. This leads to a surface charge of Nex at the two boundaries of the slab, generating an electric field E in the slab given by

$$\varepsilon_0 E = -Nex \tag{12-1}$$

where ε_0 is the permittivity of vacuum. This field exerts a force on the electrons in the slab, leading to their motion following the relationship

$$m\frac{d^2x}{dt^2} = F = -eE$$

$$\Rightarrow m\frac{d^2x}{dt^2} = -e\left(-\frac{Ne}{\varepsilon_0}\right)x = \frac{Ne^2}{\varepsilon_0}x \tag{12-2}$$

$$\Rightarrow \frac{d^2x}{dt^2} - \frac{Ne^2}{m\varepsilon_0}x = 0$$

The solution of this differential equation is

$$x = x_0 e^{i2\pi v_p t} \tag{12-3}$$

where

$$v_p = \frac{1}{2\pi}\sqrt{\frac{Ne^2}{m\varepsilon_0}} \tag{12-4}$$

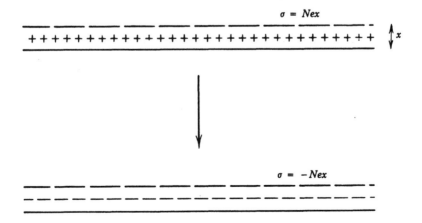

Figure 12-2. Simple example illustrating the plasma oscillation.

is called the plasma frequency, and x_0 is the oscillation amplitude. Replacing e, m, and ε_0 by their values, we find that v_p (MHz) $= 9 \times 10^{-6} \sqrt{N}$ when N is in m^{-3}.

When an electromagnetic wave of frequency v is propagating in an ionized medium, the electrons are forced to oscillate at the driving frequency v. The resulting wave equation is

$$\frac{d^2E}{dx^2} + 4\pi^2 \left(\frac{v^2 - v_p^2}{c^2} \right) E = 0 \qquad (12\text{-}5)$$

which gives a solution

$$E(x, t) = E_0 e^{i(kx - 2\pi vt)} \qquad (12\text{-}6)$$

with

$$k = 2\pi \sqrt{\frac{v^2 - v_p^2}{c^2}} = k_0 \sqrt{1 - \frac{v_p^2}{v^2}} \qquad (12\text{-}7)$$

$$k_0 = \frac{2\pi v}{c}$$

The corresponding phase velocity v_p and group velocity v_g are given by

$$v_p = \frac{2\pi v}{k} = \frac{c}{\sqrt{1 - v_p^2/v^2}} \qquad (12\text{-}8)$$

$$v_g = \frac{2\pi \partial v}{\partial k} = c \sqrt{1 - \frac{v_p^2}{v^2}} \qquad (12\text{-}9)$$

The behavior of these velocities as a function of v/v_p is shown in Figure 12-3. It is apparent that as the wave frequency approaches the plasma frequency from above, the wave slows down. At frequencies below the plasma frequency, the wave vector k is imaginary, leading to an evanescent (exponentially attenuated) wave (i.e., no propagation within the plasma). Thus, when an electromagnetic wave reaches a region in which the plasma frequency exceeds the wave frequency, the wave is reflected. In the case of the Earth's ionosphere, the electron density peak is typically about $N = 10^{12}$ e/m^3, with a corresponding plasma frequency $v_p = 9$ MHz. Thus, the ionosphere acts as a barrier for all waves with frequencies lower than 9 MHz.

When a magnetic field is present, the moving electrons will trace a helical motion, and the electron is subject to the following forces (see Fig. 12-4):

Centrifugal force $= mr\omega_H^2$
Magnetic fore $= e\nu \times \mathbf{B} = ev_t B$

Equalizing these two forces and replacing the tangential velocity v_t by $r\omega_H$, we get

$$\omega_H = \frac{|e|B}{m}$$

$$v_H = \frac{|e|B}{2\pi m} \qquad (12\text{-}10)$$

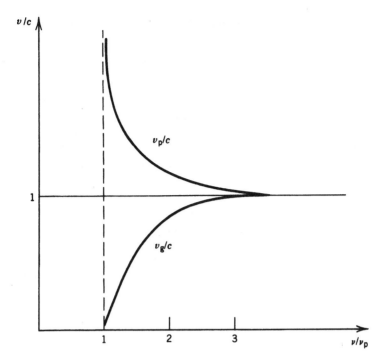

Figure 12-3. Behavior of the group velocity and phase velocity as a function of frequency.

This is called the plasma gyrofrequency. In the case of an electron,

$$v_{\mathrm{H}} = 2.8 \times 10^{10}\, B \tag{12-11}$$

where B is in Wb/m^2. To illustrate, for the case of the Earth, $B \simeq 5 \times 10^{-5}$ Wb/m^2. This gives $v_{\mathrm{H}} \simeq 1.4$ MHz.

In the presence of a magnetic field, the plasma becomes anisotropic, leading to significant complication in the wave propagation properties. The study of this case is beyond the scope of this book.

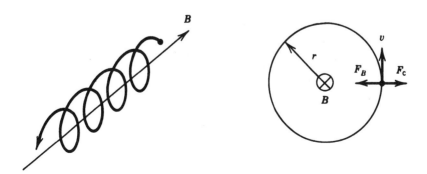

Figure 12-4. Motion of an electron in a magnetic field and forces acting on the electron. F_{c} is the centrifuge form equal to $m\omega\mathrm{H}^2 r$. F_{B} is the magnetic force equal to $ev \times \mathbf{B}$.

12-3 IONOSPHERIC PROFILE SENSING BY TOPSIDE SOUNDING

The fact that a wave of frequency v is completely reflected at the level where $v_p(z) = v$ allows the probing of different levels in the ionosphere and the measurement of the electron density profile as a function of height.

Let us consider a downlooking orbiting radio sensor that simultaneously transmits a large number of waves at neighboring, progressively increasing frequencies $v_1, v_2, \ldots, v_n$. These waves will penetrate down to different depths in the ionosphere (see Fig. 12-5).

The time delay t_i of the received echo at frequency v_i allows the determination of the location of the level where $v_p = v_i$, which, in turn, allows the derivation of the electron density at that level. The time delay t_i is given by

$$t_i = 2 \int_0^{z_i} \frac{dz}{v_g(v_i)} = \frac{2v}{c} \int_0^{z_i} \frac{dz}{\sqrt{v_i^2 - v_p^2(z)}} \tag{12-12}$$

where z_i is the location where $v_i = v_p(z_i)$. z is the distance from the sensor. It is clear from the above relationship that an interative process is necessary. The ionosphere is divided

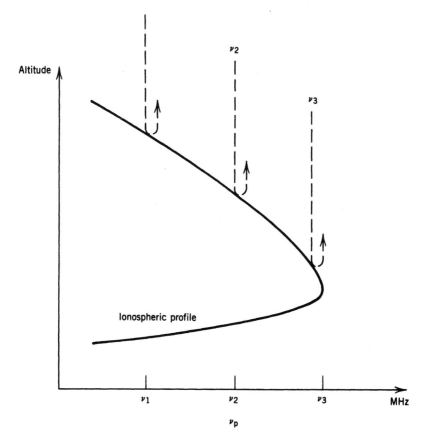

Figure 12-5. Simple sketch illustrating the depth penetration in the ionosphere of different frequency components in a swept-frequency topside sounder.

into a series of homogeneous layers (Fig. 12-6). The first (top) one starts at location $z_1 = ct_1/2$ and has a plasma frequency $v_p = v_1$. The second layer has a plasma frequency $v_p = v_2$ and starts at location z_2 such that

$$t_2 = t_1 + 2\frac{z_2 - z_1}{v_{g1}} = t_1 + \frac{2v_2(z_2 - z_1)}{c\sqrt{v_2^2 - v_1^2}} \qquad (12\text{-}13)$$

$$z_2 - z_1 = \frac{c(t_2 - t_1)}{2}\sqrt{1 - \left(\frac{v_1}{v_2}\right)^2} \qquad (12\text{-}14)$$

For the third layer, we have

$$t_3 = \frac{2z_1}{c} + \frac{2(z_2 - z_1)}{c}\sqrt{\frac{v_3^2}{v_3^2 - v_1^2}} + \frac{2(z_3 - z_2)}{c}\sqrt{\frac{v_3^2}{v_3^2 - v_2^2}} \qquad (12\text{-}15)$$

which allows the derivation of z_3, and so on. The number of layers that represent the profile cannot exceed the number of discrete frequencies used for the measurement. Thus, a very large number of neighboring frequencies allow the subdivision of the profile into

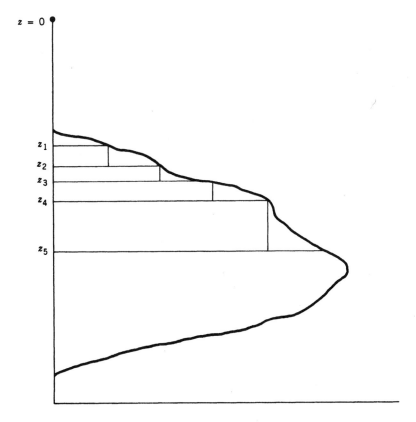

Figure 12-6. Sketch showing how an ionospheric profile is approximated by a series of homogeneous layers in order to convert the time delay frequency relationship.

smaller layers. In actuality, a continuous sweep of frequencies is transmitted, allowing the derivation of a continuous profile.

It should be noted that the above procedure works well for a smoothly increasing ionospheric density. Thus, the profile can be derived down to the ionospheric density peak. If there is a double peak, the profile in between the two peaks cannot be derived. In order to derive the bottom profile of the ionosphere, an upward-looking sensor on the surface is required.

An example of a topside sounder is the one carried by the Alouette and ISIS satellites in the 1960s. These satellites were launched into a circular polar orbit of 1000 km altitude. The sensor transmitted a frequency-swept signal pulse covering the frequency range from 0.5 to 12 MHz. By measuring the time delay of the returned echo at each frequency, the ionospheric density profile was derived on a global basis. Figure 12-7 shows an example of the data acquired with the Alouette sounder.

12-4 IONOSPHERIC PROFILE BY RADIO OCCULTATION

The index of refraction n of a plasma is given by [from Equation (12-8)]:

$$n = \sqrt{1 - \frac{v_p^2}{v^2}} \qquad (12\text{-}16)$$

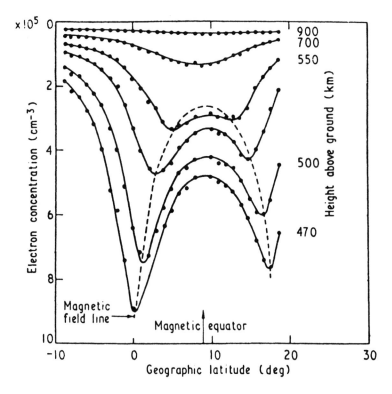

Figure 12-7. Latitude variations of the electron concentrations at different altitudes along the 110°E meridian deduced from the Alouette I sounder on June 19, 1963 at about noon local time. (From Eccles and King, 1969.)

As a spacecraft flies behind a planet, its radio signal will pass deeper and deeper into the planet's ionosphere before it reaches the atmosphere. The spherically inhomogeneous ionosphere leads to bending of the rays, similar to the case of atmospheric occultation discussed in Chapter 8. The angular bending can be derived by measuring the Doppler shift of the received signal. This in turn allows the derivation of the index of refraction profile and in turn the plasma frequency (i.e., electron density) profile.

This technique is used with the Global Positioning Satellite constellation to measure the total electron content of the Earth's atmosphere. The basic measurement uses a GPS satellite as the transmitting source and a satellite in low earth orbit as the receiver, as discussed in Chapter 9 and shown in Figure 12-8. The GPS constellation consists of 24 satellites orbiting in six different planes around the earth. This means that several GPS satellites are typically in view of any satellite in low earth orbit, and every time a GPS satellite is rising or setting relative to the receiving satellite, an occultation measurement can be made. A single receiving satellite in the proper orbit can have as many as 500 occultation opportunities per day (Kursinski et al., 1997).

The GPS occultation measurements derive the bending angle α from a measurement of the Doppler shift of the GPS signal at the receiving spacecraft. The Doppler shift, in turn, is measured as the time derivative of the phase of the GPS signals. The bending angle is related to the index of refraction, n, of the Earth's atmosphere. The index of refraction contains contributions from the dry neutral atmosphere, water vapor, free electrons in the ionosphere, and particulates, primarily in the form of liquid water as described in Chapter 9.

Each GPS satellite broadcasts a dual-frequency signal at 1575.42 MHz and 1227.60 MHz. By measuring the differential phase delay between these two signals, one can estimate the number density of the electrons, n_e, and then the contribution of the ionosphere to the bending angle can be estimated. See Chapter 9 for more details.

GPS signals have been used for ionospheric sounding by the CHAMP satellite (Beyerle et al., 2002) and the twin GRACE satellites (Gobiet and Kirchengast, 2004). The technique of ionospheric sounding using GPS signals will also be the focus of the planned COSMIC/FORMOSAT-3 project, a joint mission between the United States and Taiwan.

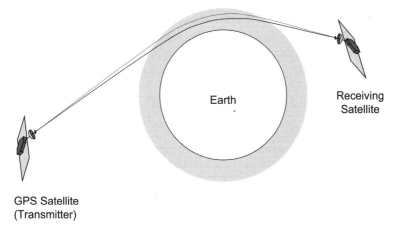

Figure 12-8. Geometry for a dual-frequency occultation measurement using a GPS satellite as a transmitter and a receiver on a different satellite in low earth orbit.

The COSMIC/FORMOSAT-3 constellation as currently planned consists of six space-craft equipped with GPS radio occultation receivers and will be launched in the December of 2005.

EXERCISES

12-1. From Figure 12-1, determine the maximum plasma frequency corresponding to the ionospheres of the terrestrial planets. If we are planning to have a topside sounder on a Venus-orbiting spacecraft, what should be the frequency sweeping range of the sensor? How about in the case of a Mars orbiter?

12-2. In order to understand the shape of the ionospheric density profile, let us consider the case in which the number density of the atmospheric atoms is given by an exponential law:

$$N(z) = N_0 e^{-z/H}$$

If F_0 is the photon solar flux in photons per square meter second normally incident at the top of the atmosphere and α is the absorption cross section in square meters of the atmospheric gases, calculate the photon flux as a function of altitude.

Let β be the photoionization cross section in square meters. Calculate the number of ionizations Q per cubic meter second. Sketch the behavior of Q as a function of z. Find the expression for the altitude where Q is a maximum.

12-3. An ionospheric sounder is in orbit around a planet whose ionospheric electron density profile is given by

$$N = \begin{cases} -6 \times 10^6 h + 1.5 \times 10^{12} & \text{for } 150 \text{ km} \leq h \leq 250 \text{ km} \\ 8 \times 10^{11} & \text{for } 75 \text{ km} \leq h \leq 100 \text{ km} \\ 0 & \text{elsewhere} \end{cases}$$

The orbit altitude is 1,000 km. A series of ten pulses at frequencies 1 MHz, 2 MHz, 3 MHz, ..., 10 MHz are transmitted simultaneously. Calculate the time delay for the echo of each one of the pulses. Plot this delay as a function of the altitude of the reflecting ionospheric layer.

REFERENCES AND FURTHER READING

Beyerle, G., K. Hocke, J. Wickert, T. Schmidt, C. Marquardt, and C. Reigber. GPS radio occultations with CHAMP: A radio holographic analysis of GPS signal propagation in the troposphere and surface reflections. *Journal of Geophysical Research,* **107**(D24), 4802, doi:10.1029/2001JD001402, 2002.

Calvert, W. Ionospheric topside sounding. *Science,* **154,** 228–232 1966.

Davies, K. *Ionospheric Radio Waves.* Blaisdell, Waltham, MA, 1969.

Eccles D., and J. W. King. A review of topside sounder studies of the equatorial ionosphere. *IEEE Proceedings,* **57,** 1012–1018, 1969.

Gobiet, A., and G. Kirchengast. Advancements of Global Navigation Satellite System radio occultation retrieval in the upper stratosphere for optimal climate monitoring utility. *Journal of Geophysical Research,* **109,** D24110, doi:10.1029/2004JD005117, 2004.

Gorbunov, M. E. Ionospheric correction and statistical optimization of radio occultation data. *Radio Science,* **37(5),** 1084, doi:10.1029/2000RS002370, 2002.

Gringuaz, K. I., and T. K. Breus. Comparative characteristics of the ionosphere of the planets of the terrestrial group. *Space Science Review,* **10,** 743–769, 1970.

Hocke, K., and K. Igarashi. Electron density in the F region derived from GPS/MET radio occultation data and comparison with IRI. *Earth Planets Space,* **54,** 947–954, 2002.

Kursinski, E. R., G. A. Hajj, J. T. Schofield, R. P. Linfield, and K. R. Hardy. Observing Earth's atmosphere with radio occultation measurements using the Global Positioning System. *Journal of Geophysical Research,* **102**(D19), 23,429–23,466, 1997.

Kursinski, E. R., G. A. Hajj, S. S. Leroy, and B. Herman. The GPS radio occultation technique. *Terrestrial Atmospheric and Oceanic Science,* **11**(1), 235–272, 2000.

Rocken, C., Y.-H. Kuo, W. Schreiner, D. Hunt, and S. Sokolovskiy. Cosmic System Description. *Terrestrial Atmospheric and Oceanic Science* **11**(1), 21–52, 2000.

Sokolovskiy, S. V. Inversions of radio occultation amplitude data. *Radio Science,* **35,** 97–105, 2000.

Sokolovskiy, S. V. Modeling and inverting radio occultation signals in the moist troposphere. *Radio Science,* **36,** 441–458, 2001.

Sokolovskiy, S. V. Tracking tropospheric radio occultation signals from low Earth orbit. *Radio Science,* **36,** 483–498, 2001.

Solheim, F. S., J. Vivekanandan, R. H. Ware, and C. Rocken. Propagation delays induced in GPS signals by dry air, water vapor, hydrometeros, and other particulates. *Journal of Geophysical Research,* **104**(D8), 9663–9670, 1999.

Vorob'ev V. V., and T. G. Krasil'nikova. Estimation of the accuracy of the atmospheric refractive index recovery from Doppler shift measurements at frequencies used in the NAVSTAR system. *Phys. Atmospheric Oceanography,* **29,** 602–609, 1993.

USE OF MULTIPLE SENSORS FOR SURFACE OBSERVATIONS

The detailed study of a planetary surface or atmosphere requires the simultaneous use of multiple sensors covering a large part of the electromagnetic spectrum. This is a result of the fact that any individual sensor covers only a small part of the spectrum in which the wave–matter interaction mechanisms are driven by a limited number of medium properties. For example, in the case of solid surfaces, x-ray sensors provide information on the content of radioactive materials, visible and near-infrared sensors provide information about the surface chemical composition, thermal infrared sensors measure the near-surface thermal properties, and radar sensors are mainly sensitive to the surface physical properties (topography, roughness, moisture, dielectric constant). Similarly, in the case of the atmosphere, in order to cover the wide range of possible chemical constituents, detect and characterize atmospheric particles (including rain), and sound the physical properties of the atmosphere, a suite of sensors covering selected bands in the visible, infrared, millimeter, and microwave spectral regions will be needed.

To illustrate how multiple sensors can be used collectively to enhance the ability of an interpreter in the study of a planetary surface, a set of data products covering the area of Death Valley in eastern California are presented. Figure A-1 shows three images of Death Valley acquired with three separate instruments in the visible/near IR (A-1a), thermal IR (A-1b), and radar (A-1c) spectral bands.

With the topography database, false illumination images can be generated to highlight the surface topography (Fig. A-2). This topography database can then be coregistered to the multispectral image data (Fig. A-1) and used to generate perspective images from a variety of observing directions, as illustrated in Figures A-3 and A-4. The observing direction, the vertical exaggeration, the spectral bands, and the color coding can be selected by the interpreter and displayed on a monitor instantaneously. This will effectively be equivalent to bringing the study site into the laboratory for detailed "dissection" and analysis. Of course, there will always be the need to do field work for direct surface observation, but the above-described database will go a long way toward developing a basic understanding of the surface properties.

Introduction to the Physics and Techniques of Remote Sensing. By C. Elachi and J. van Zyl
Copyright © 2006 John Wiley & Sons, Inc.

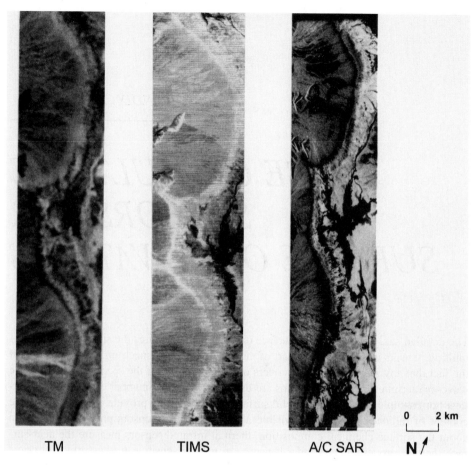

TM   TIMS   A/C SAR   N

0   2 km

Figure A-1. Images of Death Valley, California acquired in the three major regions of the electromagnetic spectrum: (*a*) Landsat image acquired in visible/NIR, (*b*) airborne TIMS image acquired in the thermal IR, and (*c*) airborne SAR image acquired in the microwave. See color section.

Figure A-2. False illumination image derived by computer processing from the topography database. See color section.

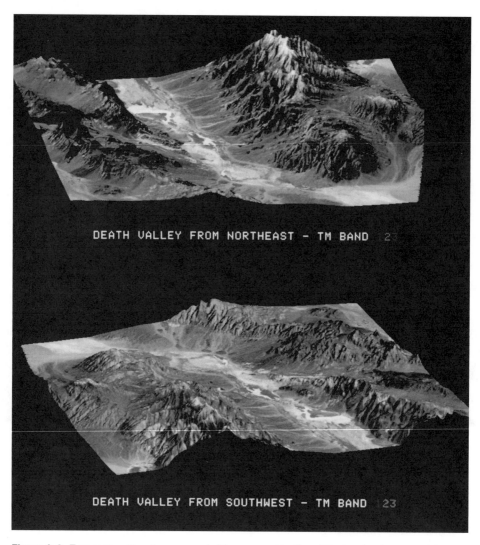

Figure A-3. Two perspective views generated from a combination of the Landsat image and the digital topography base. See color section. (Courtesy of M. Kobrick, JPL.)

COTTONBALL BASIN AND TELESCOPE PEAK

LANDSAT TM BAND
2X VERTICAL EXAGGERATION
VIEW FROM NORTHEAST

Figure A-4. Surface perspective view generated from a combination of the Landsat image and the digital topography base. See color section. (Courtesy of M. Kobrick, JPL.)

SUMMARY OF ORBITAL MECHANICS RELEVANT TO REMOTE SENSING

Orbit selection and sensor characteristics are closely related to the strategy required to achieve the desired results. Different types of orbits are required to achieve continuous monitoring, repetitive coverage of different periodicities, global mapping, or selective imaging.

The vast majority of earth-orbiting, remote sensing satellites use circular orbits. Planetary orbiters usually have elliptical orbits, which are less taxing on the spacecraft orbital propulsion system and combine some of the benefits of high and low orbits, thus allowing a broader flexibility to achieve multiple scientific objectives.

B-1 CIRCULAR ORBITS

B-1-1 General Characteristics

A spacecraft will remain in a circular orbit around a spherically homogeneous planet if the inward gravitational force F_g is exactly cancelled by the outward centrifugal force F_c. These two forces are given by

$$F_g = mg_s\left(\frac{R}{r}\right)^2 \qquad \text{(B-1)}$$

and

$$F_c = \frac{mv^2}{r} = m\omega^2 r \qquad \text{(B-2)}$$

where g_s is the gravitational acceleration at the planet's surface, R is the planet radius of the planet, r is the orbit radius ($r = R + h$), h is the orbit altitude, v is the spacecraft linear

Introduction to the Physics and Techniques of Remote Sensing. By C. Elachi and J. van Zyl
Copyright © 2006 John Wiley & Sons, Inc.

velocity, and ω is the spacecraft angular velocity. For these two forces to be equal, the spacecraft linear velocity has to be

$$v = \sqrt{\frac{g_s R^2}{r}} \tag{B-3}$$

The orbital period T of the circular orbit is then

$$T = \frac{2\pi r}{v} = 2\pi r \sqrt{\frac{r}{g_s R^2}} \tag{B-4}$$

In the case of the Earth, $g_s = 9.81$ m/sec^2 and $R \approx 6380$ km. The orbital period T and linear velocity v are shown in Figure B-1 as a function of altitude h. For instance, at $h = 570$ km, $v \approx 7.6$ km/sec and $T = 1$ hr 36 min. In such an orbit, the spacecraft will orbit the Earth exactly 15 times per day. Table B-1 gives an illustrative example for the case of some planetary orbits.

In some cases, one might be interested in calculating the orbit altitude for a given orbital period. The orbit altitude can be found by inverting Equation (B-4):

$$h = \left[\frac{g_s R^2 T^2}{4\pi^2} \right]^{1/3} - R \tag{B-5}$$

The orientation of the orbit in space is specified in relation to the Earth's equatorial plane and the vernal equinox. The angle between the orbital plane and the equatorial plane is the orbital inclination I (see Figure B-2a). The angle between the vernal equinox (direction in space defined by the Earth–Sun line at the time of day–night equality) and the node line (intersection of the orbit and equatorial plane) is the orbital node longitude Ω.

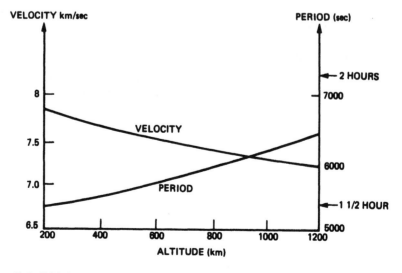

Figure B-1. Orbital velocity and period function of altitude for a circular orbit (case of the Earth).

TABLE B-1. Orbital Velocity and Period for Circular Orbits around Some of the Planets

Planet	Radius R (km)	Surface gravity g_s (m/sec²)	Orbit altitude $h = 0.2R$		Orbit altitude $h = 0.5R$	
			v (km/sec)	T (min)	v (km/sec)	T (min)
Mercury	2440	3.63	2.7	113	2.4	158
Venus	6050	8.83	6.7	114	5.9	159
Earth	6380	9.81	7.2	111	6.4	155
Moon	1710	1.68	1.55	139	1.38	194
Mars	3395	3.92	3.3	128	3.0	179
Jupiter	71,500	25.9	39.3	229	35.0	320
Saturn	60,000	11.38	23.8	316	21.2	442

The largest variation in the orbit orientation is usually due to precession, which is the rotation of the orbit plane around the polar axis and is primarily due to the Earth's oblateness. The resulting rate of change of the nodal longitude Ω is approximated by:

$$\frac{d\Omega}{dt} = -\frac{3}{2}J_2R^3\sqrt{g_s}\,\frac{\cos I}{r^{7/2}} \tag{B-6}$$

where $J_2 = 0.00108$ is the coefficient of the second zonal harmonic of the Earth's geopotential field. Figure B-3 shows the relationship between $d\Omega/dt$, I, and h for the Earth. In the case of a polar orbit (i.e., $I = 90°$) the precession is zero because the orbit plane is coincident with the axis of symmetry of the Earth.

B-1-2 Geosynchronous Orbits

An important special case of the circular orbit is the geosynchronous orbit. This corresponds to the case in which the orbit period T is equal to the sidereal day, the sidereal day being the planet rotation period relative to the vernal equinox. Replacing T in Equation

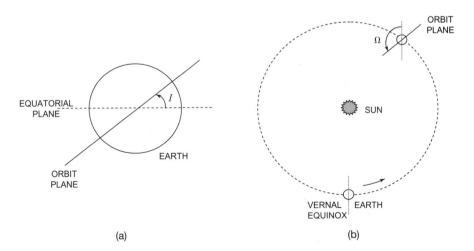

(a) (b)

Figure B-2. Orientation of the orbit in space.

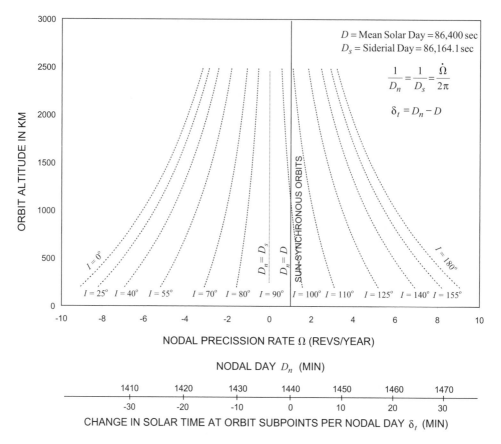

Figure B-3. Satellite orbit precession as a function of orbital altitude *h* and inclination *I*.

(B-5) by the length of the Earth's sidereal day 364/365 × 24 hr = 86,164 sec, we find that the radius of an Earth-geosynchronous circular orbit is $r \approx 42,180$ km, which corresponds to an altitude of 35,800 km. For a ground observer, a satellite at the geosynchronous altitude will appear to trace a figure-eight shape of latitude range ±*I* (see Fig. B-4). If the orbit is equatorial (*I* = 0), the satellite appears to be fixed in the sky. The orbit is then called geostationary. Geosynchronous orbits can also be elliptical. In this case, a distorted figure eight is traced relative to an Earth fixed frame (Figure B-5).

B-1-3 Sun-Synchronous Orbit

If the orbit precession exactly compensates for the Earth's rotation around the sun, the orbit is sun-synchronous (see Fig. B-3). Such an orbit provides a constant node-to-sun angle, and the satellite passage over a certain area occurs at the same time of the day.

All Landsat satellites are in a near-polar sun-synchronous orbit. Referring to Figure B-3, the Landsat orbit altitude of approximately 700 km corresponds to a sun-synchronous orbit inclination of approximately 98°. [Note from Equation (B-6) that sun-synchronous orbits require inclination angles larger than 90 degrees.]

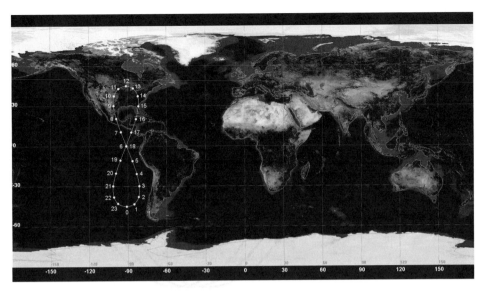

Figure B-4. Surface trace of a circular geosynchronous orbit with 45° inclination.

B-1-4   Coverage

The orbit nadir trace is governed by the combined motion of the planet and the satellite. Let us first assume that the planet is nonrotating. The nadir will resemble a sine wave on the surface map (Figure B-6*a*), with its great circle path developing a full 360° period (longitude) and covering latitudes between $\pm I$. Adding the planet's rotation causes a steady creep of the trace at a rate proportional to the ratio of the orbital motion angular rate to the planet's rotational rate (Fig. B-6*b*).

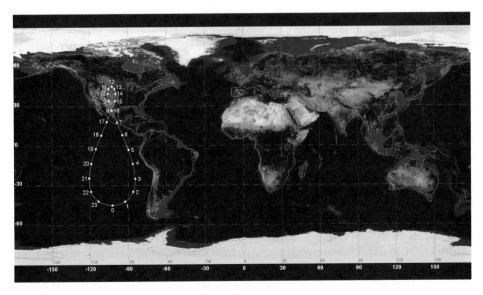

Figure B-5. Surface trace of an elliptical geosynchronous orbit with 45° inclination and 0.1 ellipticity.

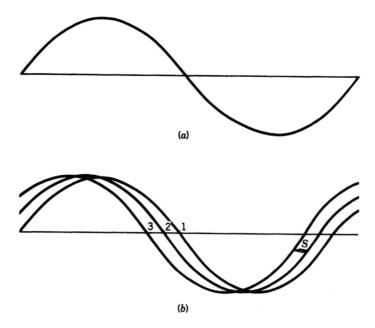

Figure B-6. (a) Orbit nadir trace on a nonrotating planet. (b) Orbit trace on a rotating planet. S is the orbit step that corresponds to the displacement of the equatorial crossing on successive orbits.

The orbit step S is the longitudinal difference between two consecutive equatorial crossings. If S is such that

$$S = 360 \frac{N}{L}$$

where N and L are integer numbers, then the orbit coverage is repetitive. In this case, the satellite makes L revolutions as the planet makes N revolutions. N is thus the cycle period (sometimes called the orbit repeat period) and L is the number of revolutions per cycle. If $N = 1$, then there is an exact daily repeat of the coverage pattern. If $N = 2$, the repeat is every other day, and so on. Figure B-7 shows the relationship between N, L, the orbital period, and the orbital altitude for sun-synchronous orbits. It clearly shows that there is a wide array of altitude choices to meet any repetitive coverage requirement.

A wide range of coverage scenarios can be achieved by selecting the appropriate orbit. Figure B-8 gives examples of four different orbital altitudes, which would allow a 16-day repeat orbit, multiple viewing of a certain area on the surface, and complete coverage with a sensor having a field of view of about 178 km. The main difference is the coverage strategy. The 542 km (241 orbits in 16 days) and 867 km (225 orbits in 16 days) orbits have a drifting coverage, that is, the orbits on successive days allow mapping of contiguous strips. The 700 km orbit (233 orbits in 16 days) has a more dispersed coverage strategy, in which the second day strip is almost halfway in the orbital step. The 824 km orbit (227 orbits in 16 days) has a semidrifting coverage strategy, in which every five days the orbit drifts through the whole orbital strip, leaving gaps that are filled up on later orbits. The Landsat satellites complete 233 orbits in a 16 day cycle, corresponding to the coverage shown for 700 km altitude orbit.

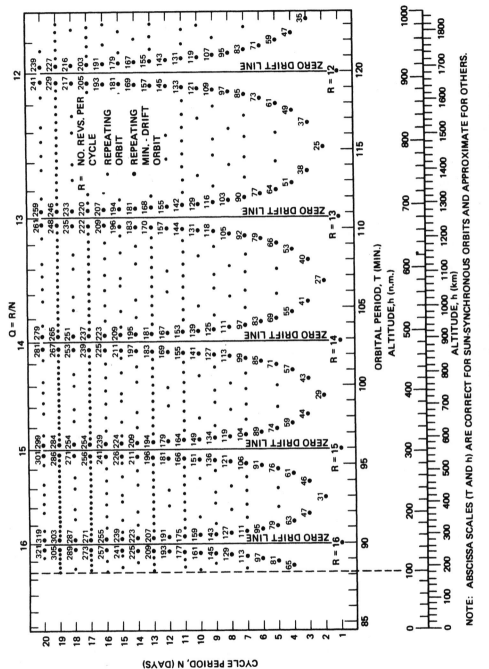

NOTE: ABSCISSA SCALES (T AND h) ARE CORRECT FOR SUN-SYNCHRONOUS ORBITS AND APPROXIMATE FOR OTHERS.

Figure B-7. Periodic coverage patterns as a function of altitude for sun-synchronous orbits (case of Earth).

531

542 km Altitude

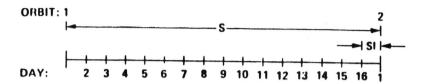

700 km Altitude

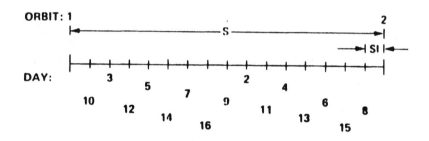

824 km Altitude

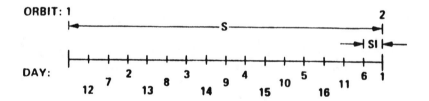

867 km Altitude

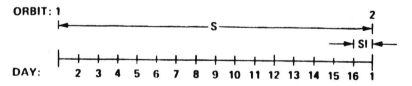

Figure B-8. Coverage scenario of four different orbital altitudes that provide an almost 16 day repeat orbit.

B-2 ELLIPTICAL ORBITS

Elliptical orbits are most commonly used for planetary orbiters. Such an orbit is desired to accommodate a wide variety of scientific objectives. For instance, high-resolution imaging, which requires low altitude, can be conducted near periapsis, whereas global meteorological monitoring can be done from near apoapsis, where the spacecraft stays for a longer time. The change in the satellite altitude will also allow the in situ measurement of the different regions of the planet's surroundings. In addition, the energy required to slow down a spacecraft to capture it in a circular orbit of altitude h_c is higher than the energy required for an elliptical orbit with a periapsis altitude h_c.

The characteristics of an elliptical orbit are derived from the solution of the two-body problem. The two most known characteristics are:

1. The radius vector of the satellite sweeps over an equal area in an equal interval of time.
2. The square of the orbital period is proportional to the cube of the mean satellite-to-planet-center distance.

The orbit is defined by

$$r = \frac{a(1 - e^2)}{1 + e \cos \theta} \tag{B-7a}$$

and

$$T = 2\pi \sqrt{\frac{a^3}{g_s R^2}} \tag{B-7b}$$

where r, θ are the satellite polar coordinates, a is the semimajor axis, and e is the orbit ellipticity (see Fig. B-9). By taking different combinations of values for a, e, and inclination, a wide variety of coverage strategies can be achieved. For example, Figure B-10 shows the ground trace for an elliptical orbit with $e = 0.75$, a 12 hour orbit ($a \approx 26{,}600$ km), and an

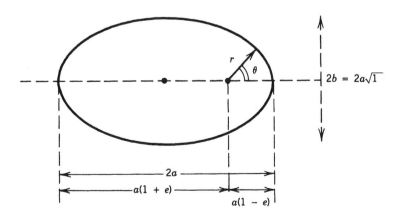

Figure B-9. Elliptical orbit.

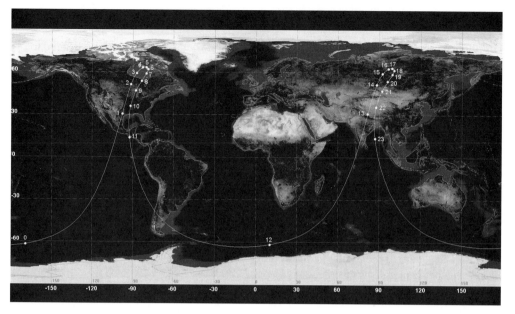

Figure B-10. Ground trace for a 12 hour elliptical orbit with an ellipticity of 0.7 and inclination of 63.4°.

inclination of 63.4°. It shows that the satellite spends most of the time over Europe and the north-central Pacific and only a small amount of time over the rest of the world. This is an example of a Molniya orbit, which has the property that rate of change of the orbit perigee is zero. For the Earth, this happens for inclinations of 63.4° and 116.6°.

B-3   ORBIT SELECTION

There are a number of factors that influence the selection of an orbit most appropriate for a specific remote sensing application. Some of these factors for an Earth orbiter are as follows:

To minimize Earth atmospheric drag—$h > 200$ km
Global coverage for Earth observation—polar or near-polar orbit
Astronomical/planetary observations from Earth orbit—equatorial orbit
Constant illumination geometry (optical sensors)—sun-synchronous orbits
Thermal inertia observations—one-day pass and one-night pass over same area
To minimize radar sensors' power—prefers low altitude
To minimize gravity anomalies perturbation—high altitude
To measure gravity anomalies—low altitude
Continuous monitoring—geostationary or geosynchronous orbit

EXERCISES

B-1. Plot the orbital velocity, velocity of projection on the surface, and orbital period for circular orbits around the Earth at altitudes ranging from 200 km to 10,000 km. Repeat the same for the case of Mercury, Venus, and Mars.

B-2. Plot the orbital velocity and orbital period for circular orbits around Jupiter and Saturn at altitudes ranging from 60,000 to 200,000 km.

B-3. Plot the nodal precession rate of circular orbits around the Earth as a function of altitude ranging from 200 km to 2500 km for orbit inclinations of 0°, 40°, 60°, 80°, 90°, 100°, 120°, 140°, and 180°. Express the precession rate in radians/year or revs/year. Indicate on the plot the sun-synchronous orbits.

B-4. Calculate the altitude of geostationary orbits for Mercury, Venus, Mars, Jupiter, and Saturn. The rotation periods of these planets are 58.7 Earth-days, 243 Earth-days, 24 hr 37 min, 9 hr 51 min, and 10 hr 14 min, respectively.

B-5. Let us assume that an Earth-orbiting sensor in polar circular orbit requires daily repeat (i.e. $N = 1$). Calculate the lowest three orbits which allow such a repeat coverage. Calculate the lowest three orbits for repeats every two days ($N = 2$) and every three days ($N = 3$).

B-6. A radar mission is designed to fly on the space shuttle with the aim of mapping as much of the Earth as possible in 10 days. Given the size and mass of the radar payload, the maximum inclination of the orbit is 57°, and the altitude range is 200 km to 250 km. Calculate the altitude of an orbit in this range that would repeat in exactly 10 days. Calculate the separation between orbit tracks along the equator.

SIMPLIFIED WEIGHTING FUNCTIONS

C-1 CASE OF DOWNLOOKING SENSORS (EXPONENTIAL ATMOSPHERE)

In the case of downlooking sensors, the weighting function is of the form

$$W(z) = \alpha_0 \exp\left[-\frac{z}{H} - \tau_{\mathrm{m}} \exp\left(-\frac{z}{H}\right)\right] \qquad \text{(C-1)}$$

where $\tau_{\mathrm{m}} = \alpha_0 H$. The first and second derivatives are given by

$$\frac{dW}{dz} = \left[-1 + \tau_{\mathrm{m}} \exp\left(-\frac{z}{H}\right)\right]\frac{W(z)}{H} \qquad \text{(C-2)}$$

$$\frac{d^2W}{dz^2} = \left[1 + \tau_{\mathrm{m}}^2 \exp\left(-\frac{2z}{H}\right) - 3\tau_{\mathrm{m}} \exp\left(-\frac{z}{H}\right)\right]\frac{W(z)}{H^2} \qquad \text{(C-3)}$$

By solving for $dW/dz = 0$ and $d^2W/dz^2 = 0$ we find the following:

$$\frac{dW}{dz} = 0 \qquad \text{for } \frac{z}{H} = \log \tau_{\mathrm{m}}$$

$$\frac{d^2W}{dz^2} = 0 \qquad \text{for } \frac{z}{H} = \log\left[\frac{2\tau_{\mathrm{m}}}{(3 \pm \sqrt{5})}\right]$$

Introduction to the Physics and Techniques of Remote Sensing. By C. Elachi and J. van Zyl
Copyright © 2006 John Wiley & Sons, Inc.

In addition:

$$\text{For } z = 0 \qquad W(z) = \alpha_0 e^{-\tau_{\rm m}} \text{ and } \frac{dW}{dz} = (\tau_{\rm m} - 1)\frac{\alpha_0}{H}e^{-\tau_{\rm m}}$$

$$\text{For } z \to \infty \qquad W(z) \to 0$$

$$\text{For } z = H \log \tau_{\rm m} \qquad W(z) = \frac{\alpha_0}{e\tau_{\rm m}} \frac{1}{eH}$$

This shows that the weighting function can have four different shapes, as shown in Figure 9-3:

For $\tau_{\rm m} < (3 - \sqrt{5})/2 = 0.382$, $W(z)$ is monotonically decreasing with positive curvature.

For $0.383 < \tau_{\rm m} < 1$, $W(z)$ is monotonically decreasing but with an inflection point.

For $1 < \tau_{\rm m} < (3 + \sqrt{5})/2 = 2.62$, $W(z)$ has a peak for $z = H \log \tau_{\rm m}$ and an inflection point after the peak.

For $\tau_{\rm m} > 2.62$, $W(z)$ has a peak at $z = H \log \tau_{\rm m}$ and has two inflection points, one on each side of the peak.

For $\tau_{\rm m} > 1$, the maximum value of $W(z)$ is constant $= 1/eH$.

C-2   CASE OF DOWNLOOKING SENSORS (LINEAR ATMOSPHERE)

In this case we have

$$W(z) = \alpha_0\left(1 - \frac{z}{H}\right)\exp\left[-\frac{\tau_{\rm m}}{2} + \tau_{\rm m}\left(\frac{z}{H} - \frac{z^2}{2H^2}\right)\right]$$

and

$$\frac{dW}{dz} = \frac{\alpha_0}{H}\left[\tau_{\rm m}\left(1 - \frac{z}{H}\right)^2 - 1\right]\exp\left[-\frac{\tau_{\rm m}}{2} + \tau_{\rm m}\left(\frac{z}{H} - \frac{z^2}{2H^2}\right)\right]$$

$$\frac{d^2W}{dz^2} = \frac{\alpha_0}{H^2}\left\{\tau_{\rm m}\left(1 - \frac{z}{H}\right)\left[-3 + \tau_{\rm m}\left(1 - \frac{z}{H}\right)^2\right]\right\}\exp\left[-\frac{\tau_{\rm m}}{2} + \tau_{\rm m}\left(\frac{z}{H} - \frac{z^2}{2H^2}\right)\right]$$

By solving for $dW/dz = 0$ and $d^2W/dz^2 = 0$, we find that

$$\frac{dW}{dz} = 0 \qquad \text{for } \frac{z}{H} = 1 - \frac{1}{\sqrt{\tau_{\rm m}}}$$

$$\frac{d^2W}{dz^2} = 0 \qquad \text{for } \frac{z}{H} = 1 - \frac{\sqrt{3}}{\sqrt{\tau_{\rm m}}}$$

In addition,

$$\text{For } z = 0: \qquad W(z) = \alpha_0 e^{-\tau_{\rm m}/2} \quad \text{and} \quad \frac{dW}{dz} = \alpha_0(\tau_{\rm m} - 1)e^{-\tau_{\rm m}/2}$$

For $z = H$: $W(z) = 0$ and $\dfrac{dW}{dz} = -\alpha_0$

For $z = H\left(1 - \dfrac{1}{\sqrt{\tau_m}}\right)$: $W(z) = \dfrac{\alpha_0}{\sqrt{e\tau_m}}$

This shows that the weighting function can have three different shapes:

For $\tau_m < 1$, $W(z)$ is monotonically decreasing with negative curvature.

For $1 < \tau_m < 3$, $W(z)$ has a peak at $z = H(1 - 1/\sqrt{\tau_m})$ and a negative curvature.

For $\tau_m > 3$, $W(z)$ has a peak and an inflection point.

C-3   CASE OF UPWARD LOOKING SENSORS

In the case of upward looking sensors, the weighting function for an exponential atmosphere is of the form

$$W(z) = \alpha_0 e^{-\tau_m} \exp\left[-\frac{z}{H} + \tau_m \exp\left(-\frac{z}{H}\right)\right]$$

The main difference from the previous case is the change in the sign inside the exponential. This leads to dW/dz always <0 and d^2W/dz always >0. Thus, the weighting function has always the shape shown in Figure 9-5.

In the case of a linear atmosphere,

$$W(z) = \alpha_0\left(1 - \frac{z}{H}\right)\exp\left[-\tau_m\left(\frac{z}{H} - \frac{z^2}{2H^2}\right)\right]$$

This gives dW/dz as always <0 and d^2W/dz^2 always >0, leading to a behavior very similar to the case of an exponential atmosphere.

COMPRESSION OF A LINEAR FM CHIRP SIGNAL

Let us assume that the radar transmits a signal that is of the form

$$v(t) = A(t) \exp\left[i2\pi(f_c t + Kt^2/2) \right]$$
(D-1)

Here, the amplitude of the signal, $A(t)$, is given by

$$A(t) = \begin{cases} 1 & \text{for} -\tau/2 \leq t \leq \tau/2 \\ 0 & \text{otherwise} \end{cases}$$
(D-2)

where τ is the length of the pulse in time. The instantaneous frequency of the transmitted signal during the pulse is

$$f(t) = \frac{1}{2\pi} \frac{\partial \phi(t)}{\partial t} = f_c - Kt \qquad \text{for} -\tau/2 \leq t \leq \tau/2$$
(D-3)

This means that the frequency varies linearly with time between values $f_c - K\tau/2$ and $f_c + K\tau/2$. The center frequency of the chirp is f_c, the chirp rate is K, and the signal bandwidth is $B = K\tau$.

If we transmit this signal at time $t = 0$, we will receive a signal from a point scatterer that is a distance R away after a time t_R, where

$$t_R = 2R/c$$
(D-4)

This received signal can be written as

$$v_r(t) = \alpha v(t - t_R)$$
(D-5)

Introduction to the Physics and Techniques of Remote Sensing. By C. Elachi and J. van Zyl
Copyright © 2006 John Wiley & Sons, Inc.

where the constant α takes into account any attenuation during propagation, as well as the radar cross section of the scatterer.

The pulse is compressed by convolving the received signal with a replica of the transmitted signal:

$$v_o(t) = \int_{-\infty}^{\infty} v^*(\xi - t)v_r(\xi)d\xi \qquad \text{(D-6)}$$

Performing the multiplications in the integrand, we find

$$v^*(\xi - t)v_r(\xi) = \alpha A(\xi - t)A(\xi - t_R) \exp[i2\pi f_c(t - t_R)] \exp[i\pi K(t - t_R)(2\xi - t - t_R)] \qquad \text{(D-7)}$$

Therefore, the output of the matched filter is

$$v_o(t) = \alpha \exp[i2\pi f_c(t - t_R)]\int_{-\infty}^{\infty} A(\xi - t)A(\xi - t_R) \exp[i\pi K(t - t_R)(2\xi - t - t_R)]d\xi \qquad \text{(D-8)}$$

The exponential outside the integral consists of the product of two terms,

$$\exp[i2\pi f_c(t - t_R)] = \exp[i2\pi f_c t] \exp[-i2\pi f_c t_R] = \exp[i2\pi f_c t] \exp[-i4\pi R/\lambda] \qquad \text{(D-9)}$$

The first term is simply a carrier frequency that will be filtered out during the down-conversion process in the receiver. The second term is the absolute phase of the radar signal because of the propagation to the scatterer and back. It is this phase that is used in interferometric radars.

Next, let us evaluate the integral in the convolution expression. We note that there are two cases we need to consider.

Case 1: $t \leq t_R$

In this case, the two signal amplitudes $A(\xi - t)$ and $A(\xi - t_R)$ are related to each other as shown in Figure D-1. The shaded area represents the values of ξ for which the integrand will be nonzero. Therefore, the integral for this case is

$$\int_{-\infty}^{\infty} A(\xi - t)A(\xi - t_R) \exp[i\pi K(t - t_R)(2\xi - t - t_R)]d\xi$$

$$= \int_{t_R - \tau/2}^{t + \tau/2} \exp[i\pi K(t - t_R)(2\xi - t - t_R)]d\xi \qquad \text{(D-10)}$$

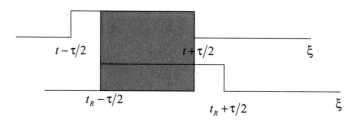

Figure D-1. The relationship between the pulse and its replica for the case in which $t \leq t_R$.

Changing variables in the integration, we find

$$\int_{t_R-\tau/2}^{t+\tau/2} \exp[i\pi K(t-t_R)(2\xi-t-t_R)]d\xi = \frac{1}{2}\int_{-\tau-(t-t_R)}^{\tau+(t-t_R)} \exp[i\pi K(t-t_R)u]du \quad \text{(D-11)}$$

Evaluating the right-hand side of this expression gives

$$\frac{1}{2}\int_{-\tau-(t-t_R)}^{\tau+(t-t_R)} \exp[i\pi K(t-t_R)u]du = \frac{\sin[\pi K(t-t_R)(\tau-(t+t_R))]}{\pi K(t-t_R)} \quad \text{(D-12)}$$

Note that for negative values of x, one can write $x = -|x|$. In this case, therefore, we can write the results as

$$\frac{\sin[\pi K(t-t_R)(\tau+(t-t_R))]}{\pi K(t-t_R)} = \frac{\sin[\pi K(t-t_R)(\tau-|t-t_R|)]}{\pi K(t-t_R)} \quad \text{(D-13)}$$

Therefore, the output of the convolution for this case, is

$$v_o(t) = \alpha\tau \exp[i2\pi f_c t] \exp[-i4\pi R/\lambda] \frac{\sin[\pi K(t-t_R)(\tau-|t-t_R|)]}{\pi K(t-t_R)} \quad \text{(D-14)}$$

This expression is valid for $t_R - \tau \le t \le t_R$.

Case 2: $t \ge t_R$

In this case, the two signal amplitudes $A(\xi-t)$ and $A(\xi-t_R)$ are related to each other as shown in Figure B-2. As before, the shaded area represents the values of ξ for which the integrand will be nonzero. Therefore, the integral for this case is

$$\int_{-\infty}^{\infty} A(\xi-t)A(\xi-t_R) \exp[i\pi K(t-t_R)(2\xi-t-t_R)]d\xi$$

$$= \int_{t-\tau/2}^{t_R+\tau/2} \exp[i\pi K(t-t_R)(2\xi-t-t_R)]d\xi \quad \text{(D-15)}$$

Changing variables in the integration, we find

$$\int_{t-\tau/2}^{t_R+\tau/2} \exp[i\pi K(t-t_R)(2\xi-t-t_R)]d\xi = \frac{1}{2}\int_{-\tau+(t-t_R)}^{\tau-(t-t_R)} \exp[i\pi K(t-t_R)u]du \quad \text{(D-16)}$$

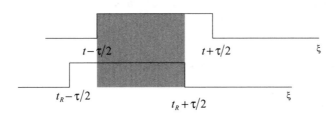

Figure D-2. The relationship between the pulse and its replica for the case in which $t_R \le t$.

Evaluating the right-hand side of this expression gives

$$\frac{1}{2}\int_{-\tau-(t-t_R)}^{\tau+(t-t_R)} \exp[i\pi K(t-t_R)u]du = \frac{\sin[\pi K(t-t_R)(\tau-(t-t_R))]}{\pi K(t-t_R)} \tag{D-17}$$

Note that for positive values of x, one can write $x = |x|$. In this case, therefore, we can write the results as

$$\frac{\sin[\pi K(t-t_R)(\tau+(t-t_R))]}{\pi K(t-t_R)} = \frac{\sin[\pi K(t-t_R)(\tau-|t-t_R|)]}{\pi K(t-t_R)} \tag{D-18}$$

Therefore, the output of convolution for this case, is

$$v_o(t) = \alpha\tau \exp[i2\pi f_c t] \exp[-i4\pi R/\lambda]\frac{\sin[\pi K(t-t_R)(\tau-|t-t_R|)]}{\pi K\tau(t-t_R)} \tag{D-19}$$

which is valid for $t_R \leq t \leq t_R + \tau$.

If we now put the two cases together, we find the final output of the matched filter as

$$v_o(t) = \alpha\tau \exp[i2\pi f_c t] \exp[-i4\pi R/\lambda]\frac{\sin[\pi K(t-t_R)(\tau-|t-t_R|)]}{\pi K\tau(t-t_R)} \tag{D-20}$$

which is valid for $t_R - \tau \leq t \leq t_R + \tau$.

For values of $|t-t_R| \ll \tau$, this expression reduces to

$$v_o(t) = \alpha\tau \exp[i2\pi f_c t] \exp[-i4\pi R/\lambda]\frac{\sin[\pi K\tau(t-t_R)]}{\pi K\tau(t-t_R)} \tag{D-21}$$

The first term on the right-hand side is just a carrier frequency, and will be filtered out once the signal is down-converted. Therefore, the base-band version of the compressed pulse is then

$$v_o(t) = \alpha\tau \exp[-i4\pi R/\lambda]\frac{\sin[\pi K\tau(t-t_R)]}{\pi K\tau(t-t_R)} \tag{D-22}$$

This is the usual expression found in many radar texts. Note that the compressed pulse has a total length in time of twice the original pulse, although most of the energy is concentrated near the center of the compressed pulse. The final question is, under what conditions is this last equation is valid. To answer this question, let us look at the width of the compressed pulse at its half-power points using this last expression. The half-power points occur where

$$\frac{\sin^2[\pi K\tau(t-t_R)]}{(\pi K\tau(t-t_R))^2} = \frac{1}{2} \tag{D-23}$$

The solution to this transcendental equation is

$$\pi K\tau(t-t_R) = \pm 1.392 \tag{D-24}$$

or

$$t = t_R \pm \frac{1.392}{\pi K \tau} \tag{D-25}$$

Therefore, the half-power time width of this compressed signal is

$$\Delta t \approx \frac{2.8}{\pi K \tau} \tag{D-26}$$

If this width is small compared to the pulse length, then we can use the approximation given in Equation (D-21). We have shown before that the bandwidth of the transmitted pulse is $K\tau$. Therefore, we can say that the approximation would be valid as along as

$$\Delta t \approx \frac{2.8}{\pi B} \ll \tau \tag{D-27}$$

Or, alternatively,

$$B\tau \gg \frac{2.8}{\pi} \tag{D-28}$$

The quantity $B\tau$ is known as the time-bandwidth product of the transmitted pulse. So, we can use the approximation as long as the time-bandwidth product is large.

Finally, let us look at the resolution of this compressed pulse. Using the Rayleigh criterion for resolution, one can discriminate two point scatterers if their echoes are separated by a minimum time such that the maximum of the second return lies in the first null of the first return. This will happen if the time difference between the two echoes is [see Equation (D-21)] $\pi K \tau \Delta t_{min} = \pi$. Since $K\tau = B$, where B is the bandwidth of the chirp signal, we can write the minimum time separation as

$$\Delta t_{min} = 1/B \tag{D-29}$$

The corresponding resolution in the slant range is then $x_r = c/2B$, and in the ground range the resolution is then $x_g = c/2B \sin \theta$.

Index

Introduction to the Physics and Techniques of Remote Sensing. By C. Elachi and J. van Zyl
Copyright © 2006 John Wiley & Sons, Inc.